REPRODUCTIVE BIOLOGY OF INVERTEBRATES

REPRODUCTIVE BIOLOGY OF INVERTEBRATES

Volume I	OOGENESIS, OVIPOSITION, AND OOSORPTION
Volume II	SPERMATOGENESIS AND SPERM FUNCTION
Volume III	ACCESSORY SEX GLANDS
Volume IV	FERTILIZATION, DEVELOPMENT, AND PARENTAL CARE (PARTS A AND B)
Volume V	SEXUAL DIFFERENTIATION AND BEHAVIOUR
Volume VI	ASEXUAL PROPAGATION AND REPRODUCTIVE STRATEGIES (PARTS A AND B)

REPRODUCTIVE BIOLOGY OF INVERTEBRATES

Edited by

K.G. and RITA G. ADIYODI

Vatsyayana Centre of Invertebrate Reproduction
Calicut University, Kerala 673635, India

VOLUME VI, PART B

Asexual Propagation and Reproductive Strategies

A Wiley-Interscience Publication

JOHN WILEY & SONS

Chichester • New York • Brisbane • Toronto • Singapore

Other Wiley Editorial Offices

John Wiley & Sons Inc., 605 Third Avenue,
New York, NY 10158-0012, USA

Jacaranda Wiley Ltd, G.P.O. Box 859, Brisbane,
Queensland 4001, Australia

John Wiley & Sons (Canada) Ltd, 22 Worcester Road,
Rexdale, Ontario M9W 1L1, Canada

John Wiley & Sons (SEA) Pte Ltd, 37 Jalan Pemimpin #05-04,
Block B, Union Industrial Building, Singapore 2057

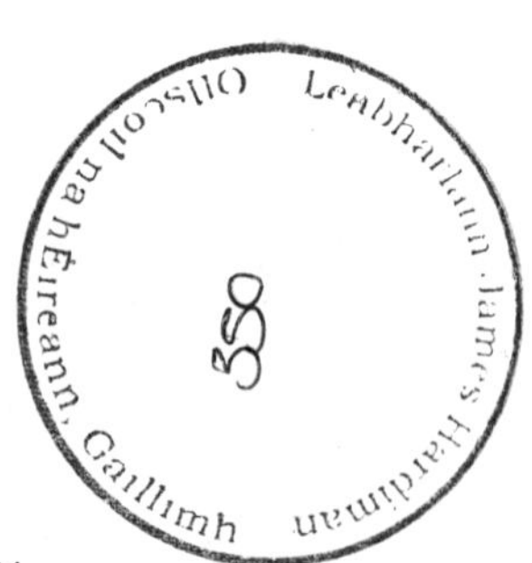

Library of Congress Cataloging-in-Publication Data

(Revised for Vol 6 Part B)
Main entry under title:
Reproductive biology of invertebrates.
"A Wiley-Interscience publication."
Includes indexes.
Contents: v. 1. Oogenesis, oviposition, and
oosorption. v. 2. Spermatogenesis and sperm function.
v. 3. Accessory sex glands. v. 4. Fertilization, development
and parental care. v. 5. Sexual differentiation and behaviour.
v. 6. Asexual propagation and reproductive strategies.
1. Invertebrates—Reproduction. I. Adiyodi, K.G., 1937
II. Adiyodi, Rita G. [DNLM: 1. Invertebrates. 2. Reproduction.
QL 364 R 425] QL 364.15.R45 592.016 88-645030

ISBN 0 471 94119 0 (v. 6)

British Library Cataloguing in Publication Data

A catalogue record for this book is available from the British Library

Reproductive biology of invertebrates,

ISBN 0 471 94119 0

Originally published in India by Oxford & IBH Publishing Co. Pvt. Ltd.,
New Delhi 110 001. Printed in India at Baba Barkha Nath Printers, New Delhi

CONTENTS

SERIES PREFACE

Invertebrates surpass vertebrates not only in the number of species (as many as 95 per cent of all known animals are invertebrates) and individuals, but also in the diversity of structure, adaptations, reproductive biology, sexual behaviour, and development. Reproductive biology is central to all biology, and that of invertebrates has great applied value in management of the myriads of species that are economically useful to man or are harmful to him and his crops and livestock, apart of course from its theoretical and phylogenetical interest. Our knowledge of various aspects of reproduction of invertebrates is, however, not very satisfactory compared to that of vertebrates. Among invertebrates, insects, crustaceans, molluscs, polychaete annelids, and echinoderms have been studied in more detail than other groups, and indeed, very little is known of sexuality, reproduction, and development of some groups such as the Kinorhyncha and Nematomorpha. The unevenness in our knowledge on reproduction of invertebrate groups seems only to become accentuated as years pass by, for the tendency is to research increasingly into areas that have been more frequently investigated. The information on invertebrate sexuality, reproduction, and development is widely scattered in a multitude of journals over a time span of approximately one century; obviously, it is difficult for the individual investigator to have access to all the literature or to benefit from comparisons with other groups.

It was these circumstances, needs, and compulsions that led us to organize and edit this multivolume treatise on invertebrate reproduction. We have preferred a thematic rather than phyletic scheme for various volumes in the series, not only because it is in keeping with the current practice of emphasis on systems and processes, but also because this arrangement, we felt, is the more suitable one to draw comparisons between different groups and for the chapters to be more authoritative and provocative. Each book in this treatise is thus independent, but paradoxically enough, also dependent on the others. When information on one group is spread over six volumes, there is a likelihood that overlap might occur between volumes. Overlap has been reduced to the minimum in this series by providing the authors in advance with a subject outline and by careful editing, but some overlap nevertheless does exist, where it has been allowed for purposes of clarity and understanding.

No book is complete, perhaps not even the scriptures. This is particularly true of books on an almost infinitely diverse and numerically large assemblage as the invertebrates. The present series, 'Reproductive Biology of Invertebrates', is no exception. In spite of our persistent efforts we did not succeed in commissioning contributions on certain groups to some volumes for various reasons — a task we have been forced to leave to the future. But what has been competently narrated, overviewed, discussed, and questioned in different volumes of this treatise encompasses a large body of very useful information, some parts of which have not been hitherto available in any form. The authors were encouraged to suggest topics of interest for future research. Many contributors did. Clearly, what is known of invertebrate propagation amounts to only a little fragment of the information that remains to be discovered. We hope that invertebrates, by virtue of their relatively small sizes, easy availability, short reproductive cycles, easy experimental manoeuvrability, great diversity, and exemption from vivisection rules will attract more investigators in the future.

A survey of the information pieced together over the past century on invertebrate sexuality, reproduction, and development shows that the edifice that has been built is based largely on glimpses of only a small number of processes in a surprisingly small number of species. Though this calls for the utmost caution in attempting any broad generalizations, it is tempting to suggest that there may be a common denominator — a common controlling system — at work in regulating sexuality, reproduction, and development in all animals, invertebrate and vertebrate, a system which differs only in details and complexity in different animals, related to their phylogenetic state and ecological needs. It is perhaps logical to assume that biological forces and laws governing propagation of species may be essentially as simple as are the physical forces and laws of nature. We hope this treatise will stimulate future investigators to test the soundness or otherwise of this unified hypothesis.

We are grateful to our contributors for their cooperation, patience, and understanding, and to scientist colleagues, too many to be listed, who offered valuable suggestions and advice and willingly and gladly peer reviewed the manuscripts. We thank most cordially the scientific and technical staff and students of the Vatsyayana Centre of Invertebrate Reproduction, Calicut University for their many kindnesses and help, which lessened a great deal the burden of the work involved, and it is to them and to our children, Nirmal and Laxmi, who missed many an evening with us which legitimately belonged to them, that this treatise is dedicated most affectionately and warmly.

Calicut University

K.G. ADIYODI
R.G. ADIYODI

PREFACE TO VOLUME VI

The emphasis in Volumes I to V of this series on "Reproductive Biology of Invertebrates" has almost exclusively been on sexual reproduction (amphimictic, via gametes) as it is the most widespread mode of propagation amongst the invertebrates and also the most studied. This volume, the sixth in the series, deals with different types of non-gametic (vegetative, somatic) propagation such as budding, fission, fragmentation, and gemmulation; the types, cytology, genetics, and evolution of parthenogenesis which is a secondary, gametic form of asexual propagation; special modes of reproduction such as polyembryony and paedogenesis; genetic, hormonal, nutritional, environmental, and social influences in fecundity and sterility; interspecific reproductive isolation; and the types of strategies adopted by various groups of invertebrates — terrestrial, marine, freshwater, brackish-water, free-living and parasitic — at the population level to ensure survival and optimum reproductive success in their respective habitats.

Regenerative ability is basic to non-gametic asexual propagation; therefore, data on restoratory, physiological, and propagative types of regeneration have been included here. The ability to regenerate lost parts and thus to propagate somatically is extremely limited in eutelic groups such as the Rotifera, Gastrotricha, Acanthocephala, and Tardigrada, in the absence of somatic divisions in the adult state. Though several groups of invertebrates such as the Porifera, Turbellaria, Annelida, and Echinodermata propagate asexually, in no group is the evolution of the phylum itself "tantamount to the evolution of asexual reproduction", as in Cnidaria.

Parthenogenesis, which involves the development of the individual from gametes — generally female, rarely male — but without fertilization, appears to have evolved from, and at least in some groups forms an alternative to, sexual reproduction. In invertebrates, this gametic form of asexual reproduction appears to be more widespread amongst the different taxa than non-gametic asexual propagation, e.g. Cnidaria, Turbellaria, Eucestoda, Rotifera, Gastrotricha, Kinorhyncha, Nematoda, Priapulida (?), Sipuncula, Mollusca, Echiura (Volume IV A, pp. 358–359), Tardigrada, Onychophora (Volume V, pp. 278–279), Crustacea, Insecta, and Echinodermata, though the cytogenetic mechanism of parthenogenetic development has not been verified in all cases.

Both non-gametic and gametic asexual reproduction are closed genetic systems with no scope for segregation and recombination that are characteristic of sexual reproduction. But there are advantages associated with asexual propagation: e.g. asexual population can escape competition with amphimictic populations, opportunistically colonize new territories, survive adversity, and escape sexual sterility caused by polyploidy (the last is questioned here in Chapter 1 on oligochaetous Annelida).

The thrust of this volume, apart from asexual propagation, is on reproductive strategies. Strategies may involve a competing or non-competing mix of asexual and sexual reproduction where both types occur, and in parasitic forms such as the trematodes and cestodes a variety of life-cycle types and ecologically different hosts. General principles, concepts, and patterns of reproductive strategies in free-living and parasitic organisms are summarized in Chapter 4 of Volume VI A. Characteristics that differ between r- and K-selection include environment (unstable or stable), fecundity (semelparity, iteroparity), sexual maturity (early or delayed), rate of development (rapid or slow), size of the egg and body size of the progeny (small or large), longevity (short or long), mortality rate (density independent or dependent), population size (variable or constant), and inter- and intraspecific competition (slight or strong), as described in Chapter 11 of Volume VI A. There is an increasing body of evidence to show that the r-K spectrum is in fact continuous with the organism positioned at some point along it and also that, based on the habitat, there is an adaptive shift along the r-K continuum even within the same species. Reproductive tactics used by animals are thus almost as infinitely diverse and individualistic as are the species themselves, and more studies are needed in this area to evolve more efficient species-specific management practices. At the organismic level, strategies involved in physiological processes associated with reproduction have only received scant attention, but departing from this practice, Chapters 3 and 4 in Volume VI B also deal with such physiological and behavioural parameters in reproductive strategies of Crustacea and Insecta.

K.G. ADIYODI
R.G. ADIYODI

CONTRIBUTORS

A. ALVARIÑO, *National Marine Fisheries Service, Southwest Fisheries Science Center, 8604 La Jolla Shores Drive, P.O.Box 271, La Jolla, California 92038-0271, U.S.A.*

JAYAPAUL AZARIAH, *Department of Zoology, University of Madras, Guindy Campus, Madras 600 025, Tamil Nadu, India*

ROBERTO BERTOLANI, *Dipartimento di Biologia Animale, Universitá di Modena, Via Università 4, 41100 Modena, Italy*

ROBERT D. BURKE, *Department of Biology, University of Victoria, Victoria, British Columbia V8W 2Y2, Canada*

B. CHRISTENSEN, *Institute of Population Biology, University of Copenhagen, Universitetsparken 15, DK-2100, Copenhagen, Denmark*

SHOU HWA CHUANG, *114 Pasir Ris Road, Singapore 1851, Singapore*

PHILIP V. MLADENOV, *Department of Marine Science, University of Otago, P.O.Box 56, Dunedin, New Zealand*

J. MUTHUKRISHNAN, *School of Biological Sciences, Madurai Kamaraj University, Madurai 625 021, Tamil Nadu, India*

CLAUS NIELSEN, *Zoological Museum, Universitetsparken 15, DK-2100 Copenhagen, Denmark*

T.J. PANDIAN, *School of Biological Sciences, Madurai Kamaraj University, Madurai 625 021, Tamil Nadu, India*

J.A. PETERSEN, *Universidade de São Paulo, Instituto de Biociências e Instituto de Biologia Marinha, Caixa Postal 11.230, 01000 São Paulo SP, Brasil (Deceased)*

JOHN RILEY, *Department of Biological Sciences, University of Dundee, Dundee DD1 4HN, U.K.*

SYSTEMATIC RÉSUMÉ OF THE INVERTEBRATES

Phylum Porifera
- Class Hexactinellida *(Farrea, Euplectella)*
- Class Calcarea
  - Subclass Calcinea *(Clathrina, Ascandra)*
  - Subclass Calcoronea *(Sycon, Grantia)*
  - Subclass Pharetronida *(Neocoelia)*
- Class Demospongiae
  - Subclass Homoscleromorpha *(Oscarella)*
  - Subclass Tetractinomorpha *(Geodia, Tetilla)*
  - Subclass Ceractinomorpha *(Halichondria, Haliclona)*

Phylum Cnidaria (Coelenterata)
- Class Hydrozoa
  - Order Hydroida *(Hydra, Obelia)*
  - Order Milleporina *(Millepora)*
  - Order Stylasterina *(Stylaster, Allopora)*
  - Order Trachylina *(Geryonia, Cunina)*
  - Order Siphonophora *(Halistemma, Physalia)*
- Class Scyphozoa
  - Order Stauromedusae *(Lucernaria, Haliclystus)*
  - Order Cubomedusae *(Carybdea, Chiropsalmus)*
  - Order Coronatae *(Nausithoë, Periphylla)*
  - Order Semaeostomeae *(Aurelia, Cyanea)*
  - Order Rhizostomeae *(Rhizostoma, Cassiopeia)*
- Class Anthozoa
  - Subclass Octocorallia *(Alcyonaria)*
    - Order Stolonifera *(Tubipora, Cornularia)*
    - Order Telestacea *(Telesto, Coelogorgia)*
    - Order Alcyonacea *(Alcyonium, Heteroxenia)*
    - Order Coenothecalia *(Heliopora)*

Order Gorgonacea *(Corallium, Gorgonia)*
Order Pennatulacea *(Pennatula, Renilla)*
Subclass Hexacorallia (Zoantharia)
Order Actiniaria *(Edwardsia, Lebrunia)*
Order Corallimorpharia *(Corynactis)*
Order Scleractinia *(Cladocora)*
Order Zoanthidea *(Zoanthus, Palythoa)*
Order Antipatharia *(Schizopathes, Dendrobrachia)*
Order Ceriantharia *(Cerianthus)*

Phylum Ctenophora
Class Tentaculata
Order Cydippida *(Mertensia, Pleurobrachia)*
Order Lobata *(Bolinopsis, Leucothea)*
Order Cestida *(Cestum, Velamen)*
Order Platyctenea *(Coeloplana, Ctenoplana)*
Class Nuda
Order Beroida *(Beroë)*

Bilateria: Protostomia

ACOELOMATA

Phylum Platyhelminthes
Class Turbellaria
Group Archoophora
Order Nemertodermatida *(Meara, Nemertoderma)*
Order Acoela *(Diopisthoporus, Childia)*
Order Catenulida *(Stenostomum, Catenula)*
Order Macrostomida *(Macrostomum, Promacrostomum)*
Order Polycladida *(Cryptocelides, Stylochus)*
Group Neoophora
Order Lecithoepitheliata *(Hofstenia)*
Order Prolecithophora *(Prolecithoplana)*
Order Neorhabdocoela (Rhabdocoela) *(Dalyellia, Typhloplana, Gyratrix)*
Order Proseriata *(Parotoplana)*
Order Tricladida *(Geoplana, Dugesia)*
Class Temnocephalida *(Temnocephala)*
Class Trematoda
Order Monogenea
Suborder Monopisthocotylea *(Gyrodactylus, Monocotyle)*
Suborder Polyopisthocotylea *(Polystoma, Hexabothrium)*

Order Aspidobothria (Aspidogastrea) *(Aspidogaster, Stichocotyle)*
Order Digenea *(Paragonimus, Fasciola, Schistosoma)*
Class Cestoidea
Subclass Cestodaria
Order Amphilinidae *(Amphilinia)*
Order Gyrocotylidea *(Gyrocotyle)*
Subclass Eucestoda
Order Caryophyllidea *(Hunterella, Glaridacris)*
Order Spathebothriidea *(Diplocotyle, Spathebothrium)*
Order Trypanorhyncha (Trypanorhynchida) *(Lacistorynchus)*
Order Pseudophyllidea *(Triaenophorus, Diphyllobothrium)*
Order Lecanicephalidea *(Lecanicephalum, Disculiceps)*
Order Aporidea *(Nematoparataenia, Apora)*
Order Tetraphyllidea *(Acanthobothrium, Oncobothrium)*
Order Diphyllidea *(Echinobothrium, Ditrachybothridium)*
Order Litobothridea *(Litobothrium)*
Order Proteocephalata (Proteocephalidea) *(Proteocephalus)*
Order Cyclophyllidea *(Hymenolepis, Taenia)*
Order Nippotaeniidea *(Nippotaenia)*

Phylum Mesozoa
Order Dicyemida (Rhombozoa) *(Dicyema, Pseudicyema)*
Order Orthonectida *(Rhopalura)*
Phylum Nemertina (Rhynchocoela)
Subclass Anopla
Order Palaeonemertini *(Cephalothrix)*
Order Heteronemertini *(Cerebratulus, Lineus)*
Subclass Enopla
Order Hoplonemertini *(Amphiporus, Prostoma)*
Order Bdellonermertini *(Malacobdella)*
Phylum Gnathostomulida
Order Filospermoidea *(Haplognathia)*
Order Bursovaginoidea
Suborder Scleroperalia *(Gnathostomula)*
Suborder Conophoralia *(Austrognatharia)*

PSEUDOCOELOMATA

Phylum Rotifera
Class Seisonidea *(Seison)*
Class Bdelloidea *(Philodina, Habrotrocha)*
Class Monogononta *(Asplanchna, Platyias)*

Phylum Gastrotricha
Order Macrodasyida *(Macrodasys, Turbanella)*
Order Chaetonotida
Suborder Multitubulatina *(Neodasys)*
Suborder Paucitubulatina *(Heteroxenotrichula)*

Phylum Kinorhyncha
Suborder Cyclorhagae *(Echinoderes, Echinoderella)*
Suborder Conchorhagae *(Semnoderes)*
Suborder Homalorhagae *(Pycnophyes, Trachydemus)*
Phylum Nematoda
Class Nematoda
Subclass Torquentia
Order Monhysterida *(Cylindrolaimus, Theristus)*
Order Desmoscolecida *(Desmoscolex, Tricoma)*
Order Areolaimida *(Camacolaimus, Bathylaimus)*
Order Chromadorida *(Paracanthonchus, Desmodora)*
Subclass Secernentia
Order Rhabditida *(Rhabditis, Bunonema)*
Order Tylenchida *(Tylenchus, Ditylenchus)*
Order Strongylida *(Strongylus, Triodontophorus)*
Order Ascaridida *(Ascaris)*
Order Spirurida *(Thelazia, Oxyspirura)*
Subclass Penetrantia
Order Enoplida *(Enoplus, Mononchus)*
Order Dorylaimida *(Dorylaimus, Actinolaimus)*
Order Trichocephalida *(Trichuris, Capillaria)*
Order Dioctophymatida *(Dioctophyme, Hystrichis)*

Phylum Nematomorpha
Order Nectonematida *(Nectonema)*
Order Gordiida *(Gordionus, Paragordius)*

Phylum Acanthocephala
Order Palaeacanthocephala *(Acanthocephalus, Gorgorhynchus)*
Order Archiacanthocephala *(Macracanthorhynchus, Moniliformis)*
Order Eoacanthocephala *(Neoechinorhynchus, Pallisentis)*

SCHIZOCOELOMATA

Phylum Priapulida *(Priapulus, Halicryptus)*

Phylum Sipuncula *(Dendrostomum, Golfingia, Aspidosiphon)*

Phylum Mollusca
Class Monoplacophora *(Neopilina)*
Class Polyplacophora *(Acanthochiton, Lepidochiton)*
Class Aplacophora *(Chaetoderma, Neomenia)*
Class Gastropoda
Subclass Prosobranchia
Order Archaeogastropoda *(Haliotis, Trochus)*
Order Mesogastropoda *(Littorina, Janthina)*
Order Neogastropoda (*Murex, Buccinum*)
Subclass Opisthobranchia
Order Tectibranchia *(Bulla, Aplysia)*
Order Pteropoda *(Spiratella, Clio)*
Order Nudibranchia *(Doris, Aeolidia)*
Subclass Pulmonata
Order Basommatophora *(Lymnaea, Planorbis)*
Order Systellommatophora *(Vaginulus, Laevicaulis)*
Order Stylommatophora *(Helix, Deroceras)*
Class Scaphopoda *(Dentalium, Cadulus)*
Class Pelecypoda (Bivalvia)
Order Protobranchia *(Nucula, Yoldia)*
Order Filibranchia *(Mytilus, Ostrea)*
Order Eulamellibranchia *(Mercenaria, Teredo)*
Order Septibranchia *(Poromya, Cuspidaria)*
Class Cephalopoda
Subclass Nautiloidea (Tetrabranchia) *(Nautilus)*
Subclass Coleoidea (Dibranchia)
Order Decapoda *(Loligo, Spirula)*
Order Octopoda *(Octopus, Argonauta)*

Phylum Echiura *(Echiurus, Urechis, Bonellia)*
Phylum Annelida
Class Polychaeta
Subclass Errantia *(Nereis, Tomopteris, Syllis, Eunice)*
Subclass Sedentaria *(Chaetopterus, Cirratulus, Sabellaria)*
Subclass Archiannelida *(Dinophilus, Polygordius, Trilobodrilus)*
Class Oligochaeta
Order Lumbriculida *(Lumbriculus, Stylodrilus)*
Order Tubificida *(Tubifex, Enchytraeus)*
Order Haplotaxida *(Haplotaxis, Alluroides, Moniligaster, Lumbricus)*
Class Branchiobdellida *(Branchiobdella)*
Class Hirudinea
Order Acanthobdellida *(Acanthobdella)*
Order Rhynchobdellida *(Glossiphonia, Piscicola)*

Order Gnathobdellida *(Hirudo, Haemadipsa)*
Order Pharyngobdellida *(Erpobdella)*

Phylum Pogonophora
Subphylum Perviata
Class Frenulata
Order Thecanephria *(Diplobrachia, Polybrachia)*
Order Athecanephria *(Siboglinum)*
Subphylum Obturata *(Lamellibrachia, Riftia, Ridgeia)*

Phylum Tardigrada
Class Heterotardigrada
Order Arthrotardigrada *(Batillipes, Halechiniscus, Lepoarctus)*
Order Echiniscoidea *(Bryodelphax, Carphania, Echiniscus)*
Class Mesotardigrada *(Thermozodium)*
Class Eutardigrada
Order Apochela *(Milnesium)*
Order Parachela *(Macrobiotus, Hypsibius)*

Phylum Onychophora *(Peripatus, Eoperipatus, Peripatopsis)*

Phylum Arthropoda
Subphylum Chelicerata
Class Merostomata *(Limulus, Tachypleus)*
Class Arachnida
Order Scorpiones *(Buthus, Palamnaeus)*
Order Uropygi *(Mastigoproctus, Thelyphonus)*
Order Amblypygi *(Acanthophrynus, Tarantula)*
Order Palpigradi *(Eukoenenia, Leptokoenenia)*
Order Araneae *(Degesiella, Araneus)*
Order Ricinulei *(Ricinoides, Cryptocellus)*
Order Pseudoscorpiones *(Chelifer, Neobisium)*
Order Solifugae *(Ammotrecha, Eremobates)*
Order Opiliones *(Phalangium, Nemastoma)*
Order Acari *(Amblyomma, Sarcoptes)*
Class Pycnogonida *(Nymphon, Callipallene)*

Subphylum Mandibulata
Class Crustacea
Subclass Cephalocarida *(Hutchinsoniella, Sandersiella)*
Subclass Branchiopoda
Order Anostraca *(Artemia, Streptocephalus)*
Order Notostraca *(Triops, Lepidurus)*

Order Conchostraca *(Limnadia, Cyclestheria)*
Order Cladocera *(Daphnia, Polyphemus)*
Subclass Ostracoda
Order Myodocopa *(Cypridina, Sarsiella)*
Order Cladocopa *(Polycope)*
Order Podocopa *(Bairdia)*
Order Platycopa *(Cytherella)*
Subclass Mystacocarida *(Derocheilocaris)*
Subclass Copepoda
Order Calanoida *(Calanus, Eurytemora)*
Order Harpacticoida *(Harpacticus, Tisbe)*
Order Cyclopoida *(Cyclopicina, Lernaea)*
Order Notodelphyoida *(Notodelphys, Demoixys)*
Order Monstrilloida *(Cymbasoma, Monstrilla)*
Order Caligoida *(Lepeophtheirus, Caligus)*
Order Lernaeopodoida *(Salmincola, Lernaeopoda)*
Subclass Branchiura *(Argulus, Dolops)*
Subclass Cirripedia
Order Ascothoracica *(Synagoga, Laura)*
Order Thoracica *(Lepas, Poecilasma)*
Order Acrothoracica *(Trypetesa, Lithoglyptes)*
Order Rhizocephala *(Sacculina, Peltogaster)*
Subclass Malacostraca
Superorder Phyllocarida
Order Leptostraca *(Nebalia, Epinebalia)*
Superorder Hoplocarida
Order Stomatopoda *(Squilla, Gonodactylus)*
Superorder Syncarida
Order Anaspidacea *(Anaspides, Koonunga)*
Order Stygocaridacea *(Stygocaris, Parastygocaris)*
Order Bathynellacea *(Bathynella, Parabathynella)*
Superorder Eucarida
Order Euphausiacea *(Euphausia, Thysanopoda)*
Order Decapoda
Suborder Natantia *(Penaeus, Pandalus)*
Suborder Reptantia *(Palinurus, Astacus, Birgus, Paratelphusa)*
Superorder Pancarida
Order Thermosbaenacea *(Monodella, Thermosbaena)*
Superorder Peracarida
Order Mysidacea *(Mysis, Praunus)*
Order Cumacea *(Diastylis, Eudorella)*
Order Spelaeogriphacea *(Spelaeogriphus)*
Order Tanaidacea *(Tanais, Heterotanais)*

Order Isopoda *(Oniscus, Porcellio)*
Order Amphipoda *(Gammarus, Orchestia)*
Group Myriapoda
Class Chilopoda
Superorder Epimorpha
Order Geophilomorpha *(Geophilus, Pachymerium)*
Order Scolopendromorpha *(Scolopendra, Cryptops)*
Superorder Anamorpha
Order Lithobiomorpha *(Lithobius, Bothropolys)*
Order Scutigeromorpha *(Scutigera, Thereuonema)*
Class Diplopoda
Subclass Pselaphognatha *(Polyxenus, Phryssonotus)*
Subclass Chilognatha
Superorder Pentazonia
Order Glomeridesmida *(Glomeridesmus)*
Order Glomerida *(Glomeris, Sphaerotherium)*
Superorder Helminthomorpha
Order Polydesmida *(Polydesmus, Oniscodesmus)*
Order Chordeumida *(Chordeuma, Abacion)*
Order Julida *(Nemasoma, Blaniulus)*
Order Cambalida *(Cambala, Leiodere)*
Order Spirobolida *(Narceus, Atopetholus)*
Order Spirostreptida *(Spirostreptus, Graphidostreptus)*
Superorder Colobognatha
Order Platydesmida *(Platydesmus, Brachycybe)*
Order Polyzoniida *(Siphonophora, Polyzonium)*
Class Pauropoda
Order Hexamerocerata *(Millotauropus)*
Order Tetramerocerata *(Pauropus, Brachypauropus)*
Class Symphyla *(Scolopendrella, Scutigerella, Geophilella)*

Group Hexapoda
Class Insecta
Subclass Pterygota

Section A. Exopterygota (Heterometabola)
Order Odonata (dragonflies, damselflies)
Order Ephemeroptera (mayflies)
Order Plecoptera (stoneflies)
Order Embioptera (web spinners)
Order Orthoptera (grasshoppers, locusts, crickets)
Order Phasmida (stick and leaf insects)
Order Dermaptera (earwigs)

Order Diploglossata (hemimerids)
Order Grylloblattoidea (grylloblattids)
Order Dictyoptera (cockroaches, mantids)
Order Isoptera (termites)
Order Zoraptera (zorapterans)
Order Psocoptera (booklice, bark lice)
Order Phthiraptera (biting and sucking lice)
Order Thysanoptera (thrips)
Order Hemiptera
Suborder Heteroptera (true bugs)
Suborder Homoptera (aphids, mealy bugs, cicadas, scale insects)

Section B. Endopterygota (Holometabola)
Order Coleoptera (beetles, weevils)
Order Strepsiptera (stylopoids)
Order Hymenoptera (ants, bees, wasps, ichneumons)
Order Neuroptera (ant-lions, lacewings)
Order Megaloptera (dobsonflies, alderflies)
Order Raphidioptera (snakeflies)
Order Mecoptera (scorpion flies)
Order Trichoptera (caddis flies)
Order Lepidoptera (butterflies, moths)
Order Diptera (true flies)
Order Siphonaptera (fleas)

Subclass Thysanura
Order Thysanura (silverfish)
Subclass Collembola
Order Collembola (springtails, snowfleas)
Subclass Protura
Order Protura (proturans)
Subclass Aptera
Order Aptera (diplurans)

Phylum Pentastomida
Order Cephalobaenida *(Cephalobaena, Raillietiella)*
Order Porocephalida *(Porocephalus, Armillifer)*

LOPHOPHORATA

Phylum Phoronida *(Phoronis, Phoronopsis)*

Phylum Bryozoa Ectoprocta
Class Phylactolaemata *(Fredericella, Plumatella)*
Class Stenolaemata
Order Cyclostomata *(Tubulipora, Heteropora)*
Class Gymnolaemata
Order Ctenostomata *(Paludicella, Sundanella)*
Order Cheilostomata *(Bugula, Dendrobeania)*

Phylum Bryozoa Entoprocta *(Loxosoma, Pedicellina, Urnatella)*

Phylum Brachiopoda
Class Inarticulata
Order Atremata *(Lingula, Glottidia)*
Order Neotremata *(Discina, Crania)*
Class Articulata *(Lacazella, Terebratulina)*

Bilateria: Deuterostomia

Phylum Chaetognatha *(Sagitta, Eukrohnia, Spadella)*

Phylum Echinodermata
Subphylum Pelmatozoa
Class Crinoidea
Order Articulata *(Antedon, Leptometra)*
Subphylum Eleutherozoa
Class Holothuroidea
Order Aspidochirota *(Holothuria, Stichopus)*
Order Elasipoda *(Deima, Elpidia)*
Order Dendrochirota *(Cucumaria, Thyone)*
Order Molpadonia *(Molpadia, Caudina)*
Order Apoda *(Synapta, Opheodesoma)*
Class Echinoidea
Subclass Regularia (Endocyclica)
Order Lepidocentroida *(Asthenosoma, Phormosoma)*
Order Cidaroidea *(Goniocidaris, Notocidaris)*
Order Aulodonta *(Diadema, Micropyga)*
Order Stirodonta *(Arbacia, Salenia)*
Order Camarodonta *(Echinus, Strongylocentrotus)*
Subclass Irregularia (Exocyclica)
Order Holectypoida *(Echinoneus, Micropetalon)*
Order Cassiduloida *(Apatopygus, Cassidulus)*
Order Clypeastroida *(Dendraster, Echinodiscus)*
Order Spatangoida *(Meoma, Spatangus)*

Class Asteroidea
Order Phanerozonia *(Porcellanaster, Archaster)*
Order Spinulosa *(Ganeria, Asterina)*
Order Forcipulata *(Asterias, Pedicellaster)*
Class Ophiuroidea
Order Ophiurae *(Ophiomyxa, Ophioscolex)*
Order Euryalae *(Asteronyx, Gorgonocephalus)*

Phylum Hemichordata
Class Enteropneusta *(Balanoglossus, Glandiceps)*
Class Pterobranchia *(Rhabdopleura, Cephalodiscus)*

Phylum Chordata
Subphylum Urochordata
Class Ascidiacea *(Ascidia, Ciona, Halocynthia)*
Class Thaliacea *(Salpa, Doliolum)*
Class Larvacea *(Oikopleura, Appendicularia)*
Subphylum Cephalochordata *(Branchiostoma, Asymmetron)*

1. ANNELIDA — CLITELLATA

B. CHRISTENSEN
Institute of Population Biology, University of Copenhagen,
Universitetsparken 15, DK-2100 Copenhagen, Denmark

I. INTRODUCTION

Asexual propagation comprises all kinds of reproduction that do not depend on a sexual or modified sexual process involving the fusion of female and male gametes. Among the Clitellata and for that matter among higher organisms in general, asexual reproduction must be a secondary phenomenon that has arisen from normal sexual outbreeding. As such, asexual reproduction represents an evolutionary novelty, or at least was so when it first made its appearance. According to conventional evolutionary thinking, such novelties must possess some kind of selective advantage in order to become established.

But which are the advantages associated with asexual reproduction? No conclusive answer can be given. Interpretations of the possible consequences of asexual reproduction are many, but they rest largely on inference from general theoretical postulations rather than on factual evidence because, until recently, there have been very few studies on these subjects.

In this chapter the main types of asexual reproduction in Clitellata are described briefly, whereas some recent observations on the possible ecological and evolutionary implications of this reproductive method are dealt with in more detail.

II. THE OCCURRENCE OF ASEXUAL PROPAGATION: AN OVERVIEW

Among clitellate annelids verified cases of asexual reproduction are only known within the class Oligochaeta. The complicated structure of the body, the fixed number of segments, and the specialized anterior and posterior ends make asexual reproduction through fission unlikely in leeches (Class Hirudinea); parthenogenesis might occur in this group but has not been established yet.

Among the oligochaetes (for classification, see Systematic Résumé), most cases of asexual propagation are obviously of secondary origin. Within most genera containing parthenogenetic species and in many cases even within the same morphological species where a parthenogenetic biotype occurs, ancestral cross-breeding species or biotypes are also known. The same pattern of sporadic, and undoubtedly secondary, occurrence is also seen in the case of architomic fission (regeneration after separation; Section III A1). This type of reproduction has been described in a few species belonging to the families Lumbriculidae, Tubificidae, Enchytraeidae, and Naididae.

Paratomic fission (regeneration before separation; Section III A2), on the other hand, would seem to represent a more fundamental feature in those taxa where it occurs. This type of reproduction is found in practically all species within the families Aelosomatidae and Naididae. Some authors (e.g. Michaelsen, 1928 and followers) consider these two families, especially the aelosomatids, among the most primitive of existing Oligochaeta, and this might indicate that this particular reproductive method is a primary feature within the entire class. However, this view is challenged by other workers, and the topic is discussed further in Section IV D.

III. MAIN TYPES OF ASEXUAL PROPAGATION

Defined as any kind of deviation from normal outbreeding in cross-fertilizing species and from inbreeding in self-fertilizing hermaphrodites, asexual propagation includes both multiplication through transverse fission and multiplication through eggs that develop parthenogenetically. This broad definition of asexual reproduction is justified, because the genetical consequences of transverse fission and parthenogenesis are in most cases identical, and, consequently, the evolutionary strategies underlying these two types of reproduction might also be more or less identical.

A. Transverse Fission

The individual divides into two or more new individuals through a transverse division of the body. The tail end of the presumptive anterior individual forms new posterior segments, and the head end of the presumptive posterior individual forms a new anterior end. Two main types of transverse fission are known: (1) In architomy the worm suddenly divides into fragments caused by vigorous contractions of the muscles in the body wall. New individuals are formed from such fragments through subsequent formation of new anterior and posterior ends. (2) In paratomy separation of the new individuals is preceded by the formation of a 'budding' or 'fission' zone, where a new anterior end and a new posterior end are formed before the individuals separate. For detailed reviews of the older literature, see Stephenson (1930), Berrill (1952), Stolte (1955), and Avel (1959).

1. Simple fission followed by regeneration

Fully grown worms divide into several fragments. The fractures are caused by vigorous contraction of the muscles in the body wall and, to a lesser extent, of muscles in the intestinal wall. *Nais paraguayensis* and *Lumbriculus variegatus* divide simultaneously into six to eight fragments (Hyman, 1938), whereas Christensen (1964) reports a mean number of four fragments in *Enchytraeus bigeminus*. The first record of asexual reproduction through fragmentation in enchytraeids was made by Christensen (1959).

In *Enchytraeus bigeminus* the fractures are always located in the middle of a segment in front of the setal bundles (Christensen, 1964). Christensen has shown that artificial fragmentation is easily induced by means of electric shocks in species reproducing in this way, whereas species that reproduce exclusively via eggs proved negative in this respect. This might indicate a special mechanism of interaction between the nervous system and muscles in species which have adopted this type of asexual reproduction.

The following outline of the regeneration of a new anterior end is mainly based upon Christensen's (1964) studies on this process in *Enchytraeus bigeminus*. Closure of the wound is accomplished by contraction of the body wall, except for a small aperture closed by damaged cells from the epidermis and intestine and detached muscle cells from the body wall (Fig. 1a). After six hours extensions from the surrounding epidermal cells cover the wound surface completely, and a new cuticle is formed. In the following hours cells of epidermal origin cover the entire wound surface, where they increase in number through mitotic divisions. Meanwhile, forward-migrating mesodermal neoblasts form two lateral aggregations immediately inside the new epidermal cells. The migratory mesodermal cells can be traced back to neoblasts present in all segments and located ventrolaterally on the posterior side of the intersegmental septal strands. Also at these early stages (Fig. 1b and c) neoblasts of endodermal origin begin to accumulate at the anterior end of the old intestine. Thus in *E. bigeminus* endodermal and mesodermal neoblasts are present in the intact worm, and, during the regeneration, they form tissues of endodermal and mesodermal derivation respectively. New tissues of ectodermal derivation may stem from dedifferentiated epidermal cells. In *Lumbriculus variegatus* mesodermal neoblasts form tissues of mesodermal derivation whereas other new tissues are produced through dedifferentiation of corresponding old tissues (Veitzman, 1937 as cited in Vorontsova and Liosner, 1960; Herlant-Meewis, 1946; Stephan-Dubois, 1956).

The main features of the organogenesis proper are shown in Fig. 1d and e. Centrally located epidermal (ectodermal) cells extend backwards as a distinct stomodaeum. Cells of epidermal origin also give rise to the new ventral nerve cord and the dorsal brain. Forward-migrating endodermal neoblasts form a new oesophagus. The mesodermal cells are separated into an outer and an inner layer by the new body coelom and form such tissues as peritoneum, blood vessels, septa, septal

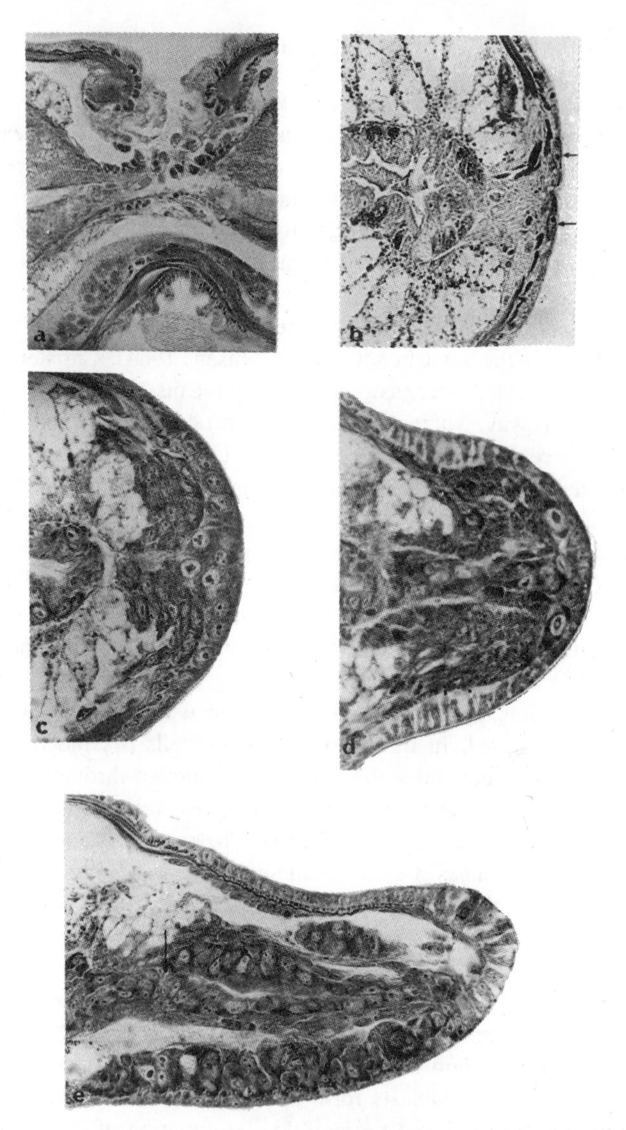

Fig. 1. Fragmentation and early stages in the formation of a new anterior end in *Enchytraeus bigeminus*. **a**: Sagittal section through a worm in the process of fragmentation. **b**: Horizontal section through the anterior end of a five to six-hour-old fragment; new epidermal cells are present on the wound surface (arrows). **c**: Horizontal section of a 12-hour-old fragment showing new endodermal cells at the anterior end of the old intestine and two lateral aggregations of mesodermal cells; the centrally located epidermal cells increase in size. **d**: Horizontal section of a 50 to 60-hour-old fragment. Stomodaeum is well developed, and circumpharyngeal connectives are distinguishable anterolaterally inside the new epidermal layer. **e**: Sagittal section of an approx. 72-hour-old fragment showing forward-migrating endodermal cells; the transition between these and the ectodermal stomodaeum is indicated by an arrow. The regeneration is complete after 144 hours. (Partly from Christensen, 1964).

glands, and musculature. In *Enchytraeus bigeminus*, regeneration of a new anterior end is complete after 144 hours, when the fragments are kept at 20–22°C (see Christensen, 1964, where a detailed time table is given).

2. Proliferation followed by fission

In the case of paratomy the regeneration is prepared and more or less completed before the individuals separate. In the fission zone a new posterior end and a new anterior end are formed, separated by the future fracture. The anterior individual forms a new tail end, the posterior individual a new head. A fission zone is always formed in the anterior part of a segment in front of the setal bundles: the division as such is therefore intrasegmental.

The simplest pattern involves a single transverse division of the parental worm into two new individuals. When these new individuals have separated and completed their regeneration, each develops a new fission zone. This type of division is found in the naidid genera *Dero* and *Ophidonais*. More frequently a new fission zone forms before division has occurred in the older one. Such delayed divisions produce chains containing more than two individuals. The position of subsequent fission zones differs between species, and, on this basis, elaborate subdivisions into a number of different types have been proposed (for details, see Stephenson, 1930; Avel, 1959).

The main outline can be summarized as follows: The first fission zone is formed in the middle region of the worm, and subsequent zones are formed in front of this, i.e. in the posterior end of the anterior individual. A second fission zone may arise in the new tissue formed immediately anterior to the first fission zone (naidian form), or the new zone may form in the most posterior original segment of the anterior individual. In this way subsequently arising zones are located one segment nearer the anterior end of the original worm (stylarian form). In the former case the proliferation can continue indefinitely, while there is a lower limit to the shortening of the body occurring in the latter case. When this limit is reached, proliferation stops for a while, and the original anterior end elongates by growth of the hind end in the usual way. When the normal size is reached, a new fission zone is intercalated in the middle of the worm, and the above process is repeated.

Aelosomatids would seem mainly to follow the naidian type described above, but superficially it looks very much like a strobilation, where new individuals are continually being formed at a fixed position in the posterior end of the original individual resulting in chains of zooids (Herlant-Meewis, 1954).

Occasionally posterior individuals may form a fission zone before they separate from the zooid, and even intermediary individuals may do so. The entire matter is furthermore complicated by the observation that individuals belonging to a species normally showing the naidian form of fission sometimes follow the stylarian pattern and vice versa. In *Aulophorus superterrenus*, Marcus (1943) reports a more or less simultaneous development of a number of nearly equidistant fission zones along the body of the worm, a situation resembling the simultaneous division into several fragments during architomy.

Organogenesis in the fission zone has been studied by several authors, e.g. DeHorne (1916), Hämmerling (1924), and Meewis (1933, 1938). According to the above described patterns, two different situations may be distinguished. In one case a new fission zone is formed within a fully differentiated segment; in the other a new zone develops in new tissues consisting of more or less undifferentiated cells stemming from the previous fission zone. The first situation is always met with in the first fission zone, and in the stylarian form also in the subsequent ones, whereas the second situation is found in connection with the majority of fission zones of the naidian form. In the following the main outlines are given of the events taking place, when a new fission zone is formed within a fully differentiated segment, and only the formation of a new head end is dealt with.

Most authors seem to favour the view that the various organs are produced from cells originating from the same germ layer as the organ in question. With respect to tissues of endodermal and mesodermal origin, organogenesis is fairly simple. At an early stage, a multiplication of embryonic peritoneal cells occurs which develop into muscles, septa, and other mesodermal elements. Later, when the gut is interrupted, certain basal cells of the gut epithelium proliferate, extend forward, and produce the endodermal portion of the new intestine. The production of new organs of ectodermal derivation is more complex, especially the formation of a new nervous system. The formation of a fission zone begins by a thickening and proliferation of the surface epithelium and a subsequent penetration of ectodermal elements into the body cavity through interruptions in the muscular coat of the body wall. Two proliferations grow up on the inside of the body wall, one on each side; later they bend inward and unite to form the new cerebral ganglion (brain). The lateral portions of these proliferations form the circumpharyngeal connectives which ventrally join the ventral nerve cord. Later, another ventral ectodermal invagination is formed, which develops into the new stomodaeum and fuses with the endodermal oesophagus.

A slight constriction already appears in connection with the proliferation of the surface epithelium, marking the anterior and posterior parts of the fission zone. This constriction gradually increases in depth, especially ventrally. The nerve cord and other internal organs are thereby divided, and ultimately the constriction progresses to a complete separation of the two individuals.

B. Parthenogenesis

Although parthenogenetic reproduction probably occurs within both leeches and oligochaetes, cytologically verified cases are only known among the latter. In the present treatment the conventional distinction between meiotic and mitotic parthenogenesis is used. In the former, meiosis still occurs in the developing oocyte, but is compensated for by a doubling of the chromosome number at some stage. In the latter, meiosis is in various ways suppressed, the maturation division(s) being mitotic in character.

1. Meiotic parthenogenesis

Depending upon the precise manner in which the somatic chromosome number is restored, several different types of meiotic parthenogenesis can be distinguished. Among annelids only two are of wider occurrence. In one of these, the second meiotic division is abortive and the original chromosome number is restored by fusion between second-division sister nuclei, i.e. between the female pronucleus and the second polar body (Fig. 2). This mode of restoration has been described in parthenogenetic species belonging to many different animal groups, but among oligochaetes it is of rare occurrence. Christensen (1961) has described it in a few enchytraeid species.

The most common type of meiotic parthenogenesis among annelids involves a premeiotic doubling of the chromosome number. Several cases have been demonstrated among earthworms of the family Lumbricidae and within the aquatic family Tubificidae (for a review, see Christensen, 1980a). Details concerning the cytology of this premeiotic doubling are given in Omodeo (1952; Fig. 2). Doubling occurs at the last oogonial division which is not completed. The abnormally short and condensed chromosomes all become included in a single nuclear membrane at a stage corresponding to telophase. Thus the oogonial nucleus contains two times the somatic chromosome number, when it enters the first meiotic division. Synapsis in the oocyte prophase is completely normal, and the number of bivalents formed corresponds to the number of somatic chromosomes (Fig. 3). The bivalents show typical chiasmata and the meiosis proceeds in an entirely regular way. Two polar bodies are given off and the female pronucleus, which contains the somatic chromosome number, becomes the developing nucleus in the embryo. Doubling also occurs in the male tissue, leading to the degeneration of most cells. But, at least in tubificids, a few morphologically normal spermatozoa are produced in parthenogenetic forms.

2. Ameiotic parthenogenesis

The cytology of ameiotic parthenogenesis is in most cases very simple. There is no synapsis, no bivalents are formed and only a single maturation division, which is an ordinary mitosis, occurs in the oocyte. Among the annelids this type of parthenogenesis is apparently of rare occurrence. In some hexaploid (and aneuploid?) forms of *Dendrobaena octaedra*, Omodeo (1955) describes a first-oocyte division where the late prophase and metaphase stages contain the somatic number of univalents. Although the details of restitution are unknown, it is probably safe to assume an ameiotic parthenogenesis following the general outline given above.

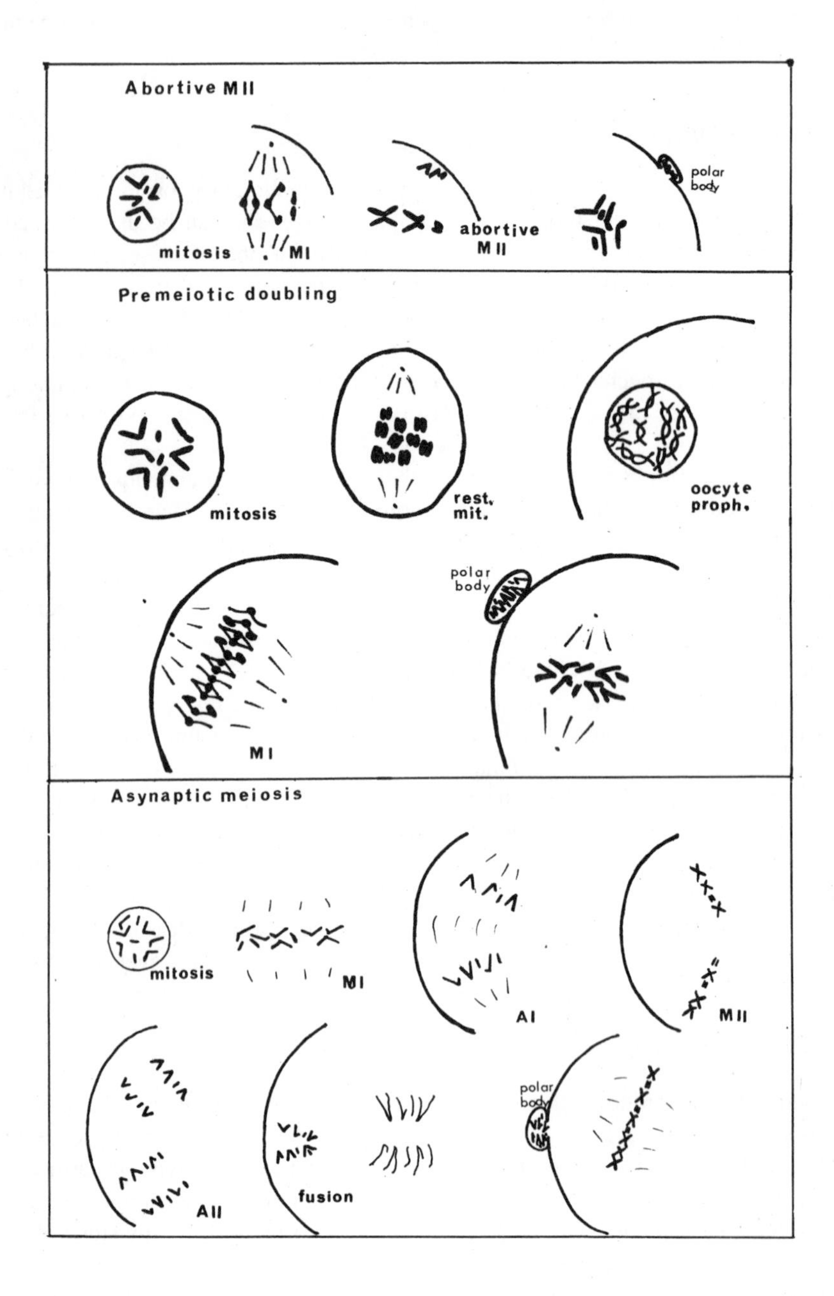

Fig. 2. Main type of parthenogenesis in oligochaetes. For further details, see text.

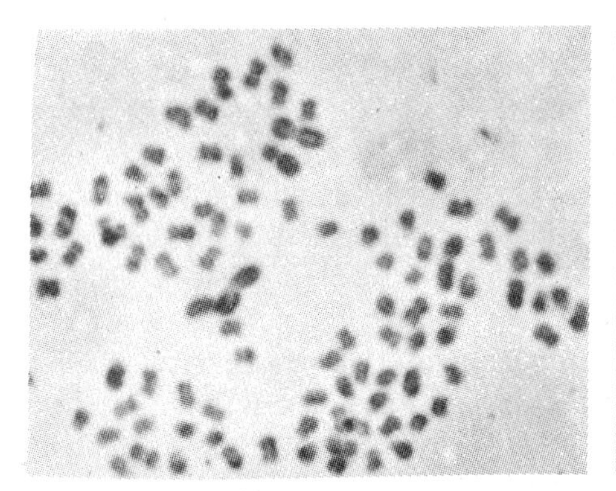
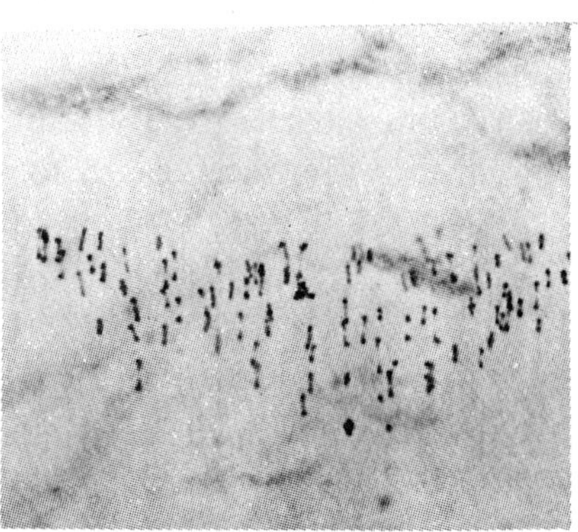

Fig. 3. *Tubifex tubifex*. To the left a mitosis and to the right a first meiotic division showing the 'diploid' number of bivalents (approx. 100) characteristic of parthenogenetic species with a premeiotic doubling of the chromosome number.

The parthenogenesis in tri-, tetra-, and pentaploid forms of the enchytraeid *Lumbricillus lineatus* follows an entirely different pattern (Christensen, 1960, 1980a; Fig. 2). In the oocyte prophase no pairing occurs between homologous chromosomes, and when the nuclear membrane disappears the chromosomes move towards the equatorial region, but no typical metaphase plate is formed. Anaphase movement follows immediately without any separation of chromatids, and the chromosomes are distributed in approximately equal numbers to the two poles, where they are arrested in a mid-anaphase stage. This lasts until the eggs are laid; then the spindle begins to elongate, and the daughter nuclei move further apart. Later the spindle bends and becomes V-shaped, with the apex of the V oriented towards the egg membrane. Anaphase movement is now arrested, and the chromosomes of the two nuclei are arranged in two second-metaphase plates. The chromosomes then divide equationally and a second anaphase stage follows, in which the chromosomes move along the still continuous spindle. This means that two of the chromosome complements, derived from each of the first-division daughter nuclei, move towards the apex of the V, and the two others towards the tips of the arms of the spindle.

The two chromosome groups, which move towards the apex of the spindle, fuse at this point and near the cell membrane. The chromosomes of the two nuclei, at the tip of the two arms of the spindle, move towards each other and fuse in the interior of the egg. As a result of these events the egg cell now possesses two nuclei, each containing the complete somatic chromosome number irrespective of the numerical distribution of the chromosomes during the first meiotic division. These two nuclei might become included in each of the two first blastomeres, or the functional nuclei of the first two blastomeres are daughter nuclei resulting from the first division of the central nucleus. In this case the peripheral nucleus is extruded as a polar body (Christensen, 1980a).

IV. GENERAL ASPECTS OF ASEXUAL PROPAGATION

Asexual reproduction, and parthenogenesis in particular, has received much attention from evolutionary biologists in the recent decades, because obligatory parthenogenesis represents an interesting alternative to the predominant sexual reproduction (see Maynard Smith, 1979, 1989). In dioecious animals parthenogenesis has a two-fold advantage in reproduction because only daughters are produced. On the other hand, parthenogenetic varieties lack the ability to bring together genetic material from two parents in a single offspring, and they therefore evolve more slowly than sexual populations and are, furthermore, in danger of accumulating harmful mutations. According to these two opposing possibilities, asexual reproduction is in some explanations considered a positive advantage, in others a defensive measure, an escape mechanism to avoid sexual sterility caused by polyploidy.

In the following, some recent observations dealing with these aspects are presented (see also Christensen, 1984).

A. Genetical Consequences of Asexual Propagation

The precise genetical principles are expected to differ in the different types of asexual reproduction. In propagation through fission and in ameiotic parthenogenesis, meiosis has been abolished; the nuclear divisions resulting in a new generation are mitotic. Segregation will not occur so that mutations and structural rearrangements of the chromosomes will tend to accumulate indefinitely. In meiotic parthenogenesis, meiosis takes place in the egg, and segregation may occur, when the mother individual is heterozygous. Heterozygosity is, however, expected to be rare in such forms, since the modes in which the somatic chromosome number is restored favour homozygosity. In time the expectation is a loss of heterozygosity, even though it might be a slow process in some cases.

All kinds of asexual reproduction constitute closed genetic systems, in which the recombination of genes from different individuals associated with sexuality is no longer possible. Consequently one might expect that particularly adaptive combinations of genes are automatically fixed and perpetuated *ad infinitum*, whereas adaptively inferior combinations are eventually completely eliminated. Furthermore, having forsaken both segregation and recombination, asexual organisms have also sacrified the two important mechanisms which accelerate the evolutionary process in sexual organisms.

The above theoretical postulations can be summarized as follows: (1) A high degree of heterozygosity in case of propagation through transverse fission and ameiotic parthenogenesis; (2) homozygosity in case of meiotic parthenogenesis; (3) strong dominance of particularly adapted genotypes; and (4) lack of genetic plasticity. However, recent observations on the genetic structure of asexual populations throw doubt upon some of these postulations (for a general review, see

Suomalainen *et al.*, 1987). The following studies on clitellate worms address these questions.

Although the mechanism for the restoration of the somatic chromosome number in the polyploid forms of the enchytraeid *Lumbricillus lineatus* represents a number of unique features (see Section III B2), the genetic consequence is that of a mitotic division and the expectation an accumulation of mutations resulting in a high degree of heterozygosity. The genetical composition of the triploid, parthenogenetic *L. lineatus* and of the diploid, cross-breeding form of the same species was studied at a number of stations along the Danish coastline. The two biotypes invariably occur together, because sperm from the diploid will have to activate the triploid parthenogenetic eggs in order for these to develop into viable embryos (Christensen and O'Connor, 1958; Christensen, 1960).

The two loci, PGM and PGI, were used as genetical markers in a study by Christensen *et al.* (1976). The results are summarized in Fig. 4. The left half of each column shows the frequencies of electrophoretic phenotypes in a sexual, diploid population; the right half gives equivalent data for the sympatric parthenogenetic, triploid form. It is clear that the expectation of a high degree of heterozygosity in triploid populations is by no means fulfilled. The frequency of heterozygotes among the latter is, in fact, in most cases very similar to that of the sympatric sexual diploid. Thus the idea of a high degree of heterozygosity as an automatic consequence of ameiotic parthenogenesis is apparently open to serious questioning. The weakness of this argument is that the genetic mechanism in question certainly makes a high degree of heterozygosity possible, but whether this is also the case depends upon the relative fitness of the genotypes in question. When the new mutations are neutral, or if heterozygotic superiority is a universal phenomenon, heterozygosity should prevail; but apparently none of these conditions is fulfilled. Studies on the adaptive differentiation among clones are discussed in Section IV C.

In the case of meiotic parthenogenesis the expectation is more or less complete homozygosity. Genetic studies on the enchytraeid worm *Fridericia striata* indicate that this situation is met with in most cases (Christensen *et al.*, 1989); 12 out of 13 polymorphic loci were homozygous for different allelic states in six different clones. Only one locus occurred in a two-banded state in two different clones; if this indicates heterozygosity, linear repetition of the locus in question may be a possible explanation of this observation.

The expectation of a strong dominance of particular genotypes in populations of asexual organisms is more or less automatically fulfilled because of their lack of segregation and recombination. However, completely monoclonal populations are rare. Observations on parthenogenetic tri- and tetraploid forms of *Lumbricillus lineatus* in two recently established environments have shown that, within both biotypes, no less than three to four different genotypes are of common occurrence, and in addition a few rare ones are usually also present (Christensen *et al.*, 1978). Observations on several other parthenogenetic organisms have revealed the same

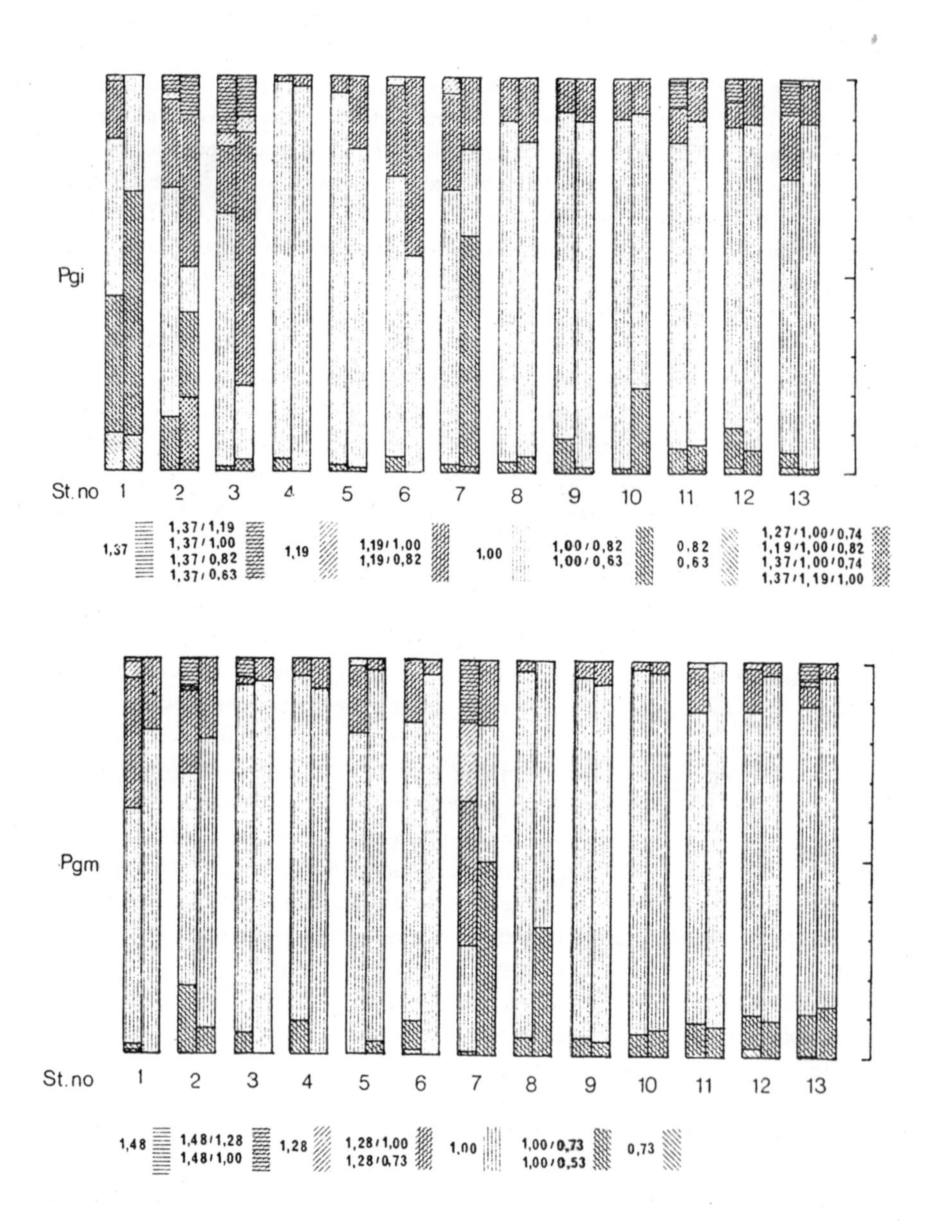

Fig. 4. Histograms showing the frequencies of phenotypes in sympatric diploid and triploid populations of *Lumbricillus lineatus* from 13 stations along the Danish coastline. The left half of each column shows the diploid population, the right half its sympatric triploid. Notice the strong similarity between the two biotypes in most stations and the fact that heterozygotes occur at fairly low frequencies in most triploid populations. The loci PGI and PGM are used as genetical markers. The symbols used are shown below each row of columns. (From Christensen *et al.*, 1976; reproduced by permission of *Hereditas*.)

degree of diversity, and the general conclusion is that clonal polymorphism is the rule rather than the exception (Noer, 1988).

The expectation regarding lack of genetic plasticity is a difficult problem to study, because it has to be measured on an evolutionary time scale whether asexual organisms are more likely to suffer extinction than sexual ones as a result of environmental changes, to which they cannot respond rapidly enough by adaptive modifications. In the studies dealt with above, which cover four to five years (Christensen *et al.*, 1978), some changes observed within a station might indicate long-term trends. If this is an adaptive response to changes in the environment, then these parthenogenetic populations respond by changes in the frequencies of pre-existing genotypes, and in this way polymorphic, parthenogenetic populations possess some evolutionary flexibility. Furthermore, the tetra- and pentaploid forms of *Lumbricillus lineatus* in question may be of very recent origin, perhaps in the order of 100 generations (Christensen *et al.*, 1978). If this holds true, and the genetic variation observed has evolved within such a small span of time, asexual organisms might not necessarily be doomed at a relatively brief evolutionary career, as is generally said.

However, this argument depends strongly upon the way in which the variation has arisen. If the polyclonal status is due to repeated origins of new parthenogenetic lineages from an ancestral sexual form (polyphyletic origin), the observed polymorphism merely reflects the genetic variation in this ancestral form. Only the situation in which the origin of asexuality is a unique event and the observed variation stems from an accumulation of mutations within this lineage (monophyletic origin) indicates the evolutionary capability of an asexual taxon. It is often difficult to distinguish between these two possibilities. In the case of *Fridericia striata*, Christensen *et al.* (1989) argue that the clonal diversity is of both mono- and polyphyletic origin. The same would seem to apply to the polyclonal, parthenogenetic form of *F. galba* (Christensen *et al.*, 1992).

B. Asexual Propagation and Polyploidy

In the few animal groups where polyploidy is of common occurrence, asexual reproduction of various kinds is also widespread. This association, although it is by no means universal, raises the question of what came first, asexual reproduction or polyploidy.

Transverse fission is apparently universal in the oligochaete families, Aelosomatidae and Naididae. In Aelosomatidae, reproduction through eggs is of rare occurrence, at least in some species, whereas in the Naididae this mode of reproduction is a regular seasonal phenomenon in most species. Knowledge of chromosome numbers is scarce in aelosomatids. Only one species, *Aelosoma hemprichi*, has been studied (Jelinek, 1974; Christensen, 1980a), and a diploid and a polyploid biotype are known. In naidids chromosome data from 12 species indicate that this is the only major oligochaete group where polyploidy is extremely rare (Jelinek,

1974; Christensen, 1980a). Thus in these taxa, where transverse fission is a fundamental feature, it is obviously a regular reproductive strategy, and not an 'escape' mechanism caused by polyploidy.

Other instances of fission have clearly arisen as a secondary phenomenon in groups with normal sexual reproduction, and the situation here could be different. The known cases include species of the genus *Lumbriculus* belonging to the family Lumbriculidae, a few tubificids (Hrabě, 1937; Loden, 1979), and four species of enchytraeids belonging to three different genera (Christensen, 1980a). *Lumbriculus variegatus* 2x and the enchytraeid *Buchholzia appendiculata* are diploids; their gametogenesis is entirely normal. In *Enchytraeus bigeminus* octo- and decaploid biotypes are known, but in spite of the high chromosome numbers gametogenesis is entirely normal, and both forms can reproduce through cross-fertilized eggs. In two polyploid *Cognettia* species (Enchytraeidae), the eggs that are occasionally produced develop parthenogenetically. In the polyploid biotypes of *L. variegatus* an abnormal meiosis has been observed, but a complete breakdown is only found in the pentaploid. In this latter biotype, fission might therefore be the only possible reproductive mechanism. However, even in the diploid form of this species sexually mature individuals are extremely rare — at least in North Zealand, Denmark — and transverse fission must consequently be the principal reproductive method also in this ancestral biotype with a normal gametogenesis. The same applies to the diploid *B. appendiculata*. The conclusion reached above that transverse fission is a reproductive strategy of its own and not an 'escape' mechanism, can, therefore, be extended also to these instances of obvious secondary origin (Christensen, 1980a).

Parthenogenesis has also arisen independently on many different occasions, and the majority of parthenogenetically reproducing oligochaetes are polyploids. This would seem to support the generally admitted view that parthenogenesis and polyploidy are strongly associated in animals. However, among the oligochaetes, this only applies to the families Tubificidae and Lumbricidae. In Enchytraeidae the majority of polyploids are sexual cross-breeders, and the high number of polyploids in this family with a normal meiosis apparently represents a unique feature among animals. This may be due to the common occurrence of bivalents with a single chiasma and the so-called achiasmatic bivalents, which completely prevent or strongly reduce the formation of multivalents even in autopolyploids. Such specialized bivalents are already present in diploid members of genera and species-groups, where polyploidy is a common phenomenon (Christensen, 1961, 1980a).

But is then the widespread occurrence of parthenogenetic reproduction among tubificids and lumbricids a consequence of polyploidy — some kind of 'escape' mechanism? Apparently not. The predominant type of parthenogenesis, involving the characteristic premeiotic doubling of the chromosome number, is also known to occur in diploids, i.e. in the earthworm *Octolasium lacteum* (see Muldal, 1952) and the eudrilid *Hyperiodrilus africanus* (see Tuzet and Vogeli, 1957). This particular reproductive strategy has, therefore, evolved because it is under some circumstances competitive, and the strong association between parthenogenesis and polyploidy in lumbricids and tubificids is rather due to the fact that this type of parthenogenesis permits the development of polyploidy, even odd-numbered polyploids. Since bivalents are undoubtedly derived from the pairing of sister-homologues (chromosomes derived from the same parent chromosome during the premeiotic doubling), the usual complications associated with the presence of more than two identical genomes are avoided. Furthermore, when a parthenogenetic biotype copulates with the ancestral sexual form, and this is known to occur, the level of ploidy may be elevated by adding a haploid chromosome set without disturbing the parthenogenetic mechanism.

Parthenogenesis first, polyploidy next, would therefore seem to be the general rule among oligochaetes. Possible exceptions are the parthenogenetic forms of *Lumbricillus lineatus* and *Fridericia galba* (see Christensen, 1980a).

Thus, the conclusion is that most types of asexual propagation in oligochaetes are reproductive strategies of their own, and the strong association with polyploidy in some groups is a secondary phenomenon caused by the fact that the particular type of asexual mechanism represents no barrier to an increase in the number of chromosome complements.

C. Ecological Implications of Asexual Propagation

It appeared from the previous section that the various types of asexual reproduction in oligochaete annelids have arisen as reproductive strategies which are able to compete successfully with sexual reproduction under natural conditions. This raises the following questions: (1) Is asexual reproduction favoured by particular environmental conditions? (2) Has asexual propagation a numerical advantage over sexual reproduction? (3) Is the fixed breeding structure in asexual populations (Section IV A) an advantage in some environments?

In order to study the first question, species with an alternation of sexual and asexual phases are the obvious choice. The naidids, where there is a regular seasonal alternation in many species, have been the subject of many studies. Such factors as quantities of nutrition and changes in temperature and oxygen contents of the water have been investigated. But none of the results can be extended to naidids in general, except perhaps that asexual propagation tends to predominate under good nutrient conditions (for reviews, see Stephenson, 1930; Lasserre, 1975). Van Cleave (1937) and Hyman (1938) relate the influence of the above factors to Child's (1915) theory of a metabolite emanating from the head, which exerts control over the developmental processes in the posterior region (see Hughes, 1989 for review).

In naidids sexual reproduction mainly occurs, if at all, after the populations have reached peak densities by fission. The proportion of individuals becoming sexual varies from zero to unity among populations of the same or different species (Timm, 1973; Learner *et al.*, 1978). The worms die after completing their sexual activity. In the parasitic species *Chaetogaster limnaei*, the proportion of sexually mature individuals varies with temperature, being higher in Russia (Vaghin, 1946) than in North Wales (Gruffydd, 1965).

The enchytraeid *Enchytraeus bigeminus* reproduces in dense laboratory cultures exclusively by transverse fission, whereas, if the worms are kept at low densities, sexual organs are developed, and a sexual reproduction via eggs commences alongside the asexual proliferation. However, worms hatched from eggs do not develop sexual organs when kept at low densities; apparently they will have to pass through a number of asexual generations before this can happen again (Christensen, 1973). In *Buchholzia appendiculata*, another enchytraeid with the same reproductive pattern, sexually reproducing individuals are found within small restricted areas, whereas transverse fission dominates completely in the intervening much larger areas. The former habitats did not differ in any consistent way from the latter, and in contrast to *E. bigeminus*, sexual reproduction occurred here in the most dense populations (Christensen *et al.*, unpublished). Thus, neither in limnic nor in terrestrial environments has any single biotic or abiotic factor which favours asexual reproduction been identified.

Naidids have relatively low individual sexual fecundity. Lochhead and Learner (1983) showed that, in *Nais variabilis*, cloning would outnumber sexual reproduction by several orders of magnitude. And in naidids, in general, asexual reproduction undoubtedly has a considerable numerical advantage. In contrast to this, the reproductive rates were very similar in these two situations, in the above-mentioned studies on *Enchytraeus bigeminus*. But it has also been shown that dense cultures of *E. bigeminus* not only suppress its own sexual reproduction but also that of some potential competitors which are obligatory sexual breeders. Thus, in this particular case, asexual reproduction may be an important factor in relation to interspecific competition. But this is probably no general mechanism.

The third question above has been addressed in some studies from our laboratory. In a polymorphic asexual population the various clones are 'frozen' genetically, and since they are reproduced unchanged from generation to generation, the same genotypes are "tested" generation after generation. Selection is in a way enhanced in asexual compared with sexual populations, and selective differentials between genotypes should turn up more readily in the former. Furthermore, in case such selective differentials exist between clones, their fixed breeding structure may result in a situation where distinct genotypes occupy particular microhabitats in a heterogeneous environment.

The first of these possibilities was studied in the naidid *Stylaria lacustris* (Christensen, unpublished). This species reproduces asexually throughout spring and summer. In late autumn all individuals become sexually mature and produce cross-fertilized eggs, which overwinter and give rise to the asexually reproducing individuals that appear next spring. Since the eggs are produced through cross-fertilization, the clones (genotypes) that give rise to the asexually reproducing summer generations are expected to occur in Hardy-Weinberg proportions. But if selection during the asexual phase picks out and propagates clones (genotypes) especially suitable for the available environment, strong deviations from the Hardy-Weinberg proportions are expected to occur among the individuals entering a new sexual phase the following autumn.

Table 1 shows the results obtained from a study of two polymorphic loci in three different lakes in North Zealand, Denmark. The observed distributions of genotypes among the sexuals in the autumn are in all cases very close to those expected, indicating that no changes have occurred in their relative frequencies during the preceding asexual phase. Thus, with respect to the present loci at least, *Stylaria lacustris* would not seem to exploit the possibility of radical genetical changes associated with asexual propagation.

This absence of response may be due to too few asexual generations between the reorganizations of the entire genetic material occurring in each sexual generation. In *Buchholzia appendiculata*, where asexual reproduction is much more dominant, Christensen *et al.* (unpublished) observed a situation more like the one expected. In this species, populations with sexual reproduction may be viewed as centres where all possible combinations of genotypes arise and eventually invade surrounding areas during the process of asexual reproduction. It was shown that the genetic diversity in asexual populations was significantly lower than that of sexual populations, so obviously the various genotypes (clones) arising in the sexual populations do not survive equally well in the surrounding asexual populations. Furthermore, constant differences in the distributions of genotypes were observed in a number of neighbouring beach and grass sites, indicating a differential distribution of clones in this heterogeneous environment.

This latter phenomenon was also observed in parthenogenetic forms of *Lumbricillus lineatus* along the shore of Hvide Sande Canal, Denmark (Christensen, 1980b). Tri-, tetra-, and pentaploid biotypes were present, and within each of these a varying number of electrophoretically distinct clones. Along a transect (4.5 m) extending from low tide level to just above the highest tide level, the three cytotypes as well as the different genotypes within them showed a highly differential distribution (Fig. 5). This distribution was constant both in space (two transects

Table 1

The genetic structure in three autumn populations of sexually mature *Stylaria lacustris* with respect to the two loci, EST and PGM

Locality	EST					PGM				
	f	m	s	N	P	f	m	s	N	P
Løgsø		0.62	0.38	106	< 75	0.02	0.98		100	< 50
Furesø		0.89	0.11	55	< 75	0.03	0.97		51	< 99
Sjaelsø	0.04	0.95	0.01	67	< 80	0.06	0.92	0.02	70	< 50

f, m, and s, Fast-, medium-, and slow-moving alleles; N, sample size; P, probability of worse fit.

15 m apart) and in time (two consecutive years), indicating the ability of genetically distinct clones in a heterogeneous environment to use distinct subniches, to which they may be specifically adapted. For the population as a whole, this probably means a highly efficient exploitation of the resources in a heterogeneous environment, and due to their fixed breeding structure multiclonal asexual species might better exploit this possiblity than sexual species.

However, not all findings lead to the above conclusions. Jaenike *et al.* (1979) and Jaenike and Selander (1985), on the basis of studies on the parthenogenetic earthworm *Octolasium tyrtaeum* in North America, are inclined to think that the two major clones, which occur over a wide variety of different habitats, are ecologically similar, or at least that any ecological diversification that might exist is completely overriden by random processes and clonal drift, occurring when populations are periodically decimated by extreme environmental conditions.

The two opposing possibilities indicated by the above examples may reflect a very real dichotomy with respect to profitable strategies to be adopted by asexual organisms. A number of narrowly adapted subniche-specialists may be an optimal strategy in a stable and strongly heterogeneous environment, whereas one or a few 'general-purpose' genotypes may be a better solution in homogeneous and/or unpredictable environments.

D. Transverse Fission and Phylogeny

As mentioned previously (Section II), some authors (e.g. Michaelsen, 1928) consider Naididae and Aelosomatidae two of the most primitive existing families of Oligochaeta. But other authors, e.g. Stephenson (1930) and followers, do not share this view. The Aelosomatidae remain an enigma. Some anatomical features are clearly primitive, but in other ways they are highly modified and specialized (Brinkhurst and Jamieson, 1971), and the view that the Naididae are primitive is also doubtful. The naidids and tubificids are closely related, but it is a matter of discussion which family is the ancestral one and which family the derived. About the only real distinction between the two families is the anterior position of reproductive organs in the Naididae. The basic argument presented here is that an apparently universal association between the number of new anterior segments formed during regeneration and the position of the gonads during the sexual phase in such species may clarify the question about the systematic position of naidids.

It is well known that there is a forward shift in the position of the testes and ovaries in those enchytraeid and tubificid species where asexual fission alternates with reproduction through eggs. In the enchytraeid species in question (i.e. *Buchholzia appendiculata* and *Enchytraeus bigeminus*), this forward shift shows a constant relation to the number of new segments in the regenerated anterior end (Fig. 6). The ovaries are always located upon the new septum formed between the posterior new segment and the anterior old segment, and they extend rearwards

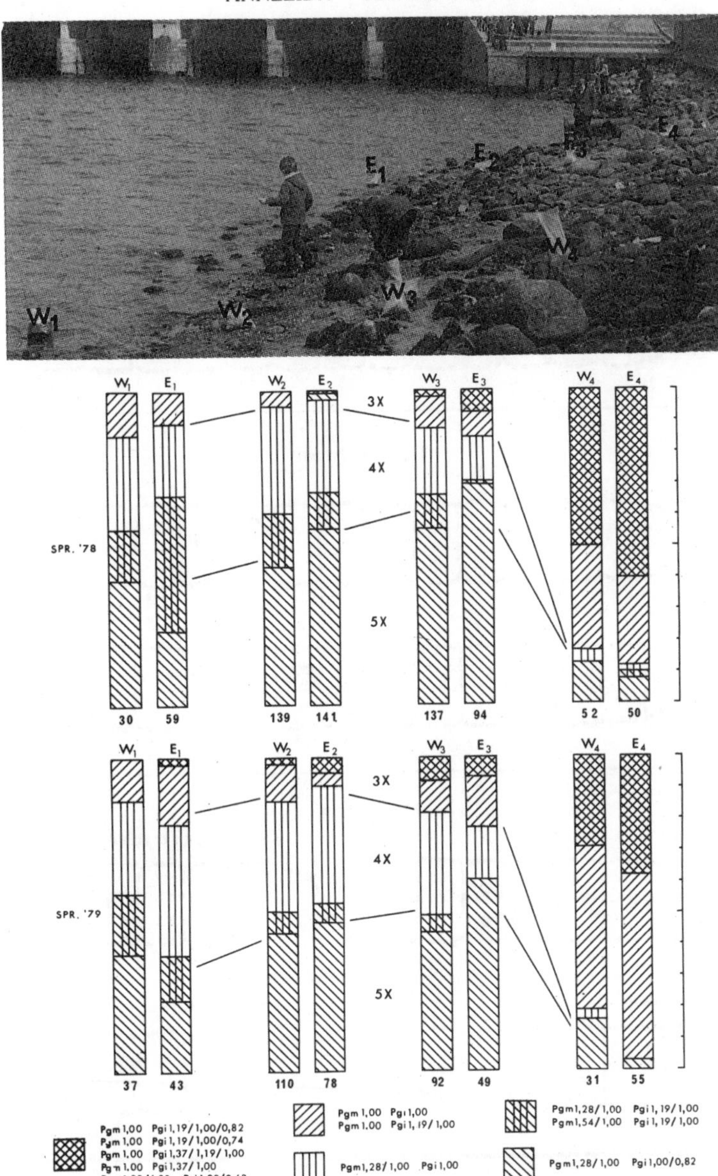

Fig. 5. Histograms showing the distribution of tri-, tetra-, and pentaploid biotypes and various electrophoretic phenotypes in *Lumbricillus lineatus* along the two transects (W and E) shown above. Observations from spring 1978 and spring 1979 are shown. In each pair of columns the one to the left shows the population at the western transect and the column to the right the population from the site at the same level in the eastern transect. Notice the strong differential distribution of the three chromosome types and the differential distribution of genotypes especially among the triploids. The number below each column gives the sample size. The symbols used for electrophoretic phenotypes are shown below. (From Christensen, 1980b).

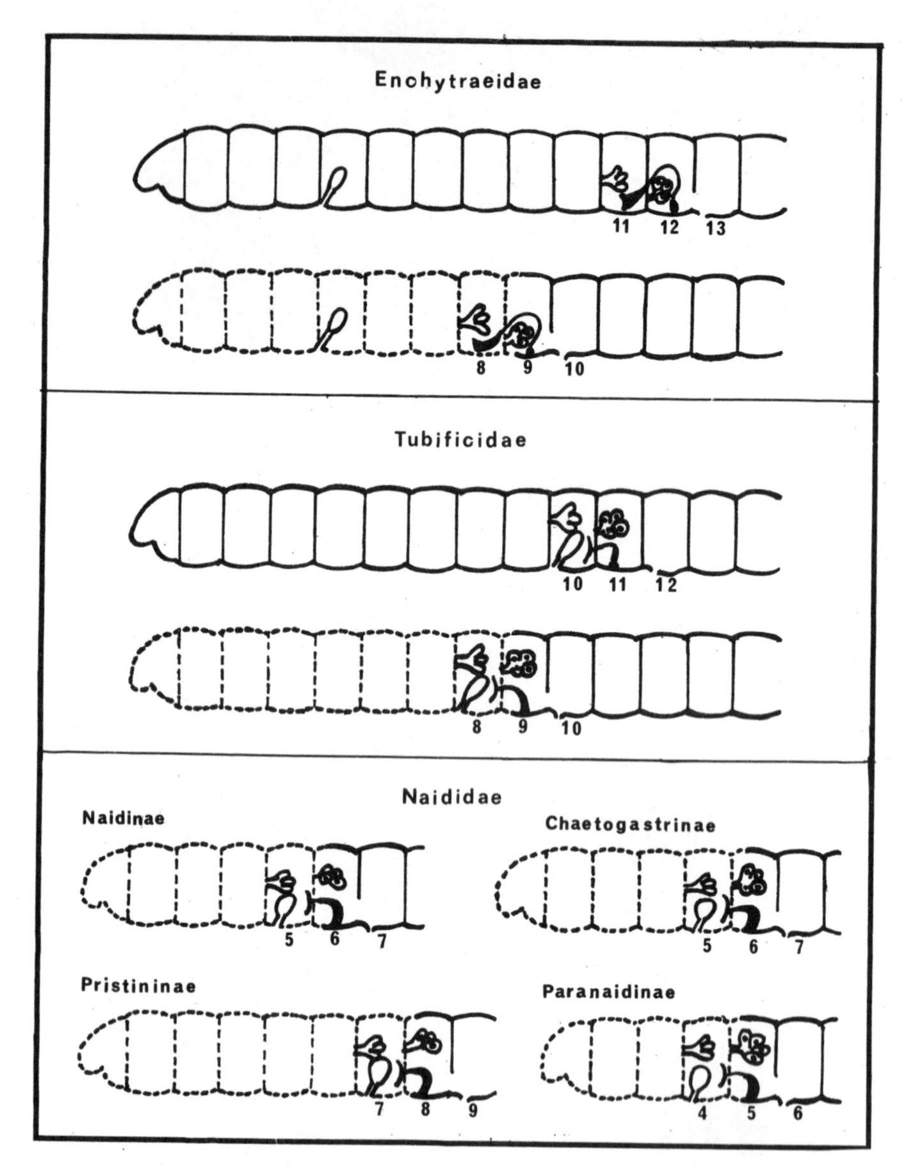

Fig. 6. The relation between the position of sexual organs and the number of new, regenerated anterior segments (broken lines) in some oligochaetes where asexual and sexual reproduction alternate.

into the latter (the anterior portion of this segment consists of new tissue). As in other enchytraeids the testes are located in the segment in front of the ovarian segment. The reproductive organs thus maintain their normal sequential arrangement, but are shifted forward in such a way that the gonads are now located on the two hindmost new septa formed during the regeneration associated with asexual reproduction.

In the few species of tubificids with both fission and sexual reproduction, the same principle would seem to apply (Hrabé, 1937). Loden (1979) reports an interesting case with respect to *Tubifex harmani*. In this species, 10 new segments are formed anteriorly after fragmentation, and, as expected, the testes are located in segment 10 and the ovaries in segment 11 in sexually mature individuals. However, this is also the normal position of the reproductive organs in the family (Tubificidae), and it shows that fragmentation as such does not automatically cause a shift to a more anterior position than that characteristic of the family in question.

The interesting fact is that, according to the informations given in Sperber (1948), exactly the same principle would seem to apply also to the naidids, where the alternation of transverse fission and sexual reproduction is a fundamental feature of the family (Fig. 6). The anterior position of the reproductive organs in naidids, compared with that of tubificids, would therefore seem to be a mere consequence of the obligatory alternation of reproductive methods in the former. Consequently, if the origin of transverse fission is a secondary phenomenon, which has arisen repeatedly among Oligochaeta, the conclusion would seem to be that naidids are merely tubificids that have adopted this reproductive method as a constant feature in their reproductive strategy. If this holds true, it raises the question whether naidids deserve the rank of a family, and whether it is a monophyletic group after all [see also Brinkhurst (1986) and Erséüs (1990)].

REFERENCES

Avel, M. (1959). 'Classe des Annélides Oligochètes (Oligochaeta)', in *Traité de Zoologie* (Ed. P.-P. Grassé), Vol. V, Pt. 1, Masson, Paris, pp. 224–470.

Berrill, N.J. (1952). 'Regeneration and budding in worms', *Biol. Rev.*, **27**, 401–438.

Brinkhurst, R.D. (1986). Guide to the freshwater aquatic microdrile of North America, *Can. Spec. Publ. Fisher. Aquat. Sci.*, **84**.

Brinkhurst, R.D., and Jamieson, B.G.M. (1971). *Aquatic Oligochaetes of the World*, Oliver and Boyd, Edinburg.

Child, C.M. (1915). *Individuality in Organisms*, Univ. of Chicago Press, Chicago.

Christensen, B. (1959). 'Asexual reproduction in the Enchytraeidae (Olig.)', *Nature. Lond.*, **184**, 1159–1160.

Christensen, B. (1960). 'A comparative cytological investigation of the reproductive cycle of an amphimictic diploid and a parthenogenetic triploid form of *Lumbricillus lineatus* (O.F.M.) (Oligochaeta, Enchytraeidae)', *Chromosoma*, **11**, 365–379.

Christensen, B. (1961). 'Studies on cyto-taxonomy and reproduction in the Enchytraeidae', *Hereditas*, **47**, 387–450.

Christensen, B. (1964). 'Regeneration of a new anterior end in *Enchytraeus bigeminus* (Enchytraeidae, Oligochaeta)', *Vid. Medd. Dansk Naturh. Foren.*, **127**, 259–273.

Christensen, B. (1973). 'Density dependence of sexual reproduction in *Enchytraeus bigeminus* (Enchytraeidae)', *Oikos*, **24**, 287–294.

Christensen, B. (1980a). 'Annelida', in *Animal Cytogenetics*, Vol. 2 (Ed. B. John), Gebrüder Borntraeger, Berlin, Stuttgart, pp. 1–79.

Christensen, B. (1980b). 'Constant differential distribution of genetic variants in polyploid, parthenogenetic forms of *Lumbricillus lineatus* (Enchytraeidae, Oligochaeta)', *Hereditas*, **92**, 193–198.

Christensen, B. (1984). 'Asexual propagation and reproductive strategies in aquatic Oligochaeta', *Hydrobiologia*, **115**, 91–95.

Christensen, B., Jelnes, J., and Berg, U. (1976). 'A comparative study on enzyme polymorphisms in sympatric diploid and polyploid populations of *Lumbricillus lineatus* (Enchytraeidae, Oligochaeta)', *Hereditas*, **84**, 41–48.

Christensen, B., Jelnes, J., and Berg, U. (1978). 'Long-term isozyme variation in parthenogenetic polyploid forms of *Lumbricillus lineatus* (Enchytraeidae, Oligochaeta) in recently established environments', *Hereditas*, **88**, 65–73.

Christensen, B., Hvilsom, M., and Pedersen, B.V. (1989). 'On the origin of clonal diversity in parthenogenetic *Fridericia striata* (Enchytraeidae, Oligochaeta)', *Hereditas*, **110**, 89–91.

Christensen, B., Hvilsom, M., and Pedersen, B.V. (1992). 'Genetic variation in coexisting sexual diploid and parthenogenetic triploid forms of *Fridericia galba* (Enchytraeidae, Oligochaeta) in a heterogenous environment', *Hereditas*, **117**, 153–162.

Christensen, B., and O'Connor, F.B. (1958). 'Pseudofertilization in the genus *Lumbricillus* (Enchytraeidae)', *Nature, Lond.*, **181**, 1085–1086.

DeHorne, L. (1916). 'Les Naidimorphes et leur reproduction asexuée', *Arch. Zool. exp. gén.*, **56**, 25–157.

Erséüs, C. (1990). Cladistic analysis of the subfamilies within the Tubificidae (Oligochaeta), *Zool. Scripta*, **19**, 57–63.

Gruffydd, L.D. (1965). 'The population biology of *Chaetogaster limnaei limnaei* and *Chaetogaster limnaei vaghini* (Oligochaeta)', *J. Anim. Ecol.*, **34**, 667–690.

Hämmerling, J. (1924). 'Über dauernd teilungsfähige Körperzellen bei *Aelosoma hemprichi* (Ehrbg)', *Biol. Zentralbl.*, **44**, 169–173.

Herlant-Meewis, H. (1946). 'Contribution a l'étude de la régeneration chez les oligochétes. Reconstitution du germen chez *Lumbricillus lineatus* (Enchytraeidae). Première Partie', *Ann. Soc. Roy. Zool. Belg.*, **77**, 5–47.

Herlant-Meewis, H. (1954). 'Etude histologique des Aelosomatides au cours de la reproduction asexuée', *Arch. Biol.*, **65**, 73–134.

Hrabé, S. (1937). 'Zur Kenntnis des *Lamprodrilus mrazeki* Hr., *Aulodrilus pluriseta* Pig. und *Aulodrilus pigueti* Kow', *Sb. Klubu prirodoved Brno*, **19**, 3–9.

Hughes, R.N. (1989). *A Functional Biology of Clonal Animals*, Chapman and Hall, London, New York.

Hyman, L.H. (1938). 'The fragmentation of *Nais paraguayensis*', *Physiol. Zool.*, **11**, 126–143.

Jaenike, J., Parker, E.D., Jr., and Selander, R.K. (1979). 'Clonal niche structure in the parthenogenetic earthworm *Octolasium tyrtaeum*', *Amer. Natur.*, **116**, 196–205.

Jaenike, J., and Selander, R.K. (1985). 'On the coexistence of ecologically similar clones of parthenogenetic earthworms', *Oikos*, **44** (3), 512–514.

Jelinek, H. (1974). 'Untersuchungen zur Karyologie von vier naidomorphen Oligochaeten', *Mitt. Hamburg. Zool. Mus. Inst.*, **71**, 135–145.

Lasserre, P. (1975). 'Clitellata', in *Reproduction of Marine Invertebrates*, Vol. III (Eds. A.C. Giese and J.S. Pearse), Academic Press, New York, pp. 215–275.

Learner, M.A., Lochhead, G., and Hughes, B.D. (1978). 'A review of the biology of the British Naididae (Oligochaeta) with emphasis on the lotic environment', *Freshwat. Biol.*, **8**, 357–375.

Lochhead, G., and Learner, M.A. (1983). 'The effect of temperature on asexual population growth of three species of Naididae (Oligochaeta)', *Hydrobiologia*, **98**, 107–112.

Loden, M.S. (1979). 'A new North American species of freshwater Tubificidae (Oligochaeta)', *Proc. biol. Soc. Wash.*, **92**, 601–605.

Marcus, E. (1943). 'Sobre Naididae do Brasil', *Boll. Fau. Fil. Ciênc. Letr. Uni. São Paulo, Brazil*, **32**, *Zool.*, **7**, 3–247.

Maynard Smith, J. (1979). *The Evolution of Sex*, Cambridge University Press, Cambridge.

Maynard Smith, J. (1989). *Evolutionary Genetics*, Clarendon Press, Oxford.

Meewis, H. (1933). 'Remarques au sujet de la scissiparité chez les Oligochètes. Lois d'apparition de la scissiparité chez *Chaetogaster diaphanus* Gruith', *Ann. Soc. Roy. Zool. Belg.*, **63**, 117–140.

Meewis, H. (1938). 'Etude de l'organogénèse lors de la reproduction asexuée chez les vers', *Ann. Soc. Roy. Zool. Belg.*, **68**, 147–194.

Michaelsen, W. (1928). 'Oligochaeta', in *Handbuch der Zoologie*, Vol. 2, Part 8 (Eds. I.W. Kükenthal and T. Krumbach), Gruyter, Berlin, pp. 1–118.

Muldal, S. (1952). 'The chromosomes of the earthworms. I. The evolution of polyploidy', *Heredity*, **6**, 55–76.

Noer, H. (1988). 'Clonal niche organization in triploid parthenogenetic *Trichoniscus pusillus*: A comparison of two kinds of evolutionary events', in *Population Genetics and the Theory of Evolution* (Eds. G. de Jong and F.R. van Dijken), Springer-Verlag, Berlin, pp. 191–201.

Omodeo, P. (1952). 'Caryologia dei Lumbricidae', *Caryologia*, **4**, 173–274.

Omodeo, P. (1955). 'Caryologia dei Lumbricidae, II Contributo', *Caryologia*, **8**, 135–178.

Sperber, C. (1948). 'A taxonomical study of the Naididae', *Zool. Bidr. Upps.*, **28**, 1–296.

Stephan-Dubois, F. (1956). 'Migration et différenciation des neoblastes dans la régénération anterieure de *Lumbriculus variegatus*', *C. R. Soc. Biol. Paris*, **150**, 1239–1242.

Stephenson, J. (1930). *The Oligochaeta*, Clarendon Press, Oxford.

Stolte, H.A. (1955). 'Ungeschlechtliche Fortplanzung und Regeneration (der Oligochaeten)', in *Klassen und Ordnungen des Tierreichs; Oligochaeta* (Ed. H.G. Bronn), Vol. 5, Akademische Verlagsgesellschaft, M.B.H., Leipzig, pp. 739–865.

Suomalainen, E., Saura, A., and Lokki, J. (1987). *Cytology and Evolution in Parthenogenesis*, RCR Press, Boca Raton, Florida.

Timm, T.E. (1973). 'The life cycle of some Oligochaeta', *Fish. Res. Board Can. Trans. Ser.*, No. 2838 (first published in *Trudy Karel'skogo Otdeleniya Gosniorkh.*, **5**, 202–204).

Tuzet, O., and Vogeli, M. (1957). 'Les chromosomes et la reproduction de deux formes d'*Hyperiodrilus africanus* forme *typica* Kinberg et forme vogelii Omodeo', *Bull. Inst. Fr. Afr. noire*, Ser. A, **19** (2), 400–411.

Vaghin, V.L. (1946). 'On the biological species of *Chaetogaster iimnaei* K. Baer', *Dokl. Akad. Nauk S.S.S.R*, **51**, 481–484.

Van Cleave, C.D. (1937). 'A study of the process of fission in the naid *Pristina longiseta*', *Physiol. Zool.*, **10**, 299–314.

Vorontsova, M.A., and Liosner, L.D. (1960). *Asexual Propagation and Regeneration*, Pergamon Press, London.

2. TARDIGRADA

ROBERTO BERTOLANI
Dipartimento di Biologia Animale, Università di Modena, Via Università 4, 41100 Modena, Italy

I. INTRODUCTION

Tardigrades (for classification, see Systematic Résumé), like other groups of small Metazoa, such as the Rotifera (see Gilbert, Volume VI A) and Nematoda, are amphimictic (dioecious or very rarely hermaphroditic) in marine habitats, and parthenogenetic in non-marine ones to the point that this may become the only type of reproduction. Parthenogenesis in tardigrades was only definitively documented about 30 years ago, and little information has been available on the distribution of parthenogenetic forms and on their relationships with the corresponding amphimictic ones. The first reports confirming parthenogenesis in tardigrades were the cytological investigation on *Hypsibius dujardini* by Ammermann (1962, 1967), rearing studies on *Milnesium tardigradum* by Baumann (1964), and karyological observations on *Macrobiotus richtersi* and *Ramazzottius (= Hypsibius) oberhaeuseri* by Bertolani (1971a, b, d). Today much more information is available and comparison with other animal groups is possible. However, tardigrades present some difficulties in interpretation due to certain unresolved problems in systematics and evaluation of the degree of intraspecies variability and thus of identification.

II. REGENERATION AND ASEXUAL PROPAGATION

Besides a physiological renewal of some cells in somatic tissues (Marcus, 1929; Bertolani, 1970a, b), regeneration is unknown among tardigrades. These animals do not appear able to repair lesions and are unable to propagate via vegetative reproduction. Reproduction in tardigrades is always linked to the production of female gametes even though fertilization is not necessarily implied. Only one animal hatches per egg and polyembryony can be ruled out.

III. PARTHENOGENESIS

A. Types

There are certainly cases of thelytokous parthenogenesis (obligatory and probably continuous) while other types of parthenogenesis, in particular heterogony (i.e. reproduction by both parthenogenesis and amphigony), have not been discovered. From a cytological point of view, parthenogenesis can be either meiotic (with a pairing of homologous chromosomes but with or without crossing-over) or ameiotic. It is not uncommon to find thelytokous populations that cannot be distinguished taxonomically from other, sympatric or not, amphimictic ones. In addition to the sex and the type of reproduction, they differ in ploidy in the majority of cases. Indeed, thelytoky is often associated with polyploidy, in particular triploidy.

B. Cytology

Parthenogenesis in tardigrades appears with a multitude of cytological modalities. Ammermann (1967) described the following processes for *Hypsibius dujardini*: The chromosome number in mitotic divisions is 2n=10. Homologous chromosomes pair at a very late stage in the first prophase of the oocyte; in fact, they pair just prior to metaphase. There is, therefore, no crossing-over. A normal metaphase is followed by an equally normal anaphase and telophase. The first division results in the formation of a polar body and an oocyte, both with a haploid number of chromosomes. In the oocyte the nuclear envelope re-forms and within it the five diads divide up into 10 chromatids, restoring the diploid number. This phase is followed by a further division which leads to the formation of the egg nucleus and of a second polar body, both diploid. In *Dactylobiotus parthenogeneticus* (cited as *Macrobiotus dispar* by Bertolani and Buonagurelli, 1975), the parthenogenetic maturation of the egg is similar in some aspects to that of *H. dujardini*. Here also the specimens are diploid (2n=10). In the first prophase of the oocyte there is a regular pairing of homologous chromosomes which display chiasmata (Fig. 1). After anaphase I the oocyte chromosomes tend to disappear, only to reappear after some time in the form of five pairs of chromosomes distributed along a plane (Fig. 2). Each chromosome consists of two partly uncoiled chromatids. The chromosomes of each pair may still be joined or lie just very close to one another. The arrangement in pairs is then lost. Another oocyte division subsequently takes place. In both *H. dujardini* and *D. parthenogeneticus*, the diploid chromosome number is restored between the first and second maturational divisions of the oocyte. The chromosomes thus undergo a total of three divisions, two which lead to the formation of polar bodies, one haploid and one diploid, and one which is endonuclear. Since the chromosomes at the first metaphase appear morphologically the same as in the oocytes of non-parthenogenetic females of other species, there must be a period of DNA synthesis between the two divisions of the oocyte. In *D. parthenogeneticus* this period may correspond to the temporary disappearance of chromosomes observed at the end of the first division of the oocyte. A similar restoration of the diploid number of chromosomes is not known in other animal groups.

As in many other animals, ameiotic parthenogenesis in tardigrades appears much more common and, for the most part, is associated with polyploidy. Since there are several cases of polyploid parthenogenetic and diploid amphimictic populations displaying the same taxonomic characteristics, the term 'biotype' and successively 'cytotype' (according to White, 1973) has been used to distinguish these populations. In fact, since genetic flow between two or more cytotypes is improbable, these forms represent a species-complex and a specific name should be used for each. Unfortunately, such precision would lead to considerable difficulties at the taxonomic level. Moreover, the possibility of breeding cannot be completely excluded (see below). Therefore, it is better to use the term 'cytotype' to designate the single components of the complex.

Observations on the various cases of ameiotic parthenogenesis do not go beyond the first metaphase, i.e. the final stage before oviposition, since they have not been made on cultured specimens. *Macrobiotus richtersi* has both the amphimictic diploid and parthenogenetic triploid cytotypes (Bertolani, 1971a, b, 1972a, b, 1975). The oocyte chromosomes of the parthenogenetic cytotype do not pair during prophase and metaphase. The presence of univalents suggests that only one division, certainly ameiotic, takes place. There are no males.

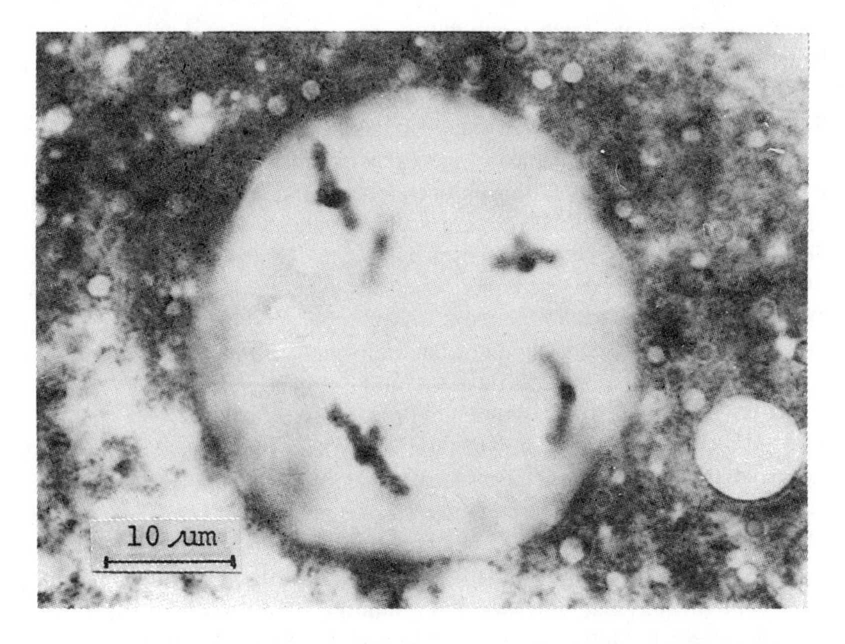

Fig. 1. Diplotene in a parthenogenetic oocyte of *Dactylobiotus parthenogeneticus* (orcein).

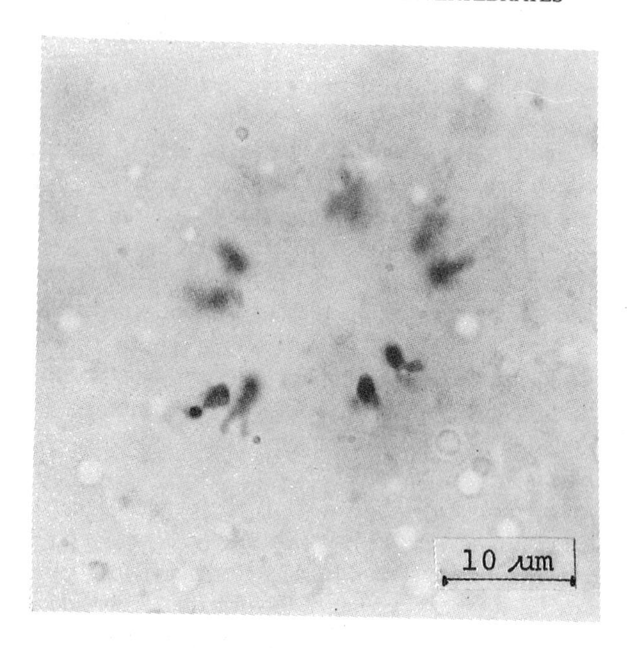

Fig. 2. Restoration of the diploid chromosome number in a parthenogenetic oocyte of *Dactylobiotus parthenogeneticus* (orcein).

Four cytotypes have been identified in *Ramazzottius oberhaeuseri*: one amphimictic diploid and three parthenogenetic thelytokous polyploid; one among the latter is tetraploid (Bertolani, 1971d, 1972b, 1973b, 1975; Rebecchi and Bertolani, 1988; Bertolani *et al.*, 1990). The diploid is less diffuse than polyploids. Oocyte division in the triploid and tetraploid cytotypes is similar to that of *Macrobiotus richtersi*, i.e. of the ameiotic type with univalents lining up on the equatorial plate during metaphase. Another cytotype showed 18 'bivalents' in the oocyte metaphase but the number of mitotic chromosomes is unknown. These different cytotypes, unlike in *M. richtersi*, may be syntopic. Note, however, that when the males were absent in a sample, only triploid females were present (and in one case tetraploid). Instead, when the males and triploid females were present, other cytotypes occurred. This finding suggests that some cytotypes could originate *in loco* by hybridization (Rebecchi and Bertolani, 1988; Bertolani *et al.*, 1990).

In *Macrobiotus hufelandi* the situation is further complicated by inadequate knowledge of the systematics. The species is currently under revision. A considerable number of cytotypes, certainly associated with several species, have been found. An amphimictic diploid cytotype and two cytotypes, one triploid and one tetraploid, which are ameiotic parthenogenetic have been reported (Bertolani, 1971c, 1972b, 1973a, 1975; Bertolani and Mambrini, 1977). One aspect of the cytology of the triploid cytotype is common to populations from different locations. Once coiling is complete, the univalents do not arrange regularly on the equatorial plane but remain scattered and tend to form two groups in a spindle-shaped vesicle. In the tetraploid cytotype the metaphase is rather regular and univalents are always present. Besides the parthenogenetic ameiotic cytotypes described for *M. richtersi, M. hufelandi* presents another triploid cytotype which can be distinguished from the former by the different cytology of parthenogenesis (Bertolani, 1975). This cytotype does not appear sympatric with the former and is characterized by a premeiotic replication of chromosome number. In fact, during oocyte prophase and metaphase the chromosomes resemble bivalents but have no chiasmata. Chromosome number is 18, as with mitotic chromosomes. These have been called 'pseudobivalents'. During metaphase they line up along the equatorial plane.

The same type of egg maturation occurs in individuals of a population of *Macrobiotus recens*, which is, however, tetraploid (Bertolani, 1975, 1982).

Univalents in prophase and metaphase have been observed in triploid cytotypes of *Pseudobiotus* (as *Isohypsibius) augusti* and *P. megalonyx* (see Bertolani, 1976), in the latter syntopic with the amphimictic diploid cytotype; in triploid *Macrobiotus pseudohufelandi* (see Bertolani *et al*., 1987); in *Diphascon granifer*, the only known case of ameiotic parthenogenesis in a certainly diploid cytotype (Bertolani, 1979); and in *Itaquascon trinacriae* (Fig. 3), whose degree of ploidy cannot be defined from the 16 univalents (Bertolani, 1979).

Egg maturation in *Echiniscus* displays certain specific cytological features (Bertolani, 1982). Males are lacking in most species of the genus. The oocytes of *Echiniscus trisetosus* and *E. testudo* have a long spindle with some coiled diads gathered at the ends. The total number of the diads in each oocyte always remains constant and is the same as the number of mitotic chromosomes, but with generally unequal distribution at the two ends (Bertolani, 1982). There is no information at all regarding the oocyte prophase. Therefore, the maturation pattern is specific but no data are available to formulate a hypothesis regarding the subsequent stages.

C. Genetics

The cytological aspects described above have various genetic consequences. Ameiotic parthenogenesis lacks chromosome pairing and therefore is genetically the same as mitosis. The descendants of one female, thus, constitute a clone. In this case heterozygosity, if it exists, is maintained and may increase as a result of the accumulation of mutations (White, 1973). The parthenogenesis with pseudobivalents of *Macrobiotus hufelandi* and *M. recens* may be interpreted in two ways since we do not know if non-disjunction follows chromosome duplication or if disjunction occurs and then pairing. If the former were true (Fig. 4A), the consequences would be the same as for ameiotic parthenogenesis; in the second case (Fig. 4B) there would be a tendency towards homozygosity, since there is no crossing-over. In fact, the homologous chromosomes which remain in the egg after the maturation divisions may be either accidentally heterozygotic or completely homozygotic. In the next generation the homozygotes remain as such, while some of the heterozy-

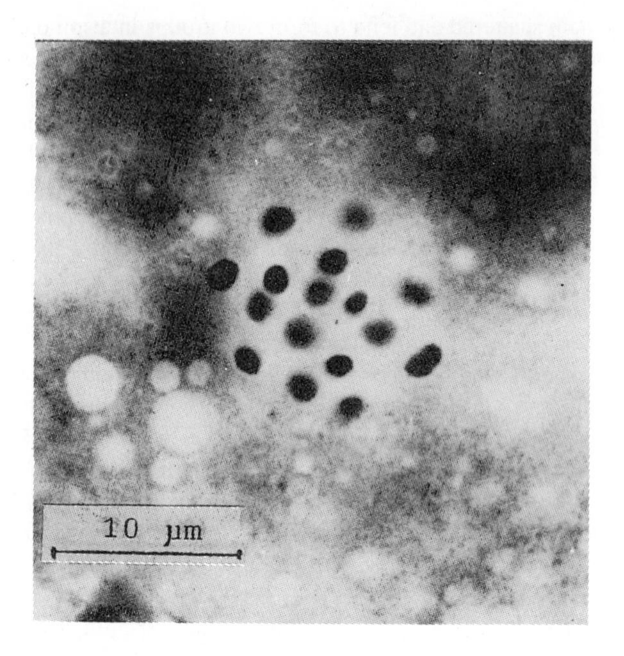

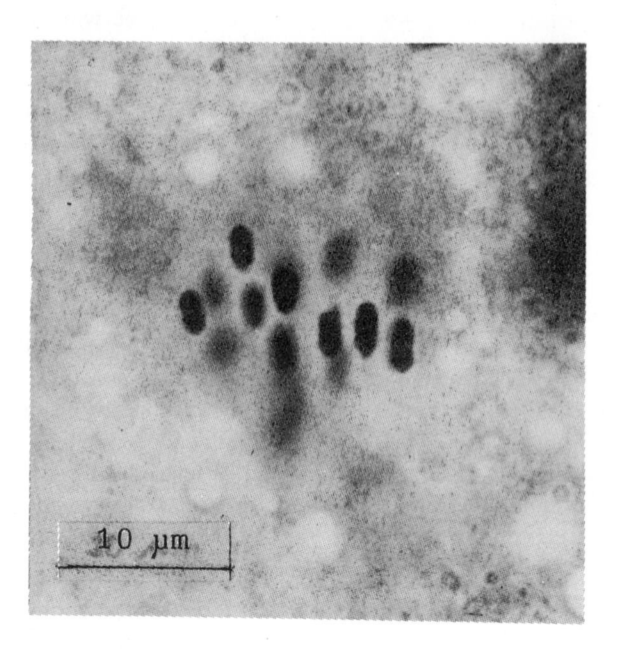

Fig. 3. Metaphase with univalents of ameiotic parthenogenetic oocyte in *Itaquascon trinacriae*: polar (**A**) and lateral (**B**) view (orcein).

gotes will accidentally become homozygotes. Thus little by little full homozygosity is reached. The hypothesis of non-disjunction seems the more tenable one because it can explain the loss of crossing-over, which would be pointless between two genetically identical chromosomes.

Complete and immediate homozygosity is brought about by the nature of egg maturation in *Hypsibius dujardini* since during prophase there are no chiasmata and restoration of the diploid number is a result of the separation of two sister chromatids from each chromosome (Fig. 5A). In *Dactylobiotus parthenogeneticus* the situation is different, even if the cytology has some aspects in common with *H. dujardini*. The presence of crossing-over in prophase I allows recombination. The two chromatids which separate to restore the diploid numbeı may be genetically different (Fig. 5B). If, however, parthenogenesis survives as time passes, in this case also there will be a tendency towards homozygosity. To avoid this, amphimictic generations should appear, but the phenomenon of heterogony was not observed in the studied population (Bertolani and Buonagurelli, 1975).

D. Evolution

A recent study by Bertolani *et al.* (1990) points out that the parthenogenetic strains of a given morphospecies, *Ramazzottius oberhaeuseri*, collected from the same area have a distribution different from that of amphimictic ones. Considering the same substratum, viz. the lichen *Xanthoria parietina*, it was found that lichen clumps collected from tree trunks contained only females (ameiotic), whereas those collected from large rocky outcrops almost always contained males in addition to diploid females or polyploid females with ameiotic oocytes. The second finding indicates that different strains can live together and thus there is no habitat difference. Since semiterrestrial tardigrades undergo passive transport, the different distribution has been explained by a greater fitness of amphimictic over parthenogenetic strains. Parthenogenetic and amphimictic females produce a similar number of eggs. Parthenogenetic reproduction requires only one individual to colonize a new territory and can more rapidly populate substrata of relatively recent origin (e.g. tree lichen). On the other hand, amphimictic reproduction cannot easily colonize recent formations, whereas it populates large substrata of ancient origins.

These data support the assertions by Cuellar (1977), Bell (1982), Lynch (1984), and Went (1984) that low-stability environments are more easily populated by parthenogenetic than amphimictic strains. This is confirmed by the lack of parthenogenesis in the more stable marine environment and its presence in freshwater and terrestrial habitats, an observation valid not only for tardigrades but also for other animal groups with parthenogenesis.

In particular, parthenogenesis in tardigrades is very common among moss-dwelling, wood-litter-dwelling and freshwater species, as in nematodes and rotifers. Pilato (1979) suggests that cryptobiosis (anhydrobiosis, anoxibiosis, encyst-

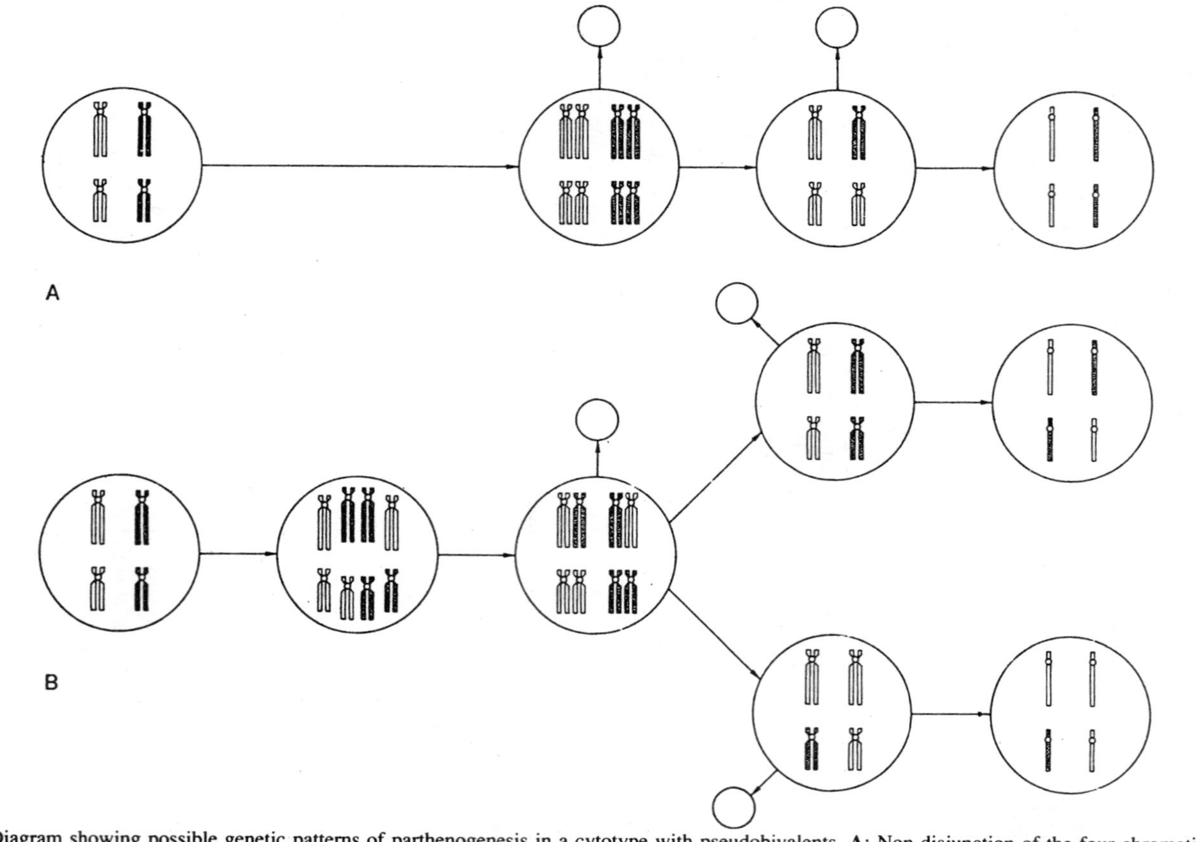

Fig. 4. Diagram showing possible genetic patterns of parthenogenesis in a cytotype with pseudobivalents. **A**: Non-disjunction of the four chromatids. **B**: Disjunction and subsequent pairing. (Different patterns of chromosome represent heterozygous condition.)

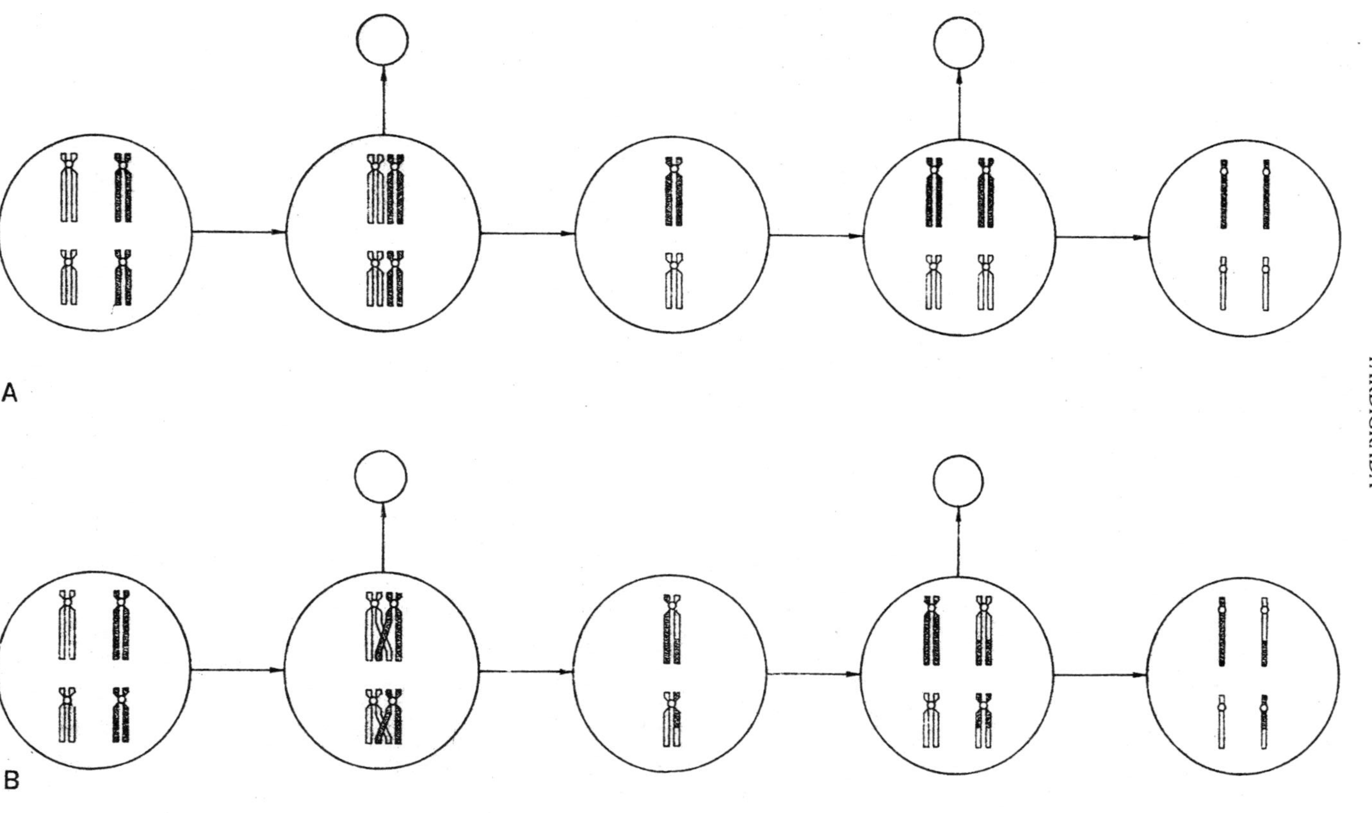

Fig. 5. Meiotic parthenogenesis. Diagrams of oocyte maturation in *Hypsibius dujardini* **(A)** and in *Dactylobiotus parthenogeneticus* **(B)** showing the different genetic characteristics. (Different patterns indicate heterozygous condition.)

ment, cryobiosis), adopted by these animals to resist adverse conditions, allows parthenogens to minimize selective pressures in their environment. This capability together with the rapid colonization after passive dispersal has permitted the survival of parthenogenetic strains and allowed their evolution, as indicated by the diversification of the bdelloid Rotifera, which lack amphimixis. For the tardigrades, *Echiniscus* (a moss-dwelling heterotardigrade) is the only genus presenting numerous species but only very few with males, suggesting a possible parthenogenetic evolution. In the other heterotardigrades, parthenogenesis is less frequent or absent, whereas in eutardigrades it is common and almost always correlated with amphimictic forms. In the latter, the parthenogenetic forms seem to be of recent origin. Specifically, an *in loco* origin of some strains was hypothesized for the previously cited case of *Ramazzottius*, suggesting a possible polyphyletic origin of the polyploid cytotypes (Rebecchi and Bertolani, 1988).

IV. LIFE-HISTORY TACTICS

In tardigrades, the sex ratio varies over a wide range of values. In marine species, males may be more common than females or are sometimes found in about equal numbers. The former case can be explained in *Batillipes pennaki* by the longer life span of the males (Grimaldi de Zio *et al.*, 1977). On the contrary, many non-marine species show the predominance of females. The presence of only females can be accounted for by parthenogenesis. On the other hand, the presence of males in small numbers represents a more complicated situation. My experience reveals sex ratios which vary greatly from the 1:1 value only in cases of mixed parthenogenetic and amphimictic populations. Without chromosome analysis the two populations are sometimes indistinguishable.

The reproductive period is variable. Marine tardigrades show reproductively mature and newly hatched individuals in samples collected at various times during the year (Pollock, 1970). Semi-terrestrial and freshwater tardigrades exhibit similar behaviour (von Wenck, 1914; Bertolani, 1971c, 1975, 1976). In *Dactylobiotus grandipes* mature females are found only in August and September but young specimens are also found in April and July (Schuster *et al.*, 1977).

Population density is also different. Some marine species are more plentiful in the spring and autumn and others in summer and autumn (see Pollock, 1975). Moss-dwelling tardigrades can be found, in variable numbers, throughout the year, while freshwater ones, particularly those from rivers (lotic waters), are absent during the summer (Bertolani, 1976, 1988). My observations over various collections, in addition to data by other authors, substantiate these findings. Nevertheless, both *Dactylobiotus grandipes* and *Murrayon pullari* have been found in a North American lake throughout the year and in their greatest numbers during the summer months (Schuster *et al.*, 1977). In any case, tardigrades are certainly *r*-strategists.

V. INTERSPECIFIC REPRODUCTIVE ISOLATION

Various species of tardigrades are cosmopolitan, and it is therefore difficult to sustain geographic isolation as the main mechanism of speciation. The available data suggest that reproductive isolation is unlikely due to the different maturation times of the gametes. Speciation may occur whenever obligatory parthenogenesis begins which isolates the carriers of this characteristic from the main strain. Such speciation may even be instantaneous. In *Isohypsibius granulifer*, centric fusion of some chromosomes is known in some populations but is absent in others having the same morphology (Bertolani, 1975, 1982, 1987). This fact may also be responsible for reproductive isolation. Other modes of reproductive isolation are unknown to date.

VI. CONCLUSIONS

Reproduction via gametes, either amphimictic or parthenogenetic, is the only known method of propagation among tardigrades.

Parthenogenesis is a common reproductive strategy in non-marine tardigrades. Parthenogens escape severe competition with amphimictic populations since they can better colonize new territories. Parthenogenesis may be meiotic but is more often ameiotic, revealing different cytological patterns and genetic characteristics. No cases of facultative or occasional parthenogenesis or of heterogony are known among tardigrades. Only in *Dactylobiotus parthenogeneticus* does parthenogenesis appear not obligate and continuous, but in the populations whose cytology has been studied males have never been observed (Bertolani and Buonagurelli, 1975).

Little additional information on reproductive strategies is available. We do not know if the hermaphroditism observed in a limited number of cases is cross-fertilizing or self-fertilizing. Moreover, there is a complete lack of data on the factors influencing fertilization and on the hormonal activities associated with reproduction. Lastly, we are still missing specific data on the possible presence of diapause, which in any case should be present in some environments such as marine and fresh water, where tardigrades can be absent even for months during the span of the year.

In any case, much still remains to be done in the field of tardigrade reproductive biology. The topic generates remarkable interest due to the systematic position of tardigrades, close as they are to annelids and arthropods, and for the biological peculiarities, such as cryptobiosis, that allow these animals to live like aquatic organisms in terrestrial environments.

REFERENCES

Ammermann, D. (1962). 'Parthenogenese bei dem Tardigraden *Hypsibius dujardini* (Doy)', *Naturwissenschaften*, **49**, 115.

Ammermann, D. (1967). 'Die Cytologie der Parthenogenese bei dem Tardigraden *Hypsibius dujardini*', *Chromosoma*, **23**, 203–213.

Baumann, H. (1964). 'Über den Lebenslauf und die Lebensweise von *Milnesium tardigradum* Doyere (Tardigrada)', *Veröff. Uberseemus. Bremen*, **3**, 161–171.

Bell, G. (1982). *The Masterpiece of Nature: The Evolution and Genetics of Sexuality*, University of California Press, Berkeley.

Bertolani, R. (1970a). 'Mitosi somatiche e costanza cellulare numerica nei Tardigradi', *Atti Accad. Naz. Lincei, Rend.*, Ser. VIII, **48**, 739–742.

Bertolani, R. (1970b). 'Variabilitá numerica cellulare in alcuni tessuti di Tardigradi', *Atti Accad. Naz. Lincei, Rend.*, Ser. VIII, **49**, 442–445.

Bertolani, R. (1971a). 'Rapporto-sessi e dimorfismo sessuale in *Macrobiotus* (Tardigrada)', *Atti Accad. Naz. Lincei, Rend.*, Ser. VIII, **50**, 377–382.

Bertolani, R. (1971b). 'Partenogenesi geografica triploide in un Tardigrado *(Macrobiotus richtersi)*', *Atti Accad. Naz. Lincei, Rend.*, Ser. VIII, **50**, 487–489.

Bertolani, R. (1971c). 'Contributo alla cariologia dei Tardigradi. Osservazioni su *Macrobiotus hufelandi*', *Atti Accad. Naz. Lincei, Rend.*, Ser. VIII, **50**, 772–775.

Bertolani, R. (1971d). 'Osservazioni cariologiche su biotipi bisessuati e partenogenetici in *Hypsibius oberhaeuseri* (Tardigrada)', *Atti Accad. Naz. Lincei, Rend.*, Ser. VIII, **51**, 411–413.

Bertolani, R. (1972a). 'Sex ratio and geographic parthenogenesis in *Macrobiotus* (Tardigrada)', *Experientia*, **28**, 94.

Bertolani, R. (1972b). 'La partenogenesi nei Tardigradi', *Boll. Zool.*, **39**, 577–581.

Bertolani, R. (1973a). 'Presenza di un biotipo partenogenetico e suo effetto sul rapporto-sessi in *Macrobiotus hufelandi* (Tardigrada)', *Atti Accad. Naz. Lincei, Rend.*, Ser. VIII, **54**, 469–473.

Bertolani, R. (1973b). 'Primo caso di una popolazione tetraploide nei Tardigradi', *Atti Accad. Naz. Lincei, Rend.*, Ser. VIII, **55**, 571–574.

Bertolani, R. (1975). 'Cytology and systematics in Tardigrada', *Mem. Ist. Ital. Idrobiol.*, **32**, Suppl., 17–35.

Bertolani, R. (1976). 'Osservazioni cariologiche su *Isohypsibius augusti* (Murray, 1907) e *I. megalonyx* Thulin, 1928 (Tardigrada) e ridescrizione delle due specie', *Boll. Zool.*, **43**, 221–234.

Bertolani, R. (1979). 'Parthenogenesis and cytotaxonomy in Itaquasconinae (Tardigrada)', *Zes. Nauk. Uniw. Jagiellon., Prace Zool.*, **25**, 9–18.

Bertolani, R. (1982). 'Cytology and reproductive mechanisms in tardigrades', in *Proceedings of the Third International Symposium on the Tardigrada* (Ed. D.R. Nelson), East Tennessee State University Press, Johnson City, Tennessee, pp. 93–114.

Bertolani, R. (1987). 'Sexuality, reproduction and propagation in tardigrades', in *Biology of Tardigrades* (Ed. R. Bertolani), Selected Symposia and Monographs UZI, **1**, Mucchi, Modena, pp. 93–101.

Bertolani, R. (1988). 'Tardigradi delle acque dolci, con riferimento ai corsi d'acqua della Lunigiana e della Garfagnana', *Boll. Mus. St. Nat. Lunigiana*, **6–7**, 133–138.

Bertolani, R., and Buonagurelli, G.P. (1975). 'Osservazioni cariologiche sulla partenogenesi meiotica di *Macrobiotus dispar* (Tardigrada)', *Atti Accad. Naz. Lincei, Rend.*, Ser. VIII, **58**, 782–786.

Bertolani, R., Garagna, S., Manicardi, G.C., and Redi, C.A. (1987). '*Macorobiotus pseudohufelandi* Iharos as a model for cytotaxonomic study in populations of eutardigrades', *Experientia*, **43**, 210–213.

Bertolani, R., and Mambrini, V. (1977). 'Analisi cariologica e morfologica di alcune popolazioni di *Macrobiotus hufelandi* della Valsesia', *Atti Accad. Naz. Lincei, Rend.*, Ser. VIII, **62**, 239–245.

Bertolani, R., Rebecchi, L., and Beccaccioli, G. (1990). 'Dispersal of *Ramazzottius* and other tardigrades in relation to type of reproduction', *Inv. Reprod. Dev.*, **18**, 153–157.

Cuellar, O. (1977). 'Animal parthenogenesis', *Science*, **197**, 837–843.

Grimaldi de Zio, S., D'Addabbo Gallo, M., and Morone De Lucia, R.M. (1977). 'Ciclo biologico di *Batillipes pennaki* Marcus (Heterotardigrada)', *Atti IX Congr. Soc. Ital. Biol. Mar.*, 285–292.

Lynch, M. (1984). 'Destabilizing hybridization, general-purpose genotypes and geographic parthenogenesis', *Q. Rev. Biol.*, **59**, 257–290.

Marcus, E. (1929). 'Tardigrada' in *Klassen und Ordnungen des Tierreichs* (Ed. H.G. Bronn), Vol. 4, Akademische Verlagsgesellschaft, Leipzig, pp. 1–608.
Pilato, G. (1979). 'Correlations between cryptobiosis and other biological characteristics in some soil animals', *Boll. Zool.*, **46**, 319–332.
Pollock, L.W. (1970). 'Reproductive anatomy of some marine Heterotardigrada', *Trans. Am. Microsc. Soc.*, **89**, 308–316.
Pollock, L.W. (1975). 'Tardigrada', in *Reproduction of Marine Invertebrates* (Eds. A.C. Giese and J.S. Pearse), Vol. 2, pp. 43–54.
Rebecchi, L., and Bertolani, R. (1988). 'New cases of parthenogenesis and polyploidy in the genus *Ramazzottius* (Tardigrada, Hypsibiidae) and a hypothesis concerning their origin', *Inv. Reprod. Dev.*, **14**, 187–196.
Schuster, R.O., Toftner, E.C., and Grigarick, A.A. (1977). 'Tardigrada of Pope Beach, Lake Tahoe, California', *Wasmann J. Biol.*, **35**, 115–136.
von Wenck, W. (1914). 'Entwicklungsgeschichtliche Untersuchungen an Tardigraden *(Macrobiiotus lacustris* Duj)', *Zool. Jb., Abt. Anat. Ontog. Tiere*, **37**, 465–514.
Went, D.F. (1984). 'Parthenogenetic strategies in insect reproduction', *Adv. Inv. Reprod.*, **3**, 303–315.
White, M.J.D. (1973). *Animal Cytology and Evolution*, Cambridge University Press, Cambridge.

3. ARTHROPODA — CRUSTACEA*

T.J. PANDIAN
School of Biological Sciences, Madurai Kamaraj University, Madurai 625 021, India

I. INTRODUCTION

The vast majority of crustaceans are disexual (dioecious) and dimorphic and reproduce sexually. In many crustaceans (Cladocera, Decapoda, Stomatopoda, Copepoda, and Euphausiacea; for classification of Crustacea, see Systematic Résumé), spermatophores containing non-flagellated, non-motile spermatozoa are transferred by some specialized appendages onto the female body and hence fertilization is external or more precisely epizoic (for details of fertilization, see Hinsch, Volume IV B of this series). Fertilization by placement of spermatophore(s) around the genital pore or onto the sternum of the female can be regarded as neither internal nor external; the term 'epizoic' is considered more appropriate here (see spawning in dragonflies: Corbet, 1962). In the other crustaceans (Cirripedia, Mysidacea, Cumacea, Isopoda, Amphipoda, Ostracoda), internal fertilization is ensured with intromittent structures, which transfer flagellated spermatozoa whose motility is motivated only at the proximity of the ova (see Pochon-Masson, Volume II of this series and, Jamieson, 1991 for spermatology). Thus, the characteristic strategy of Crustacea is to ensure fertilization with non-motile or slightly motile spermatozoa (Table 1). Most crustaceans brood their eggs either fastened to certain appendages (e.g. Decapoda) or held in marsupium (e.g. Isopoda), or carried in attached (e.g. Cladocera) or projected sacs (e.g. Copepoda), or retained in the mantle cavity (e.g. Cirripedia). Hence egg carriage may be considered the hallmark strategy of crustacean reproduction (see Section I B below). Development in most crustaceans is indirect, involving one or more, sometimes a series of, larval stages; thus, post-embryonic development involves different degrees of metamorphosis, ranging from anamorphic to strongly metamorphic. Comparative studies on reproductive strategies of crustaceans may indeed greatly enhance our appreciation of the different

*Dedicated to my guru, Prof. O. Kinne, on his 70th birthday (30 August 1993).

Table 1

Strategy of sperm transfer in major crustacean groups

Group	Nature of sperm	Transmitting structure	Fertilization	Sperm storage
1. Flagellate sperm strategy				
Ostracoda	Long, flagellated, motile	Sclerotized penis near caudal furca	Internal	?
Cirripedia	Flagellated, not highly motile	Long protrusible penis	Internal	Activated only at the proximity of ova
Isopoda	Flagellated, motile	Genital apophyses/appendix masculina	Internal	Sperm fertilize eggs in oviduct
Amphipoda	Flagellated, not highly motile	Penis papillae	Marsupial	Activated only at the proximity of ova; storable for four days
Mysidacea	Flagellated	Penis + fourth pair of pleopod	Marsupial	?
Cumacea	Flagellated	?	?	?
Anostraca	?	Eversible copulatory processes	Internal	Not storable
2. Non-flagellate sperm strategy				
Cladocera	Non-flagellated, non-motile	Modified postabdomen	?	?
Copepoda	Non-flagellated, non-motile	Fifth thoracic legs	Epizoic	Transmitted as firm spermatophore; storable
Branchiura	?	?	?	Transmitted as spermatophore
Anaspidacea	?	First two pairs of pleopods	Epizoic	Transmitted into sternal pouch-like seminal receptacle

Euphausiacea	Non-flagellated, non-motile	Petasma	Epizoic	Transmitted as firm spermatophore; storable
Stomatopoda	Non-flagellated, non-motile	Penis attached to the last thoracic legs	Epizoic/internal	Transmitted as sperm and into sternal pouch-like seminal receptacle
Decapoda	Non-flagellated, non-motile, sometimes star-like	Gonopods + first two pleopods e.g. Natantia, some macrurans and anomurans	Epizoic	Transmitted as firm spermatophore into sternal pocket-like receptacle or to pleopods
		e.g. Brachyura, crayfish and lobsters	Internal	Transmitted as sperm plug; plug; storable in internal seminal receptacle

ecological niches inhabited by them, the adaptive changes in the kinds, numbers and sizes of crustacean eggs, and the varying morphological stages at which their larvae hatch.

A. Strategy of Sperm Transfer

1. Motile sperm

In the sessile cirripedes, the transfer of spermatozoa is characteristically by a copulatory act. However, the precopulatory activity, which involves movements of either partner, must necessarily be absent. At the time of copulation, the penis of balanoid barnacles, much extended by turgor pressure (Klepal *et al.*, 1972), makes searching movements to locate a functional female and on acceptance by the latter, the penis is inserted into the mantle cavity and there follows a whole series of responses and behavioural patterns associated with copulation (H. Barnes and M. Barnes, 1956). There is multiple insemination and a mass of semen is thus transferred. The ova are shed into a sac, produced by oviducal atrium (Walley, 1965) and it is around this sac that the mass of semen is deposited. The spermatozoa of cirripedes display very limited active movement (velocity 30 μm/sec) in the seminal vesicle (Munn and H. Barnes, 1970) but those collected from the mantle, an hour after multiple insemination, were motile, the velocity being 127 μm/sec (H. Barnes *et al.*, 1971). Walley *et al.* (1971) have shown that the clear, fluid secretion of the oviducal gland is a powerful activating agent, inducing motility in the spermatozoa of *Semibalanus balanoides* after the semen has been transferred to the functional female, but they found that spermatozoa from the vesiculae seminales could not be activated in the same way; these observations suggest that passage of spematozoa through the penis followed by further activation from oviducal gland secretion is essential for the attainment of their full motility in this species (H. Barnes *et al.*, 1971). H. Barnes (1962, 1963) found that the level of ATP of the spermatozoa is low in *Balanus balanus* and is likely to be even lower in *S. balanoides* so that its delayed activation conserves its energy until the semen is in the proximity of the ova.

Similarly, suggestive evidence for the delayed activation of the sperm is also available in amphipods. In Amphipoda and Mysidacea, fertilization is reported to occur in the marsupium (e.g. *Gammarus duebeni*: Kinne, 1954; *Neomysis vulgaris*: Kinne, 1955). Beck (1977) observed the presence of unfertilized eggs (without fertilization membrane) in the marsupium of the freshwater mysid *Taphromysis bowmani* and concluded that fertilization occurs in the marsupium. In the amphipod *Talitrus saltator*, the sperm is relatively large, about 400 μm long. Moulting, copulation, and ovulation all are well synchronized: immediately following a moult, copulation takes place; ovulation occurs anytime between moulting and day 4 following moulting. The sperm are motivated by a secretion of the unfertilized eggs or a secretion released by the female during ovulation (Williamson, 1951).

The strategy of these crustaceans is thus to ensure fertilization by flagellated sperm, whose motility is not initiated until they are in the proximity of ova, such delayed motivation affording them considerable scope for energy conservation.

2. Nonmotile sperm

In the Copepoda, the spermatophore is an elongated, flask-shaped object containing spermatozoa and associated secretions. The fifth pair of pleopods is modified as a secondary sexual character, whose function is the transfer of the spermatophore from the male to the female (Table 1; Hopkins *et al.*, 1978). The male must attach the spermatophore to the female genital segment in such a position that the opening of the neck of the spermatophore, through which the spermatozoa flow out, is in contact with the opening of the female cavity, presumably for storage in seminal receptacles (Blades, 1977). In some calanoid copepods (e.g. Centropagidae: Lee, 1972), the opening of the spermatophore neck is situated amidst a coupling device composed of one or more plates. In others, a specialized coupling device is absent and the unmodified spermatophore neck is attached to the female cavity by a cement-like secretion excreted from the spermatophore itself. In the calanoids with couplers and those without such devices (e.g. *Clausocalanus* spp. : Frost and Fleminger, 1968), the site and mechanism of attachment is precise (e.g. *Pareuchaeta (Euchaeta) norvegica*: Hopkins and Machin, 1977), species-specific, and in the case of *Labidocera jollae* (see Fleminger, 1967), racially-specific.

The mode of sperm transfer in decapods may be epizoic or internal. Epizoic fertilization involves the transfer of spermatophores onto the sternum, as in *Panulirus*, or to the pleopodal region (e.g. *Emerita asiatica*: Subramoniam, 1977). In *E. asiatica*, the spermatophores are held together in a row on a strip of membrane by a peduncle (i.e. pedunculate) and are structurally species-specific (Subramoniam, 1991a), as in many anomurans (Mouchet, 1931), or they may be in the form of a tubular mass, i.e. non-pedunculate (see Subramoniam, 1991a), as in macrurans (Mathews, 1954; Berry and Heydorn, 1970). The absence of intromittent organ is partly compensated by the sticky nature of the spermatophore, which, on adhesion to the female body, undergoes structural changes leading to the hardening of the outer layer (e.g. *Panulirus interruptus*: Martin *et al.*, 1987). In Brachyura and most Macrura with a seminal receptacle, fertilization is internal; the spermatophores are non-pedunculate and are transmitted in fluid medium to the female genital opening by well-developed intromittent organs, aided by the anterior pleopods. In Macrura, the tips of the first pleopods are inserted into the seminal receptacle during copulation and the liquid semen flows along the grooves of the pleopods. In brachyurans the first pleopod is in the form of a cylinder, into which fits the piston-like second pleopod. Sperm emitted from the gonopods are pumped through the conducting first pleopod.

3. Efficiency and cost of the strategies

Of the two different strategies of sperm transfer described above for crustaceans, the strategy of copepods appears more precise, efficient, and economic than that of cirripedes. It is, perhaps, for this reason that the higher crustaceans not only retained the copepod strategy but also elaborated it further during the course of evolution. Observing 300 randomly chosen mating couples of the calanoid *Centropages typicus* in the laboratory, Blades (1977) estimated that some 80 per cent of the couples performed pheromonally guided (see Uchima and Murano, 1988 for *Oithona davisae*) 'optimum' mating behaviour, resulting in proper placement of the spermatophore. In the calanoid *Pareuchaeta norvegica*, Hopkins and Machin (1977), who sampled 10,500 females, found that the first spermatophore was attached to the exact position in the genital segment in 78.5 per cent of the cases. The records of Parrish and Wilson (1978) on laboratory cultures of the calanoid *Acartia tonsa* maintained over seven years and extending for 32 filial generations show that on the first day after mating, 84 per cent females laid fertile eggs; likewise, Zurlini *et al.* (1978) found that the mean egg fertility of the harpacticoid *Euterpina acutifrons* was 83 per cent and the level of fertility remained unaffected in eggs from successive egg sacs. The percentage of successful hatching of *Calanus helgolandicus* was more or less equal in the laboratory-spermatophored and ocean-spermatophored females (81 per cent) (Paffenhofer, 1970). These observations clearly show that the spermatophore is correctly placed by the males and that sufficient sperm are available to ensure fertilization of all the eggs, even in the prodigious calanoids (see Paffenhofer, 1970; Ianora *et al.*, 1989). Among the euphausiids, in *Thysanoëssa raschii*, 85 per cent of the females carry spermatophores during April-May in the Clyde Sea (Mauchline and Fisher, 1969); the high fertility of *T. raschii* is comparable to that of the copepods. The decapods too fix the spermatophores precisely. Of the 10,684 female *Penaeus merguiensis* sampled by Crocos and Kerr (1983), 78 per cent were inseminated. Lumare (1981) observed that over 68 per cent of eggs of *P. japonicus* were successfully fertilized from ocean-spermatophored females. Even the diminutive male *Emerita asiatica*, which wanders in the gill region and anterior part of the female, deposits the spermatophores only in the vicinity of egg-carrying pleopods (Subramoniam, 1977).

Male copepods ejaculate a maximum of three, but usually two, spermatophores during their life (e.g. *Euterpina acutifrons*: Haq, 1972; *Diaptomus nevadensis*: as cited by Daborn, 1973). A male *Pareuchaeta norvegica* transfers 0.9 per cent of its body weight into each ripe spermatophore (Hopkins, 1978). In other words, by transferring spermatophores equal to 1.8 per cent of the body weight, a male copepod is able to ensure a blanket fertilization of over 80 per cent eggs in a fertilized female. Thus, the crustacean strategy of transferring non-motile spermatozoa, packed in a spermatophore, appears to be precise and efficient.

On the other hand, in *Semibalanus balanoides* and *Balanus balanus* the mass of semen transferred amounts to as much as 50 per cent the body weight (H. Barnes *et al.*, 1963). Addition of traces of ascorbic acid (14 μl/l) to water induces multiple copulation in *Balanus* sp. (Collier *et al.*, 1956). The high ascorbic acid content (21 μg/ml) in the seminal plasma of *B. balanus* (see H. Barnes, 1962) and its release into water contained in the mantle cavity ensures multiple and mass inseminations. Briefly, the fertility, ensured by the copepod at a cost of only 1.8 per cent of its body weight, is ensured by the cirripede at a cost of 50 per cent of its body weight. While such comparisons are open to criticism, they are of some use to assess the economy and efficiency of different strategies of sperm transfer among crustaceans.

Another possible source of criticism is the reported occurrence of uneconomic, multiple placements of spermatophores in a number of copepod species: for example, Wolf (1905) found, in *Diaptomus gracilis*, 10 to 15 spermatophores in one female; Gibbons (1936), 15 on a single female *Calanus finmarchicus*. However, Marshall and Orr (1955) considered the occurrence of more than two or three spermatophores on a female *C. finmarchicus* as rare. From their studies on seasonal variation in the number of adult females carrying different number of spermatophores, Hopkins and Machin (1977) estimated that single, double, and triple, spermatophore placements accounted respectively for 71.6, 19.8, and 6.5 per cent of the attached spermatophores in *Pareuchaeta norvegica*. The remaining placements of four, five, or six spermatophores on a female accounted for 2.1 per cent of the attached spermatophores. This fact strongly suggests that single spermatophore placements are by far more frequent in the Copepoda.

In a way, multiple placements of spermatophores appear to reflect the adult male to unmated female ratio in a copepod population. With many males searching for dwindling numbers of unfertilized females, it becomes likely that a number of males will simultaneously attempt to mate with a single female. Cases of a number of males attempting to copulate at the same time with one female have been reported for a number of copepods (e.g. *Cyclops americanus*: Hill and Coker, 1930; *Pseudodiaptomus coronatus*: Jacobs, 1961).

Such an economy of sperm-transfer strategy of copepods is perhaps made possible at the energy costs of spermatophore construction and precopulatory activity of the partners including that expended on pheromone system of communication (Katona, 1973; Uchima and Muranó, 1988). Energy expenses on these heads may be very little in *Balanus* spp., which, however, heavily invest in the development of a complex penis annually (Klepal *et al.*, 1972). Thus, the copepods appear to invest less in production and transfer of sperm to fertilize a maximum of about 2,000 eggs (e.g. *Calanus helgolandicus*: Paffenhofer, 1970). Conversely, the cirripedes make a far heavier investment of semen production to fertilize a maximum of about 6,000 eggs (H. Barnes and M. Barnes, 1968). It is tempting to suggest that the concept of r- and K-strategy, now widely considered from the point of egg production (see Pianka, 1974), should also be extended to sperm production. Even though the gonadal weight-based values, reported in Table 2, may not give a correct picture of the total number of sperm produced by the species concerned, they do emphasize the need for the extension of the concept of r- and K-strategy to the males as well.

Table 2

Sperm-production potential of some crustaceans. Data represent the highest gonadal index values

Species	Testis as percentage of body weight	Reference
Emerita asiatica	2.5	Subramoniam (1977)
Pareuchaeta norvegica	1.8[1]	Hopkins (1978)
Metapenaeus affinis	1.8	Pillay and Nair (1971)
Uca lactea annulipes	1.0	Pillay and Nair (1971)
Portunus pelagicus	1.0	Pillay and Nair (1971)
Sesarma intermedium	0.7[2]	Kyomo (1988)
Penaeus indicus	0.8	Subrahmanyam (1963)
Jasus edwardsii	0.2	MacDiarmid (1989)

[1]Value based on the transferred spermatophore alone.
[2]Spermatheca weight as percentage of body weight.

4. Intromittent organs

Table 1 also lists the different structures used as intromittent organs by crustaceans. The cirripede penis is highly complex (Klepal *et al.*, 1972) and of taxonomic importance (H. Barnes and Klepal, 1971; Jamieson, 1991). Crustacea with flagellate sperm strategy use specialized copulatory organs to transfer liquid semen into the female, whereas those with non-flagellate sperm strategy use one or the other appendages to facilitate epizoic fertilization. Among decapods only in brachyurans and some macrurans is fertilization internal; the first two pairs of pleopods actually transfer the liquid semen from gonopods into the seminal receptacle.

5. Sperm storability

Low motility obviously lengthens the period of sperm survival (K.G. Adiyodi and R.G. Adiyodi, 1974). That most crustaceans transfer spermatophores containing non-motile sperm or motile sperm whose motility is not activated until they are in the proximity of the ova (e.g. *Semibalanus balanoides*: H. Barnes *et al.*, 1971), indicates that the crustacean sperm may survive over a relatively long period. The structural and functional aspects of spermatheca, correlated with moult, suggest three types of sperm storage: type 1, non-storable sperm; type 2, storable sperm at the cost of male; and type 3, storable sperm at the cost of female. At the population level, the storage of mature sperm of a species, either in the male or in the female, in itself costs little extra energy, but must be of much temporal and spatial relevance, and holds significant genetic consequences.

Separate mating, obligatorily required for each spawning in type 1, must cost time and energy on communication and premating behaviour, but offers more scope to select genetically better mates. Maintenance of the female in types 2 and 3 ensures fertilization of more than one clutch of eggs, as and when the female is ready to extrude eggs, but limits the chances of mate selection and genetic variability. However, investigations on these interdisciplinary fields are totally wanting in the Crustacea.

(a) Type 1: Non-storable sperm

Moulting, mating, and fertilization are highly organized processes and occur in that sequence in most cases, as most crustaceans carry their eggs. In primitive forms such as Anostraca, in which the sperm-storage facility is totally wanting, each clutch of eggs is fertilized by a separate copulation—a factor that might have driven them to resort to parthenogenesis.

(b) Type 2: Storable sperm at the cost of male

In the absence of a highly secretory spermathecal gland (for accessory sex glands of Crustacea, see K.G. Adiyodi and G. Anilkumar, Volume III of this series), the sperm may be retained in the female for a brief period; the seminal fluid must keep the sperm viable until fertilization. In such forms, repeated matings occur, perhaps to accumulate the required substances from the accessory gland (e.g. *Oratosquilla (Squilla) holoschista*: Deecaraman, 1980). In diecdysic forms with epizoic fertilization, spermatophores are lost along with the exuvia; another impregnation is required before the next batch of eggs could be spawned. The seminal products within the spermatophore of lobsters (e.g. Radha and Subramoniam, 1985) and prawns (e.g. Sasikala and Subramoniam, 1987) not only yield energy but also help maintain micro-environment within the spermatophore. Of particular interest in this respect is the presence of a large quantity of acid mucopolysaccharides in the spermatophore, which may prevent dehydration and microbial infection (see also Subramoniam, 1991b).

(c) Type 3: Storable sperm at the cost of female

The presence of a highly secretory spermatheca (see e.g. Krishnakumar, 1985) may help retain the sperm for a longer period. For instance, in anecdysic forms, such as *Callinectes sapidus*, the female mates only once immediately after the terminal moult and the spermatozoa remain viable for more than one year and are used for repeated spawnings (Tagatz, 1968). In terrestrial isopods, such as *Trioniscus pusillus, Oniscus asellus*, and *Porecellio scaber*, Heeley (1941) observed that the sperm remained viable for over two years and that females required only a single copulation to produce a series of broods. In the anecdysic *Cyclops*, as many as seven to 13 pairs of ovisacs full of eggs are produced, successively at intervals of one to six days, all fertilized by sperm retained from a single copulation, i.e. the sperm are viable for a period of three months (Pennak, 1953). Some diecdysic forms appear to use the sperm obtained from a single male sparingly for three or more spawnings, and others may even retain the excess sperm over one or more moults. Claims that a single impregnation suffices for two to eight successive broods have frequently been made for decapods (e.g. *Doelea gracilipes*: Gore, 1971). Transmoult retention of sperm occurs in *Portunus sanguinolentus* (see Ryan, 1965), *Menippe mercenaria* (see Cheung, 1968), and a few other species including *Paratelphusa hydrodromous* (see R.G. Adiyodi, 1988, p. 174), all of which successfully brood without further mating after an ordinary moult. In others, such as *Cancer* and *Pachygrapsus*, the presence of sufficient spermatozoa in the spermatheca may even postpone the next moult (Hiatt, 1948). The storage of sperm from more than one male in a female that copulated at different ecdyses has been demonstrated in two species of *Armadillidium* by using colour genes (Leuken, 1963).

Development of techniques for *in vitro* fertilization of females using sperm from the same spermatophore of a male is desirable to produce genetically pure strains of crustaceans in aquaculture. A successful device of electroejaculation of spermatophore has been described for prawns (Perysn, 1977; Sandifer and Smith, 1979). However, the percentage of fertilized eggs is much higher in the naturally impregnated prawns than in the artificially inseminated ones. For instance, Lumare (1981) reported that the artificially inseminated *Penaeus japonicus* yielded only 8 per cent fertilized eggs, as against 68 per cent from naturally fertilized females. It is not clear whether the quality of spermatophores used in these experiments was good. More recently, Leung-Trujillo and Lawrence (1987) observed a steady decline in sperm quality of *P. setiferus* held under laboratory conditions. Cryogenic storage of spermatophores for later use in artificial insemination for stockpiling and exchanging selected male genetic material has been attempted for *Macrobrachium* (see Sandifer and Lynn, 1980) and *Oratosquilla (Squilla) holoschista* (see Deecaraman, 1980). Jeyalectumie and Subramoniam (1989) investigated the effects of temperature and cryoprotectants, and recommended the combination of −79°C and glycerol for long-term preservation of *Scylla serrata* sperm.

Two important recent observations pertain to the processes of capacitation and anaerobic metabolism of decapod sperm during storage in the female. In the shrimp *Sicyonia ingentis*, the sperm was capacitated only after its transfer to the female, as evidenced by its capacity to undergo normal acrosomal activation (Clark *et al.*, 1984). After mating, the spermathecal contents of the crab *Scylla serrata* are enriched with organic substances derived from the semen. Enzyme studies of the spermathecal contents of the crab showed low, moderate, and high levels of activity for succinate dehydrogenase, lactate dehydrogenase, and fumarate reductase respectively (Jeyalectumie and Subramoniam, 1991). These observations suggest that the sperm metabolism of *S. serrata* is anaerobic.

B. Strategy of Brood Protection

1. Egg-carriage strategy

Protection, ventilation, and in certain cases, nutrient supply are the important aspects of crustacean maternal care. Besides, the mother animals effectively control (e.g. Crisp and Spencer, 1958; Katre and Pandian, 1972) and time (Ennis, 1973, 1975) the events of hatching and release of neonates at conditions favourable for feeding (H. Barnes,

1962) and dispersal (Lucas, 1975; DeCoursey, 1979). Even after the release, the young may secure maternal protection by clinging onto the mother's body.

The eggs of a number of crustaceans, including most calanoid copepods (e.g. *Calanus*: Conover, 1967), branchiurans (e.g. *Argulus*: Schöne, 1961), syncarids (e.g. *Anaspides*: Green, 1971), some euphausiids (e.g. *Euphausia*: Mauchline and Fisher, 1969), and penaeids (e.g. *Penaeus*: Badawi, 1975), are shed freely into the water (Table 3). There are exceptions to either category. Some calanoids, e.g. *Pseudocalanus elongatus* (see Paffenhofer and Harris, 1976) and *Pareuchaeta norvegica* (see Hopkins, 1977), carry eggs in masses. Among the stomatopods, some (e.g. *Oratosquilla holoschista*: Deecaraman, 1980) carry eggs in their chelate thoracic legs, but in the others like *O. investigatoris* (see Losse and Merrett, 1971) and *Gonodactylus*, the eggs are not carried; the female not only guards the egg mass in her 'chamber' (Fig. 1), but also palpates and turns it over regularly for aeration until hatching (Dingle and Caldwell, 1972; also see Giesbrecht, 1910). Since protection and ventilation are two important functions of crustacean maternal care (see Balasundaram and Pandian, 1981), the stomatopods are considered along with those that carry the eggs. Of the 26,000 species of crustaceans, only about 1,650 species (6 per cent) shed their eggs freely into the water; hence, egg carriage may be considered as a hallmark strategy of crustacean maternal care.

2. Sites of egg carriage

Egg carriage among crustaceans thus far has been little studied, but seems to have been a major event in the evolutionary history of crustaceans, resulting in

Fig. 1. Egg-carrying behaviour of stomatopods. *Squilla mantis* carrying the egg mass (left); *Gonodactylus oerstedi* (right) guarding the egg mass in its burrow (redrawn from Schmitt, 1965).

Table 3

Sites of egg carriage in crustaceans

A. Taxonomic groups retaining eggs			Total (No.)	Species (per cent)
Class	Crustacea		26,000[1]	
	1. In mantle cavity		2,980[1]	11.5
Order	Conchostraca (Subclass Branchiopoda)	(180)		
Subclass	Ostracoda	(2,000)		
Subclass	Cirripedia	(800)[1]		
	2(a) In attached pouch		725	2.8
Order	Cladocera (Subclass Branchiopoda)	(425)		
Order	Notodelphyoida (Subclass Copepoda)	(300)		
	2(b) In projected pouch		3,219	12.3
Subclass	Cephalocarida	(2)		
Order	Anostraca (Subclass Branchiopoda)	(175)		
Order	Notostraca (Subclass Branchiopoda)	(15)		
Subclass	Copepoda (excepting Calanoida and Notodelphyoida)	(3,000)		
Order	Thermosbaenacea (Subclass Malacostraca)	(4)[1]		
Order	Euphausiacea (Subclass Malacostraca)	(23)		
	3. In marsupium		8,725	33.6
Order	Mysidacea (Subclass Malacostraca)	(450)		
Order	Cumacea (Subclass Malacostraca)	(425)[1]		
Order	Teniidacea (Subclass Malacostraca)	(250)		
Order	Isopoda (Subclass Malacostraca)	(4,000)		
Order	Amphipoda (Subclass Malacostraca)	(3,600)		
	4. In thoracic and abdominal legs			
	(a) In thoracic legs		8,190	31.5
Order	Leptostraca (Subclass Malacostraca)	(7)		
Order	Stomatopoda (Subclass Malacostraca)	(180)		
	(b) In abdominal legs			
Order	Decapoda (Subclass Malacostraca except Penaeidea)	(8,003)		
B. Taxonomic groups shedding eggs			1,656	6.4
Order	Calanoida (Subclass Copepoda)	(1,200)		
Subclass	Branchiura	(75)		
Superorder	Syncarida (Subclass Malacostraca)	(6)		
Order	Euphausiacea (Subclass Malacostraca)	(57)		
Section	Penaeidea (Subclass Malacostraca)	(318)		

[1] Denotes the subsequent inclusion of more species (see R.D. Barnes, 1974). Values in parentheses indicate the number of species in the respective categories.

interesting morphological, functional (e.g. ventilation: Pandian and Balasundaram, 1980; inhibition of somatic growth during egg carriage: Kurup and R.G. Adiyodi, 1987), and behavioural (e.g. combing brood plates: McIntyre, 1954; tending and grooming eggs: Schmitt, 1965; Ansell and Robb, 1977; Bauer, 1981) adaptations.

The problems posed by the egg-carriage strategy of crustaceans are focused on here for the first time and some of the following considerations are partly speculative.

There are a number of exceptions and questionable inclusions to the types of egg carriage listed in Table 3. For instance, the Anostraca carry a ventral egg pouch, which is soon shed along with the eggs (Fig. 2A). Some ostracods are reported to shed their eggs freely into water (R.D. Barnes, 1974), but detailed information is not available. Some sphaeromids possess a special pouch-like invagination of the thoracic sternites and the oostegites may be absent (R.D. Barnes, 1974); however, these sphaeromids are also included under marsupial type of brooding.

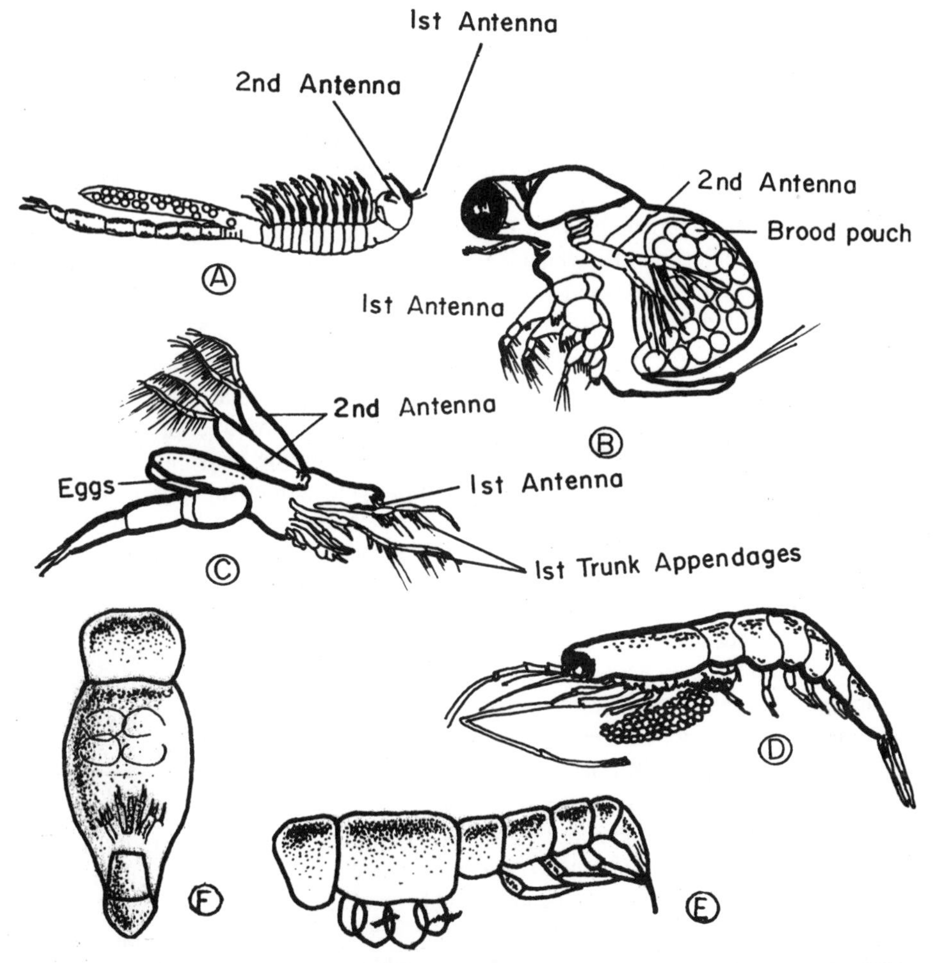

Fig. 2. Position of egg sacs in **A**: *Branchinecta*, **B**: *Polyphemus*, **C**: *Leptodora* (from Pennak, 1953); and **D**: *Nematoscelis difficilis* (redrawn from Nemoto *et al.*, 1972). **E**: Lateral and **F**: ventral views of the marsupium in *Parathemisto gaudichaudi* showing the egg-carrying site (redrawn from Sheader, 1977b).

Retaining eggs in the mantle cavity should pose no carriage problem but the egg-holding capacity is limited by the available space and the ability of the animal to ventilate the eggs. A prerequisite for this type of carriage is the presence of a mantle cavity in the anatomy, which occurs in about 11.5 per cent of crustaceans.

For want of a special ventilation mechanism, holding too many eggs in a pouch or sac may not be possible. Besides, a closed dorsal pouch, such as in Cladocera, may hinder movement and may alter posture (Fig. 2B and C). Notodelphyoid copepods are commensals in tunicates, which may ventilate their dorsal egg pouches. Clearly, this type of egg carriage is not prevalent and occurs only in about 2.8 per cent of crustaceans.

Projected egg sacs of Anostraca, Copepoda, and Euphausiacea (Fig. 2D) are better exposed dorsally, laterally, and ventrally, respectively; these crustaceans are active swimmers and hence their sacs are better ventilated. Projected sacs may retard the speed of locomotion but considerably reduce sinking rate in most of these pelagic forms (Clarke, 1967). Some 12.3 per cent of crustaceans adopt this type of egg carriage.

A distinct characteristic of most Peracarida is the presence of a ventral brood pouch or marsupium (Fig. 2E), which is formed by large plate-like processes, called 'oostegites', on certain thoracic coxae. The oostegites project inward horizontally and overlap with one another to form the floor of the marsupium. The marsupium may pose no resistance to movement but suffer space limitation and restricted ventilation (see Shillaker and Moore, 1987). The longer the female, the larger the available space in the marsupium and the number of eggs carried. A process from the basal portion of the maxilliped projects backward into the marsupium; the vibrations of this process create an anteroposterior water current (in contrast to the posteroanterior water current produced by the Caridea) through the brood chamber facilitating ventilation of developing eggs (see Sheader, 1977b); besides, the pellucid lobes of the second gnathopod, peculiar to the females, frequently turn and change the position of eggs in the marsupium (see Hurley, 1968). The marsupium of most gammarideans bears plumose marginal setae which aid in preventing the eggs from falling out. In comparison to the projected egg sac or open carriage on the appendages, the marsupium appears to afford a more effective protection as borne out from the fact that the largest number of crustaceans (nearly 33.6 per cent) brood their eggs within the marsupium.

Perhaps the appendages of decapods provide better ventilation and more space. Fastening the eggs to the thoracic legs appears to hinder the feeding activity. For example, *Oratosquilla (Squilla) mantis* carries a walnutsized, agglutinated egg mass on the smaller, chelate thoracic appendages and constantly turns and cleans it. The female does not feed while brooding (Schmitt, 1965). Only a small number of Euphausiacea, Stomatopoda, and Leptostraca incubate their eggs on thoracic legs (see Table 4, Figs. 1 and 2D).

A large number of crustaceans (approximately 32 per cent) carry eggs on their appendages. In decapods, brooding is accomplished on setae of the endopodite

Table 4

Maximum egg-carrying capacity of some crustaceans

	Taxonomic groups retaining eggs	Egg No. (per brood or spawn)	Reference
1.	In mantle cavity		
	e.g. *Balanus amphitrite* (Cirripedia)	300	Pillay and Nair (1972)
2(a)	In attached pouch		
	e.g. *Scapholeberis kingi* (Cladocera)	21	Murugan and Sivaramakrishnan (1976)
(b)	In projected pouch		
	e.g. *Artemia salina* (Anostraca)	110	Hentig (1971)
	Tisbe gracilis (Copepoda)	78	Battaglia (1957)
	Nematoscelis difficilis (Euphausiacea)	405	Nemoto *et al.* (1972)
3.	In marsupium		
	e.g. *Praunus neglectus* (Mysidacea)	80	Mauchline (1971a)
	Diastylis polita (Cumacea)	162	Corey (1981)
	Caecidotea intermedia (*Asellus intermedius*) (Isopoda)	230	Ellis (1961)
	Gammarus wilkitzki (Amphipoda)	250	D.H. Steele and V.J. Steele (1975a)
4(a)	In thoracic legs		
	e.g. *Squilla mantis* (Stomatopoda)		Schmitt (1965)
	Nematoscelis difficilis (Euphausiacea)	405	Nemoto *et al.* (1972)
(b)	In abdominal legs		
	e.g. *Crangon crangon* (Decapoda)	8,000	Clarke (1979)
	Jasus verreauxi (Decapoda)	2,000,000	Kensler (1967)
	J. edwardsii (Decapoda)	422,000	MacDiarmid (1989)
	Emerita analoga (Decapoda)	5,000	Efford (1969)
	Mutata lunaris (Decapoda)	65,000	Perez (1990)
	Callinectes sapidus (Decapoda)	2,750,000	Hines (1982)
5.	Groups shedding eggs		
	e.g. *Calanus hyperboreus* (Copepoda)	2,000	Conover (1967)
	Euphausia superba (Euphausiacea)	2,800	Harrington and Ikeda (1986)
	Penaeus merguiensis (Decapoda)	450,000	Crocos and Kerr (1983)
	P. setiferus (Decapoda)	1,000,000	Williams (1965)

of pleopods by which the surface area available for egg attachment is enormously increased. The adhesiveness of the eggs, developed after spawning, is specific to the endopodite setae; the bonds are stronger with the setae of endopodites than with the setae of the exopodites of the pleopods or with any other surface (Cheung, 1966). Ventilation is afforded by a current of water produced actively (see Pandian and Balasundaram, 1980) by forward movements of the pleopods. However, the safety requirements of the eggs become increasingly important: the abdomen, with the egg-carrying pleopods, is tightly flexed beneath the thorax in Brachyura as well as hippid and galatheid Anomura.

3. Energy cost of egg carriage

The fraction of energy expended on carrying the brood of eggs, weighing 2 to 40 per cent of the female's own body weight, is considerable, as shown in lizards (Tinkle, 1969; Tinkle and Hadley, 1975). Egg carriage may appreciably affect hydrodynamics (Spaargaren, 1979), increase the metabolic costs of swimming and walking, make the mother animals predation-prone, and penalize them in competitive interactions. Decrease in pyloric caecae reserve suggests that the cost of brooding eggs in the starfish *Leptasterias hexactis*, which does not feed while brooding eggs, amounts to about 70 per cent of the brood weight (Menge, 1974). Such estimations for crustaceans are totally wanting, but are urgently required. Brood capacity and incubation period are the two important factors that influence the energy cost of egg carriage; of these, the first one is considered at some length below.

4. Brood capacity

The relation between the site of egg carriage and the brood capacity has been analysed by the author. Lack of special ventilation mechanism and critical dorsal position of the pouch appear to have limited the egg-holding capacity of Cladocera to a maximum of 21 only (Table 4). Projection of egg sacs enhanced the capacity to about 100 in Anostraca, 75 in Copepoda, and about 400 in Euphausiacea. The fact that isopods carry usually 30 to 50 (R.D. Barnes, 1974), the amphipods about 50 (D.H. Steele and V.J. Steele, 1975a), and the mysids 25 to 75 eggs (Mauchline, 1973) suggests that space is the critical limiting factor. Carrying eggs on the setae of the pleopods enormously increases brooding capacity: even the terrestrial decapods carry a minimum of 100 eggs (Bliss, 1968); from this minimum, the values range to thousands and millions (Balss, 1955; Kensler, 1967; Hines, 1982). Thus, in crustaceans the pleopod is perhaps the most ideal site to incubate the largest number of eggs.

5. Brood weight

Table 5 summarizes the available data on brood size expressed as percentage of body weight as well as of body volume. Mean density of Crustacea is 1.052 g/ml for eggs and 1.175 g/ml for adults (see Table 6); for instance, the values for *Pareuchaeta norvegica* are available in terms of weight and volume; a conversion of brood volume into brood weight revealed 11 per cent difference between the calculated and observed values; age-related changes in water content of developing eggs may introduce considerable error.

Because of the difference in density between sea water and fresh water (Table 6), it may be energetically costlier to carry an equal volume or weight of brood for the freshwater planktonic crustaceans than for their marine counterparts;

Table 5

Brood size as percentage of body size in some crustaceans

Species		Brood as		Reference
		Percentage of live body weight	Percentage of live body volume	
I.	In mantle cavity			
	Balanus perforatus	9		Crisp (1954)
	Chirona (Balanus) hameri	9		Crisp (1954)
II.	In attached pouch			
	Daphnia carinata	25		Santharam (1979)
III.	In projected pouch			
	Branchinecta gigas	9		Daborn (1975)
	Pareuchaeta norvegica	6	8	Hopkins (1977)
	Pareuchaeta norvegica	12		Nemoto *et al.* (1976)
	Palaemonetes pugio	17		Anderson (1974)
	Canuella perplexa	21		Ceccherlli and Mistri (1991)
	Meganyctiphanes norvegica		10	Mauchline (1988)
	Thysanoëssa raschii		10	Mauchline (1988)
	Nematoscelis difficilis	18		Nemoto *et al.* (1972)
	Copepods		23	Mauchline (1988)
IV.	In marsupium			
	British mysids		10	Mauchline (1973)
	Ligia oceanica	12		Pandian (1972)
	Mysids		20	Mauchline (1988)
V.	On appendages			
	Macrobrachium nobilii	13		Pandian and Balasundaram (1982)
	Portunus pelagicus	11		Pillay and Nair (1971)
	Uca lactea annulipes	19		Pillay and Nair (1971)
	Deiratonotus tondensis	21	17	Fukui and Wada (1986)
	Ptychognathus ishii	21	16	Fukui and Wada (1986)
	Decapods		10	Mauchline (1988)

in fact, if time for the brooding period is also considered total energy cost of brood carriage may become even more. Yet, the freshwater planktonic crustaceans, such as *Cyclops* and *Daphnia*, carry broods equal to 18 to 25 per cent of their body weight, against the 6 to 12 per cent brood weight carried by the marine copepod *Pareuchaeta*.

Recently, Mauchline (1988) reviewed the scattered literature on the brood size of marine crustaceans and drew some very interesting conclusions, which are briefly summarized here. Brood volume, relative to body volume, of these crustaceans tends to remain constant as body size increases and represents 10 to 15 per cent of the body volume. The constraints on body volume devolve from the

Table 6

Density of egg, larva, and adult of some crustaceans at 20°C

Life stage	Density (g/ml)	Reference
Eggs	1.052	Herring (1974b)
Larvae	1.045	Herring (1974a)
Adults	1.175	Spaargaren (1979)
Sea water (35‰)	1.024	Riley and Skirrow (1975)
Pure water	0.998	Reid (1961)

possession of a rigid exoskeleton but some variation independent of body volume is still possible. The options open to the small crustaceans are, therefore, more limited than those available for the larger ones. There may be a minimum viable size of eggs. Consequently, small crustaceans may increase fecundity through the production of successive broods, whereas larger crustaceans can increase the number of eggs by decreasing the egg size. For instance, the large mysid *Gnathophausia ingens* (151 mm in length) produces a single brood of 150 to 250 eggs but the smaller *Metamysidopsis elongata* (5 mm in length) produces some 14 or more broods totalling about 340 eggs (Childress and Price, 1978). Likewise, the golden kingcrab *Lithodes aequispina* produces, by crab standards, fewer ($< 40,000$) large (> 2.2 mm) eggs (Jewett *et al.*, 1985), whereas *Paralithodes* spp. produce many ($> 380,000$) small (> 1.2 mm) eggs (Sasakawa, 1975).

6. Brood functions

(a) Protection

Several factors render the maternal protection afforded by crustaceans ineffective to varying extents; consequently, some 12 to 43 per cent eggs in a brood may suffer mortality (Table 7). Throughout the incubation period, egg loss is common among many ovigerous female crustaceans due to (1) fouling epibionts some of which, by occluding developing eggs from gaseous exchange, suffocate and kill them, while others live on the egg material, and (2) severing of eggs as a result of volumetric expansion by imbibition of water, abrasion of peripheral eggs while the animal moves around or ventilates, and erosion of individual territories leading to shocking encounters (see below).

In the absence of mucus, as in fish (see Bauer, 1978), the surface of eggs and body of crustaceans serve as excellent substrata for fouling organisms. Table 8 lists the incidence of microbial epibionts on the surface of eggs of some crustaceans. Bauer (1975, 1977, 1978, 1979, 1981) demonstrated the presence of cleaning brushes on chelipeds and maxillipeds in representatives of 13 out of 15 surveyed caridean families. When the cleaning chelipeds were ablated, *Hepacarpus pictus* was fouled with 20 times more epizoites and particulate debris (Bauer, 1978).

Table 7

Brood mortality in some crustaceans (adapted from Balasundaram and Pandian, 1982a)

Site of egg carriage	Brood mortality (per cent)	Reference
1. In pouch/sac	12	
Daphnia pulex	17	Buikema (1973)
D. carinata	21	Santharam (1979)
Pseudocalanus elongatus	12	Corkett and Zillioux (1975)
Tachidius discipes	7	Heip and Smol (1976)
Paronychocamptus nanus	11	Heip and Smol (1976)
Pareuchaeta norvegica	4	Hopkins (1977)
2. In marsupium	24	
Armadillidium vulgare	4	Paris and Pitelka (1962)
Idotea pelagica	10	Sheader (1977a)
Jaera nordmanni	27	Jones (1974)
J. albifrons	50	Jones and Naylor (1971)
Dynamene bidentata	36	Holdich (1968)
Corophium volutator	26	Fish and Mills (1979)
C. arenarium	26	Fish and Mills (1979)
Hippomedon whero	14	Fenwick (1984)
Protophoxus australis	35	Fenwick (1984)
Mysids	10	Mauchline (1973)
3. In appendages	43	
Macrobrachium nobilii	46	Balasundaram and Pandian (1982a)
M. rosenbergii	30	Wickins and Beard (1974)
Palaemon serratus	50	Reeve (1969)
Homarus americanus	36	Perkins (1971)
Nephrops norvegicus	45	Morizur *et al.* (1981)
Clibanarius chapini	47	Ameyaw-Akumfi (1975)
C. senegalensis	27	Ameyaw-Akumfi (1975)
Cancer magister	55	Wickham (1979)
Paratelphusa hydrodromous	53	C.K. Pillai (personal communication)

On accumulation, these fouling non-pathogenic microbionts effectively occlude gaseous exchange and suffocate and kill the eggs (e.g. Fisher, 1976; Fisher *et al.*, 1978; Balasundaram and Pandian, 1981).

Reports widely scattered in the literature on the tending, grooming, and cleaning of eggs by crustaceans (e.g. Schmitt, 1965; Ansell and Robb, 1977; Bauer, 1979) suggest that the decapod strategy to minimize the fouling load of epibionts is purely mechanical and not chemical.

Phycomycete fungi, such as *Lagenidium callinectes* and *Haliphthoros milfordensis*, are known to penetrate into the egg mass; the infected eggs do not hatch, while the development of the uninfected eggs on the same egg mass is not affected (Table 8).

Table 8

Incidence of epibionts on some crustacean eggs

Epibiont	Host	Suggested treatment	Reference
NON-PATHOGENIC EPIBIONTS			
Protozoa			
Vorticella	Many decapods		
Epistylis	Penaeid shrimps	Quinacrine hydrochloride (1 mg/l)	S.K. Johnson (1976)
Cyanophyta			
Oscillatoria	Many decapods		
Anabaena	Many decapods		
Fungi			
Aphanomyces	*Macrobrachium*	Malachite green	Balasundaram and Pandian (1981)
Bacteria			
Leucothrix mucor	*Cancer irroratus*	Streptomycin (4 mg/l)	P.W. Johnson *et al.* (1971)
	Cancer magister	Penicillin (1 mg/l) + streptomycin (1 mg/l)	Fisher (1976)
	Artemia salina	Decapsulation	Sorgeloos *et al.* (1977)
PATHOGENIC EPIBIONTS			
Fungi			
Lagenidium callinectes	*Homarus americanus*	Malachite green	Nilson *et al.* (1976)
	Cancer magister		Couch (1942)
Haliphthoros milfordensis	*Homarus americanus*	Malachite green	Fisher *et al.* (1975)
	Pinnotheres	Furnace	Vishniac (1958)
	Callinectes sapidus		Tharp and Bland (1977)
	Artemia salina		Tharp and Bland (1977)

There are reports of the occurrence of one or more 'commensal' species among the eggs carried by ovigerous females (e.g. *Gammarinema ligiae* in *Ligia oceanica*: Pandian, 1972; *Carcinonemertes errans* on *Cancer magister*: Coe, 1902). These commensals are now suspected to be egg predators (Kuris, 1978; Subramoniam, 1979). Wickham (1979), who observed *Carcinonemertes errans* feed on *Cancer magister* eggs in the field, even reared *Carcinonemertes errans* on this diet in the laboratory. He calculated that 55 per cent of the eggs carried by *Cancer magister* population in central California are consumed by the epibiont *Carcinonemertes errans*. High-density aquaculture practices, particularly those involving addition of nutrients (Fisher, 1976), enhance the scope of fouling epibionts; intensive research on these aspects is urgently needed.

In crustaceans, imbibition of water during incubation leads to a volumetric expansion of individual eggs (Pandian, 1970a, b; see also Pandian, 1972). As this process is linked to the simple osmotic hatching mechanism in crustaceans (Davis, 1968), there is hardly any exception. That the rate of volumetric expansion of crustacean eggs ranges from 1.2 to 5.4 (Table 9) results in a struggle for space among the developing eggs, especially in those carried in sac and marsupium; those which are not well placed or attached and those which are not allowed to volumetrically expand by the 'elbowing' neighbours degenerate or get lost easily. Since water imbibition is known to occur during the initial (from 50 to 56 per cent) and terminal (60 to 85 per cent) phases of incubation (Pandian, 1970a, b), the more significant egg loss during these phases than during the interim period in decapods such as *Macrobrachium nobilii* may be related to the imbibition of water during the said periods (Balasundaram and Pandian, 1981). In crustaceans which incubate eggs in sac and marsupium, abrasion may not be an important consideration, but accommodation of eggs that are doubled in volume is; the 10 to 30 per cent egg loss, suffered by these crustaceans, may solely be due to this problem. Abrasion can be an important cause for egg loss, especially in burrowing (Perkins, 1971; Clarke, 1979) and rock-dwelling carideans.

When an individual's territory is invaded by an intruder, or eroded by a neighbour, encounters occur at times; such encounters are shocking to sensitive ovigerous females, which react by rejecting the eggs in lumps. Rejection of eggs in lumps is frequently reported in literature (e.g. Costlow and Bookhout, 1968; Phillips, 1971; Balasundaram and Pandian, 1982a; Kurup and R.G. Adiyodi, 1987). In aquaculture practices, high-density culture is often resorted to in order to maximize production efficiency within the available resources. High-density culture practices certainly result in the erosion of individual territories. For instance, egg loss suffered by *Macrobrachium nobilii* increased from 6 to 46 per cent of the initial clutch size, when the rearing density was raised from 1 male + 1 female to 1 male + 7 females per aquarium (Balasundaram and Pandian, 1982a). In the field too, *M. nobilii* suffered over 36 per cent egg loss. Therefore, it appears that egg loss occurs in almost all egg-carrying crustaceans, the loss being low (around 12 per cent) in the fully closed sac of Copepoda, moderate (around 24 per cent) in the partly closed marsupium of Peracarida, and high (around 43 per cent) in Decapoda, which carry fully exposed eggs on the pleopods (Table 7).

(b) Ventilation

As development advances, the rate of gaseous exchange is also accelerated. Copepods are likely to swim faster to comply with the requirement for faster ventilation. No information is available on the rate at which ventilation current of water is passed through the eggs in marsupium. Ovigerous decapods exhibit a very characteristic behaviour of raising the abdomen to aerate their eggs by fanning the pleopods (Ansell and Robb, 1971). Underwater photography revealed the usual occurrence of males and non-berried females of *Nephrops norvegicus* in tunnels with a single opening, while ovigerous females occupied invariably tunnels with front and rear openings, which facilitated the unmixed current of water aerating the eggs (Rice and Chapman, 1971).

Table 9

Increase in egg volumes of certain crustaceans

Species	Volume increase	Reference
	Natantia	
Potimirim glabra	1.6	Davis (1964)
Caridina weberi	1.6	Rao *et al.* (1980)
Macrobrachium nobilii	1.6	Balasundaram (1980)
Acanthephyra spp.	2.0	Herring (1974a)
Palaemonetes vulgaris	2.2	Davis (1965)
Palaemon serratus	2.25	Wear (1974)
Crangon crangon	2.5	Pandian (1967)
	[acrura	
Panulirus homarus	1.2	Berry (1971)
Jasus lalandii	2.0	Silverbauer (1971)
Homarus gammarus	2.7	Pandian (1970a)
H. americanus	2.7	Pandian (1970b)
Nephrops andamanicus	2.8	Berry (1969)
	Anomura	
Galathea dispersa	1.7	Wear (1974)
Pagurus (Eupagurus) bernhardus	1.8	Pandian and Schumann (1967)
P. prideauxii	2.45	Wear (1974)
Petrolisthes aramatus	2.3	Davis (1966)
P. elongatus	5.4	Greenwood (1965)
Porcellana platycheles	2.5	Wear (1974)
Galathea squamifera	2.6	Wear (1974)
Pisidia longicornis	2.75	Wear (1974)
	Brachyura	
Cancer pagurus	1.5	Wear (1974)
Maja squinado	1.65	Wear (1974)
Carcinus maenas	1.8	Wear (1974)
Inachus dorsettensis	1.85	Wear (1974)
Eurynome aspera	1.95	Wear (1974)
Goneplax rhomboides	2.0	Wear (1974)
Ebalia tuberosa	2.2	Wear (1974)
Macropodia rostrata	2.2	Wear (1974)
M. longirostris	2.35	Wear (1974)
Liocarcinus (Macropipus) depurator	2.2	Wear (1974)
L. (M.) pusiuus	2.3	Wear (1974)
L. (M.) holsatus	2.35	Wear (1974)
	Copepoda	
Diaptomus siciloides	2.8	Davis (1959)
Cyclops bicuspidatus	2.55	Davis (1959)

Contd.

Table 9 Contd.

Species	Volume increase	Reference
	Cirripedia	
Balanus balanus	1.75	H. Barnes (1965)
Semibalanus (*Balanus*) *balanoides*	1.6	H. Barnes (1965)
	Isopoda	
Cirolana sp.	2.0	Davis (1969)
	Amphipoda	
Parathemisto gaudichadi	1.9	Sheader (1977b)

In decapods such as *Palaemon serratus* (see Phillips, 1971) and *Macrobrachium idae* (see Pandian and Katre, 1972), fanning frequency of pleopods is accelerated with advancing development. The frequency of fanning in *M. nobilii* increases from 2,676 times/hr during the initial incubation period to 8,760 times/hr during the terminal period. On hatching the last batch of eggs, the female ceases to sway the pleopod. Pandian and Balasundaram (1980) also estimated the optimum of current of water that ensured successful incubation in a newly designed incubator during different developmental stages. By plotting this pleopod-beating frequency against the current of water, they estimated the optimum quantity of water required for ventilation (Fig. 3). It increased from 0.5 to 2.5 l/hr during the incubation period. The ovigerous females ensure flow of the required current of water by changing the frequency of pleopod activity. Thus, the crustacean maternal care, which involves (1) egg carriage to the tune of about 13 per cent of the animal's body weight, (2) continuous cleaning of the surface of eggs, and (3) incessant pleopod fanning, constitutes an important component of energy expenditure in their reproductive energetics.

(c) Ovoviviparity

Ovoviviparity only occurs in a few crustaceans and is not a common strategy. Early workers were curious to confirm its occurrence in Cladocera and Peracarida. Some cladocerans, such as *Moina* and *Polyphemus*, secrete a nutritive fluid into the brood pouch (Gravier, 1931) and the embryos of these genera are claimed to be incapable of development outside the brood pouch. But the others, e.g. *Daphnia, Chydorus*, undergo normal development under *in vitro* conditions (Rammer, 1933).

With the advancement of *in vitro* development technics some interesting experiments have been conducted to understand the processes of adoption, replacement, and identification of embryos/young ones in the marsupium by the mother animals. Wittmann (1978) was perhaps the first to successfully induct adoption of (equal-sized) eggs of *Gammarus mucronatus* by the mysid *Leptomysis lingvura*. However, many gammarids (e.g. *G. palustris*: Borowsky, 1983) can discriminate their eggs from those of other species. *Corophium bonnellii* can distinguish eggs of its species from similar-sized eggs of *C. volutator* and *Lembos websteri* and

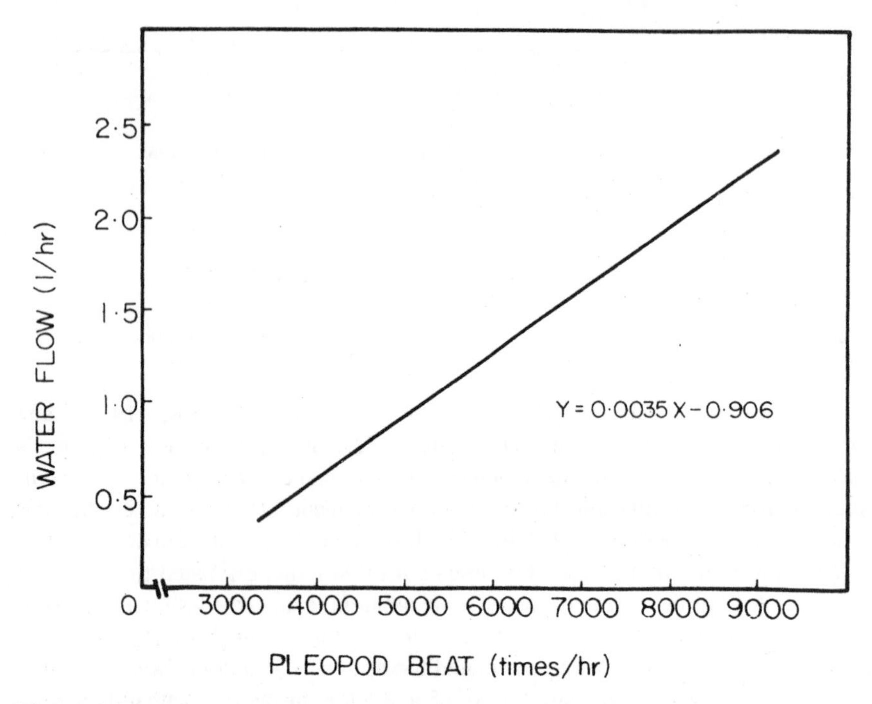

Fig. 3. Relationship between the frequency of pleopod beats of *Macrobrachium nobilii* and water flow rate in the incubator (from Pandian and Balasundaram, 1980).

replace only eggs of its own species into its marsupium (Shillaker and Moore, 1987). *Corophium bonnellii* has narrow oostegites and hence a less rigid brood pouch; consequently, it is predisposed for occasional loss of brooded eggs (see also Krishnan and John, 1974 for *Melita zeylanica*) and appears to have evolved a behavioural mechanism of replacing displaced eggs. To behavioural ecologists and functional morphologists, the study of adoption of eggs of conspecific and interspecific females by the peracarids may prove to be a fruitful area of research.

That over 78 per cent incubated eggs of *Marinogammarus marinus* successfully complete *in vitro* development at optimum temperature (15°C) (Vlasblom, 1969) suggests that most amphipods do not receive any maternal nutrients in the marsupium. The work of Saudray and Lemercier (1960) indicates that the embryos of the isopod *Ligia oceanica* receive maternal nutrients. Comparative studies on energetics of yolk utilization showed that the energy content of a freshly hatched young *L. oceanica* is 73 per cent of that of the egg, as against the usual value of 62 per cent observed in a number of decapods studied by us (Pandian, 1967, 1970a, b; Pandian and Schumann, 1967; Pandian and Katre, 1972). Pandian (1972) considered that this extra 11 per cent energy of the neonate *L. oceanica* may represent the maternal contribution (however, see also Green, 1965). Though small in fraction, the receipt of this moiety may be obligatory.

7. Control of hatching

During development, as mentioned above (Section I B 6a), crustacean eggs imbibe water (Pandian, 1970a, 1972; Herring, 1974a; Wear, 1974) and as they swell, the egg membrane breaks. This simple mechanism is the usual means of egg hatching in Crustacea (Davis, 1968). In addition to this osmotic mechanism of hatching, a proteolytic enzyme is released by embryos of the crabs *Neopanope sayi, Uca pugilator*, and *Sesarma cinereum* (see DeVries and Forward. 1991). Yet, the exact timing of hatching of the eggs, which are osmotically ready for the process, appears to be under the control of the ovigerous female in most Crustacea (Pandian, 1970a; Branford, 1978). An extract of the 'shell' (i.e. tissues within the shell and hypodermis lining, together with attached muscle ends) of the ovigerous individual stimulates mass hatching of barnacles and, in general, of all cirripedes (Crisp and Spencer, 1958). This hatching substance represents a product of barnacle tissue metabolism, and is a relatively stable, easily diffusible molecule, which forms lactone at low pH. It activates hatching not enzymatically, but by stimulating the movements of the embryos. The hatching substance is released into the egg mass in the mantle cavity by the mother at a time conducive for larval feeding.

The female chooses an appropriate time not only for feeding but also for larval dispersal. Monitoring the hatching time of the intertidal fiddler crabs *Uca pugnax, U. pugilator*, and *U. minax* under laboratory and field conditions, DeCoursey (1979) found that the hatching profile lasting only for a 10-minute period for 1,500–94,000 eggs in a female was highly synchronized to the nocturnal high tide. The female aided hatching by vigorous abdominal contractions. Such tidal timing may reduce predation of the ovigerous female approaching the water mark to deposit the zoeae, and serve to flush them.

In several decapods, eggs in a clutch, all apparently fertilized and berried at the same time and ready to hatch, are not however hatched in a single burst, but in batches spread over a period of time (Table 10); hatching is also restricted to nights and for a brief period only. This rhythm in the hatching profile is under control of the mother animal, as all eggs in a clutch, if incubated *in vitro*, hatch simultaneously (Balasundaram and Pandian, 1981). Larvae of the last batch contain 8 and 30 per cent less reserve yolk energy than those of the first batch in *Homarus americanus* (Pandian, 1970b), *Macrobrachium idae* (see Pandian and Katre, 1972), and *M. nobilli* (Balasundaram, 1980). Besides, they swim so slowly it is difficult for them to escape from predators. Yet, for some unknown ecological advantage, a number of decapods choose to hatch their eggs in batches over a period of time.

C. Post-embryonic Development

In most crustaceans, development involves one or more larval phases. The occurrence or absence of specialized ontogenetic digressions labels the development as metamorphic or anamorphic (Snodgrass, 1956). Nauplius is the typical free-living

Table 10

Batching effect of hatching in some decapods

Species	No. of batches	Reference
Homarus gammarus	22	Pandian (1970a)
H. americanus	5	Ennis (1973)
Acanthephyra pelagica	9	Herring (1974a)
Jasus lalandii	5	Silverbauer (1971)
J. edwardsii	4	MacDiarmid (1989)
Macrobrachium nobilii	5	Balasundaram and Pandian (1981)
M. idae	3	Pandian and Katre (1972)
Clibanarius chapini	4	Ameyaw-Akumfi (1975)
C. senegalensis	3	Ameyaw-Akumfi (1975)

larval phase with which a crustacean commences its post-embryonic development. Passing through a series of 10 or more moults, it develops into an adult by progressively differentiating more segments and their appendages. The simple, stepwise transformation from hatching to maturity, observed in over 15 per cent crustaceans such as Anostraca, Ostracoda, and Euphausiacea (Table 11), is reminiscent of the ontogeny of hemimetabolous insects.

Deviation from the basic anamorphic development (anamorphic group, Table 11) occurs in over 53 per cent of crustaceans (metamorphic group, Table 11), in which one (or more) of the pelagic free-swimming larval phases is involved in dispersal of benthic (Brachyura, Anomura, Scyllaridea, and Stomatopoda) or sessile benthic (Cirripedia) forms. In the parasitic crustaceans, the larval phase is not only dispersive but also infective.

There is a tendency to extend the period of embryonic development and to exclude larval phases in different crustaceans, particularly the decapods. Thus naupliar, zoeal, and later stages may be passed within developing eggs. Barring the parasitic gnathiideans and epicarideans, the Peracarida retain their eggs in marsupium until a complete, or nearly complete (manga), or a juvenile stage is reached. The involvement of a direct or nearly direct development may partly account for the brood capacity, which is far lower than that of most decapods (Fig. 4, Table 4).

1. Density problems

In comparison to that of eggs carried by crustaceans, the density of sea water and fresh water is low (Table 6). Unless carried by the mother animals, these spherical eggs may sink slowly in sea water and rapidly so in fresh water. The occurrence of

Table 11

Post-embryonic development and larvae of crustaceans (from Waterman and Chace, 1960; modified and added)

Post-embryonic development type	Larva(e)	Total species	
		No.	Per cent
Anamorphic		2,545	9.8
Ostracoda (2,000)	Nauplius		
Branchiura (75)	Nauplius		
Euphausiacea (90)	Nauplius → Calyptopis (Protozoea) Furcilia (zoea)→ Cryptopia (postlarva)		
Cephalocarida (2)	Metanauplius		
Branchiopoda (375)	Metanauplius		
Mystacocarida (3)	Metanauplius		
Mildly metamorphic		4,021	15.5
Calanoida (1,200)	Nauplius → Copepodid (postlarva)		
Harpacticoida (1,200)	Nauplius → Copepodid		
Cyclopoida (1,000)	Nauplius → Copepodid		
Penaeidae (318)	Nauplius → Protozoea → Mysis (zoea) → Mastigopus (postlarva)		
Nephropsidae (313)	Mysis → Postlarva (zoea)		
Metamorphic		8,214	31.3
Thoracica (550)[1]	Nauplius → Cypris		
Sergestodae (15)	Nauplius → Elaphocaris (protozoea) → Acanthosoma (zoea) → Mastizopaus (postlarva)		
Caridea (1,590)	Protozoea → Zoea → Postlarva		
Stenopodidea (22)	Protozoea → Zoea → Postlarva		
Scyllaridea (84)	Phyllosoma → Puerulus, nisto or, Pseudibaccus (postlarva)		
Anomura (1,270)	Zoea → Glaucothoe (in pagurids) or Grimothea (postlarva)		
Brachyura (4,428)	Zoea → Megalopa (postlarva)		
Stomatopoda (180)	Pseudozoea → Alima → Synzoea → Stomatopodid (postlarva)		
Gnathiidae (75)	Larva → Praniza (postlarva)		

Contd.

Table 11. Contd.

Post-embryonic development type	Larva(e)	Total species	
		No.	Per cent
Strongly metamorphic in females alone		1,350	5.2
Notodelphoida (300)	Nauplius → Copepodid (postlarva)		
Caligoida (400)	Nauplius → Copepodid		
Lernaepodida (300)	Nauplius → Copepodid		
Epicaridae (350)	Epicaridium → Microniscus → Cryptoniscium → Bopyridium (postlarva)		
Strongly metamorphic		273	1.1
Monstrilloida (35)	Nauplius → Parasitic → larva		
Acrothoracica (12)	Nauplius → Cypris		
Aseothoracica (25)	Nauplius → Cypris		
Rhizocephala (200)	Nauplius → Cypris → Kentrogon		
Epimorphic		852	30.2
Cladocera (425)	(In *Leptodora*: metanauplius)		
Anaspidacea (6)			
Spelaegriphacea (1)			
Mysidacea (420)[1]			
Epimorphic with manga		7,861	30.2
Neabliacea (7)	Manga		
Thermasbaenacea (4)[1]	Manga		
Cumacea (425)[1]	Manga		
Tanaidacea (250)[1]	Manga		
Amphipoda (3,600)[1]	Manga		
Isopoda (3,575)[1]	Manga		
(exceptions: Gnathiidae, Epicaridae)			

[1]Denotes the inclusion of more species (see R.D. Barnes, 1974).
Values given in parentheses refer to the number of species.

calanoid (Kasahara *et al.*, 1975a) and cladoceran (Onbe, 1977) resting eggs in marine sediments (see also Table 40) shows that these dense, freely shed, resting eggs sink through the less dense sea water. On the other hand, sea water is denser than most crustacean larvae which therefore are able to drift at the desired level without much effort. Low density of the fresh water reduces the scope for such drift by planktonic larval stages, making it energetically expensive. Hence colonization of

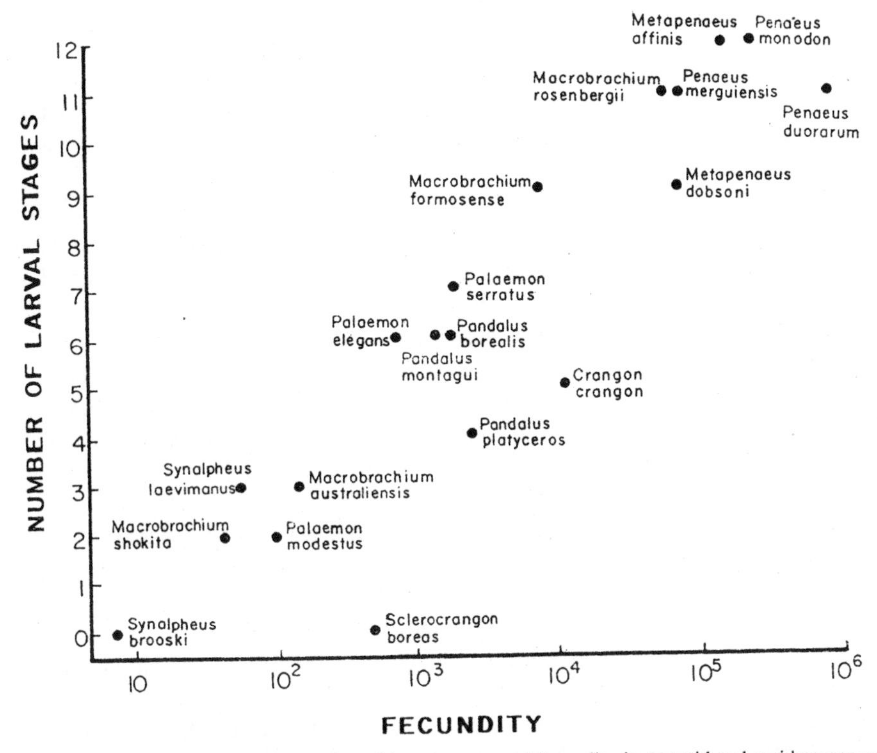

Fig. 4. The relation between the number of larval stages and fecundity in penaeid and caridean prawns (from Wickins 1976).

fresh water by crustaceans as well as other invertebrates involved extended embryonic development, with a few or no larval stages. Larger size contracts this reduced scope further, especially in crustaceans such as decapods. Thus, decapods provide contrasting examples for indirect or almost direct development even among very closely related species inhabiting the marine or freshwater environment. For instance, the marine *Halicarcinus ovatus* and the estuarine *H. australis* pass through three zoeal stages each, whereas the development of the freshwater *H. lacustris* is direct, and involves no free larval stage (Lucas, 1971).

In decapods that spawn in the sea, there is a protracted larval life, with a number of free-swimming stages between the egg and the post-larva (see Table 11). In freshwater species, the young one hatches out as a miniature adult. Direct or almost direct development involves large and yolky eggs, and consequently, there is a reduction in fecundity too. Thus, some natantians spawn only a few, large, yolky eggs, which hatch at advanced juvenile stage, while other nearly identical species with many small eggs, poor in yolk, have a series of larval stages. Dobkin (1969) extended this idea and noted that abbreviated development in Carridea occurs largely in forms from Arctic waters and fresh water, and in those living on inquilines.

2. *K*- and *r*-strategies

During crustacean development, small, dense eggs decrease in density, largely as a result of increased volume (Pandian, 1970a; Herring, 1974b) and the less dense larvae are effortlessly drifted to the planktonic mode of existence. As the larvae grow and approach the adult size, they become denser and are recruited to the benthic population. Thus, the egg size (Dobkin, 1969) and larval density (Herring, 1974b) are the two key factors determining the larval duration. According to Thorson (1950), the number of eggs spawned by marine invertebrates with planktotrophic larvae ranges from 11×10^2 to 5×10^8, the species with a few large eggs with extended embryonic development occupying the lower extreme and those spawning many, small eggs with protracted larval life, the other extreme of the recorded range. In other words, these two extremes represent the K- and r-strategies; but the fact that fecundity values of crustaceans extend over nearly the entire range indicates that each species occupies a point in what may be more appropriately called the 'K–r continuum' (see Giesel, 1976; Hamer and Appleton, 1991).

Approximately 50 per cent of the benthic marine invertebrates pass through a larval period of four weeks (Thorson, 1961). Thorson (1950) assessed that less than 0.1 per cent of the total quantity of invertebrate larvae, released annually into the plankton, achieved settlement. In most r-strategists, which suffer heavier mortality than K-strategists, the pelagic period is longer and involves more larval stages. Apparently, the longer the pelagic life, the greater the larval mortality. In general, K-strategists suffer a much lower mortality and achieve a higher percentage of recruitment from the pelagic to benthic populations. The problem of larval mortality is particularly severe for all small decapods, in which the larvae hatch as zoea; hence the decapods must have some mechanism to achieve higher recruitment to benthic populations (Thorson, 1966). A common strategy of several decapods is to reduce the pelagic period and moult, a critical event, during which the larvae are more vulnerable to predation (e.g. *Palaemonetes*, Table 12).

Table 12

Pelagic period and number of moults in *Palaemonetes* (from Hubschmann and Broad, 1974)

Species	Size (mm)		Pelagic period (day)	No. of moults
	at hatching	at metamorphosis		
P. vulgaris	2.3	6.3	15–30	7–11
P. pugio	2.6	6.3	15–30	7–11
P. intermedius	3.5	7.0	13	6–8
P. kadiakensis	4.4	7.5	16–30	5–8
P. paludosus	3.8	4.5	5–10	3
P. cummingi	4.8	5.5	9	3

One of the most active areas of theoretical ecology is the study of life-history strategies, and the two theories which have enjoyed much support are: (1) r- and K-selection (MacArthur and Wilson, 1967; Pianka, 1970) and (2) bet-hedging (Murphy, 1968; Stearns, 1976). Table 13 (see Fenwick, 1984) summarizes the expected predictions of these theories. Many of these predictions of life-history strategies are difficult to apply to invertebrates because of the uncertainties in collecting reliable age-specific data on mortality and natality. Consequently, much of the work on these animals has concentrated on larval development (e.g. Todd and Doyle, 1981). Several studies have been undertaken to test these theories with reference to crustaceans (e.g. Lynch, 1980; Nelson, 1980; Van Dolah and Bird, 1980). A detailed long-term study undertaken by Fenwick (1984) on gammarid amphipods and an ostracod indicated that the r- and K-selection theory is inadequate, especially with reference to the comparison of strategies between species (or phylogenetic groups: see e.g. Reaka, 1979; Lynch, 1980). Fenwick (1984) found that his results provided little support to the bet-hedging theory. The principal reason for the failure of these theories is that they were devised with little regard to the organisms to which they were intended to apply. The approach of Lynch (1980), which recognizes the possibility of several equally successful combinations of strategies for a given situation, appears better suited for crustaceans. His ideas on optimal body size have brought together information on size-specific foraging efficiencies, size-specific mortality (due to predation), size-specific growth rates, and size-specific reproductive efforts. Briefly, species-growth patterns have evolved to balance optimal foraging size against size-specific mortality on one hand and the demand for reproduction on the other. Lynch's (1980) approach has demonstrated that there is far more to life-history strategies than simply age at maturity, longevity, and number of broods (see also Fenwick, 1984).

3. Dispersal strategies

About 70 per cent of the decapods complete their planktonic life within five weeks. The larvae of *Emerita* and *Panulirus* groups pass through an extended pelagic life (Thorson, 1961). *Emerita analoga* is one of the few decapods for which there is sufficient literature on larval dispersal (M.W. Johnson, 1940; M.W. Johnson and Lewis, 1942); it combines a low fecundity with one of the longest pelagic-development periods (ca. 130 days). Therefore, *E. analoga* represents an extreme case, in which a species that produces only a few thousand eggs a year has an extended pelagic life and yet ensures greater probability of being recruited to the coastline. Efford (1970) suggested that countercurrent is the mechanism by which the pelagic larvae disperse non-randomly and by which late zoeal stage tends to remain near adult habitat. In other words, as *E. analoga* inhabits such coastal areas alone, where the currents move parallel to or along the coast, the chances of its larvae being transported to far-off places are reduced. This mechanism also offers

Table 13

The combination of life-history strategies predicted by r- and K-selection and by bet-hedging (from Fenwick, 1984)

Stable environment	Unstable environment
1. r- and K-selection and bet-hedging with high adult mortality	
a. Slow development	b. rapid development
late maturity	early maturity
iteroparity	semelparity
fewer young	more young
small reproductive effort	larger reproductive effort
long life	short life
2. Bet-hedging with high juvenile mortality	
a. Early maturity	b. late maturity
iteroparity	iteroparity
more young/brood	fewer young/brood
larger reproductive effort	smaller reproductive effort
fewer broods	more broods
shorter life	long life

an explanation for the pattern of distribution of *E. analoga* in the northern Pacific coast of America.

Exactly the reverse is true for the *Panulirus* group. The larvae pass through perhaps the longest period of pelagic life (180 days). The majority of the phyllosoma larvae of *Panulirus* are held in eddy flows and are carried beyond the shelf; at the end of their long planktonic existence, the late phyllosoma larvae or the early puerulus are drifted by gyral current towards the coast, where they settle in shallow reef areas. An extended pelagic life together with the 'true long distance' nature of the phyllosoma explains the wide and often circumtropical distribution of *Panulirus* (Chace and Dumont, 1949).

There are two basic mechanisms of recruitment of juveniles to estuarine decapod populations: (1) recruitment by retention of larvae and (2) recruitment by immigration of juveniles and adults (Sandifer, 1975); these more or less correspond to the nursery-area hypothesis and the rearing-current hypothesis of Efford (1970), respectively. According to the nursery-area hypothesis, the recruitment is from larvae that have avoided dispersal by currents and get recruited near the adult habitat. For instance, the larvae of this group of decapods in the York river estuary are more abundant in the lower layer of the water column, where net transport is upstream, than near the surface and thus tend to be retained within the estuary (Sandifer, 1975). To cite another example, the crabs *Halicarcinus ovatus* and *H. rostratus* are

K-strategists with low fecundity; the pelagic larval life lasts for 23 days at 20°C. They are abundant in semi-enclosed water masses. A major factor for this abundance is that these localities act as 'nursery areas', restricting dispersal of larvae away from the adult habitats, compensating for their low fecundity with relatively high rates of recruitment (Lucas, 1975).

The rearing-current mechanism is the one in which the adults migrate to a spawning area to breed and the larvae develop, while drifting back to the adult habitats (e.g. *Penaeus*: George, 1969; *Macrobrachium*: Balasundaram, 1980).

II. PATTERNS OF SEXUALITY

The patterns of sexuality in crustaceans have been extensively reviewed and discussed in Volume V of this series by Charniaux-Cotton *et al.* and, therefore, only points relevant to this chapter are included here. The vast majority of crustaceans are disexual and dimorphic and reproduce sexually as mentioned in Section I, though, in certain groups, parthenogenesis plays an important role in the life cycle (Table 14).

A. Parthenogenesis

Geographic parthenogenesis (e.g. see Table 15) and cyclic parthenogenesis occur in crustaceans. The former is characterized by the existence of disexual and parthenogenetic forms in a single, or in two closely occurring, species; the two forms generally have different geographic distribution (Charniaux-Cotton, 1960). Among the Cladocera, short-day photoperiods induce an increase in females, whereas males prevail in long-day photoperiods (Stross, 1969, 1971). The males thus become a rarity among several crustaceans inhabiting high latitudes (e.g. *Ilyocypris*: Pennak, 1953). Hence, geographic parthenogenesis, mostly found in high latitudes, is accompanied by polyploidy. For instance, two different races exist in the terrestrial isopod *Trichoniscus elisabethae* (see Vandel, 1940). The first, composed of triploid females (3n = 24), reproduces by ameiotic parthenogenesis, the second is disexual and diploid (2n = 16). In northern Europe, only the first race is found; in France, the two races are sympatric, but there is an absolute reproductive isolation between them. Females of the diploid race do not exhibit any tendency toward parthenogenesis. In the triploid race, males may appear sporadically and may even mate with triploid females, but fertilization does not occur; likewise, fertilization of the eggs does not actually take place, in spite of copulation with a very long intromittent organ in certain freshwater ostracods (Lowndes, 1935). Cyclic parthenogenesis is characterized by heterogony and occurs mostly among cladocerans.

Analysing the geographic patterns of distribution of parthenogenetic populations, Cuellar (1977) and Glosener and Tilman (1978) concluded that most parthenogenetic species exist in natural disclimax communities and that within these communities, they exist in isolation from the closely related congeneric species.

Table 14

Sexuality and kinds of males and eggs in some crustaceans

Taxonomic group	Sexuality	Kind of male	Kind of egg
Cephalocarida	Simultaneous hermaphodites with common gonoduct and opening		
Anostraca	Dioecious	Male not known in some species	Parthenogenetic; produce sexual and asexual resting eggs
Ostracoda	Dioecious	Male not known in some species	Parthenogenetic; produce resting sexual eggs
Cladocera	Dioecious	Seasonal male	Parthenogenetic; produce sexual resting eggs
Copepoda	Dioecious	Dwarf male in parasitic forms: dimorphic males in some species switching sex from male to female in *Euterpina* sp.	Produce resting eggs; any life stage may get encysted
Cirripedia	Some dioecious; most others simultaneous hermaphrodites; cross-fertilization obligatory; prodigious (***r***-strategy) sperm producers	Complemental males; with simplified anatomy	
Euphausiacea	Dioecious	Dimorphic males in some species	
Tanaidacea	Hermaphrodite		
Isopoda	Dioecious; some protandrously hermaphrodites	Dwarf male parasitic forms	Resting stage, monogeny
Amphipoda	Dioecious; intersexuality not uncommon		Resting stage; monogeny
Decapoda	Dioecious; parasitic castration occurs; pandalids are protandrous hermaphrodites; primary and secondary females in *Emerita*		Diapausing eggs

Table 15

Geographic parthenogenesis in crustaceans (from Charniaux-Cotton, 1960)

Taxonomic status	Species
Branchiopoda	
Anostraca	*Artemia salina*
Notostraca	*Lepidurus*
Conchostraca	*Limnadia*
Cladocera	*Daphnia pulex*
Ostracoda	
Podocopa	*Cypris fuscata*
	Candona
	Cypricercus
	Ilyocypris
	Potamocypris
Copepoda	
Harpacticoida	*Elaphoidella*
Cyclopoida	*Attheyella*
Cirripedia	*Sylon*
Malacostraca	
Isopoda	*Trichoniscus elisabethae coelebs*
Amphipoda	*Rhabdosoma*
	Stygobromus

Without this isolation, parthenogens will be outcompeted by the sexual species. Browne (1980a, b) tested this hypothesis by allowing competitive interactions between the obligate parthenogenetic strains of *Artemia salina* from India and the obligate dioecious outbreeders from America. He established two alternative competitive regimes, the first one with the initial high density and high food density, and the second one with low initial density and low food density. Even with the inherent advantage of not having to pay for 'the cost of meiosis' (Williams, 1975), the sexual strains of *A. salina* eliminated the parthenogenetic strains, which lends experimental support to the hypothesis of Glosener and Tilman (1978).

B. Hermaphroditism

The relative advantages of gonochorism and simultaneous or consecutive hermaphroditism are poorly understood (Emlen, 1973). Ghiselin (1969) suggested that natural selection favours sequential hermaphroditism over dioecy, when an individual's reproductive success as a male or a female is closely related to age or size, and where the relationship is different for each sex. For example, if large size increases the egg output of a female but does not aid a male in competition to fertilize eggs, selection favours genes which cause an individual to operate as a male, when small, then switch to a female at a later age or larger size. The reverse order of sex change would be favoured, if larger size or old age were relatively more important for male reproductive success (see also Wenner, 1972). A series of papers employing population-genetic techniques have generally supported Ghiselin's (1969) suggestion (R.R. Warner, 1975; Leigh *et al.*, 1976; Charnov, 1979).

1. Sequential hermaphroditism

Sequential hermaphroditism appears more prevalent among crustaceans. Wenner (1972) summarized and classified the different types of sequential hermaphroditism (sex reversal) under reversal, intermediate, and anomalous patterns of sex ratio; since then evidences in support of Wenner's concept have been reported frequently (e.g. Ahmed and Mustaquim, 1974; Swartz, 1976; Subramoniam, 1981). Moreover, the sex-reversal events are now considered important switch mechanisms in population regulation (Wildish, 1977).

Almost all species of cymothoids are protandrous hermaphrodites (Weinstein and Heck, 1977). The occurrence of protogyny in *Clibanarius* (see Wenner, 1972) is regarded as adaptive; it is selectively advantageous for a female to be smaller in size, since her brood must also be accommodated within the shell (Ameyaw-Akumfi, 1975). Large males of the harpacticoid copepod *Euterpina acutifrons*, found in Brazilian waters, appear to be genetic females that had switched sex in response to some epigenetic factor (Stancyk and Moreira, 1988).

2. Simultaneous hermaphroditism

Thoracican barnacles are mostly hermaphrodites and represent the only large group of hermaphroditic crustaceans (Table 14). Tiny dwarf males, which attach themselves to the female or to hermaphroditic individuals, are found in some pedunculate genera like *Scalpellum* and *Ibla*, in a few species of *Balanus*, and in the boring Acrothoracica (e.g. *Trypetesa*). In some species of *Scalpellum*, the males are miniatures of the females of hermaphroditic individuals. When attached to a hermaphrodite, these dwarf males are called 'complemental males' (Table 14). The body of the dwarf males consists of little more than a sac containing testes, sperm ducts, and an enormous penis—all housed in a mantle. A single female *Trypetesa* may contain as many as 14 dwarf males attached to the outside of her mantle. Similarly, a single female *Clypeoniscus hanseni* may carry as many as four diminutive males attached to her (Sheader, 1977a).

Ecologically speaking, males are a luxury for a population because they utilize resources which otherwise can be used more productively by females. As the continued presence of males intensifies, it would be a great 'ecological luxury' (Ricklefs,

1973) for some of the cirripedes to maintain complemental males and that too in such great numbers as is the case with *Trypetesa*. Nevertheless, maintenance of a 'harem' of complemental males may be an adaptive strategy to ensure multiple and mass inseminations rather than a luxury. The complemental males have no digestive tract and possibly can maintain themselves on dissolved nutrients from the medium (Lawrence, 1973; Lawrence *et al.*, 1977; for a review, Pandian, 1975). They may not therefore compete with the female for the same source of food.

C. Dwarf Males

Some parasitic (Gnathiidae among Isopoda and certain families of Copepoda) and sessile crustaceans ensure fertilization by epizoic dwarf males, which apparently acquire sexual maturity without somatic growth. The strategy of size reduction is ubiquitous among the parasites (e.g. *Ione thoracica, Peltogaster paguri*). Reduction in size may mitigate 'immediate competition' and minimize the energy cost of carriage of these 'incipient parasites' (Wharton, 1942). In *Emerita asiatica*, the epizoic dwarf male, weighing more than 5 mg (= 5 mm carapace length), stop hanging on to the female and become 'free-living' (Subramoniam, 1977). Through his careful study on the chronology of sexualization in *E. asiatica*, Subramoniam (1981) has since convincingly demonstrated that the freed male gradually gets redifferentiated into a secondary female. In the flabelliferid isopod *Anilocra physodes*, individuals in male phase living attached to the female redifferentiate into the female phase, later than the isolated male-phase individual; the attached males do not pass to the female phase until the female partner has been removed. The sexually undifferentiated first copepodid larva of *I. thoracica* differentiates, on reaching the host *Callianassa tyrrhena (laticaudata)*, into a female and the subsequent one into a male that lives attached to the female.

If young males are removed from females and placed on unparasitized individuals of *Callianassa tyrrhena*, the parasite, in the majority of cases, becomes a female. Conversely, a juvenile female may develop into a typical dwarf male, if it is placed on a female (Reverberi and Pittoti, 1942).

D. Dimorphic Males

An unusual characteristic of *Euterpina acutifrons* is that the males are dimorphic, and appear to perform different functions (Haq, 1965, 1972). Distinct size and structural differences in the antennae, antennules, and second pair of legs have been observed; 'small' males range from 0.53 to 0.60 mm in body length, large ones from 0.65 to 0.70 mm. A similar intrasexual dimorphism has also been described for the *Labidocera jollae* group (Fleminger, 1967) and *Cyclops* (see Lowndes, 1929). Both laboratory and field studies have shown that small males are reproductively more active, more cold-adapted, and more numerous than the large ones (Haq,

1972; Moreira and Vernberg, 1968; Vernberg and Moreira, 1974; D'Appolito and Stancyk, 1979). Recently, Stancyk and Moreira (1988) observed the presence of 20 per cent large males in Brazilian populations of *E. acutifrons*; the large males, in turn, sired large males and they did not occur in the small-male-sired clutches. Thus, there seems to be a genetic basis for this male dimorphism in males.

Despite this marked difference in body size among dimorphic males, the size of the spermatophore is about the same (Haq, 1972). In an attempt to explain this intrasexual dimorphism, Haq (1973) crossed each type of male with virgin females and found that the proportions of different sexes and of dimorphic males produced in two successive broods at 16° and 20°C remained unaltered. This observation strongly suggests that some mechanism, closely associated with genetic factors, may determine different sexes and dimorphic males. That the large males are more warmth-adapted (Moreira and Vernberg, 1968) and occur in great numbers during the summer months (D'Appolito and Stancyk, 1979) indicates that a higher temperature induced the production of dimorphs and large males. Structurally, the small male resembles the female more closely than the large male (Haq, 1965).

Recently, Fleminger (1985) described the presence of female dimorphism in 14 species of calanoid copepods. Studies on commensal and parasitic copepods have illustrated the polygenic nature of sex determination in this group. These studies have also shown that factors such as host age (Hippeau-Jacquotte, 1984) or presence of conspecifics (Do and Kajihara, 1986) may influence sex determination and even morphotype within a single sex. It appears that the female is the fundamental sex and developmental programmes of the two males are turned on by specific ecological stimuli. For instance, sex determination in the parasitic poecilostomatid copepod *Pseudomysicola spinosus* depends largely on the presence of an adult at the second copepodid stage; individuals generally became large crawling males, if a female was present; at least about 50 per cent became females and the rest differentiated into small, actively swimming males (Do and Kajihara, 1986). The easily culturable *Euterpina acutifrons* provides an excellent opportunity to learn about the mechanism of sex determination in copepods, especially the epigenetic and other ecological factors in the determination of sex dimorphism in male and the costs/benefits of being female, large male, or small male under various environmental conditions. A different kind of intrasexual dimorphism is prevalent among the euphausiids. Males of *Nematoscelis* from different geographic regions are known to differ remarkably with respect to the saddle-shaped thickenings (see James, 1973).

E. Mating Morphism

Mating morphism refers to the changes in morphic features following moult in a single individual; it differs from polymorphism in that individuals with different morphic characters occur simultaneously. In the astacoid decapod *Orconectes immunis*, two distinct forms of males have been observed: the copulatory individuals of the spring and autumn moult into non-copulatory form in summer and winter, respectively. The first pleopods of the copulatory form are corneous, hard, and sculptured; they aid in transmission of the spermatophores during spring and autumn. The soft pleopods of the non-copulatory form are uniramous and unsculptured and incapable of transmitting spermatophores. Just as the absence of pleopodal hairs inhibits berrying (see Sandifer and Lynn, 1980) the loss of sculpture and biramous nature of the first pleopod can also effectively prevent mating (Pennak, 1953). Such mating morphism also occurs in Tanaidacea (e.g. *Leptochelia dubia*) and Brachyura (e.g. *Macropodia rostrata*) (Teissier, 1960).

F. Sex Ratio

Sexual processes allow genetic recombination in each generation; as such, sex is an important way in which genetic variability is maintained. However, hereditability of an individual is halved (Pianka, 1974). According to Fisher's theory of sex ratio, natural selection favours a 1:1 parental expenditure on offspring of the two sexes (Kolman, 1960).

Differential mortality between the sexes and other factors which may create a significant difference in the cost of producing each sex may lead to variously skewed sex ratios (Wilson and Pianka, 1963). Sperm are small, energetically inexpensive to produce, and produced in large numbers. In *Aratus pisonii* (see Conde and Diaz, 1989) and many other crustaceans, the males grow bigger than females as the latter divert their energy towards reproduction and thus concentrate in intermediate size classes. Energy investment in male-gamete production being lower, the males can mate frequently, though semen production could prove costly and offset the advantage partly. As a male can fertilize a very large number of females, it may not be desirable to maintain equal number of males in a population, a fact that has pushed the sex ratio in favour of female production to such an extent that in several species males are rare or unknown (Table 16). *Cancer magister* appears to represent the near 'ideal' situation, in which the parity in sex ratio is maintained from the larval to adult stage. Besides differential mortality, a number of other contributory factors interacting with locality, season, and ontogeny of the species are responsible for the observed alternations in sex ratio.

1. Sex ratio and size

Since age- or size-dependent differential mortality is the prime factor responsible for the observed skewed sex ratios, a few authors have attempted to study sex ratio as a function of size (e.g. W.S. Johnson, 1976a, b). Wenner (1972) observed that male-female relationships in Crustacea fall into four discrete patterns:

(1) Standard: male-female ratios equal at smaller sizes, males predominate at the largest sizes, e.g. *Portunus sanguinolentus*.

Table 16

Sex ratio of some aquatic crustaceans (from Darnell, 1962)

Species	Identity	Sex ratio (Percentage of females)	Remarks
		MALES UNKNOWN	
Caenestheriella gynecia	Freshwater conchostracan	100	Females presumably parthenogenetic
Candona caudata	Freshwater ostracod	100	Females presumbly parthenogenetic
Stygobromus spinosus	Subterranean amphipod	100	Females presumably parthenogenetic
		MALES RARE	
Potamocypris smaragdina	Freshwater ostracod	ca. 50–100	Males normally absent
Oithona similis	Marine copepod	87–98	Males almost absent in late winter
Stenhelia	Deep-sea copepod	93	Males few as in other deep-sea copepods
Artemia parthenogenetica	Asexual population	98	Males few
Pleuroxus hamulatus	Freshwater cladoceran	ca. 99	Males almost unknown
		EQUAL SEX RATIO	
Pseudocalanus minutus	Marine copepod	ca. 50	Sex ratio nearly equal through copepodite stages; subsequently, most males die abruptly
Mysis relicta	Freshwater mysid	51–93	Sex ratio nearly equal; females may outlive males
Cancer magister	Marine brachyuran	ca. 50	Sex ratio nearly equal in larval, postlarval, and adult populations
		MALES PREDOMINATE	
Atya scabra	Freshwater caridean	21–39	Males predominate in primary habitat with females inhabiting marginal areas
Panulirus guttatus	Marine macruran	ca. 10	Predominance of males due to differential mortality

Contd.

Table 16 Contd.

Species	Identity	Sex ratio (Percentage of females)	Remarks
		SEX RATIO VARIES SPATIALLY	
Trichoniscus biformatus	Hydrophilic isopod	ca. 50–100	Dioecious in favourable habitats, parthenogenetic and tetraploid in unfavourable habitats
Cyprinotus incongruens	Freshwater ostracod	ca. 50–100	In Germany, parthenogenetic; In Hungary, males frequently recorded; In North Africa, males as many as females
Triops longicaudatus	Freshwater notostracan	? –100	Parthenogenetic or hermaphroditic in California and Pacific islands; disexual over the other range
Eubranchipus vernalis	Freshwater anostracan	36–60	Males dominate in some populations, females in others
Callinectes sapidus	Estuarine brachyuran	4–85	Males inhabit more saline waters, especially at low temperatures
		SEX RATIO VARIES TEMPORALLY	
Penaeus setiferus	Estuarine penaeid	43–83	Males increase in brackish water during spawning
Temora stylifera	Marine copepod	19–64	More males during favourable season and more females during unfavourable season
		SEX RATIO CHANGES DURING ONTOGENY	
Eubranchipus oregonus	Freshwater anostracan	43–100	Females outlive males
Dynamene bidentata	Intertidal isopod	50–29	Females live for one year, while males live for the second year

(2) Reversal: smaller individuals are all females and the larger ones all males as in the hermit crab, *Clibanarius zebra*. Conversely, smaller individuals may all be males and later become females as in *Pandalus* (see Butler, 1964).

(3) Intermediate: sex ratios are intermediate between the standard and reversal patterns and may begin as male-biased (e.g. *Pleuroxus hypsinotus*) or female-biased (e.g. *Carcinus latens*).

(4) Anomalous: the male : female sex ratio for small individuals is equal, then becomes (a) male-biased in larger and finally female-biased in the largest individuals (e.g. *Emerita analoga*: W.B. Barnes and Wenner, 1968; see also Subramoniam, 1977, 1981 for *E. asiatica*) or (b) first female-biased and finally male-biased (e.g. *Carcinus laevimanus*).

Briefly, the standard pattern represents a male : female equal secondary sex ratio with differential mortality, longevity, or growth rate in one sex; the reversal pattern represents sex reversal of individuals; the intermediate pattern is a mixed category including cases of sex reversal and those which commence with biased secondary sex ratios; and the anomalous pattern results from sex reversal or a complex interaction of differential life span (longevity), migration, mortality, and growth rates (Wenner, 1972). Sex reversal is unknown in the Brachyura (Swartz, 1976). Dismissing almost all these explanations of Wenner (1972), Swartz (1976) showed that it is the preferential predation of one sex that is responsible for the anomalous pattern of sex ratio in the xanthid crab *Neopanope sayi*. Differential predation of one sex has also been observed in *Diaptomus shoshone* and *D. coloradensis* (see Maly, 1970). Predation, however, is not an important cause of mortality in the mangrove crab *Aratus pisonii* (see Diaz and Conde, 1989) in which the female-biased sex ratio has been attributed to different growth rates and moulting mortality.

2. The Wildish model

The population homeorheostat model, proposed by Wildish (1977), is sufficiently generalized to account for all the cases of biased sex ratios as well as monogeny. In r-strategists, arrhenogeny is a means of decelerating and thelygeny a means of accelerating the intrinsic rate of increase towards an optimum population control, via genetic mechanism, of the sex ratio, which is responsive to environmental fluctuations. For instance, food scarcity is known to affect sex ratios and to favour the appearance of a greater proportion of males among several species of copepods (e.g. *Calanus hyperboreus*: Conover, 1967). It assumes the occurrence of monogeny and individual-based selection. In K-strategists, the homeorheostat is a social-behaviour one. Negative feedback controls, which regulate population size, are due to direct density effects on survivorship, individual productivity, and selective female mortality. Positive feedback controls are the reverse of the above, associated with low densities.

3. Breeding structure

Over evolutionary time, natural selection operates to produce a correlation between male fitness and female preference, because those females preferring the fittest male associate their own genes with the best male genes and therefore produce the fittest male offspring (Pianka, 1974). As a result of such mating preferences, populations have breeding structures. At one extreme is homogamy (inbreeding), at the other heterogamy (out-breeding). Both these extremes represent non-random breeding structures and can occur in some crustaceans. For instance, the European and American strains of the copepods *Tisbe lagunaris, T. bulbisetosa, T. holothuriae*, and *T. battagliai* can be panmictic and heterogamous, as they are intercrossable and the F_1 hybrids are fertile giving rise to viable offspring (Table 17). *Tisbe clodiensis* is homogamous, as the survival of interpopulation hybrids to adult is zero

Table 17

Crosses between *Tisbe* populations from Europe and Atlantic coast of the United States (from Battaglia and Volkmann-Rocco, 1973)

Species	Type of cross female (F) × male (M)	Mean F_1 offspring/female	Viability to adulthood
T. lagunaris	F Venice × M Beaufort	66	1.10
	Venice control	60	1.00
	F Beaufort × M Venice	51	1.03
	Beaufort control	49	1.00
T. bulbisetosa	F Venice × M Beaufort	60	1.13
	Venice control	52	1.00
	F Beaufort × M Venice	71	1.13
	Beaufort control	65	1.00
T. holothuriae	F Helgoland × M Beaufort	63	1.03
	Helgoland control	61	1.00
	F Beaufort × M Helgoland	48	1.10
	Beaufort control	43	1.00
T. battagliai	F Anzio × M Beaufort	47	1.11
	Anzio control	42	1.00
	F Beaufort × M Anzio	41	1.02
	Beaufort control	40	1.00
T. clodiensis	F Venice × M Beaufort	33	0.00
	F Beaufort × M Venice	37	0.00
	F Banyuls × M Beaufort	29	0.00
	F Beaufort × M Banyuls	40	0.00

(Battaglia and Volkmann-Rocco, 1973). Experimental studies of Menzies (1972) on interbreeding of geographically separated populations have shown that *Limnoria tripunctata* is heterogamous with reference to strains of Europe and America (except for the cross between Beaufort × Chatham, U.S.A.).

4. Mating systems

Mating systems, when present, can considerably alter the quarternary sex ratio; an analysis of the sex ratio in conjunction with mating system operating in a population may reveal the functional aspects of sex ratio in greater detail. In other words, mating system may represent a behaviour mechanism that may alter the functional aspects of sex ratio, perhaps in response to environmental factors including population density. Description of mating systems for Crustacea is almost totally wanting. What follows is the simple extrapolation of our knowledge of vertebrate mating systems to crustaceans. Since the crustacean females have much more at stake during reproduction, they tend to exert much stronger mating preferences than males and to be more selective of acceptable mates. The outcome of this mating preferences is polygyny (one male having pair bond with two or more females) or polyandry (one female with two or more males). Monogamy (one male having pair bond with one female) is a compromise between these two extremes. Under a monogamous mating system, both the male and the female expend time and energy on raising their offspring (Pianka, 1974). In almost all crustaceans, it is the female that undertakes the responsibility of incubation, and monogamy is not likely to be practised by any free-living and free-swimming crustaceans. Of course, the occurrence of a male attached with a female throughout his life is not uncommon among parasitic crustaceans (e.g. *Ione thoracica*: Reverberi and Pittoti, 1942). However, to depict this type of monogamy, in which a male does not contribute time and energy on raising the offspring, a suitable term may be required. Polygyny occurs in a few crustaceans (e.g. *Cancer irroratus*: Elner and Elner, 1980). Such species, in which a female maintains more than one dwarf male (e.g. *Emerita asiatica*: Subramoniam, 1977; *Clypeoniscus hanseni*: Sheader, 1977a; *Scalpellum* and *Ibla*: R.D. Barnes, 1974), may be described as polyandrous. The sexual relationship is said to be promiscuous, when no pair bond beyond copulation is formed (Emlen, 1973) and most crustaceans may come under this category.

G. Parasitic Castration

An analysis of the life-history features of insect parasitoids and crustacean castrators suggests that these are similar trophic phenomena, distinct from parasitism and predation. A parasitoid consumes only one host during its life time; parasitic castrators cause the reproductive death of only one host. Since population densities of many insect species are regulated by parasitoids (Varley, 1947; van den Bosch *et al.*, 1966), parasitic castrators may also play an important role in the regulation of crustacean host population (Kuris, 1974). But only a few host species have been sampled well enough to allow accurate density estimates; even among these, the population dynamics of the parasitic castrators and their hosts have not been studied concomitantly to establish a correlation between the changes in host and parasite density.

An estimated 3 per cent of crustacean species are castrators of other Crustacea; the corresponding value for insects is 12 per cent. A number of instances of castration within the crustacean orders is known (e.g. *Balanus improvisus* by *Boschmaella balani*: Bocquet-Vedrine and Parent, 1973; *Idotea pelagica* by *Clypeoniscus hanseni*: Sheader, 1977a) and there is at least one example within the same family (Bopyridae: *Gyge branchialis* by *Pseudione euxinica*: Caroli, 1946; however, see also Cattalano and Restivo, 1965). Densities of many castrators of crustaceans are very low (Bourdon, 1960, 1964). Occasionally, incidence of the castrators may range from 25 to 75 per cent (Gifford, 1934; Bourdon, 1963; Hartnoll, 1967; Sloan, 1984). In some populations of the entoniscid isopod *Portunion conformis* in *Hemigrapsus oregonensis*, the incidence reaches the level of 92 per cent (Kuris, 1974). Most bopyrids seem to infect only young hosts (Pike, 1960); the others are restricted to adult hosts (Pike, 1961). Sacculinids appear to selectively parasitize juvenile hosts (Veillet, 1945), whereas entoniscids infect hosts of all ages (Kuris, 1971).

Parasitic castrators are typically highly host-specific. Epicaridean isopods are classically regarded as strictly host-specific, although it is likely that much of this specificity is ecologically influenced (Kuris, 1974). A few species have a low degree of host specificity in the field; for example, *Hemiarthrus abdominalis* has been recorded from as many as 22 host species (Pike, 1960).

Multiple infection by a single species neither speeds up castration effects nor produces additional deleterious effects (Veillet, 1945). In fact, the strategy of some crustaceans (e.g. *Ione thoracica*) is to masculinize the second infecting larva. Though not so common, mixed infection of parasitic castrators does occur (Altes, 1962). Usually, the second arrival often causes the death of the first castrator. Caroli (1946) reports that the young bopyrid *Pseudione euxinica* spends the early part of its life on the bopyrid *Gyge branchialis*. Upon gradual destruction of a previously infecting sacculinid by entoniscid, *P. euxinica* becomes the primary parasite on the host *Upogebia littoralis*. Likewise, *Cryptoniscus paguri*, a hypercastrator of *Septosaccus rodriguezi* may sometime become a primary parasite of the host *Clibanarius erythropus* (see Altes, 1962). Thus, this sort of hyperparasitic interaction is akin to multiple infection. The portunid crabs *Petrolisthes boscii* and *P. rufescens* are attacked by bopyrid and rhizocephalan parasites: the bopyrid parasitized 5.2 per cent *P. rufescens* and 0.6 per cent *P. boscii* and the rhizocephalan infected 6.4 per cent *P. boscii* and 0.3 per cent *P. rufescens*; evidently, these parasites almost excluded each other from their hosts (Ahmed and Mustaquim, 1974).

Since key features of crustacean life histories, such as moulting and metamorphosis, are under the control of endocrine systems, the castrators can 'read' and 'direct' these systems; for instance, moulting of *Bopyrus fougerouxi* is in synchrony with its host *Palaemon (Leander) serratus* (see Tchernigovtzeff, 1960). Parasitic castrators may cause precocious development (e.g. *Cryptoniscus paguri* on *Septosaccus rodriguezi*: Altes, 1962; *Probopyrus pandalicola*: Anderson, 1977) or prolonged life and gigantism (e.g. *Pinnotherion* sp. on *Pinnotheres pisum*: Mercier and Poisson, 1929). It appears that the replacement of ovary by the parasite results in an energetically less expensive total system (e.g. *Hemigrapsus oregonensis* castrated by *Portunion conformis*). Hyperfeminization, the appearance of adult characteristics in juvenile female hosts, is an oft-reported effect of sacculinization (Reinhard, 1956). Polyembryony, resulting in the production of many progeny from a single infection, is a regular feature of the life cycle of some Rhizocephala (Bocquet-Vedrine and Parent, 1973; see also Sloan, 1984). Drastic modifications in host behaviour sometimes accompany parasitization by castrators; sacculinized *Carcinus maenas* behave as reproductive females, migrating to shallow water and displaying egg-care behaviour (Rasmussen, 1959); a similar situation has also been reported for the golden king crab *Lithodes aequispina*, infected by the rhizocephalan parasite *Briarosaccus callosus* (see Sloan, 1985).

Non-crustacean parasitic castrators of Crustacea include the coccidian *Aggregata inachi* (see Smith, 1905), the fecampid turbellarian *Fecampia spiralis* on the Antarctic isopod *Serolis schytiei* (see Baylis, 1949), the enigmatic Ellobiopsidae (e.g. *Thalassomyces fagei* on the euphausiid *Thysanoëssa raschii*: Mauchline, 1966), Nematomorpha (e.g. *Nectonema gracile* on the decapod *Palaemonetes vulgaris*: Born, 1967), echinostome fluke cercaria on the tadpole shrimp, *Lepidurus packardii* (see Ahl, 1991), and fungus (parasite in ova of the thoracican cirripede *Chthamalus fragilis denticulata*: T.W. Johnson, 1958). For unknown reasons, the crustaceans are subjected to castration by members of several different taxa.

H. Monogeny

Another aspect of the host-parasite relationship pertinent here is the feminizing role played by some protozoans during the process of sex determination in some amphipods and by some bacteria in isopods. Whereas the sex castrators cause the reproductive death of their male hosts, these transovarially transmitted microsporidians (in *Gammarus*) or haplosporidians (in *Orchestia*) are responsible for the production of only females (thelygeny) or only males (arrhenogeny) in their hosts. A list of transmitted parasites likely involved in monogeny in crustaceans is presented in Table 18. *Triops newberryi* (Branchiopoda) (see Sassaman, 1991), *Tisbe reticulata* and *Tigriopus japanicus* (Copepoda), *Trioniscus provisorius, Armadillidium vulgare, Jaera albifrons*, and *Asellus aquaticus* (Isopoda), *G. duebeni, Orchestia gammarella*, and *O. mediterranea* (Amphipoda) are some crustaceans in which monogeny has been experimentally verified (see Wildish, 1977). It is very likely that some epigenetic factor is responsible for the occurrence of monogeny in these crustaceans.

Observations on infected intersexes with rudimentary androgenic glands and on experimentally infected males demonstrate that this organ (or its analogue) is not attacked by the parasite. Hence, the substances or by-products of the microsporidians, excreted during their multiplication, appear to inhibit the differentiation of the androgenic gland (Bulnheim, 1975) and the feminization is caused by a progressive reduction of the titre of the androgenic hormone (Veillet and Graf, 1959). The feminizing influence of the parasites may be affected by superimposing en-

Table 18

Microsporidian and other parasites responsible for monogeny in some Peracarida

Parasite species	Host species	Reference
Octosporea effeminans	*Gammarus duebeni*	Bulnheim (1967, 1969, 1970)
F bacteria	*Armadillidium vulgare*	Juchault and Legrand (1968)
Epigenetic factor	*Porcellio dilatatus*	Juchault and Legrand (1975)

vironmental factors such as temperature and salinity. An increase of the ambient salinity to 25–30‰ results in disappearance of the parasite *Octosporea effeminans; Thelohania hereditaria* is more resistant to high salinity level. Consequently, infected *Gammarus duebeni* may produce eggs, all of which are not parasitized by *O. effeminans*, and a mixed progeny may arise.

In oniscoids (e.g. *Armadillidium vulgare*), females are heterogametic (ZW) and males homogametic (ZZ). The complex phenomena of monogeny in oniscoids have been explained as resulting from a modification of the sex differentiation by symbiotic feminizing T_f bacteria or masculinizing Ar_m bacteria. ZZ males, when infected by T_f bacteria, become functional neofemales; likewise, the ZW females, on infection by Ar_m bacteria, differentiate into functional males. In numerous populations, WZ genetic females have disappeared and all individuals are ZZ males, some of which, on infection, by T_f bacteria function as females. It appears that the level of feminizing or masculinizing factor is not stabilized. Consequently, the sex ratio of *A. vulgare* ranges from 100 per cent females to 100 per cent males in different populations (Juchault and Legrand, 1968).

III. EGG PRODUCTION

All reproductive strategies are considered as compromises, reflecting balance of selecting pressure on parental survival, survival of offspring, and fecundity (Emlen, 1973). Maturing age, breeding frequency, and clutch size are some 'gears' available to an organism to regulate its fecundity and crustaceans represent one of the prosperous aquatic groups which have successfully applied these 'gears', usually in combination but rarely singly, to effectively regulate population size for optimal utilization of resources. Life span of crustaceans may be as short as 13 days (e.g. *Moina micrura*: Murugan, 1975b, Table 19) or as long as over 20 years (e.g. *Cancer pagurus*: Bennett, 1974). It is a general norm that with increasing size, the life span is extended (McNaughton and Wolf, 1974) and crustaceans fit into this norm. Organisms invest considerable time and energy on reproduction. Since most crustaceans carry the developing eggs until hatching, body size may profoundly alter reproductive potential and quantitative aspects of investment and commitments of these two components. Reproductive energetics has begun to receive some attention in recent years (e.g. Enders, 1976; Muthukrishnan and Pandian, 1987; Vernberg, 1987).

A. Time Cost of Juvenile Stages

Our knowledge of the life span of crustaceans is very limited and is based on economically important species that are easily reared in the laboratory and aquaculture systems (Kinne, 1977). Temperature is a potent factor which determines the life span of invertebrates (Kinne, 1963) and crustaceans are no exception. For instance. the life span of some temperate cladocerans is about 76 days at 10°C (Table 19).

Table 19

Life span (LS) in days and juvenile period (JP) in per cent of LS and number of adult instars (AI) of some temperate cladocerans (from Bottrell, 1975, modified and added). For comparison, values available for the tropical cladocerans are also given (source as indicated)

Species	Temperature (°C)											
	5			10			15			20		
	LS	JP	AI	LS	JP	AI	LS	JP	AI	LS	JP	AI
Temperate cladocerans (11 ± 5°C)												
Eurycercus lamellatus	162	62	3	105	62	3	63	44	6	42	45	6
Alona affinis	144	51	4	90	44	5	62	41	6	37	46	5
Pleuroxus uncinatus	132	59	3	79	48	4	50	45	5	31	44	6
Acroperus harpae	119	59	3	74	51	4	45	54	4	29	53	5
Simocephalus vetulus				76	48	3	57	39	5	38	38	6
Sida crystallina				74	65	2	60	43	5	40	40	6
Chydorus sphaericus	96	40	4	59	35	5	34	32	6	24	37	6
Graptoleberis testudinaria	95	47	3	54	44	3	34	46	4	23	41	5
Tropical cladocerans (28 ± 3°C)												
Simocephalus acutirostratus				44	9	18	Murugan and Sivaramakrishnan (1973)					
Daphnia carinata				24	21	8	Navaneethakrishnan and Michael (1971)					
D. lumholtzi				42		12	Sharma *et al.* (1984)					
Scapholeberis kingi				21	10	17	Murugan and Sivaramakrishnan (1976)					
Moina micrura				13	15	11	Murugan (1975b)					
Ceriodaphnia cornuta				21	10	18	Murugan (1975a)					

while it is about a third (25 days) in the tropical cladocerans. Temperature also accelerates associated events such as the time invested on juvenile stage and the number of moultings by adult. The tropical cladocerans undergo as many as 15 adult moults, as compared to 5.5 moults by their temperate counterparts. Since over 80 per cent of the adult moults are followed by the release of neonates (see Murugan and Sivaramakrishnan, 1973, 1976), the egg-bearing moults (12) of the former are over 2.5 times as much as the latter (4.4 moults). Secondly, the temperate cladocerans appear to invest about half of their life span on the juvenile period, whereas the time cost on this investment is only 13 per cent for a tropical cladoceran. That higher temperatures prevailing in the tropics reduce the time investment on juvenile stage is also indicated by the fact that among the temperate cladocerans, a rise in temperature from 5° to 20°C resulted in a decrease of this value from 53 to 43 per cent. Thus, temperature may greatly alter egg production by modifying the time cost of juvenile period and breeding frequency.

Reports on other groups of crustaceans too confirm this observation. For instance, in Lake Kinneret (Israel, 33°N), where the annual temperature fluctuations range from 14° to 28°C, the dominant plankton *Mesocyclops leukkarti* invests 41 per cent of its life span (128 days) to pass through nauplius and copepodite stages and to produce 27 eggs in two clutches at 15°C; the corresponding values for the copepods at 27°C are 31 per cent of 61 days and 93 eggs (Gophen, 1976). That crustaceans exposed to low temperature prevailing in temperate zones tend to incur heavy time cost on juvenile period is only true not of lower crustaceans such as Cladocera and Copepoda with short life span but also of higher crustaceans such as Amphipoda and Isopoda with a longer life span of over one to four years (Table 20). New Zealand, European, and American isopods and amphipods invest about half of their lifetime on juvenile stages; unfortunately, corresponding data for the tropical crustaceans are unavailable.

Available data for the arctic copepods indicate that they invest between 75 and 90 per cent of their lifetime on juvenile stages (e.g. *Limnocalanus johanseni*: Comita, 1956; *Diaptomus sicilis*: Edmondson *et al.*, 1962). This suggests that the time investment on juvenile stages by crustaceans may increase with increasing latitude. The need to accommodate the major events (e.g. breeding) within a short adulthood during the short spell of summer has enforced monocyclic breeding or univoltinism among several arctic crustaceans (e.g. *Arctodiaptomus, Acanthodiaptomus*: Edmondson *et al.*, 1962) and all the broods produced by these arctic copepods are composed of resting eggs alone (e.g. *D. shoshone, D. coloradensis*: see Maly, 1973); conversely, with more or less uniform high temperature prevailing throughout the year, most tropical crustaceans invest a major percentage of their life span on breeding and repeatedly reproduce almost throughout the year. The breeding season of *Emerita talpoida* in Beaufort, U.S.A. (36°N), lasts from June to September (Wharton, 1942), whereas *E. portoricensis* in Jamaica (18°N) reproduces almost throughout the year (Goodbody, 1965).

B. Generation Time

From the point of age at which reproduction commences, generation time (the time lapsed from egg to egg stage) is of great significance. Our knowledge of generation time of crustaceans is based mostly on experimentally cultured copepods and cladocerans. The generation time of tropical cladocerans is about one-seventh of that of their temperate counterparts (Table 21). It lasts for about six weeks at 10°C for the temperate copepods for which data are reported (Table 22). McLaren (1978) estimated it to range from three weeks in *Pseudocalanus minutus* to seven weeks in *Calanus finmarchicus* in their habitat at about 11°C. He is of the opinion that experimental estimation of generation time of copepods on the basis of isochronal development may or may not hold good for those inhabiting seas at high latitudes and in some temperate fresh waters, for there is a delay or even retardation of

Table 20

Life span and juvenile period as a percentage of life span in some temperate isopods and amphipods (from Carefoot, 1973)

Species	Life span (month)	Juvenile period (percentage of life span)	Reference
Isopoda			
Idotea emarginata	12	58	Naylor (1955)
Dynamene bidentata	24	50	Holdich (1968)
Jaera albifrons	12	50	Jones and Naylor (1971)
Caecidotea intermedia *(Asellus intermedius)*	12	25	Ellis (1961)
A. aquaticus	12	13	E.A. Steele (1961)
A. eridianus	12	13	E.A. Steele (1961)
A. tomalensis	12	83	Ellis (1961)
Mancasellus macrourus	12	67	Markus (1930)
Ligia oceanica	28	43	Nicholls (1931)
L. pallassii	21	57	Carefoot (1973)
Ligidium japonica	24	50	Saito (1965)
Cylisticus convexus	24	50	Hatchett (1947)
Porcellio scaber (U.S.A.)	24	50	Hatchett (1947)
P. scaber (U.K.)	36	66	Heeley (1941)
Armadillidium vulgare	48	25	Paris and Pitelka (1962)
Trichoniscus pusillus	36	33	Heeley (1941)
Oniscus asellus	42	57	Heeley (1941)
	Mean	46	
Amphipoda			
Corophium insidiosum	7.4	45	Nair and Anger (1979a)
Jassa falcata	9.4	56	Nair and Anger (1979b)
Gammarus oceanicus	9.0	50	V.J. Stelle and
G. mucronatus	5.0	50	D.H. Stelle (1972) in Dauvin (1989)
G. (Marinogammarus) marinus	7.3	60	Vlasblom (1969)
Hippomedon whero	7.7	46	Fenwick (1984)
Patuki repori	8.5	77	Fenwick (1984)
Diagodias littoralis	6.5	50	Fenwick (1984)
Protophoxus australis	5.5	51	Fenwick (1984)
Ampelisca sarsi	21.0	57	in Dauvin (1989)
Ampharete acutifrons	24.0	50	in Dauvin (1989)
Nephtys hombergi	60.0	70	in Dauvin (1989)
Abra alba	24.0	50	in Dauvin (1989)
	Mean	54	

post-embryonic development in some species and the timing of their life cycle, especially the resting stage, is attuned to food availability.

Generation time of males has a different pattern. Loss of males following mating tends to reduce intraspecific competition (Ricklefs, 1973). It is not known how ubiquitous this strategy is among crustaceans. An extreme example is perhaps provided by the meiobenthic copepod *Asellopsis intermedia* (see Lasker *et al.*,

Table 21

Generation time of some cladocerans (from Bottrell, 1975, modified and added)

Species	Generation time (day)	Reference
Temperate cladocerans (10°C)		
Eurycercus lamellatus	52	Bottrell (1975)
Sida crystallina	48	Bottrell (1975)
Alona affinis	39	Bottrell (1975)
Pleuroxus uncinatus	38	Bottrell (1975)
Acroperus harpae	38	Bottrell (1975)
Simocephalus vetulus	37	Bottrell (1975)
Graptoleberis testudinaria	24	Bottrell (1975)
Chydorus sphaericus	21	Bottrell (1975)
Mean	37	
Tropical cladocerans (28°C)		
Daphnia carinata	7	Navaneethakrishnan and Michael (1971)
S. acutirostratus	6	Murugan and Sivaramakrishnan (1973)
Moina micrura	3	Murugan (1975b)
Ceriodaphnia cornuta	3	Murugan (1975a)
Scapholeberis kingi	3	Murugan and Sivaramakrishnan (1976)

1970) in which, following copulation, males died in good numbers and became scarce, and egg maturation was completed in about seven months after insemination.

Maintenance of sterile post-reproductive females in a population may be an ecological burden, as are the males after accomplishing mating. Very little attention has been paid to these aspects of population biology of crustaceans. Female *Euterpina acutifrons* (Copepoda) spend at least 28 per cent of their adult life as sterile individuals (Table 23). When fed more than two-thirds of the maximum food supply, the post-reproductive period is extended to over 50 per cent of the respective adult life span. Reduced food supply extends both pre- and post-reproductive periods and depresses egg production. Food supply at more than optimal levels (6.10^3 cells/ml; see Table 23), ensured greater egg production, but led to the continued presence of reproductively exhausted females, which competed for resources intensely; it may be a built-in endogenous mechanism to regulate population size. Temperature is another important environmental factor that determines the duration of the reproductive period; for instance, the reproductive period of *Artemia parthenogenetica* (French strain) lasts for 55 per cent of its life span at 24°C; it is reduced to 20 and 30 per cent at 30 and 15°C, respectively (Browne *et al.*, 1988).

Table 22

Generation time and reproductive potential (measured as intrinsic rate (r_m/day) and doubling time t_2 (in days) of some crustaceans

Species		Mean generation time (day)	r_m/day	t_2 (days)	Reference
Ostracoda					
Cyprideis torosa			0.015	46.2	Heip (1977)
Loxoconcha elliptica			0.030	30.0	Heip (1977)
Copepoda					
Tachidius discipes	5°C	83.0	0.039	17.8	Heip and Smol (1976)
	10°C	41.5	0.085	8.2	
	15°C	25.1	0.134	5.2	
	20°C	17.3	0.186	3.7	
	25°C	13.9	0.239	1.8	
Paronychocamptus nanus	5°C	150.0	0.028	24.8	Heip and Smol (1976)
	10°C	48.0	0.057	12.8	
	15°C	29.1	0.083	8.4	
	20°C	22.0	0.113	6.1	
	25°C	18.6	0.142	4.9	
Tisbe reluctans	18°C	39.8	0.102		Volkmann-Rocco and Fava (1969)
T. persimilis	18°C	39.0	0.088		Volkmann-Rocco and Fava (1969)
T. clodiensis	18°C		0.187		Volkmann-Rocco and Battaglia (1972)
T. dobzhanskii	18°C		0.291		Volkmann-Rocco and Battaglia (1972)
Euterpina acutifrons		26.5	0.161		Zurlini *et al.* (1978)

Table 23

Influence of food concentration on adult life span, post-reproductive period, and egg production in *Euterpina acutifrons* (from Zurlini *et al.*, 1978; modified and recalculated)

Food concentration (cells/ml)	Adult life span (day)	Post-reproductive period		Egg production	
		(day)	(as percentage of life span)	(sacs/female)	(total eggs/female)
Starved	21	13.5	64	2.8	52
6.10^2	20	11.0	54	2.3	44
6.10^3	47	13.0	28	6.5	68
6.10^4	47	24.3	51	8.8	150
6.10^5	38	18.4	48	9.8	170

Egg fertility is another built-in mechanism in endogenous population regulation. Zurlini *et al.*(1978) observed that about 13 per cent of the eggs of *Euterpina acutifrons* were not fertile. The egg fertility remains unaltered throughout the adult life span. Examination of late stage of *Panulirus longipes cygnus* revealed that some 4.5 per cent eggs are infertile and that the fertility level is the same in mother animals of different sizes (Morgan, 1972). No information is available on other major groups of Crustacea. In all such studies, careful attention should be paid to distinguish infertile eggs from fertile eggs that could not get attached to the mother animal. For instance, the presence of an ectoparasite (?) *Hemiarthrus abdominalis* on the first pleopod of the shrimp *Dichelopandalus bonnieri* eliminated the possibility of nearly a third of the fertilized eggs getting normally attached to the pleopod setae (Al-Adhub and Bowers, 1977).

C. Reproductive Potential

The reproductive potential of a species can be measured with intrinsic rate of natural population increase (r_m); it is a parameter which measures the speed with which population increases in the absence of limiting factors. It is thus a measure of production of neonates in an unlimited environment. The net reproduction rate (*Ro*) and intrinsic rate of increase are determined by combining data on neonate production with those on sex ratio and survival obtained under unrestricted feeding and other environmental conditions (for details, see Heip and Smol, 1976; Zurlini *et al.*, 1978). Table 22 shows that the reproductive potential of copepods is very high; for instance, the value reported for *Tachidius discipes* is higher than those obtained for a few ostracods and nematodes from the same area; at 25°C, the time required for doubling the population of *T. discipes* is just 1.8 days, as compared to 30 days for the ostracod *Loxoconcha elliptica* (see Heip, 1977) and 24 days for the nematode *Oncholaimus oxyuris* (see Heip *et al.*, 1978).

Secondly, the influence of temperature on the reproductive potential of *Tachidius discipes* and *Paronychocamptus nanus* is profound. The time required to double the population numbers, calculated from *Ro*, decreases from 18 to 2 days in the former and from 25 to 5 days in the latter, when temperature is elevated from 5 to 25°C. Extrapolation of the data with the habitat temperatures indicates that more than half the potential of these species is realized during spring bloom and thus most of its breeding duration is attuned to the best part of the year (Heip and Smol, 1976). When compared to the reproductive potential of the copepods estimated under unrestricted laboratory conditions, only a third of the potential is actually realized in the field by *T. discipes* (see Heip, 1972).

Before breeding frequency is considered, a brief reference must be made to the variations observed in the (age) size at maturity among populations of a species, and among individuals of a population. Describing a standard procedure for fixing the minimum size of sexual maturity and possibly the growth pattern of crustacean population, Wenner *et al.*(1974) pointed out that owing to the difference in food availability, *Emerita analoga* population from the Santa Cruz Island attained sexual maturity at a carapace length of 16 mm. But the population at Santa Barbara (only about 30 km away) only matured at a carapace length of 21 mm.

D. Spawning Frequency

Maximum body weight, attained by crustaceans, ranges from a few milligrams (e.g. 2 mg, *Pareuchaeta norvegica*: Nemoto *et al.*, 1976) to a few kilos (e.g. 1.6 kg, *Cancer pagurus*, 20-year-old male: Bennett, 1974). But most crustaceans are small and are subjected to heavy predation; such crustaceans, prone to heavy adult mortality, tend to display a reproductive pattern in which breeding begins early and is distributed (iteroparous) to different size classes (Giesel, 1976). Thus even in the egg shedders, like the calanoids, egg laying is spread over a long period (e.g. 74 days, *Calanus finmarchicus*) and tends to occur in a series of bursts, each burst lasting for a week (Marshall and Orr, 1952).

1. Spawning interval

Egg size and yolk quality play a decisive role in determining the length of incubation period (e.g. for copepods, see McLaren, 1965). A consequence of producing larger eggs and/or egg mass is that spawning causes an enormous drain of energy resources of the female, but this may be offset to some extent by the prolonged incubation characteristic of larger eggs (see also Herring, 1973, 1974a, b). Consequently, intervals between successive egg layings in crustaceans range from a few hours (24 hours) in *Moina micrura*, brooding about five small eggs (Murugan, 1975b) to a few days (19 days) in *Macrobrachium nobilii* carrying a few hundred eggs (Pandian and Balasundaram, 1982) and to a few (6 to 12) months in *Panulirus longipes cygnus*, producing 500,000 large eggs (Chittleborough, 1976).

At species level, population density and food scarcity are known to prolong the interval between successive egg layings and the extended interval serves to procure sufficient nutrients. Walker (1979) reared the harpacticoid copepod *Amphiascoides* sp., collected from a high-density culture, in fresh medium and conditioned medium. The interval between successive egg layings is extended from 1.3 days in fresh culture medium to 3.4 days in conditioned medium. Fed on 3×10^5 *Isochrysis* cells/ml, it lasts between 1 and 10 hours in *Pseudocalanus minutus*, but when the cell density is decreased to 3×10^4 and 3×10^3/ml, the interoviposition period is prolonged to 100 and 150 hours, respectively (Corkett and McLaren, 1969). Similar observations have also been reported by Sindhukumari and Pandian (1991), who fed *Macrobrachium nobilii* on different rations.

2. Effect of egg carriage

Moulting and reproduction are two major events, involving cyclic mobilization of organic reserves from storage depots to the epidermis and gonad, respectively. In most diecdysic crustaceans, spawning is obligatorily preceded by moult. The presence of eggs on the pleopods of decapods inhibits (e.g. *Palaemon serratus*: Panouse, 1947) and postpones moulting (e.g. *Orconectes (Cambarus) propinquus*: Scudamore, 1948); as a matter of fact, even injection of ecdysterone into ovigerous *Palaemonetes kadiakensis* did not induce moulting (Hubschmann and Armstrong, as cited in Hubschmann and Broad, 1974), due to the presence of moult-inhibiting hormone (MIH: K.G. Adiyodi and R.G. Adiyodi, 1970). Though the eggs are mature and ready for extrusion, spawning is deferred if eggs are present on pleopods in some decapods that spawn more than once during an intermoult period. Likewise, no spawning occurs in the presence of eggs in egg sacs in some copepods (e.g. *Amphiascoides* sp.: Walker, 1979) or in the mantle cavity of some cirripedes (e.g. *Elminius modestus*: Crisp and Davies, 1955). By putting off the next moult and spawning, the egg carriage evidently reduces the frequency of breeding. The energy cost of egg carriage in crustaceans has been discussed in Section I B 3 above. When females of *Macrobrachium nobilii* are relieved from the task of incubation, the frequency of berried moults increases from 61 per cent in normal females to 81 per cent in the relieved ones (Table 24). Consequently, the egg production in these relieved females increases by 1.5 times compared to the normal ones (Pandian and Balasundaram, 1982). In *M. idella*, however, the duration of the moult cycle is not determined by the presence or absence of brood, but by the (vitellogenic or non-vitellogenic) state of the ovary (Narayanan, 1990).

Many crustaceans invest considerable time and energy on egg carriage; egg-carrying crustaceans therefore produce fewer eggs than their counterparts that spawn freely. Some 140,000 eggs are incubated by *Macrobrachium acanthurus* (see Dugan *et al.*, 1975), a species known to carry the largest number of eggs among the species of *Macrobrachium* (see Balasundaram, 1980). This value amounts to less than 15 per cent of the total eggs (one or two million eggs) spawned by some penaeids (Table 4). Likewise, the number of eggs incubated by some egg-carrying

Table 24

Egg production in normal and relieved females of *Macrobrachium nobilii* (from Pandian and Balasundaram, 1982)

Parameter	Normal females	Relieved females
Duration of the moult cycle (day)	19.3	17.7
Moult (No./annum)	18.9	20.6
Berried moult (per cent)	61.0	81.0
Berried moult (No./annum)	11.5	16.7
Mean clutch size (No./clutch)	2,161.0	2,194.0
Total eggs produced (No./annum)	24,852.0	36,640.0

euphausiids is in the range of about 100, whereas free-spawning euphausiids produce as many as 300 eggs (Table 25). The number of eggs laid by pelagic copepods is also often large, when the eggs are shed freely in water. For instance, the number of eggs shed by the calanoids ranges from 1,340 for *Calanus hyperboreus* (see Conover, 1967) to 2,000 for *C. finmarchicus* (see Marshall and Orr, 1955) and *C. helgolandicus* (see Paffenhofer, 1970). On the other hand, the egg-carrying harpacticoids tend to lay only a small number of eggs (e.g. 100 to 300, *Euterpina acutifrons*: Neunes and Pongolini, 1965). *Pseudocalanus* is an exceptional calanoid that carries most of its eggs in sacs. Sazhina (1971) computed that *P. elongatus* produces 567 eggs, and carries them in 21 egg sacs, each containing 27 eggs. Others report even lower reproductive performance by *P. elongatus* (e.g. Paffenhofer and Harris, 1976). A comparative study of egg-carrying and egg-spawning calanoids by Corkett and Zillioux (1975) shows that the inclusion of eggs in a sac and the duration of egg carriage are mostly responsible for the reduced fecundity of *Pseudocalanus*.

Much of our knowledge on the endocrine control strategies of moulting and reproduction in decapods has been gained from crabs and shrimps subjected to eyestalk ablation or multiple limb autotomy. Crustacean eyestalks contain hormones that inhibit moult (MIH) and the gonad (GIH) but the course of events

Table 25

Egg production in some euphausiids (from Mauchline and Fisher, 1969; modified and added to from Barange and Stuart, 1991)

Species	Egg number/brood
Egg-carrying euphausiids	
Nyctiphanes australis	60
N. simplex	50
N. capensis	80
N. couchii	50
Stylocheiron suhmii	3
S. carinatum	430
Nematoscelis megalops	170
Pseudeuphausia sinica	40
Mean	95
Free-spawning euphausiids	
Meganyctiphanes norvegica	530
Euphausia superba	555
E. tricantha	310
E. lucens	80
E. pacifica	80
E. eximia	175
E. hanseni	198
Thysanoëssa raschii	300
T. inermis	300
Mean	296

initiated by ablation varies with the taxonomic group, age of the individual, and season (R.G. Adiyodi, 1985). In decapods, moulting and reproduction are two major energy-demanding processes; hence, an interrelationship between them determines the pattern of reproductive and/or somatic growth. Brachyurans have a long intermoult period, during which all gonadal activities including spawning, fertilization, and hatching of eggs from the pleopods are accommodated. Hence, the somatic and the reproductive growths are programmed as antagonistic events in Brachyura. Conversely, palaemonid natantians are diecdysic and their premoult period is longer than the intermoult period; however, there are exceptions to this generalization (e.g. *Macrobrachium idella*: Narayanan, 1990); consequently, the gonadal activities are also extended to the premoult period. This kind of synergism between moulting and reproduction demands the partitioning of nutrients from the storage organs, if both activities are to occur at the same time (see K.G. Adiyodi and R.G. Adiyodi, 1970, 1974).

Through a series of carefully designed research investigations, the Adiyodis have shown that in brachyurans, such as *Paratelphusa hydrodromous*, the long intermoult stage is clearly demarcated into two alternating physiological phases—a reproductive phase and a somatic phase. During the reproductive phase, the physiology of the animal is tilted in favour of gonadal development; hence, eyestalk ablation or mutiple limb autotomy during this phase results in precocious ovarian (e.g. Kurup and R.G. Adiyodi, 1981, 1984; Gupta *et al.*, 1987) or testicular (Gupta *et al.*, 1989; Mathad and K.G. Adiyodi, 1990) growth (see also Wilber, 1989). During the somatic phase, the physiology of the animal is, however, poised in favour of body growth; therefore eyestalk ablation or multiple limb autotomy (e.g. Kurup and R.G. Adiyodi, 1984) accelerates somatic growth, often culminating in ecdysis. The recent work of these authors has shown that this antagonism is maintained primarily by alternating responsiveness of tissues to hormones involved in growth and reproduction (K.G. Adiyodi and R.G. Adiyodi, 1985).

Eyestalk ablation leads to different endocrine pathways, through which somatic and/or reproductive growth processes are regulated, in the Decapoda. (1) In palaemonids, the ablation precipitates moult, which is mostly followed by a single spawning within a moult cycle (e.g. Sindhukumari and Pandian, 1987). (2) In penaeids, though repeated spawnings occur within one moult cycle, no spawnings were observed during late premoult and early postmoult (Emmerson, 1980; Browdy and Samocha, 1985; Shlagman *et al.*, 1986; Choy, 1987). Oosorption was noticed with the onset of premoult (Browdy, 1988). Essentially the somatic and reproductive growth processes are antagonistically programmed in penaeids as in Macrura (e.g. Quackenbush and Herrnkind, 1981; Rahaman and Subramoniam, 1989) and Brachyura (R.G. Adiyodi, 1985).

Calcification of the exoskeleton of crustaceans has a direct impact on the duration and timing of the moult cycle. Hence the division of decapod crustaceans into those with calcified and those with poorly calcified exoskeleton (Drach and Tchernigovitzeff, 1967). In the poorly calcified natantians (e.g. *Macrobrachium rosenbergii*: Fieber and Lutz, 1984; *Aristeus antennatus*: Sarda *et al.*, 1989), the difference in Ca content of the old carapace before and during stage D_2 is much more pronounced, as it is energetically cheaper to reabsorb the crystallized form of Ca available in the old exoskeleton; the energy thus saved can be utilized for growth and reproduction. The calcified crustaceans are characterized by long intermoult period and K-strategies, the poorly calcified ones by short intermoult period and r-strategies (Sarda *et al.*, 1989). The relation between calcification process and endocrine events following eyestalk ablation is a promising area for future research.

E. Clutch Size

Brood or clutch size denotes the number of eggs produced by a female from a single ovulation and fertilization though not necessarily a single copulation. Breeding may be restricted to a given season, or a succession of broods may be produced continuously: the rate at which eggs are produced is referred to as 'fecundity'; more precisely, fecundity is defined as the number of eggs produced per given increase in weight of the adult (the slope of the adult size-egg number regression) per unit time (H. Barnes and M. Barnes, 1968). Enormous data have been accumulated on egg production as a function of female size in almost all the common species of crustaceans (see Mauchline, 1988), but very little effort has been made to relate this to time scale.

1. Size-dependency of clutch

Since most crustaceans carry their developing eggs, the size of the mother animal is an important factor that regulates the number of eggs in a brood. Working on several marine Malacostraca, Jensen (1958) concluded that the 'absolute number' of eggs (i.e. the total number of eggs carried in all the broods of a female) is determined by environmental factors; however, 'the relative number' (i.e. the total number of eggs carried in a single brood at any one time) exhibits a linear relationship to the volume of the mother and hence is dependent upon the mother herself. Considering a larger number of taxonomic groups of crustaceans, Mauchline (1988) confirmed Jensen's (1958) conclusions. This has also been discussed elsewhere (Section I B 5).

Table 26 gives the mean, range, and standard deviation values obtained for the clutch size as a function of maternal body size of *Pareuchaeta norvegica*. The wide range in egg number within individual size classes and the large degree of overlap between the ranges indicate that the egg-producing performance of females can be highly variable, perhaps owing to a host of environmental factors, which decisively also alter the relative number of eggs. In general, a linear trend explaining the clutch-mother size relationship is obtained for almost all the major taxonomic groups of crustaceans (e.g. Cladocera: Fig. 5; Copepoda: Fig. 6; Amphipoda: Fig. 7), but the slope may significantly be shifted to different levels by one or the other environmental factors. Thus, food supply (e.g. *Tigriopus brevicornis*: G.W. Comita and J.T. Comita, 1966, Fig. 6) and associated factors such as tidal levels altering feeding durations, tidal cycle (e.g. *Semibalanus balanoides*: H. Barnes and Powel, 1953; H. Barnes and M. Barnes, 1968), temperature-related seasons (e.g. *Corophium insidiosum*: Sheader, 1978, Fig. 7), and latitudes (e.g. *B. improvisus*: H. Barnes and M. Barnes, 1968 and *Scapholeberis mucronata*: Green, 1966; Fig. 5) are known to alter remarkably the slope of the clutch-mother size relationship. However, the trend obtained for the clutch-age (e.g. *Daphnia pulex*: Richman, 1958) or instar number of adults (e.g. *Ceriodaphnia cornuta*; Murugan, 1975a) is distinctly different from those hitherto described. In *C. cornuta*, the egg number steadily increases with increasing age or instar of mother animal only up to a particular limit, beyond which the number begins to dwindle (Fig. 8), and thus the trend obtained is very similar to the one mathematically arrived at by Giesel (1976). Food supply too does not change the pattern but only the level of the relationship (Fig. 8). Data reported for egg number in successive broods of *Acanthocyclops viridis* also suggest an initial increase followed by a decline. The records of Parrish and Wilson (1978) show that in most female *Acartia tonsa*, egg production declined during the last days between the 30th and 40th day of adult age. It is not clear whether the linear relation obtained for all the major classes of crustaceans does include the declining segment, representing the oldest age or the largest size groups.

It is difficult to make further analysis of the linear relation obtained for clutch-mother size among the major groups of crustaceans, as has been done for the relationship between egg size and incubation period. First, clutch size of crustaceans

Table 26

Observed variations in egg numbers within and between different maternal body-size groups *Pareuchaeta norvegica* (from Hopkins, 1977; modified)

Body size (prosoma length) (mm)	No. of observations	Mean	No. of eggs/sac	
			Range	S.D.
4.65	12	23.17	17–31	3.99
4.75	2	25.50	25–26	
4.85	10	26.60	19–40	6.31
4.95	9	27.55	19–37	6.40
5.05	23	29.17	20–37	3.94
5.15	3	29.00	25–33	4.00
5.25	7	30.43	23–37	4.61
5.35	13	29.15	17–36	5.10
5.45	54	30.43	17–39	4.96
5.55	26	30.46	18–40	5.76
5.65	27	32.48	15–41	6.50
5.75	10	32.30	22–37	4.97
5.85	9	32.22	25–40	5.93

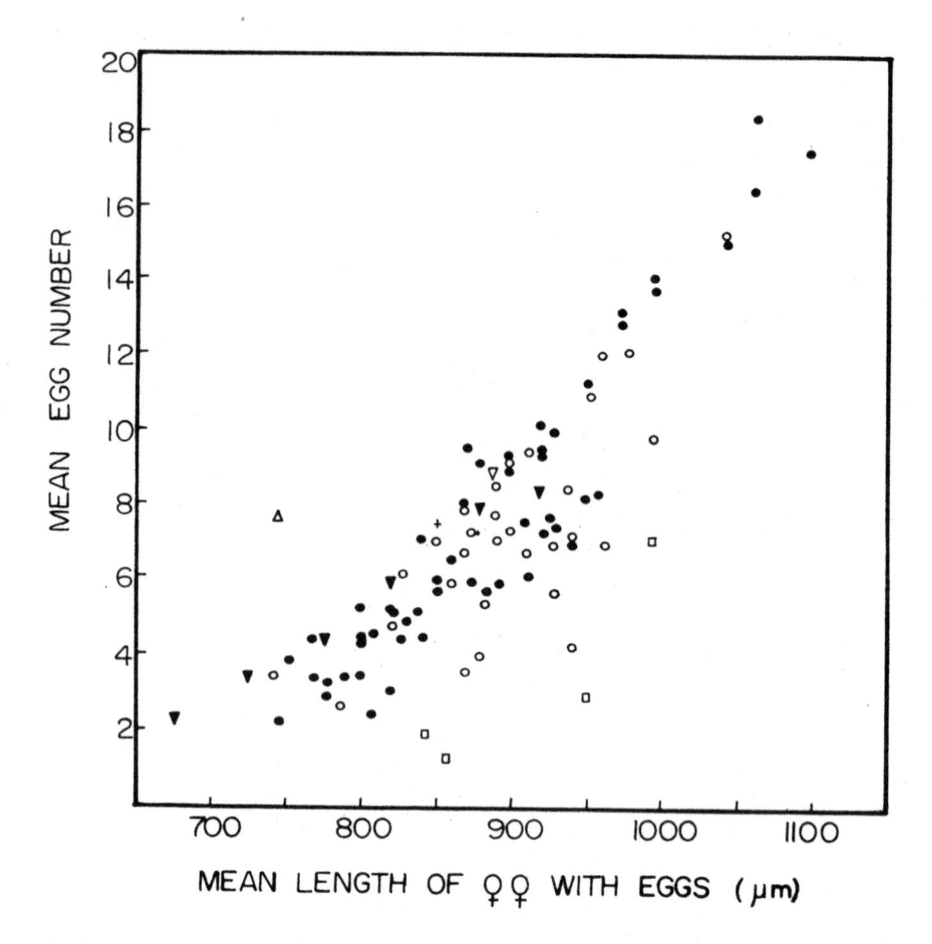

Fig. 5. Relationship between egg number and maternal size in the cladoceran *Scapholeberis mucronata* as altered by latitudinal variations. •, Females from Hampton Court Long Water (U.K.), 1960 and 1961; ○, females from Hampton Court Long Water (U.K.), 1963; +, females from Rethwaite (U.K.); □, females from Greenland; ▽, females from Ager So (Denmark); ▼, females from Log So (Denmark); △, females from Fondo Toce, Lago Maggiore (Italy) (from Green, 1966).

ranges from two eggs (e.g. interstitial copepods *Intermediosyllus, Remanae, Psammotopa*: Harris, 1972) to two million eggs (e.g. *Callinectes sapidus*: van Engel, 1958). Second, a host of parameters has been used to represent the mother's size: (1) weight (e.g. Cirripedia: H. Barnes and M. Barnes, 1968), (2) prosoma length (e.g. Copepoda: Hopkins, 1977), (3) body length (e.g. Cladocera: Green, 1966; Peracarida: Sheader, 1978), (4) body volume (e.g. Decapoda: Katre, 1977a), (5) carapace length (e.g. Brachyura: Mauchline, 1988), (6) age (e.g. Cladocera: Richman, 1958), and (7) instar number (e.g. Cladocera: Murugan, 1975a).

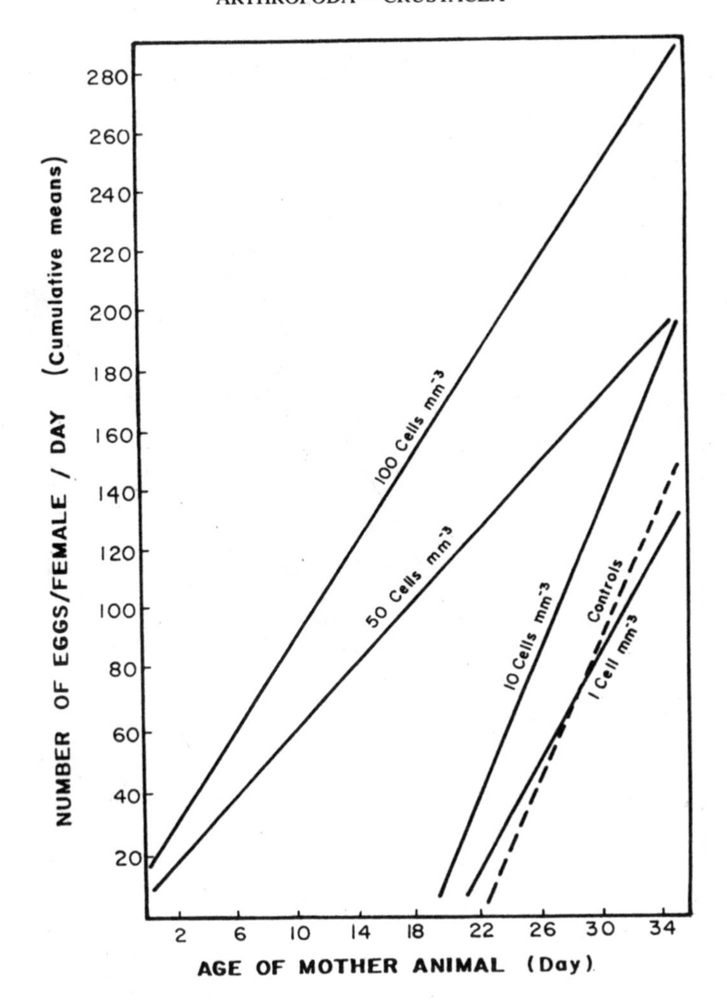

Fig. 6. Fecundity-adult age relationship in the copepod *Tigriopus brevicornis* fed on different algal (*Phaedactylus tricornutum*) rations (from Comita and Comita, 1966; modified).

2. Environmental factors

Nutritional requirements of only a few crustaceans have been studied. Administration and estimation of the consumed feed pose problems. The need for tearing, kneading, and mastication of the feed by most crustaceans (see Pandian, 1975), the scope for a fraction being lost from the buccal cavity (Wickins, 1976), and the possibility of the soluble feed being lost in water have together deterred research work in this area. Vitamin B_{12} substantially increases the fertility of *Daphnia magna* (see D'Agostino and Provasoli, 1970); supplementation of thiamine and/or pantothenic acid to B_{12} ensures fertility of *D. magna* for at least 200 generations

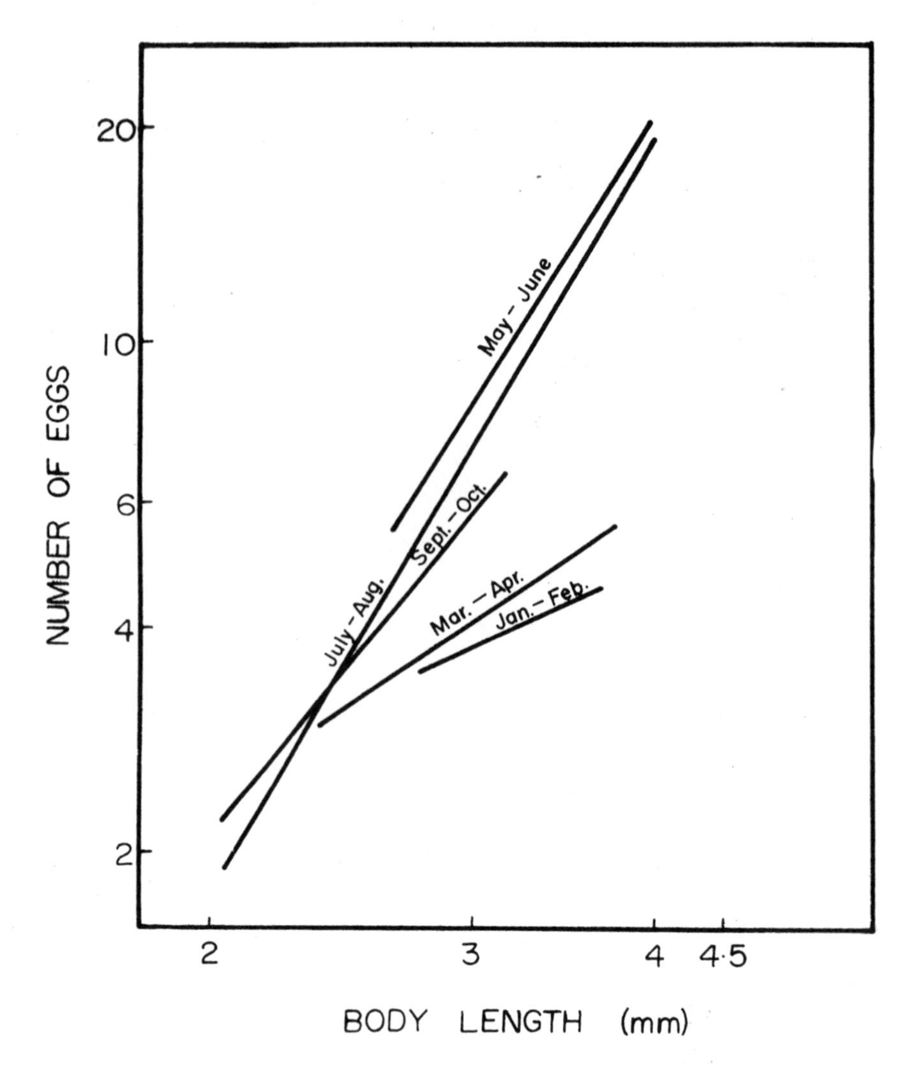

Fig. 7. Relationship between egg number and maternal size in the amphipod *Corophium insidiosum*. Note the linear trend shifts from one level to the other in females collected during different temperature-related seasons from the same area (from Sheader, 1978).

(Conklin and Provosoli, 1977). Middleditch *et al.* (1980), who successfully induced spawning in *Penaeus setiferus* in the laboratory, consider that dietary lipid is primarily responsible for the promotion of ovarian development; cholesterol is the dominant sterol present in the gonad; polyunsaturated fatty acids, obtained through the feed *Glycera dibranchiata*, enable the prawn to mature and spawn the eggs.

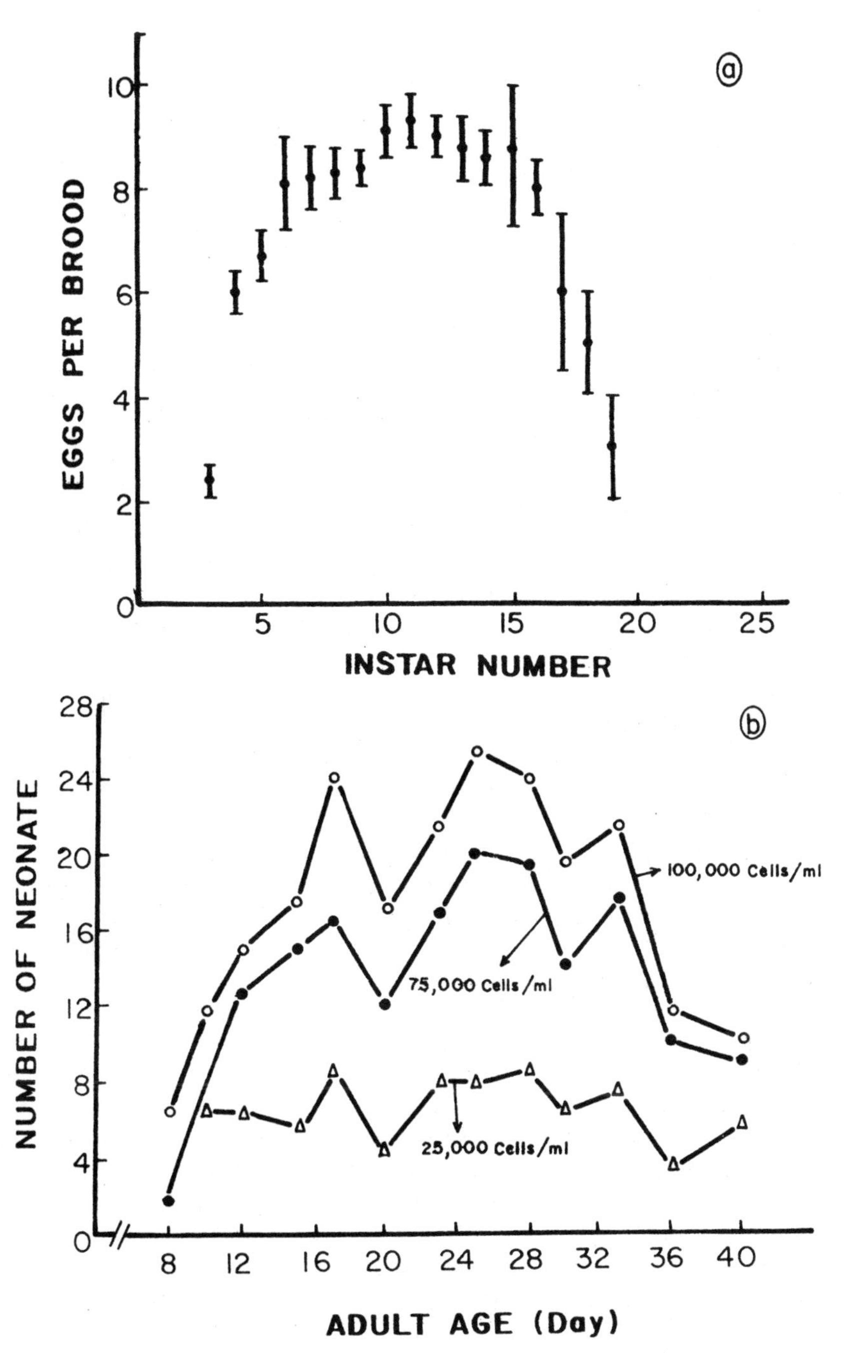

Fig. 8. **a:** Egg number as a function of successive adult instars in *Ceriodaphnia cornuta* (from Murugan, 1975a). **b:** Neonate number as a function of adult age in *Daphnia pulex* (from Richman, 1958).

Table 27

Effect of food quality on egg production in *Calanus finmarchicus* (from Marshall and Orr, 1952; modified)

Algal species	Egg laying (per cent females)	Total eggs (No.)	Eggs (No./sac)	Eggs (No./female/day)
Starved	51	83	16.3	1.5
Chlorella stigmatophora	47	71	15.3	0.8
Hemiselmis rufescens	77	145	14.3	1.7
Diragteria inornata	92	200	16.8	2.5
Coscinodiscus centralis	87	261	11.5	3.0
Rhizosolenia delicatula	80	463	22.6	6.6
Skeletonema costatum	93	787	19.5	9.4
Peridinium trochoideum	100	865	24.8	11.5
Chlamydomonas sp.	94	1117	26.1	11.7
Syracosphaera carterae	100	1016	28.9	11.8
Ditylum brightwelli	100	1422	24.2	16.7
Lauderia borealis	100	1779	33.4	21.2
Cymnodinium sp.	100	1974	29.9	21.9

Many crustaceans are able to utilize a wide range of food sources to produce eggs. For instance, *Calanus finmarchicus* maintains its high fecundity in 50 per cent of the algae on which test cultures were made (Table 27); of the dozen algae thus tested, *Chlorella* alone was totally unsuitable. In some species, prey size is remarkably important: egg production is doubled in *Acanthocyclops viridis* fed on larger specimens of *Daphnia obtusa*, in comparison to that fed on smaller ones; yet, *A. viridis* too produced eggs at all the tested feed (Table 28). Detritus food web adds stability to an ecosystem by making energy, fixed seasonally by primary producers, available to the consumers over a very long time. Seasonal variations in body size and number of eggs laid per female *Pseudocalanus minutus* in Bedford Basin give indirect evidence that assimilation of a diet composed of mainly non-living particles is channelled into growth and reproduction (Poulet, 1976). Hence, some crustaceans may be speculated to draw energy and nutrients from algal and detritus sources. Experimental studies reveal that, to the estuarine copepod *Eurytemora affinis*, detritus-algal (*Isochrysis gablana*) mixture is a more adequate diet than either component fed separately. Rate of egg production on mixed diet is significantly higher than on the algal or detritus diet alone (Table 29). The detritus in this situation meets most of the energy requirements of the copepod, while the algal cells contribute some trace metabolites necessary for egg production (Heinle *et al.*, 1977). Feeding protozoans, possibly ciliates, singly or in combination with orange bacteria resulted in the highest egg production. It is very likely that *E. affinis* utilizes algal-detritus source during the summer, and detritus-protozoan/detritus-protozoan-bacteria source during the winter, in view of the abundance of detritus (see Heinle and Flemer, 1975). Impressive data on the role of bacteria on egg production of a few Copepoda have been reported by Reiper (1978).

Table 28

Effect of food quality on fecundity of *Acanthocyclops viridis* (from Smyly, 1970; modified)

Feed	Adult instar duration (day)	Maximum no. of broods	No. of eggs/brood	Fecundity (no. of eggs)
Algae	30	1	19	19
Protozoa	57	4	34	137
Ceriodaphnia quadrangula	89	4	64	255
Chydorus sphaericus	106	6	58	347
Daphnia obtusa (large)	99	8	60	478
D. obtusa (small)	94	12	74	888
Artemia salina	68	10	92	924

Table 29

Effect of different natural feed combinations on egg production in the copepod *Eurytemora affinis* (from Heinle *et al.*, 1977; modified)

Feed combination	No./brood	Egg production		
		No. of broods	Total eggs	Length of adult life in days
Starvation control	12	0.9	15	5
Algal control (0.375×10^4 cells/ml)	11	3.7	42	10
Unautoclaved detritus (10 mg/l) + algal control)	10	4.7	50	11
Unautoclaved detritus (100 mg/l)	7	0.9	10	8
Autoclaved detritus (100 mg/l)	5	0.8	5	5
Natural protozoa	18	5.5	103	12
Natural protozoa + bacteria	13	5.4	73	12

Ovarian development is dependent upon adequate food supply. Interruption of the food supply leads to a regression of ovarian tissue. However, resumption of feeding, even after five months, as in *Semibalanus balanoides* (see H. Barnes and M. Barnes, 1967), renews development of the ovary; once initiated, ovarian development depends upon adequate food supply and continuous food supply ensures normal breeding. As indicated elsewhere (Section III D), *Panulirus longipes cygnus* may breed twice annually. Some 67 per cent of the females supplied with abundant food in the aquaria bred twice, but only 11 per cent of the females observed in field could be confirmed as having spawned twice during the breeding season; thus, abundance and scarcity of food can decisively affect breeding frequency (Chittleborough, 1976). The correlation between algal content and egg production is so significant that some workers read the egg production of cladocerans (within the biokinetic range of the species) from an estimation of chlorophyll content (e.g. *Simocephalus vetulus*: Green, 1966).

Temperature may act as a time-setter of reproduction (e.g. *Menippe mercenaria*: Cheung, 1968) and affect fecundity. In most cases, it alters egg number per brood, brood number, and the length of the interlaying period; for instance, the first two characters increased from 25 and 2.4 at 5°C to 35 and 8 at 25°C in *Tigriopus brevicornis*, while the interlaying period was reduced from about 10 to three days (Harris, 1973). Incidentally, latitudinal variations profoundly alter the size of an egg as well as brood size in response to the ambient temperature. Green (1966) made a very detailed study on this aspect in some cladocerans, especially *Scapholeberis mucronata*. Egg number increased with increasing size (body length) of females collected from different geographic localities (Fig. 5). At a given body length, Greenland populations produce fewer eggs than the British and Danish populations. The Italian populations, exposed to appreciably higher summer temperatures than those of Britain and Denmark, produce a larger number of eggs than the British and Danish populations. Green (1966) also noted geographic variations in egg size of *S. mucronata*; the eggs produced in Greenland are much larger than those produced in Britain even as the British populations reproduced at temperatures (ca. 6°C) lower than the summer temperature in Greenland (11°C). A definite inverse relation becomes apparent when the seasonal variations in egg size of *Simocephalus vetulus* (British populations) are plotted against temperature (Fig. 9); a very similar trend also emerged for the egg size-temperature relationship in *Scapholeberis mucronata* populations from different geographical localities. Evidently, temperature determines both egg and clutch size.

In crustaceans inhabiting estuarine and coastal waters, salinity plays an important role in determining the egg number per brood and the brood size (e.g. *Eurytemora herdmani*), besides the species' abundance pattern (Conde and Diaz, 1989).

F. Energy Budget

Retention of matter and energy within an organism results in growth, which is expressed in volumetric expansion and/or energy concentration. Beyond a critical level of accumulation of matter and energy, organisms choose to reproduce, rather than grow further, or grow and reproduce simultaneously (see Slobodkin, 1960). Most crustaceans grow and reproduce on attaining sexual maturity but a few such as copepods and some crabs cease to grow and only reproduce as adults like pterygote insects. The energy budget in relation to egg production, body growth, and metabolism for various crustaceans is given in Table 30. Most commendable is the efficiency of the parthenogenetic cladoceran, for which the values range from 52 to 69 per cent. Similarly, the parthenogenetic *Artemia salina* too allocates about 28 per cent of the assimilated energy for reproduction. Energy invested on reproduction by *Penilla* is 2.5 times more than that allocated for somatic growth. *Artemia* and *Daphnia* invest seven and 10.5 times more energy on reproduction than on body growth. An all-out effort must be made to cultivate them as feed organisms, in view of their

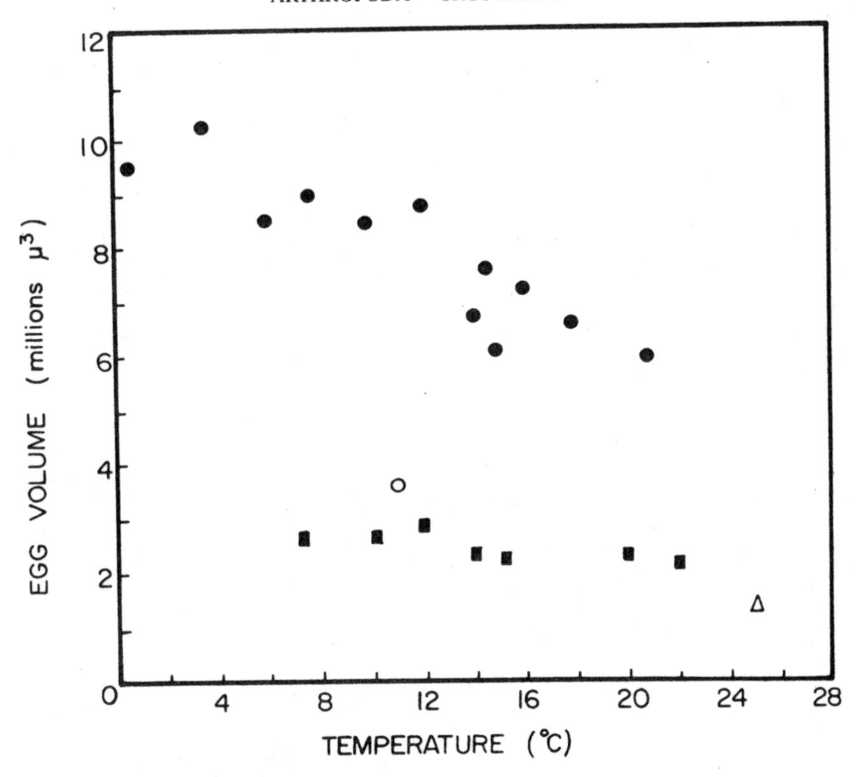

Fig. 9. Relationship between egg volume and water temperature in *Simocephalus vetulus* (●●) and *Scapholeberis mucronata* (■■); the values indicated by symbols ○ and △ belonged to the Greenland and Italian population of *S. mucronata*, respectively (from Green, 1966; modified).

ability to allocate such a large fraction of the assimilated energy on reproduction. It will be of great interest to know whether parthenogenesis obligatorily involves such a heavy investment or such a heavy investment is peculiar to the Cladocera alone.

Among gonochoristic crustaceans, copepods make the heaviest investment on reproduction, a strategy to compensate loss due to heavy predation. Several insects which are prone to heavy predation are also known to make heavy investment on reproduction. As a group, the copepods commit about 16 per cent of the assimilated energy on reproduction, which in most cases is about four times more than that utilized for body growth. Of the 11.3 per cent assimilated energy invested on growth in *Balanus glandula*, 4.2 per cent is spent on body growth proper and the remaining 7.1 per cent on reproductive growth (Wu and Levings, 1979). For crustaceans, such as the barnacles, which suffer heavy mortality during their planktonic stages and their early settlement, it is indeed an adaptive strategy to invest a lion's share of the assimilated energy on reproduction (Wu and Levings, 1979). Likewise, the parasite *Probopyrus pandalicola* has to pass through the epicardium-microniscus-

Table 30

Energy budget of some curstaceans. All values are given as percentage of assimilated energy

Species	Egg production	Exuvia	Body growth	Meta-bolism	Basis	Reference
Anostraca						
Artemia salina	28	4	4	54	Calorie	Khmeleva (1967)
Branchinecta gigas	8	11			Calorie	Daborn (1973)
Cladocera						
Daphnia pulex	69		6	25	Calorie	Buikema (1975)
D. pulex	67		7	26	Calorie	Richman (1958)
Thames cladocerans			10–30		Dry weight	Bottrell (1975)
Copepoda						
Asellopsis intermedia	11	1.7	5	82	Carbon	Lasker *et al.* (1970)
Acartia tonsa	58		10	38	Nitrogen	Kiorboe *et al.* (1985)
Tigriopus brevicornis	23	0.4	4	73	Nitrogen	Harris (1973)
Calanus finmarchicus	12	0.9	25	62	Nitrogen	Corner *et al.* (1967)
C. helgolandicus	19			81	Dry weight	Gaudy (1974)
Temora stylifera	14			86	Dry weight	Gaudy (1974)
Pareuchaeta norvegica	19	0.9	27	53	Calorie	Bamstedt (1979)
Centropages typicus	17			83	Dry weight	Gaudy (1974)
Cirripedia						
Balanus glandula	13	2.5	11	73	Calorie	Wu and Levings (1979)
Mysidacea						
Metamysidopsis elongata	13	7	19	55	Calorie	Clutter and Theilacker (1971)
Isopoda						
Cirolana harfordi	8		32	59	Calorie	W.S. Johnson (1976a, 1976b)
Amphipoda						
Crangonyx richmondensis	0.5	3.5	17	79	Calorie	Mathias (1971)
Hyalella azteca	2	7	16	75	Calorie	Mathias (1971)
Calliopius laeviusculus	7	4.7	16	47	Carbon	Dagg (1976)
	7	5.3	18	38	Carbon	Dagg (1976)
	8	5.9	19	35	Carbon	Dagg (1976)
Euphausiacea						
Euphausia pacifica	1	7	6	86	Carbon	Lasker *et al.* (1970)
Deep sea euphausiids		10			Dry weight	Jerde and Lasker (1966)
Decapoda						
Macrobrachium nobilii	14	7	7		Calorie	Pandian and Balasundaram (1982)

cryptoniscus-bopyridium stages and must still find the right hosts at the appropriate time; to compensate for these non-feeding and highly risky stages, this species invests as much as 63 per cent of its assimilated energy on reproduction, which is perhaps the highest recorded in literature (G. Anderson, 1977).

All these values only represent the energy investment on egg production. No information is yet available on the energy cost of mating and egg carriage, which are essential components of the reproductive cost. Interestingly, in peracaridean *K*-strategists such as *Metamysidopsis elongata, Cirolana harfordi, Crangonyx richmondensis*, and *Hyalella azteca*, energy allocated for reproduction is just equal to, or less than what is committed for body growth. Some relatively large insects are known to allocate 13 to 14 per cent of the assimilated energy on reproduction (e.g. *Bombyx mori*: Hiratsuka, 1920; *Oxya velox*: Delvi and Pandian, 1971); their investment on somatic growth is far more than what is relegated for reproduction.

G. Reproductive Effort

The term 'reproductive effort' is often used to refer to the parental investment of energy in gamete production over some biologically meaningful period of time and is expressed as a fraction or percentage of total assimilated energy (see Pandian, 1987). Available data on reproductive effort of crustaceans for their entire life span are listed in Table 31. Parthenogenetic curstaceans invest 14–39 per cent of their assimilated energy on reproductive effort. As expected, those copepod species which suffer heavy predation show the highest values of reproductive effort (13–39 per cent); in fact, the daily reproductive-effort values reported for copepods range from 7 per cent of respective body weight to 106 per cent (see Ambler, 1985). Likewise, for sessile Cirrepedia and parasitic Isopoda, whose larvae undertake the risk of reaching suitable habitat or host, the reported values for the reproductive effort are high. Therefore, these data on reproductive efforts confirm the generalization made from the data on energy budget.

IV. YOLK UTILIZATION

Yolk is a product accumulated in the egg as a result of vitellogenesis (see R.G. Adiyodi and Subramoniam, Volume I of this series). It is a special kind of reserve substance containing almost the entire spectrum of nutrients to ensure the normal development of the embryo; in most crustaceans, a smaller or larger fraction of the yolk remains in the neonate until it is able to subsist on its own. Quality and quantity of yolk accumulated in the eggs may remarkably alter the mode of development. For instance, accumulation of trehalose and glycerol, in the place of glycogen, shifts the eggs of *Artemia salina* from viviparous, subitaneous mode of development to the one interrupted by a period of dormancy (Clegg, 1965). The quantity of yolk in the egg is known to determine the direct or indirect mode of development of the hybrid embryos of the two sibling species (?) of the crab *Halicarcinus*. The freshwater form of *H. lacustris* broods 50 large (0.7 mm diameter) eggs that develop directly into young ones in about 55 days; the marine form produces over 400 small eggs (0.35 mm in diameter), which undergo indirect development (in about 25 days) involving three zoeal stages. When crossed, a few

Table 31

Gross egg production efficiencies of some crustaceans

Species	Per cent	Basis	Reference
Anostraca			
Artemia salina	14	Calorie	Khmeleva (1967)
Copepoda			
Acartia clausi	2	Calorie	Petipa (1967)
A. tonsa	39	Nitrogen	Kiorboe *et al.* (1985)
Calanus glacialis	5	Carbon	Hirche (1989)
C. helgolandicus	4	Dry weight	Gaudy (1974)
C. finmarchicus	14	Nitrogen	Corner *et al.* (1967)
C. marshallae	29	Dry weight	Peterson (1988)
C. pacificus	32	Dry weight	Runge (1985)
Paracalanus parvus	33	Dry weight	Checkley (1980)
Pseudocalanus elongatus	3	Dry weight	Harris and Paffenhofer (1976)
Temora stylifera	7	Dry weight	Gaudy (1974)
Eurytemora affinis	13[1]	Carbon	Heinle *et al.* (1977)
Pareuchaeta norvegica	18	Calorie	Bamstedt (1979)
Centropages typicus	2	Dry weight	Gaudy (1974)
Asellopsis intermedia	6	Carbon	Lasker *et al.* (1970)
Tigriopus brevicornis	22	Nitrogen	Harris (1973)
Diaptomus siciloides	7	Calorie	G.W. Comita (1964)
Cirripedia			
Balanus glandula	13	Calorie	Wu and Levings (1979)
Isopoda			
Cirolana harfordi	7[2]	Calorie	W.S. Johnson (1976b)
Probopyrus pandalicola	63	Calorie	Anderson (1977)
Amphipoda			
Calliopius laeviusculus	7	Carbon	Dagg (1976)
Decapoda			
Macrobrachium nobilii	12	Calorie	Sindhukumari and Pandian (1991)

[1]For those producing one egg sac; the value for those producing two sacs is 30 per cent.
[2]K_2.

females of either form retain at least some hybrid eggs and hatch them successfully. Among the hybrids there is no intergradation of direct and indirect development. Hybrid embryos (F_1) develop as well as hybrid females produce eggs (F_2), according to the female parental form, and the amount of yolk in the eggs determines the mode of development, according to the original female parental form (Lucas, 1970; for *Menippe* hybrids, see Wilbur, 1989).

A. Energy Values

Presenting the first survey on calorific value of a variety of species ranging from ciliates to fish, Slobodkin and Richman (1961) showed that most of their results are between 5,400 and 6,100 g cal/g dry weight. In Crustacea, most of the reported values fall within the limits of 6,000–6,500 g cal/g dry weight for eggs and 4,200–4,700 g cal/g dry weight for neonates (Table 32). The high calorific value indicates that almost all crustaceans accumulate a large quantity of energy-rich lipids in their eggs. The higher values recorded for the eggs and neonates of *Probopyrus* and *Macrobrachium* are due to the relatively greater concentration of energy-rich lipid in the eggs and the retention of a greater fraction of yolk in the neonate. Availability of such unusually large fraction of yolk in the neonates is useful to land the non-feeding, actively searching epicaridean larva of *Probopyrus* on to its host (G. Anderson, 1977; for copepod parasites, see Izawa, 1973), or to take the passively migrating zoea to the estuarine waters, where alone it can moult and attain the feeding, zoeal stage (Balasundaram and Pandian, 1982b).

B. Efficiency Values

In general, developing embryos are known to convert the yolk with the highest efficiency, and crustaceans are no exception. The reported values for yolk utilization efficiency range from 54 per cent for *Pagurus (Eupagurus) bernhardus* (see Pandian and Schumann, 1967) to 74 per cent for *Macrobrachium nobilii* (see Balasundaram, 1980; also see Balasundaram and Pandian, 1982 b) and average 63 per cent. This value is comparable to those reported for different groups of vertebrates (e.g. fishes—*Solea solea*—68 per cent: Fluchter and Pandian, 1968; Pandian, 1987; amphibians: Seale, 1987; reptiles: Waldschmidt *et al.*, 1987; birds—*Gallus*—56 per cent: Brody, 1945). Environmental factors such as temperature and salinity alter significantly the efficiency of some crustaceans. At any tested salinity-temperature combination, *Artemia salina* utilizes the yolk with an efficiency which is comparable to or even higher than those reported for other crustaceans (Table 33). This observation of Hentig (1971) suggests that *Artemia* is indeed one of the very few euryplastic animals that can flourish under widely differing habitats. At any tested temperature, the efficiency of *A. salina* decreases from 82 per cent at the lowest salinity (5‰) to 64 per cent in the brine water (70‰). Increase in temperature at any tested salinity resulted in 4 to 9 per cent decrease in the efficiency. To produce the most energy-rich nauplius feed, *Artemia* eggs should be hatched at 30°C in dilute sea (salt) water of 5‰.

C. Composition of Eggs

Major constituents of crustacean eggs are: water (55–65 per cent), protein (15–25 per cent), lipids (20–30 per cent), and salts (2–5 per cent). Unusually high or low values have also been recorded: 72 per cent water and 10 per cent protein in *Ligia oceanica* (see Pandian, 1972); 67 per cent lipids in *Parapasiphae sulcatifrons* (see Herring, 1974b). Water, ash, and lipids are the three important variables during

Table 32

Calorific values of eggs and neonates of some arthropods

Species	Energy content (g cal/g dry weight		Reference
	Eggs	Neonate	
CRUSTACEA			
Branchiopoda			
Artemia salina	5,549	5,565	Hentig (1971)
Copepoda			
Calanus finmarchicus	5,478[1] 4,136[1]		G.W. Comita *et al.* (1966)
Isopoda			
Ligia oceanica	5,956	4,175	Pandian (1972)
Probopyrus pandalicola	7,858	7,426	Anderson (1977)
Decapoda			
Macrobrachium nobilii	7,020	5,753	Balasundaram (1980)
M. lamarrei	6,325	6,026	Katre (1977b)
M. idella	6,231[1]	4,410[1]	Vijayaraghavan and Easterson (1974)
Crangon crangon	5,915	4,544	Pandian (1967)
Homarus americanus	6,636	4,292	Pandian (1970b)
H. gammarus	6,172	4,524	Pandian (1970a)
Pagurus (Eupagurus) bernhardus	6,053	4,783	Pandian and Schumann (1967)
Caridina nilotica	1,497[1]	587[1]	Ponnuchamy *et al.* (1979)
Emerita holthusi	4,288[1]	992[1]	Vijayaraghavan *et al.* (1976)
INSECTA			
Poecilocerus pictus	5,509		Delvi (1972)
Brachythermis contaminata	6,607	4,874	Mathavan (1985)
Orthetrum sabina	6,473	4,888	Mathavan (1985)
ARACHNIDA			
Argiope pulchella	6,161		Prakash and Pandian (1978)
A. aurantia	6,343		J.F. Anderson (1978)
Cyrtophora cicatrosa	6,440	5,460	Palanichamy (1980)

[1]Estimated following organic matter method.

embryonic development. Absorption of water and salts from the surrounding water raises the levels of these constituents, while lipids are heavily depleted. Lipid depletion is reflected in the decreased calorific content of the neonates (Table 32). Table 34 shows that during development, marine crustaceans depend heavily on the surrounding medium for water and salts. Eggs of freshwater crustaceans already contain a high dose of salts and they have therefore reduced the magnitude of their dependence on the medium.

Table 33

Effects of different salinity and temperature combinations on yolk utilization of *Artemia salina* in terms of dry weight and energy (from Hentig, 1971; modified from his tables 4, 5 and 6)

Salinity (‰)	Temperature (°C)	Weight of nauplius (μg)	Nauplius as percentage of gastrula	Energy in nauplius (g cal)	Nauplius energy as percentage of gastrula
	Gastrula	2.52	100	0.0146	100
5	10	1.96	78	0.0112	80
	15	2.10	84	0.0117	80
	20	2.16	86	0.0112	83
	30	2.19	87	0.0123	84
15	10	1.86	74	0.0104	71
	15	1.97	79	0.0110	75
	20	2.06	82	0.0115	79
	30	2.09	83	0.0116	80
30	10	1.76	70	0.0097	67
	15	1.83	73	0.0102	69
	20	1.89	75	0.0106	72
	30	1.92	76	0.0106	72
70	10	1.63	65	0.0089	61
	15	1.69	67	0.0091	62
	20	1.72	68	0.0095	65
	30	1.78	71	0.0098	67

Table 34

Water and salt requirements of eggs of some crustaceans for successful completion of their embryonic development. Values based on mg wet weight per egg (from Pandian, 1970b; modified and added)

Species	Required water (mg)	Quantity of salt (mg)	Reference
Marine crustaceans			
Homarus gammarus	2.1	61.1	Pandian (1970a)
H. americanus	1.3	50.4	Pandian (1970b)
Crangon crangon	0.8	5.5	Pandian (1967)
Pagurs (Eupagurus) bernhardus	0.8	15.7	Pandian and Schumann (1967)
Ligia oceanica	0.1	67.5	Pandian (1972)
Freshwater crustaceans			
Caridina nilotica	0.5	4.7	Ponnuchamy *et al.* (1979)
C. weberi	0.2	2.3	Rao *et al.* (1980)

1. Proteins

Suyama (1959) reported the amino acid composition of the ovarian eggs of *Panulirus japonicus*. Glutamic acid and aspartic acid are the major constituents, each forming 12 per cent of the total. Arginine, alanine, glycine, histidine, tyrosine, and valine are all present in the range of 4 to 7 per cent. The possibility of changes in them during development has not been studied in detail except in *Balanus balanus* and *Semibalanus balanoides* (see H. Barnes and Evens, 1967) and *Artemia salina* (see Emerson, 1967).

Studies on *Balanus balanus* and *Semibalanus balanoides* have shown that protein is lost throughout development (H. Barnes, 1965). Table 35 presents the amino acid composition of the eggs of these two species and the percentage of loss of amino acid nitrogen during development; the loss is much the same for most amino acids. The loss of both alanine and glycine is low. On the whole, 26 to 30 per cent of the amino acid nitrogen is lost during development. Given allowance for the loss of non-protein nitrogen, the fraction of protein nitrogen expended on embryonic metabolism is likely to be around 20 per cent. Interestingly, the values reported for the percentage of egg protein utilized for embryonic metabolism are 21 per cent for *Pagurus (Eupagurus) bernhardus* (see Pandian and Schumann, 1967), 17 per cent for *Crangon crangon* (see Pandian, 1967), 12 per cent for *Homarus gammarus* (see Pandian, 1970a), and 10 per cent for *Ligia oceanica* (see Pandian, 1972).

From available data, the amount of protein metabolized during the conversion of yolk substances equivalent to 1 g cal from the egg into the larva or neonate in some animals has been calculated, and the data are presented in Table 36. Non-cleidoic eggs of freshwater animals (e.g. salmon) oxidize as much as 122 μg protein in order to convert yolk substances equivalent to 1 g cal, while those of cleidoic terrestrial eggs of birds and insects use only about 20 μg and those of marine demersal crustacean eggs about 30 μg protein. In other words, suppression of protein metabolism is one-sixth in terrestrial cleidoic eggs, and one-fourth in marine demersal crustacean eggs, as compared to the eggs of non-cleidoic freshwater animals. The suppression of protein metabolism among the marine eggs becomes increasingly acute as the habitat of the chosen crustaceans is closer and closer to the coast. The hermit crab, data for which are given in Table 36, was trawled at 60 m depth (off Helgoland); the shrimp was netted at about 10 m depth; the lobster was caught by divers in the coastal waters of Helgoland at about 5 m depth, and *Ligia* was collected at supratidal zone of Helgoland harbour. To convert yolk substances equivalent to 1 g cal, the hermit crab oxidizes as much as 41 μg protein, the shrimp 31 μg protein, the lobster 16 μg protein, and *Ligia* 9 μg protein. The fact that in *Ligia* the amount of protein oxidized during the conversion of yolk substance equivalent to 1 g cal is only less than half of that oxidized in the cleidoic terrestrial eggs of birds and insects is of great interest; it suggests that developing eggs of *Ligia* are adapted to the semiterrestrial life primarily through suppressing protein metabolism rather than converting end product of protein metabolism into urea or

Table 35

Amino acid composition of eggs and the loss incurred by them during development (from H. Barnes and Evens, 1967; modified)

Amino acid	*Balanus balanus*		*Semibalanus balanoides*	
	Initial level (mM/10^6 eggs)	Total loss (per cent)	Initial level (mM/10^6 eggs)	Total loss (per cent)
Aspartic acid	0.66	35	0.46	29
Threonine	0.45	38	0.35	41
Serine	0.56	40	0.44	43
Glutamic acid	0.99	35	0.70	34
Proline	0.39	34	0.27	19
Glycine	0.56	9	0.50	1
Alanine	0.42	9	0.35	5
Valine	0.64	37	0.4S	33
Methionine	0.17	40	0.14	55
Iso-leucine	0.42	44	0.29	39
Leucine	0.62	40	0.42	34
Tyrosine	0.30	17	0.27	35
Phenylalanine	0.26	33	0.20	34
Lysine	0.50	30	0.42	38
Histidine	0.14	33	0.12	37
Arginine	0.41	30	0.36	38
Tryptophan	0.06		0.06	
Cystine	0.05		0.04	
Total N (mg)	149	Mean = 30	108	Mean = 26

uric acid at considerable energy cost (for details, see Pandian, 1975). A general suppression of nitrogen metabolism by the semiterrestrial (e.g. *Ligia oceanica*) and terrestrial (e.g. *Armadillidium vulgare*) isopods has been considered by Dresel and Hoyle (1950) as the chief adaptation for life on land, rather than a conversion of ammonotely to urea- or uricotely.

2. Lipids

Triglycerides and phospholipids are the major lipid components of eggs of several crustaceans (e.g. Cirripedia: *Semibalanus balanoides*, see Dawson and H. Barnes, 1966; Decapoda: *Heterocarpus*, R.J. Morris, 1973; see also Herring, 1973). In boreo-arctic species of *Semibalanus*, lecithin and phosphatidyl ethanolamine make up a large proportion of phospholipid fraction; eicosapentaenoic and docosahexaenoic acids predominate the fatty-acid spectrum (Dawson and H. Barnes, 1966). The fatty-acid composition, particularly the concentration of polyunsaturated acids, of eggs of the crustaceans vary; the high degree of polyunsaturated acids in barnacles may protect them against the deleterious effects of extremely low temperatures (see Allen, 1960).

Table 36

Estimation of protein metabolized during the conversion of yolk substances, equivalent to one gram calorie, in developing eggs of some animals (from Pandian, 1972; modified)

Species	Habitat	Amount of protein metabolized (μg)	Reference
Chick *Gallus*	Terrestrial	18	Needham (1931)
Silkworm *Bombyx*	Terrestrial	22	Needham (1931)
Salmon	Freshwater	122	Needham (1931)
Hermit crab *Pagurus (Eupagurus) bernhardus*	Marine, 60 m depth	41	Pandian and Schumann (1967)
Shrimp *Crangon crangon*	Marine, 10 m depth	31	Pandian (1967)
Lobster *Homarus gammarus*	Marine, 5 m depth	16	Pandian (1970a)
Isopod *Ligia oceanica*	Marine, supratidal	9	Pandian (1972)

Over 75 per cent of the triglycerides contained in yolk is utilized to meet the energy demand of embryonic metabolism (Dawson and H. Barnes, 1966). Several marine crustaceans draw about 80 per cent of the energy required for embryonic metabolism by oxidizing fat. Upon oxidation, fat releases larger quantities of metabolic water (1 g fat: 1.07 g water; 1 g carbohydrate: 0.56 g water; 1 g protein: 0.41 g water; Baldwin, 1964); unlike protein, oxidation of fat and carbohydrate does not result in ammonia production, the removal of which may require water. These two properties of fat, namely production and conservation of water, obviously represent a strategic advantage for eggs of marine crustaceans, to which water is not as readily available as to freshwater inhabitants.

Needham (1950) has attributed the following three metabolic properties to terrestrial cleidoic eggs: (1) independence from environment of water and salts; (2) suppression of protein metabolism, and (3) 'gearing up' of fat metabolism. There are a few exceptions to Needham's concept: some terrestrial cleidoic eggs, such as those of insects, depend upon their environment for water; in aquatic habitat, non-cleidoic eggs of freshwater animals do not depend upon the environment for salts. The non-cleidoic marine demersal eggs, while heavily dependent on the surrounding sea water for water and salts, display considerable suppression of protein metabolism and a remarkable enhancement of fat metabolism. Non-cleidoic eggs of freshwater animals utilize 66 and 27 per cent energy by oxidation of protein and fat, respectively, while cleidoic, terrestrial eggs use 61 per cent fat energy and 7 per cent protein energy (Pandian, 1970a, b). The marine demersal eggs of crustaceans such as *Pagurus bernhardus* too draw as much as 67 per cent energy from fat oxidation and only 28 per cent energy from protein oxidation (Table 37). Therefore, as regards these two metabolic properties, namely suppression of protein metabolism and 'gearing up' of fat metabolism, marine demersal crustacean eggs are similar to the terrestrial cleidoic eggs.

Table 37

Yolk constituents utilized as energy source for embryonic metabolism in some crustaceans

Species	Protein (per cent)	Fat (per cent)	Reference
Marine crustaceans			
Pagurus (Eupagurus) bernhardus	28	67	Pandian and Schumann (1967)
Crangon crangon	21	75	Pandian (1967)
Homarus gammarus	13	87	Pandian (1970a)
Ligia oceanica	10	87	Pandian (1972)
Emerita holthusi	80	20	Vijayaraghavan *et al.* (1976)
Freshwater crustaceans			
Macrobrachium lamarrei	8	92	Katre (1977b)
Caridina nilotica	19	81	Ponnuchamy *et al.* (1979)
C. weberi	59	41	Rao *et al.* (1980)

Studies on yolk utilization in the freshwater prawn *Caridina weberi* (see Rao *et al.*, 1980) and the marine crab *Emerita holthusi* (see Vijayaraghavan *et al.*, 1976) indicate that, at least in these two crustaceans, there is no apparent correlation between the major energy source of embryonic metabolism (Rao *et al.*, 1980); these authors suggested that the eggs weighing less than 25 μg (dry weight) tend to utilize protein as the major energy source, while in the larger ones fat replaces the protein. Much credence cannot be attached to the basic data collected by these authors; for instance, protein values summarized by Rao *et al.* (1980) in their table 3 include non-protein nitrogen too; secondly, energy values reported for the eggs of *Caridina nilotica* (1,497 g cal/g dry weight: Ponnuchamy *et al.*, 1979) and the freshly hatched larvae of *Emerita holthusi* (992 g cal/g dry weight: Vijayaraghavan *et al.*, 1976) are not theoretically acceptable as they are not comparable to values reported for eggs and neonates of crustacean and other arthropods (Table 32).

Nevertheless, the suggestion made by Rao *et al.* (1980) deserves some consideration. Unfortunately, they have not consulted a pertinent publication by Herring (1974b) listing an excellent correlation between egg size (wet weight), density, and lipid content in some crustaceans (Table 38). The high concentration of lipid (38 per cent) in large eggs (> 3.00 mm length) inevitably decreases the density, because the lipid fraction is the least dense major fraction of the egg constituent. The lipid is mostly composed of triglycerides (Herring, 1973) and/or phospholipids (W.V. Allen, 1972). The triglycerides have a density of 0.92 g/ml (Lewis, 1970). Lipid accumulation is a strategy to decrease density and to reduce energy cost of egg carriage in crustaceans characterized by abbreviated development coupled with extended period of incubation (see also Herring, 1973). Necessarily, lipid is utilized as the main energy source during embryonic development. Hence, the density of these large, buoyant eggs tends to increase during embryonic development; however, the imbibition of water and the expansion of egg volume compensate and reduce the density. An alternate strategy adopted by some crustaceans such as *Emerita holthusi* and *Macrobrachium idella* (see Vijayaraghavan and Easterson, 1974) is to reduce the density of small but dense eggs, by imbibing water and expanding the volume (e.g. *Acanthephyra purpurea*: Herring, 1974a). In these small (< 1.25 mm in length), dense (due to about 14 per cent lipid content) eggs, the importance of lipid as main energy source may be reduced to a certain extent. However, the scope for the occurrence of a large, low-lipid and high-protein type of crustacean egg, implying fundamental difference in the pattern of embryonic metabolism, is little. For, the size/density relations of the eggs of the ostracod *Gigantocypris*, the copepod *Valdivella* and the mysid *Eucopia* (Table 38) also tend to conform to one of the alternate strategies suggested above. The respiratory quotient values reported for the developing embryo of *Simocephalus vetulus* too suggest greater fat utilization (Hoshi, 1950). Hence, the size/density/lipid relationship is an inevitable consequence of the development pattern in crustaceans.

3. Carbohydrates

Only about 3 per cent of the required energy for embryonic metabolism is drawn from the oxidation of carbohydrates in some marine decapods (Pandian, 1970a). In *Artemia salina*, carbohydrate, especially trehalose, provides 24 to 35 per cent of the metabolic energy; as explained elsewhere (Section V F), trehalose accumulation before cryptobiosis and its subsequent utilization during development are an adaptive strategy of *A. salina*. Hentig (1971) studied the effects of different salinity and temperature combinations on utilization of yolk substances for embryonic metabolism. Temperature does not significantly alter the proportion of different yolk substances utilized for embryonic metabolism (Table 39). On the other hand, with increasing salinity from 5‰ to 70‰, the percentage contribution of metabolic energy progressively decreased from 35 to 24 for cabohydrate and increased from 30 to 37 for protein and from 35 to 39 for fat.

4. Pigments

Haem pigments are present in the form of cytochromes but their occurrence and function in crustacean embryos have not been studied, as their concentration is not sufficient to cause any colouration (Green, 1971). Some non-malacostracans have readily developed the necessary pigments to colonize hostile waters with poor oxygen supply. Haemoglobin has been detected in the parthenogenetic eggs of *Daphnia* (see Fox, 1948) and haem pigment in the egg shells of *Artemia* (see J. Needham and D.M. Needham, 1930). More haematin is deposited in shells of eggs shed in concentrated brines with low oxygen content than in more dilute media with higher oxygen content (Gilchrist and Green, 1960). In *Daphnia* too, haemoglobin is maternal in origin (Dresel, 1948) and the quantity passed into

Table 38

Size, density, and lipid content of some crustacean eggs (from Herring, 1974b; modified)

Species	Egg dimension (mm)	Density (g/ml)	Lipid (g wet weight)
Natantia			
Acanthephyra purpurea	0.90 × 0.62	1.065	14.4
A. pelagica	1.00 × 0.68	1.063	16.5
A. curtirostris	0.94 × 0.67	1.061	9.4
A. acutifrons	0.85 × 0.75	1.058	12.7
A. acanthitelsonis	0.92 × 0.64	1.068	16.1
Meninigodora vesca	0.95 × 0.70	1.056	
M. miccylus	0.94 × 0.72	1.054	
Notostomus auriculatus	0.98 × 0.64	1.054	13.0
N. elegans	0.76 × 0.76	1.053	18.5
Hetarocarpus grimaldii	0.63 × 0.53	1.073	18.7
H. ensifer	0.54 × 0.40	1.081	13.5
Nematocarcinus cursor	0.54 × 0.42	1.075	14.5
N. exilis	0.64 × 0.50	1.082	24.9
Plesionika edwardsi	0.44 × 0.34	1.079	
Pontophilus talismani	0.74 × 0.52	1.076	12.0
Systellapis debilis	3.40 × 1.88	1.030	
S. crisata	4.12 × 2.72	1.028	40.4
S. braueri	4.64 × 3.12	1.027	27.6
Oplophorus spinosus	3.04 × 1.95	1.029	41.1
Ephyrina bifida	4.70 × 3.52	1.026	42.2
E. hoskynii	4.00 × 3.12	1.027	36.0
Hymenodora gracilis	2.68 × 2.17	1.027	22.4
Pasiphaea hoplocerca	2.24 × 1.60	1.041	27.5
P. multidentata	2.52 × 1.76	1.038	19.7
Pasiphaeid sp.	4.92 × 3.32	1.027	52.8
Parapasiphaea sulcatifrons	4.20 × 2.68	1.026	67.0
Reptantia			
Galtheid A	0.36 × 0.36	1.091	
Galtheid B	1.74 × 1.60	1.043	
Geryon sp.	0.66 × 0.66	1.074	13.6
Polycheles granulatus	0.66 × 0.58	1.069	9.4
Ostracoda			
Gigantocypris mulleri	1.85 × 1.75	1.030	
Copepoda			
Valdivella insignis	1.92 × 1.40	1.028	
Mysidacea			
Eucopia unguiculata	1.90 × 1.40	1.040	

each egg depends on the nutritional state of the mother and the oxygen content of the waters in which she lives. In poorly aerated water, *D. magna* produces fewer eggs with a high dose of haemoglobin (Green, 1956). Haemoglobin is considered

Table 39

Energy expenditure on embryonic metabolism in *Artemia salina* and the percentage of energy contribution from fat, protein, and carbohydrate (from Hentig, 1971; recalculated from his tables 4, 6, and 11)

Salinity (‰)	Temperature (°C)	Yolk energy expended (g cal)	Fat (per cent)	Protein (per cent)	Carbohydrate (per cent)
5	10	0.0034	37	31	32
	15	0.0029	40	27	33
	20	0.0024	32	31	37
	30	0.0023	29	32	39
	Mean		35	30	35
15	10	0.0042	39	34	27
	15	0.0036	40	32	28
	20	0.0031	34	34	32
	30	0.0030	33	36	31
	Mean		37	34	29
30	10	0.0048	40	35	25
	15	0.0044	39	35	26
	20	0.0040	37	36	27
	30	0.0040	36	37	27
	Mean		38	36	26
70	10	0.0057	40	36	24
	15	0.0055	40	36	24
	20	0.0051	38	38	24
	30	0.0046	36	40	24
	Mean		39	37	24

primarily to supply protein rather than oxygen in embryos of *Simocephalus vetulus* (see Hoshi, 1957).

The eggs of Crustacea are pigmented by carotenoids which may be free or linked to protein. When free, the colour ranges from yellow to red, but when linked to protein, the colour range is vastly increased (Green, 1971). Information on the occurrence of a range of carotenoid pigments in crustacean eggs is scattered in literature (Goodwin, 1960; Zagalsky *et al.*, 1967; Green, 1971; Herring, 1973). The carotenoid pigments in eggs are localized in the yolk protein at the time of laying but as development proceeds, the chromatophores appear and the relative volume of yolk decreases considerably. However, no major quantitative change occurs in the pigment content of a single egg (e.g. *Homarus*: Goodwin, 1951; *Balanus* and *Semibalanus*: H. Barnes, 1965; *Emerita*: Gilchrist and Lee, 1972; see Ceccaldi, 1968 for general discussion), but the carotenoproteins may be esterified or broken down (see Green, 1971) towards the end of embryonic metabolism. Possibly, linking a protein with a carotenoid shields the protein from certain enzymes. Green and blue egg caroteno-pigments alone occur in decapods inhabiting shallow waters, but in the eggs of abysso-benthic decapods, dark colours like deep brown and purple occur (Herring, 1973). The strategic role played by carotenoid pigments during development is still a subject of much speculation (see Gilchrist and Lee, 1972).

D. Rate of Development

Barring the case where development is interrupted by diapause, the incubation period is determined by the rate of yolk utilization. In general, the larger the egg, the longer the development period. McLaren (1966) considers that egg diameter—an index of egg size—reflects the surface-to-volume restrictions on gas exchange and therefore on metabolism of the egg. Temperature profoundly alters the yolk utilization rate and hence the development time. With increasing temperature, the development time is known to decrease curvilinearly (e.g. Copepoda: McLaren, 1966; Cirripedia: Patel and Crisp, 1960; Amphipoda: D.H. Steele and V.J. Steele, 1975d; Decapoda: Branford, 1978). Eggs of *Pseudocalanus* sp., collected at high latitudes during spring, require a longer development period than those from warm waters of summer and autumn. This is related to the larger size of females and their eggs in colder seasons (however, see also Hart and McLaren, 1978). Considerable attention has been paid by copepodologists to predict the duration of embryonic development as a function of temperature. Following McLaren (1965, 1966), most workers have used Belehradek's equation:

$$D = a(T - \alpha)^b$$

where D represents the development time in days, T is the temperature in degrees Celsius, a and b and α are fitted constants. Briefly, the formula is the simplest of several equations describing the three ways in which monotonic response to temperature may differ: a accounts for the differences in elevation by defining shifts along the Y or development axis; b depicts the degree of curvilinearity of response over the entire biokinetic range; and α, sometimes referred to as the 'biological zero', is the theoretical temperature at which development time becomes indefinite and accounts for shifts of the response along the temperature or X axis. McLaren (1966) and McLaren *et al.* (1969) have shown that while there is no correlation between egg size and a with the fitted parameters, there is a correlation if the parameters are 'regularized'; i.e. if a constant b (real b) is assumed, and new values of a and α are calculated, a is correlated with egg size. Wear (1974) used a constant b but found no correlation between egg size and a in decapods. Finding no correlation between a and egg size of amphipods, isopods, cumaceans, and mysids, D.H. Steele and V.J. Steele (1975a) stated that there is no reason to assume that different species will respond in the same way to changes in temperature.

McLaren's (1965, 1966) view has also been criticized by some copepodologists (e.g. Heip, 1974), as the b values obtained for the copepods range widely: on an average 1.97 for calanoids, 1.40 for cyclopoids, 1.72 for all copepods. Cooley and Minns (1978) found that in absolute terms, the greatest differences occur between the predicted and actual developmental times for certain freshwater calanoids at certain temperature and that there is a great deal of difference even at certain temperatures. For cyclopoids, lower a values are associated with species whose subitaneous eggs are likely to develop in cooler water at some time in their life cycle. However, Cooley and Minns (1978) observed that in calanoids, and to a lesser extent in cyclopoids, a and α are predictably related and prediction may be possible if the *in situ* egg hatching temperature (T_H) is known. Thus, McLaren's view may have a limited use for the prediction of development time for calanoids.

V. LIFE-HISTORY STRATEGIES

The vast majority of crustaceans are aquatic; they display an array of interesting life-history strategies to tide over unfavourable situations and to quickly repopulate the habitat, when conditions become suitable. Several crustaceans successfully survive through unfavourable seasons as resting eggs (e.g. Anostraca, Cladocera, and some species of Ostracoda and Copepoda) or as cysts (e.g. *Artemia*, some cyclopoid copepods); developing eggs of a number of decapods may undergo diapause and the adult females of several isopods and amphipods as well as a few prawns enter a reproductively quiescent resting period, to time the release of their progenies at the most favourable season.

A. Heterogonic Strategy

Heterogonic cycles in daphnids include a parthenogenetic period and an amphigonic period. Briefly, for a certain period, parthenogenetic females produce only females. Large numbers of such unfertilized eggs rapidly complete development in a brood chamber. After a number of generations, certain females produce male and female individuals; a few large eggs rich in yolk are laid in the brood chamber; unless fertilized either in the ovary or brood chamber, such eggs degenerate. On fertilization, the walls of the brood undergo great changes and form epihippium, which is separated from the carapace at the next moult. The resting eggs hatch out into parthenogenetic females, after a period of dormancy. The high reproductive potential owing to parthenogenesis is a remarkable biological and ecological feature by which the Cladocera are distinguished from all other planktonic crustaceans. An increase of any cladoceran population in a planktonic community occurs at a rapid, almost explosive rate (r-strategists) (Onbe, 1977). The reproductive potential of a parthenogenetic cladoceran individual is theoretically twice that of its amphigonic counterpart (see Glosener and Tilman, 1978). Monocyclic (e.g. *Moina*), dicyclic, and polycyclic races (e.g. *Daphnia*) or species have been described in daphnids and in other Cladocera correlated with more or less frequent appearance of amphigonic individuals in the populations. In limnetic population of large lakes, reproduction may be entirely parthenogenetic or acyclic, and in smaller lakes several complete cycles occur during a year. In the arctic *Daphnia*, production of resting eggs in the absence of males has been recorded (Edmondson, 1955); these eggs are called 'pseudosexual eggs'.

1. Paedogenesis in parthenogenesis

A very interesting type of parthenogenesis, known as 'paedogenesis', occurs in podonid cladocerans. The advanced embryos of *Evadne nordmanni*, which have developed parthenogenetically within a brood pouch, already carry their own eggs in their embryonic brood space; these embryos develop further, and by the time they are liberated from the brood pouch as free-swimming young females, their eggs have segmented to blastula stage with a large cleavage cavity (Kuttner, 1911). This phenomenon has been frequently reported to occur in a number of Polyphemoidea (Dolgopolskaya, 1958; Le Tourneau, 1961; Gieskes, 1970; Bosch and Taylor, 1973; Onbe, 1974).

2. Cytogenetic aspects

In 1886, Weismann established that a single polar body is formed in the parthenogenetic egg of *Daphnia*, in which the diploid number of chromosomes is maintained, presumably due to the abolition of a second maturation division. A close re-examination of the whole process in *D. pulex* by Bacci (1965) revealed the occurrence of an entire sequence of meiotic process. During the first maturation division, a transient pairing of chromosomes occurs; however, neither is the achromatic spindle formed nor the nuclear membrane dissolved; consequently, no division takes place. With increasing size and the appearance of a large vacuole in the cytoplasm, the undivided nucleus is shifted to a corner of the oocyte. Subsequently, the second maturation division occurs and thus the single polar body formed corresponds to the second maturation division and not to the first one, as was presumed by Weismann (1886).

Recombination can give rise to genetic variability within a single parthenogenetic strain of *Daphnia*. This explains why variability in thelygenous parthenogenesis of daphnids is high as in the offspring of amphigonic reproduction. Endomeiosis in *Daphnia* also indicates the possibility of obtaining different sex genotypes via recombination within parthenogenetic lines of daphnids and consequently the possibility of a not purely phenotypic sex determination. Thus, the occurrence or absence of endomeiosis in an egg is the key cytological mechanism that decides whether the offspring is to become parthenogenetic or amphigonic. A number of workers have made elaborate studies to identify the factor(s) responsible for triggering the switch from parthenogeny to amphigony.

3. Environmental factors

Banta (1939) reared several common cladocerans for 800 to 1,600 successive parthenogenetic generations (27 years) by frequently changing the culture medium and reducing the population. He considered such results a proof of the complete control of environmental factors upon the sex cycle. However, field observations do not suggest that the quantitative abundance of food is always followed by corresponding increase in egg production. In *Simocephalus vetulus*, for example, the egg production remains lower at lower temperatures than at higher temperatures, so that even in the presence of abundance of food, the egg production is restricted at lower temperatures (Green, 1966). The food quality, especially a specific nutrient procurable from a particular food source alone, may strongly influence the switch from parthenogeny to amphigony. Von Dehn (1955) reared cultures of *Moina rectirostris* fed on yeast or defatted yeast. Yeast-fed cultures produced about 30 per cent males and epihippial eggs, whereas those maintained on defatted yeast produced all female progeny with only a few exceptional males. The addition of fatty extract to the defatted yeast restored the high incidence of males. She further demonstrated that a mixture of ergosterin and olive oil in 1 : 1 ratio produced the same effect as that of fatty extract of yeast. She concluded that the formation of resting eggs depends upon the presence of ergosterin (Von Dehn, 1955). Ergosterol is present in several species of phytoplankton and hence easily available to these crustaceans (Conklin and Provosoli, 1977). In all her experiments, Von Dehn (1955) employed only parthenogenetic progenies of a single female with the aim of working on a genetically homogeneous material; this careful experimental design, however, appears now questionable, after the demonstration of endomeiosis (Bacci, 1965). Secondly, Green (1966) reported the seasonal occurrence of male epihippial eggs in 19 most frequent species of cladocerans in Hampton Court Long Water. Though the same phytoplanktons served as a food source to almost all of them, nine species continued to reproduce parthenogenetically throughout the year and three other species partly switched to amphigony during January-February or February-March, whereas the remaining seven species were amphigonic for different durations. It is not clear why the nutrient factor, which according to Von Dehn induces amphigony in, say, *S. mucronata* from December to March, has not induced amphigony in *S. vetulus*.

Group effect is another important factor causing the appearance of mictic females and of males in cyclical parthenogenesis. The mechanism by which it induces amphigony has not yet been understood. Reared in an aseptic medium, *Daphnia magna* reproduces parthenogenetically; however, on being transferred to a medium where amphigonic individuals had been produced by a different strain in the presence of bacteria, it switched over to amphigony (Treillard, 1925). Bacterial factors, like the external metabolites of the ectocrine kind, are considered to induce the production of male and epihippial eggs in *D. magna*.

In cladocerans, the switch to amphigony is also under the control of photoperiod interacting with population density and other extrinsic factors. The transitional reversal from parthenogenesis to disexual reproduction in *Daphnia pulex* is promoted by a critical photoperiod of LD 12.75 : 11.25 at 19°C (Stross and Hill, 1968). Interestingly, the critical day length is influenced only slightly by temperature; at 12°C, it is LD 13 : 11. At 19°C, all broods were parthenogenetic at LD 12 : 12 when only three adults were reared in 50 ml medium, but all broods were sexual at a density of 30 adults/50 ml. The density-related stimulus is suggested to be volatile. Furthermore, the extent of sexual induction is related to the age of the grandparent when the mother was born. Thus the inherent genetic factor, physiological (photoperiodic) rhythms, and ecological density may act in combination to trigger the time and extent of switch-over from parthenogeny to amphigony. Hence, any attempt to identify a single causative factor or agent that induces amphigony in Cladocera could be futile.

4. Marine cladocerans

In contrast to the prosperity of the order Cladocera in fresh water with more than 400 recorded species, only eight cladoceran species belonging to three genera (*Evadne*, *Podon*, and *Penilla*) are truly marine; besides, a couple of species occur in the coastal waters (Purasjoki, 1958; Shirgur and Naik, 1977). The cladocerans survive on the sea bottom in the form of resting eggs after completely disappearing from the plankton, and the planktonic populations of the next season arise from the parthenogenetic females to be hatched from these eggs. In sediments of the Inland Sea, Japan, the maximum density of 122,000 *Penilla* resting eggs/m^2 has been recorded during September (Table 40). These eggs hatch most successfully at temperatures of 17 to 20°C, which interestingly coincide with the sea temperature of the season, when this species first appears in the plankton.

Returning to environmental factors that trigger the appearance of sexual morphs, a different mechanism appears to operate in induction of sexual males in the marine dicyclic cladoceran *Podon polyphemoides* (see Bosch and Taylor, 1973). The species occurs over a temperature range of 0.2° to 27°C and a salinity range of 2.5‰ to 31.5‰. In Chesapeake Bay (39°N), its highest density (60,000 podonids/m^3) is recorded (in June) at 19.6°C and 10.4‰. Parthenogenetic females dominate the population through most of the seasons. With the onset of spring, males, sexual females, and resting eggs appear at 11°C and are found until the temperature remains less than 23°C. They reappear during the fall, when the temperature falls to 17°C. The sexual morphs of *P. polyphemoides* appear at both ends of the thermal cycles within 11° to 17°C limits, when photoperiods are around 13L : 11D during the spring and 10L : 14D during the fall. It may be recalled that the critical photoperiods which induce the development of sexual morphs are 13L : 11D for *Daphnia pulex* (45°N: Stross and Hill, 1968) and 22L : 2D for *D. middendorffiana* (71°N: Stross, 1969). The observation reported for *P. polyphemoides* shows no correlation between the spring and fall photoperiods, and also no correlation to that obtained for *Daphnia* spp. However, a definite

Table 40

Density of resting eggs of some calanoid copepods and cladocerans

Species	Egg density (No/m^2)	Reference
Calanoida		
Eurytemora affinis and *Acartia clausi*	3,200,000	Naess (1991)
A. clausi	3,200,000	Kasahara *et al.* (1975a)
A. erythreae	780,000	Kasahara *et al.* (1975a)
Centropages abdominalis	148,000	Kasahara *et al.* (1975a)
Tortanus forcipatus	62,000	Kasahara *et al.* (1975a)
Calanopia thompsoni	24,000	Kasahara *et al.* (1975a)
Labidocera bipinnata	3,500	Uye *et al.* (1979)
Cladocera		
Penilla avirostris	122,000	Onbe (1977)
Evadne tergestina	20,000	Onbe (1977)
Podon polyphemoides	4,000	Onbe (1977)

correlation between the production of sexual morphs and temperature readily becomes apparent for *P. polyphemoides*. Thus, the induction and termination of sexual morphs in different cladocerans appear to be under the control of multiple factors that act together in a complex manner.

B. Resting Eggs

Resting eggs are developing sexual eggs whose development is arrested at some variable point within a stage or range of stages. Almost all the major groups of lower crustaceans are capable of producing resting eggs. Eggs of branchiopods usually have hard external membranes and frequently suffer desiccation before development is completed. They have the capacity to delay the development, while resisting desiccation (e.g. *Streptocephalus*: Bernice, 1970) or freezing (e.g. *Branchinecta*: Daborn, 1975). Among ostracods, some are known to produce sexual resting eggs (e.g. *Elofsonia baltica, Cytherura gibba, Cytherois fischeri*, and *C. arenicola*: Theisen, 1966).

In temperate anostracans *Chirocephalus diaphanus* (see Mathias, 1937) and *C. stagnalis* (see Cottareli, 1966), two kinds of eggs are produced; one kind, the cyst, may hatch after drying and the other without predrying, provided the osmolarity of the medium is sufficiently low (Khalaf and Hall, 1975a, b). The embryonic development in the dried cyst is completed up to a 'breaking point' before the dormancy but the completion of development and hatching requires drying.

The tropical anostracans produce drought-resistant eggs, in which hatching occurs normally only after drying (e.g. *Streptocephalus seali*: Moore, 1967). Bernice (1970) found that "without drying, *S. dichotomus* eggs do not hatch even though they are kept immersed up to a period of four months in water". However, a few authors have also reported the hatching of these resting cysts without drying (see Prophet, 1962). Sam and Krishnaswamy (1979) think *S. dichotomus* produces two kinds of eggs, like the temperate anostracans *Chirocephalus diaphanus* and *C. stagnalis*. The second kind of undried eggs hatch and establish a new population in a drying pond, refilled and diluted (specific conductance below 80 μmoles/cm^2) with fresh rain water. It is perhaps the second kind of eggs with which Bernice (1970) did not succeed in hatching and that too with water of osmolarity higher than 80 μmoles/cm^2. Yet, the factors which initiate the production of the second kind of eggs remain puzzling.

Certain species of calanoid copepods produce two kinds of eggs: subitaneous and resting. Widely scattered information on the occurrence of resting eggs of copepods in freshwater as well as marine environments is summarized in Table 41. It has been generally accepted that most freshwater calanoid copepods lay two different types (subitaneous and resting) of eggs that differ from each other in morphology and/or physiology, the type of egg produced being dependent upon seasonal environmental changes (Hutchinson, 1967). A similar concept is now being also extended to some marine copepods. Kasahara *et al.* (1974) described four types of resting eggs in *Tortanus fasciatus*. No morphological difference was visible between subitaneous and resting eggs of *Acartia tonsa* collected from the east coast of the United States (Zillioux and Gonzalez, 1972). However, scanning electron microscopic studies revealed distinct differences in the ornamentation between summer and winter eggs of *Labidocera aestiva* from the same area (Grice and Gibson, 1975). Likewise, resting mechanisms may also differ from species to species at the same locality and in the same species from different localities. Therefore, it is important to investigate the mechanism over a wide geographical area and in more species.

Eggs of calanoids, shed freely, sink to the bottom in shallow waters, where no vertical turbulence occurs. In the Inland Sea (Japan), for instance, *Tortanus forcipatus* appears early summer and produces subitaneous eggs, which hatch readily in a day, and the unique structure of the egg membrane retards sinking rate. Before sinking to the bottom, most eggs hatch into nauplii and are recruited into the planktonic population. Those eggs that sink to the bottom also hatch, when washed free of mud by turbulence. After the annual maximum temperature, *T. forcipatus* population begins to decline; while thus declining, it produces resting eggs, which do not hatch immediately even when the temperature is high, and sink to the bottom, where they overwinter (Kasahara *et al.*, 1975b). Embryonic development is not suppressed within the mud at a certain fixed developmental stage as is the case in diapause but is arrested at some variable points within a stage or range of stages (Uye *et al.*, 1979). By spring, most of the resting eggs are capable of hatching the subitaneous way. Thus, the resting eggs serve as a means to overwinter. *T. forcipatus* provides a mechanism to repopulate the area with the species (Kasahara

Table 41

Occurrence and hatching of resting eggs in some calanoid copepods

Species	Remarks	Reference
Freshwater Calanoida		
Epischura lacustris	O_2 deficiency inhibits hatching; in American waters	Main (1962)
Diaptomus laciniatus	Physiologically different resting eggs	Hacher (1902)
D. stagnalis		Brewer (1964)
D. oregonensis	Reddish-brown resting eggs hatch after overwintering at 40°C for 4 weeks	Cooley (1971)
Limnocalanus macrurus		Roff (1972)
Marine Calanoida		
Pontella mediterranea	Winter resting eggs in the Black Sea; hatching of resting eggs only after exposure to low temperature for 1 or 2 months	Sazhina (1968)
P. meadi	In American waters	Grice and Gibson (1977)
Tortanus discaudatus	Resting eggs in the coastal Atlantic	M. Johnson (1967)
T. forcipatus	Subitaneous and resting eggs; 15°C stimulates hatching in resting eggs; in the Inland Sea, Japan	Kasahara *et al.*(1975b)
Centropages ponticus	Winter resting eggs in the Black Sea	Sazhina (1968)
C. hamatus	Resting eggs in the White Sea	Pertzova (1968)
C. abdominalis	Resting eggs hatch 10° and 15°C during winter-spring; in the Inland Sea, Japan	Uye *et al.*(1979)
C. yamada	Resting eggs hatch 10° and 15°C during summer-fall; in the Inland Sea, Japan; 15×10^5 resting eggs/m^2	Kasahara *et al.* (1975a)
Labidocera aestiva	Resting eggs hatch at 11°C in the Woods Hole Harbour	Grice and Gibson (1975)
L. bipinnata	Resting eggs hatch at 15°C; in the Inland Sea, Japan	Uye *et al.* (1979)
Calanopia thompsoni	Resting eggs hatch at 15°C; in the Inland Sea, Japan	Uye *et al.* (1979)
Acartia erythreae	Resting eggs hatch at 20°C; in the Inland Sea, Japan	Kasahara *et al.* (1974)
A. clausi	Resting eggs at extreme salinities; hatch at 20°C in the coastal Pacific	Uye and Fleminger (1976)
A. tonsa	Temperature-dependent resting eggs in the Atlantic	Zillioux and Gonzalez (1972)
A. tonsa	Resting eggs from the Pacific; over 50 per cent hatch at 14°C	Marcus (1990)

et al., 1975b). The resting eggs also permit the occurrence of a species in different latitudes (Grice and Gibson, 1975).

Temperature is an important factor affecting the development and hatching of resting eggs. Resting eggs of the warm-water species *Calanopia thompsoni* and *Labidocera bipinnata* are stimulated to hatch at 15°C, and of *Acartia erythreae* at 20°C; those of the cold-water species *A. clausi* hatch at 25°C and of *Centropages abdominalis* and *C. yamada* between 10 and 15°C in the Inland Sea, Japan (Uye *et al.*, 1979). The alternating cycles of abundance of *A. clausi* in cold seasons and of *A. tonsa* in warm seasons in the Atlantic coast of the United States are correlated to this temperature-dependent hatching of the resting eggs (Uye and Fleminger, 1976). Thus, the species-specific thermal requirements of the resting eggs may, at least in part, explain the succession of different copepod species in the plankton. Data on density of the eggs of marine calanoids and cladocerans are presented in Table 40.

C. Resting Cysts

It is well known that *Artemia salina* releases resting cysts that survive an obligate, intensive period of desiccation and successfully hatch, when water with optimum concentration of oxygen is available. Offspring of *A. salina* are released as nauplii and cysts during the same brooding period. Hentig (1971) reared *A. salina* at different temperature (10°, 15°, 20°, and 30°C) and salinity (5‰, 15‰ and 70‰) combinations and estimated the production of offspring. Reconsideration of his data (Table 42) permits the following interesting conclusions:

(1) In none of the combinations are either cysts or nauplii alone produced; at low salinity (5‰) as well as in certain combinations of low temperature and low salinity, neither egg nor cyst is produced.

(2) The lowest temperature-salinity combination at which *Artemia* is able to release offspring is 15°C at 32‰ and 20°C at 15‰.

(3) Production of offspring in terms of either number of eggs or number of broods increased with increasing temperature as well as salinity.

(4) With increasing salinity, cyst production decreased at 10°C but increased at 30°C; i.e. elevated temperature-brine concentration and lower temperature-salinity dilution represented possibly the extreme situations. Thus, the artemian strategy is to devote a certain percentage of offspring production as cysts and to enhance the percentage at either extreme salinity-temperature combination.

Working on other species and strains of *Artemia*, Browne (1980a) and Browne *et al.* (1988) confirmed the findings of Hentig (1971). All the eight strains belonging to the obligate dioecious outbreeding and parthenogenetic species utilize the combination of viviparous and encysted zygotic production, and viviparity is the more favoured mode of reproduction than encystment (Table 43). The strain CR is characterized by large-scale encystment early in the reproductive span (broods 1 to 6), SF in the middle life (broods 6 to 10), and MA very late in the life cycle (broods 11 to 15: see Browne, 1980a). The mean value for encystment for five strains obtained by Browne (1980a) and Browne *et al.* (1988) ranged from 6 to

Table 42

Cyst production in *Artemia salina* as functions of temperature and salinity (from Hentig, 1971; modified)

Temperature (°C)	Salinity (‰)	Interspawning interval (day)	No. of broods	Number of offspring			No. of cysts/female
				per female	per brood	per cent as cysts	
10	70	26	1.0	110	110		
15	32	11	2.8	74	207	22	45
	70	12	2.8	118	330	8	25
20	15	7	2.4	124	283	35	99
	32	7	3.0	132	396	31	123
	70	8	2.8	148	428	23	98
30	15	4	2.5	108	270	14	38
	32	4	3.6	130	470	29	136
	70	4	3.4	143	473	39	184

30 per cent; therefore the production of cyst appears to entail many disadvantages. Cysts have only a low hatch rate, ranging from 10 to 69 per cent (Hentig, 1971). Cyst production appears to be costlier than viviparous reproduction; the data obtained by Browne (1980a) and Browne *et al.* (1988) show that a typical cyst brood is only 74 per cent the size of viviparous brood; i.e. the cyst brood is smaller. Hentig (1971) determined that 22 per cent of the dry mass of the cyst was utilized for encapsulation, which may explain the smaller size of the encysted brood. When conditions are favourable, viviparously produced offspring will probably dominate the growth of future population. On the other hand, production of cysts would be selectively advantageous, if the environment tended towards instability. Thus, there is a trade-off between the advantages and disadvantages involved in the production of encysted and viviparous offspring.

D. Resting Adults

Ovarian diapause or resting stage is the condition in which oogonia do not enlarge and either the oostegite setae, as in Amphipoda (e.g. *Amphiporeia lawrenciana*: Downer and D.H. Steele, 1979), or the oostegites themselves, as in Isopoda (e.g. *Jaera ischiosetosa*: D.H. Steele and V.J. Steele, 1972b), are lost at the next moult and thus the production of young ones is effectively prevented during the resting stage. Inclusion of resting stage is the strategy adopted by Amphipoda and Isopoda (and possibly by other Peracarida too) to tide over the unfavourable season and to time the release of young ones with the onset of optimum conditions.

The onset as well as termination of the resting state is not determined directly by temperature but rather by photoperiod (e.g. *Gammarus setosus*: V.J. Steele *et al.*, 1977). The seasonal release of young ones during spring-summer in most

Table 43

Encystment of eggs of *Artemia parthenogenetica* and *A. franciscana* as a function of temperature

Strain	Egg encysted (per cent) Temperature (°C) 15	24	30	Reference
Obligate parthenogenetic, *A. parthenogenetica*				
Madras, India (MA)		13		Browne (1980a)
Kutch, India (KU)	0	24	30	Browne *et al.* (1988)
Cadz, Spain (CA)	34	28	45	Browne *et al.* (1988)
Salin de Giraud, France (GI)	6	30	18	Browne *et al.* (1988)
Obligate dioecious outbreeder, *A. franciscana*				
San Francisco, U.S.A. (SF)	0	28	0	Browne *et al.* (1988)
Cabo Rojo, Puerto Rico (CR)		22	0	Browne (1980a)
Megrine, Tunisia (TU)	85	99		Browne *et al.* (1988)
Larnaca Lake, Cyprus (CYP)	73	100	100	Browne *et al.* (1988)

northwestern Atlantic species of *Gammarus* is correlated with the appearance and growth of filamentous and leaf-like algae and abundance of organic matter in them (D.H. Steele and V.J. Steele, 1975c). Thus, the resting stage serves as an important strategy in regulation of the timing of release of young ones to coincide with optimum conditions. Obligate and long period of resting stage is characteristic of arctic forms such as *G. wilkitzki*, *G. setosus*, and *Gammarellus homari* (see D.H. Steele. 1972), which produce a few but large, slow-developing eggs (Table 44). *Gammarus oceanicus* and *G. finmarchicus* undergo resting stage during August to November. and produce medium-sized eggs in three to eight broods (Table 44). Yet others such as *G. tigrinus* and *G. lawrencianus* are small gammarids which produce large numbers of small eggs in about 10 broods. The egg size is correlated with the occurrence of the resting stage to ensure that the release of the young is well timed and placed in the optimum season.

Among the decapods, shrimps, lobsters, and crabs may enter resting stage to time the release of their young and, in some species, possibly to reduce intraspecific competition among the propagated larvae. The European shrimp *Crangon crangon* breeds from May to September, with peaks in June and August; the breeding season commences with an ecdysis and the appearance of egg-carrying setae on pleopods. The adult shrimp moults once every 30 days (at 12°C) during most of the year except from November to March. In sexually mature females, two kinds of intermoult periods occur: (1) egg-carrying and (2) resting (neuter) without oviposition (Lloyd and Yonge, 1947). The resting phase of the ovary, after the termination of the breeding season, is induced by Ovary (Gonad)-Inhibiting Hormone (GIH). This hormone was detected in the eyestalks during the period from November to April. It appears not only to halt vitellogenesis but also to cause the disappearance of

Table 44

Annual production of neonates by a female of some species of *Gammarus* and *Eulimnogammarus* (from D.H. Steele and V.J. Steele, 1975c; modified)

Species	Generations (No.)	Maximum broods (No.)	Resting stage	Production of neonates (No.)	Reference
G. wilkitzki	1	1	obligatory	170	D.H. Steele and V.J. Steele (1975a)
G. setosus	1	1	obligatory	85	V.J. Steele and D.H. Steele (1970)
G. oceanicus	1	3	Aug. to Nov.	270	V.J. Steele and D.H. Steele (1972)
Eulimnogammarus (Gammarus) obtusatus	1	3	Aug. to Nov.	50	D.H. Steele and V.J. Steele (1970a)
G. duebeni	1+	5	Aug. to Nov.	200	D.H. Steele and V.J. Steele (1969)
G. finmarchicus	1	8	Sept. to Oct.	240	D.H. Steele and V.J. Steele (1975b)
G. lawrencianus	2+	10	Sept. to Nov.	2,100	D.H. Steele and V.J. Steele (1970b)
G. tigrinus	2+	10	Sept. to Feb.	2,100	D.H. Steele and V.J. Steele (1972a)
G. stoerensis	2+	9	Jan. to May	450	D.H. Steele and V.J. Steele (1975a)

egg-carrying setae, thereby preventing egg carriage (see Sandifer and Smith, 1979) and incubation. Also in *Palaemon squilla*, cessation of vitellogenesis is accompanied by the disappearance of temporary secondary sex characters connected with incubation, namely the egg-bearing hairs (Callan, 1940). Thus, the release of larvae during the winter in these prawns is completely avoided.

In the Australian rock lobster *Panulirus longipes cygnus*, Chittleborough (1976) recognized two kinds of females based on their moulting cycle. After the breeding season, the females of the first kind (nine individuals) moulted in February-March to a resting neuter stage characterized by short pleopod setae (< 2 mm); they moulted again between June and October, the pleopods then having the long (> 10 mm) setae. The second kind of females (15 individuals) moulted only once between successive breeding seasons and remained setose at this moult, so that there was no resting phase through an intermoult period. It is very likely that the onset and termination of resting stage in *P. longipes cygnus* is also regulated by GIH. As the inclusion of resting stage is not synchronized in this species at the population level — as is the case with *Crangon crangon* — it appears that the release of GIH is induced by density-dependent factors. Those females which are nutritionally impoverished enter into resting stage.

The Indian shrimp *Macrobrachium nobilii* moults and breeds once every 19 days throughout the year. The two kinds of moult cycles described above were also observed in this species and a few females even continued the resting stage over three successive moults. At any time during the year, 39 per cent females have moulted to the resting stage (Pandian and Balasundaram, 1982). Maternal care involving the carriage of eggs which amount to 13 per cent of its body weight, continuous cleaning of the surface of eggs, and incessant pleopod-fanning appear to constitute an important component of energy expenditure in the reproductive energetics of *M. nobilii*. That only 19 per cent of the females relieved of their eggs in our experiments entered resting stage indicates that, if some extra energy were available (e.g. by saving the energy cost of incubation or possibly from food sources), a greater proportion of females would avoid entering the neuter stage (Pandian and Balasundaram, 1982). In *Gammarus finmarchicus*, the resting stage occurs during September-October at the population level, but a few females with non-setose oostegites and apparently in a resting stage have been observed thoughout the year (D.H. Steele and V.J. Steele, 1975b). It may be that there are two different types of resting stages, one in the entire population and the other only in a certain fraction of the population. The second type of resting stage, observed in the iteroparous shrimp *M. nobilii* and the prodigal rock lobster *Panulirus longipes cygnus*, seems to serve as an escape from the severe energy drain due to breeding (65 per cent of the converted energy: see Pandian and Balasundaram, 1982) in nutritionally improverished females, and to reduce intraspecific competition among the progenies that are propagated more or less simultaneously.

E. Diapause

In insects, diapause is characterized by a markedly lowered metabolic rate, as measured by oxygen uptake, and by a relatively enhanced ability to survive anaerobiosis; it is induced by photoperiod at some developmental stage. Diapause is a fascinating strategy successfully also employed by developmental stages of crustaceans to tide over unfavourable living conditions.

1. Diapausing post-embryonic stages

In many cyclopoid copepods, the copepodid instar or even adult secretes an organic cyst-like covering and remains inactive within under unfavourable conditions like drought and extreme cold. Such cysts, buried in mud, are particularly adapted to withstand desiccation and enable the copepod to aestivate during temporary drying of the pool and ponds in which it lives. They also provide a means of dispersal in the muddy feet or bodies of birds and other animals. Observations on the initiation, development, and termination of diapause in encysted larvae and adults of cyclopoid copepods have been recorded for *Diacyclops bicuspidatus* (see Birge and Juday, 1908), *Canthocamptus staphylinoides* (see Cole, 1953a), *C. staphylinus* (see Roen,

1957), *Cyclops strenuus strenuus* (see Elgmork, 1959, 1964), *C. leuckarti* (see Smyly, 1962), *C. vicinus* (see Einsle, 1967), *Cyclopoida* spp. (see Wierzbicka, 1962), and *Diacyclops navus* (see Watson and Smallman, 1971a).

Cole (1953b) found that encysted fourth copepodites of *Diacyclops bicuspidatus* were able to survive anaerobiosis and relatively high concentrations of potassium cyanide. He rightly postulated that the arrested development in *D. bicuspidatus* was similar to the 'diapause' which has been described for many insects. All stages of *D. navus* are obligate aerobes, even though during diapause, oxygen uptake may be so low as to be scarcely detected (Watson and Smallman, 1971b). Einsle (1967) described the combinations of day length and temperature which are required to initiate and terminate arrested development in *Cyclops vicinus* from temporary ponds.

Experimental studies undertaken by Watson and Smallman (1971a) have shown that *Diacyclops navus* is sensitive to photoperiodic stimulation only during the first copepodid instar; day length and temperature are the environmental cues which interact to cause an arrest in development at the fourth copepodid instar. For instance, at 15°C, the day length at which 50 per cent of *D. navus* enter arrest is just below 12 hr and is shifted to 13.5 hr at 10°C. This shifting of critical day lengths with lowering temperature is considered an evidence for the primary role of photoperiod and the modifying role of temperature as causative factors in the induction of arrest in *D. navus* and, possibly, in other cyclopoids too. When placed at 25°C under 16 hr day lengths with adequate food supply, the diapause is terminated in about 10 days and 100 per cent of *D. navus* complete development within 40 days (Watson and Smallman, 1971a). Our present knowledge on photoperiodic control of diapause in cyclopoids has been gathered from only American and European species; in view of the fact that the timings of the inductive and initiative stages are species-specific responses, information on the tropical cyclopoids is urgently wanted.

2. Diapausing embryonic stage

An alternate strategy, adopted by certain decapods, is to carry diapausing eggs and to enhance larval survival by broadcasting them, when food is abundantly available. The presence of developing eggs on the pleopods inhibits moulting (e.g. *Palaemon serratus*: Panouse, 1947; *Paratelphusa hydrodromous*: Kurup and R.G. Adiyodi, 1987) and spawning (e.g. *Clibanarius* spp: Ameyaw-Akumfi, 1975). Hence, the presence of diapausing eggs on the pleopods for a period effectively eliminates the subsequent spawning and moulting, and thus aids the semelparous decapods to accumulate necessary reserve energy.

At temperatures between 11° and 15°C, the eggs of some decapods cease to develop beyond the gastrula stage, which is achieved in three to four days after spawning. Subsequently, the eggs undergo diapause, while being carried by the females: 16 weeks in *Hyas coarcticus*, 14 weeks in *Corystes cassivelaunus*, eight weeks in *Cancer pagurus*, and six weeks in *Maja squinado* (see Wear, 1974). The diapause of these species is not broken by simply raising the temperature but it is possible to do this with short photoperiod. In *H. coarcticus* and *Corystes cassivelaunus*, the incubation lasts for a period of 10 months, which is the longest known for Brachyura (Hartnoll, 1963, 1972a, b). In these two species, eggs berried in May-July rest in diapause until October-Novermber and develop at a slow rate during the winter, and the larvae are broadcast in early spring (March and April), when planktonic food is abundant. It is tempting to suggest that some decapods have evolved diapause of incubating eggs as a strategy to enhance the chances of larval survival (Wear, 1974) and to become more and more semelparous.

In summary, development in normal, subitaneous eggs of Crustacea is rapid, and hatching takes place while the eggs are being carried. Production of dormant eggs or cysts is stimulated by a variety of external factors such as population density, temperature, photoperiod, and salinity. Inclusion of a resting stage during the adult life span and of diapause in the incubated eggs may serve to time the release of young ones to ensure maximum larval survival. In other words, the eggs are prepared for rapid hatching, when the precise environmental conditions that are required are present. The controlling factors—oxygen, salinity, temperature, and/or photoperiod—that break dormancy vary with species and are related to the type of habitat for which each species is adapted.

F. Biochemical Aspects of Cryptobiosis

Artemia salina is a biological oddity in every sense of the word. It is one of the few crustaceans which have attracted extensive investigations by reproductive biologists, cytogeneticists, biochemists, and ecologists. Some races of this species are parthenogenetic, others disexual; thin-shelled subitaneous eggs and/or thick-shelled dormant cysts are produced (Lochead, 1961; Hentig, 1971) through parthenogenetic and sexual reproduction; the thin-shelled eggs are retained in the brood pouch and give rise to nauplii after a few days. *Artemia salina* has a diploid amphigonic race and different parthenogenetic races, which are diploid with 2n = 42 chromosomes, triploid, tetraploid (Artom, 1931), pentaploid, or octaploid (Barigozzi, 1958) (see also Table 45). This species thrives in salt lakes and survives a wide range of temperature (10° to 35°C) and salinity (5 to 70‰). Perhaps it is the only animal species to occupy osmotically the most hostile environment, the salt lakes. *Artemia salina* has been chosen here to summarize the physiological and biochemical aspects of cyst dormancy for (1) the reasons described above, (2) availability of extensive literature (which, in turn, is due to the availability of the cysts in such large quantities), and (3) occurrence of diguanosine tetraphosphate in the resting eggs of *Daphnia magna* (see Oikawa and Smith, 1966) and *Eubranchipus vernalis* (Warner, unpublished data) and of glycerol in the eggs of *Chirocephalus diaphanus* (see Hall and MacDonald, 1975), which were hitherto considered unique to *Artemia*. This suggests the possibility that one or more of the artemian adaptive strategies is shared by the crustaceans during their dormancy.

Covered by a complex shell (J.E. Morris and Afzelius, 1967; Anderson *et al.*, 1970), encysted *Artemia* enter a period of obligate dormancy and arrested development at the gastrula stage (Dutrieu, 1960; Benesch, 1969). The dormant embryos, composed of a partial syncytium of about 4,000 nuclei (Table 45; Nakanishi *et al.*, 1962), are released from the ovisac into the environment, in which they undergo dehydration. The abilities of the dried cysts are impressive: (1) they can withstand almost total desiccation with only traces of 'residual water' remaining (0.7 μg/g dried cysts: Clegg *et al.*, 1978) when the cysts are placed over strong desiccants or under reduced pressure (Whitaker, 1940; Clegg, 1967, 1974, 1978a, b); (2) they resist temperatures exceeding 100°C for over an hour (Hinton, 1954) and can be exposed to near absolute zero, apparently for an indefinite period (Skoultchi and Morowitz, 1964); (3) they exhibit considerable resistance to bombardment by various forms of high-energy radiation (Snipes and Gordy, 1963; Iwasaki, 1965, 1973; Iwasaki *et al.*, 1971); (4) little, if any, mortality is observed in cysts that have been embedded within crystals of NaCl, a result of their ability to withstand distortion and mechanical damage (Clegg, 1974); and (5) the durable nature of the shell is indicated by the fact that the cysts can be soaked for a month or more in acetone (J.E. Morris, 1968), a wide variety of alcohols, and other organic solvents (Ewing, 1968) with little or no decrease in viability. Although hydrated cysts are less tolerant of these environmental hazards, they can survive exposure to anoxia for periods exceeding five months, an ability not frequently observed in the Metazoa

Table 45

Number of nuclei per cyst in various populations of *Artemia salina* (from Olson and Clegg, 1976; modified)

Locality	Ploidy	Mode of reproduction	Year of collection	No. of nuclei ($\bar{X} \pm$ SD)
San Francisco (U.S.A.)	2N	Disexual	1938	3,164 ± 239
			1951	3,462 ± 182
			1961	3,086 ± 356
			1965	2,959 ± 192
				4,004 ± 192
Great Salt Lake (U.S.A.)	2N	Disexual	1951	3,543 ± 192
			1967	3,900 ± 150
Inagua (Bahamas)	2N	Disexual	1969	3,534 ± 134
Aio-machi (Japan)		Parthenogenetic	1968	3,769 ± 171
Jamnagar (India)	3N	Parthenogenetic	1968	3,221 ± 156
Comacchio (Italy)	4N	Parthenogenetic	1960	3,468 ± 209
Sete (France)	2N	Parthenogenetic	1962	3,415 ± 243

(Dutrieu and Christia-Blanchine, 1966). Almost all the hydrated cysts retain their ability to be redesiccated until emergence of the embryo occurs (J.E. Morris, 1971). The following are some of the important components that are considered to have endowed *Artemia* cysts with the impressive abilities to survive the magnitudes of environmental insults.

1. Glycerol

The presence of glycerol in substantial amounts (2–6 per cent of the dry weight) in the cysts of *Artemia* but in no other stage of the life cycle (Clegg, 1962, 1965) is of potential significance. When an *Artemia* cyst undergoes dehydration, the glycerol content per cyst remains the same, but its concentration increases enormously. Based on its physical and chemical properties, glycerol appears to alleviate many of the dehydration problems. These properties may be listed as follows: (1) its relatively high electric constant (of 43, about half that of water) has considerable solvent properties for macromolecules and inorganic electrolytes; (2) its ability to stabilize a variety of proteins and to prevent their aggregation and precipitation protects many enzymes and other proteins against denaturation due to factors such as temperature extremes, pH ultra-treatment; (3) its least compressibility (exhibits only 10 per cent decrease in volume at 4,000 lbs/in^2) could bear the enormous stress generated in structures whose dimensions are at the Ångstrom-level, by removal of water. These and the other non-toxic, low volatility, anti-oxidant, and radio-protectant properties suggest a potential adaptive role for glycerol in desiccation-resistant cells; (4) by virtue of its structure, it might replace the hydration lattice of certain proteins of low-water activities (D.T. Warner, 1969), preventing their aggregation and/or inactivation activities. If such a possibility could be extended to other molecular species, glycerol may play the role of water substitute in the 'glycerolated' cysts (see Crowe, 1971; Crowe and Madin, 1974).

2. Trehalose

Trehalose occurs in very small amounts in a few crustaceans (Fairbrain, 1958; Telford, 1968). Barring the encysted embryo, which contains as much as 17 per cent of its dry weight as trehalose (Dutreiu, 1960; Clegg, 1962), it does not occur in any other stage in the life cycle of *Artemia*. Clegg (1965) demonstrated the absence of trehalose in the haemolymph of female *Artemia* fed on yeast, which contained about 12 per cent of its dry body weight as trehalose. The results, presented in Table 46, show that trehalose is synthesized and accumulated in the cells of only those embryos that are destined to enter dormancy. It is synthesized largely at the expense of stored glycogen, but is rapidly metabolized, when dormancy is terminated, being used for respiration and for the synthesis of glycogen (see Table 46; see also Clegg, 1964). Thus, trehalose-glycogen interconversion in the dormant egg should involve considerable expenditure of energy; but it suggests that the storage of carbon and energy in the form of trehalose is of some adaptive significance. Reducing sugars, especially glucose, interact with the amino groups of free amino acids and the side chains of proteins to form a variety of insoluble 'melanoidans' (Browning reaction). As this reaction is to occur particularly at very low water concentrations, there is the danger of significant amounts of reducing sugar becoming insoluble in the dried cells of the cyst. By virtue of its non-reducing character, owing to the stable 1–1 linkage, trehalose holds the highest possibility of minimizing Browning reactions, which can destroy the biological activity of proteins in the cells. Besides, trehalose is not susceptible to aminolyses, as is the case with other non-reducing sugars (see El-Nockrashy and Frampton, 1967). Taken together, these two properties of trehalose suggest that a large amount of substrate for energy metabolism and biosynthesis could be stored in the dried embryo with greater stability and integrity. Incidentally, appreciable amounts of trehalose have been found in the desiccation-resistant structure of several invertebrates (Fairbrain, 1958) and other lower organisms (Sussman and Halvorson, 1966). Hence, an urgent search for trehalose in dormant stages of other crustaceans is recommended.

3. Diguanosine tetraphosphate

Finamore and Warner (1963) reported the occurrence of p^1, p^4-diguaniosine 5′-tetraphosphate (diGDP), which is synthesized by the developing ovarian eggs (Warner and Finamore, 1965; Warner and McClean, 1968; Warner *et al.*, 1972)

Table 46

Composition of embryos and nauplii produced by *Artemia salina* (disexual, diploid, California strain) under laboratory and natural conditions (Clegg, 1965; modified)

Stage	Mean concentration (μg/1,000 embryos or nauplii)		
	Glycogen	Trehalose	Glycerol
From laboratory population			
Presumptive non-dormant			
(i) 10 hr embryos	442	9	20
(ii) Newly released nauplii	402	6	92
Presumptive dormant			
(iii) 10 hr embryos	336	80	36
(iv) Newly shed dormant embryos	92	334	139
From field population			
(v) Dormant embryos	28	375	123
(vi) Newly emerged nauplii	374	6	104

and is found in the encysted embryos and in stages leading to the newly hatched nauplius (Warner and Finamore, 1967; McClean and Warner, 1971). Several proposals have been advanced for the functions of diGDP, but only one adaptive significance is considered here relevant. In its natural habitats, the cysts are often found amidst decaying debris in wind-rows along the shores and hence undergo successive periods of hydration due to humidity and rain, followed by redesiccation, rehydration, etc. Clearly, they must have devices to prevent exhaustion of their endogenous supplies and premature emergence. Anaerobic conditions bring carbohydrate metabolism to a standstill (Dutrieu and Christia-Blanchine, 1966; Ewing, 1968; Ewing and Clegg, 1969), the opposite of the usual response of Metazoa to anaerobiosis. However, diGDP becomes the major energy source, when conventional carbohydrate-based energy metabolism is stopped during anaerobiosis (Stocco *et al.*, 1972).

4. DNA

Data presented in Table 45 indicate that, within a given population of *Artemia*, dormancy is initiated at essentially the same developmental stage with respect to the number of nuclei present. The relative constancy of nuclei at the time of embryonic dormancy in different populations that reproduce parthenogenetically or disexually as well as in those that are separated by thousands of miles indicates that the onset of dormancy is rather rigidly programmed into the development. It is acquired and maintained in each population, independent of geographic and environmental conditions.

Development of the encysted gastrula into prenauplius (nauplius I) occurs in the absence of cell division (Nakanishi *et al.*, 1962). Following emergence and during further development up to nauplius II, an increase of 25 per cent in the number of cells occurs. However, when the freshly hatched larva is exposed to FUdR (10 μg/ml), cell division and hence DNA synthesis (see Warner and McClean, 1968) are largely inhibited (see Finamore and Clegg, 1969), but the observable development including moulting nevertheless proceeds normally (Olson and Clegg, 1978). Evidently, the processes of morphogenesis and cell differentiation involved in development of the gastrula into nauplius II can occur in the absence of cell division and DNA synthesis (see Benesch, 1969). Once again, this exceptional behaviour shown by *Artemia salina* can perhaps be accounted for by the problems this species faces in nature, such as the exposure of hydrated cysts to anoxia and desiccation at any point up to the emergence from the shell. Desiccation itself is known to be mutagenic in prokaryotic cells (Zamenholf *et al.*, 1968). Hence, it is not surprising that sensitive processes, like gene replication and cell division, are shut down during this critical period.

5. Hydration

One of the major characteristics of living systems is the presence of large amounts of intracellular water. The induction of dormancy in *Artemia* involves dehydration either by osmotic withdrawal or by evaporation (Clegg, 1974). Though dehydration is an essentially complete process, a small amount of tightly bound residual water remains in the cyst (Clegg *et al.*, 1978). The cyst resumes metabolism and development when rehydrated (Clegg, 1978a); the process of rehydration is intimately related to the reduction in glycerol concentration (Table 47). As rehydration proceeds, a variety of biochemical activities are initiated at about 0.3 g water/g cyst (Table 48); further increase in water content up to 0.6 g/g cyst does not result in the initiation of any additional metabolic activity. However, at 0.65 g/g cyst level, a host of metabolic events begin and developmental processes are also resumed.

G. Breeding Aggregation

A population is considered aggregated when individuals are not randomly distributed thoughout the area or volume in which they are found. A breeding aggregation is a grouping together for the purpose of mating (Clutter, 1969). This behavioural strategy provides the breeders a wide choice for selection of their partners. Some species of the marine cypridinid ostracod *Philomedes* possess benthic females and planktonic males. There is a period of aggregation, during which the females swim to join the males. Following copulation, the females sink to the bottom and cut off their swimming setae (see R.D. Barnes, 1974). Among the Mysidacea, breeding aggregation occurs during spring (e.g. *Gastrosaccus normani, Erythrops elegans, Hemimysis lamornae*), summer (e.g. *H. lamornae, Paramysis*

Table 47

Glycerol-water relationships as a function of hydration in *Artemia* cysts (from Clegg, 1978a; modified)

Cyst hydration (g water/g dry cysts)	Cyst water in cells		Weight of glycerol*
	(g)	(per cent)	(per cent)
0.05	0.039	77	98
0.10	0.079	79	47
0.20	0.166	83	22
0.30	0.255	85	15
0.40	0.352	88	11
0.50	0.450	90	8
0.60	0.546	91	7
0.70	0.644	92	6

*The glycerol content of these cysts is 3.72 per cent of the dry weight and the location is exclusively in the cells.

Table 48

Hydration-dependence of cellular metabolism in *Artemia* cysts (from Clegg, 1978a; modified)

Cyst hydration (g water/g cyst)	Metabolic events initiated
0–0.10	None observed
0.10	Decrease in ATP concentration
0.11–0.30	No additional event
0.31	Metabolism involving several amino acids
	Krebs cycle and related intermediates
	Short-chain aliphatic acids
	Pyrimidine nucleotides
	Slight decrease in glycogen concentration
0.31–0.60	No additional event
0.61	Cellular respiration
	Carbohydrate synthesis
	Mobilization of trehalose
	Net increase in ATP
	Major changes in amino acid pool
	Hydrolysis of yolk protein
	RNA and protein synthesis
	Resumption of embryonic development
0.61–1.40	No additional event

arenosa, Schistomysis ornata), or winter (e.g. *Anchialina agilis, Pseudomma affine, Mysidopsis didelphys*). The total number of individuals aggregated in these species ranges from 114 in *P. affine* to 3,122 in *Paramysis arenosa* (see Mauchline, 1971b). The occurrence of breeding aggregation of euphausiids has been known for some time (Mauchline and Fisher, 1969). Populations of *Meganyctiphanes norvegica* begin to form breeding aggregations in late autumn, the highest population densities occurring at 150 m depth or more (Mauchline, 1960). The aggregation behaviour of the 'deep species' of mysids appears to be closely similar to that of the euphausiids. Negative photoperiodism of individuals is suggested as a mechanism initiating aggregation of species at the bottom of deep-water basins (Mauchline, 1971a). Reaction of one individual to another or group of individuals is considered necessary to intensify the initial degree of aggregation and to maintain the integrity of breeding aggregations once formed.

H. Breeding Migration

For the egg-shedding euphausiids (e.g. *Thysanoëssa longicaudata*: Williams and Lindley, 1982) and calanoid copepods (e.g. *Calanus helgolandicus*: Williams *et al.*, 1987), ontogenetic vertical migration, i.e. developmental ascend, has been described. Decapods carry their eggs attached to the inner branches of the abdominal appendages until hatching occurs. The egg attachment, following spawning, may require a layer of soft sediment and in certain species this may involve spawning migration to the continental shelf. Broekhuysen (1936) observed that the portunids *Carcinides maenas* held in bare aquaria fail to attach their eggs after spawning; normal attachment occurs only when the crabs are provided with a thick layer of sand. Observation of Norse (unpublished) on several Jamaican *Callinectes* spp. held in bare and sand-layered aquaria also confirmed the need for the soft sediments for attachment of eggs, for the egg mass of the portunids are huge, often forcing their abdomen 90° or more from the sternum, and exceed the boundaries of sternum, abdomen, and pleopodal exopodites. Extrapolated, these observations may explain the unusual discrepancy in sex ratios of oceanic and shelf samples of the portunid crab *Euphylax dovii*. The crab, so highly adapted for life in the pelagic realm, has not evolved a mechanism of egg attachment independent of the substratum. Therefore, the females must undertake extensive migration until they reach waters sufficiently shallow for spawning in the soft sediments. Ovigerous females of this pelagic species remain near the bottom in shallow waters because their large egg masses may increase drag, temporarily raising the energetic cost of staying in the water column. Second, hatching in coastal waters may provide larvae with a richer food supply than can be obtained from the shore (Norse and Fox-Norse, 1977).

There are over 100 species belonging to the genus *Macrobrachium* (see Williamson, 1972), some of which are important from the point of aquaculture. An essential component in *Macrobrachium* breeding is the obligate necessity of exposing the non-feeding first instar zoea to dilute sea water; only on such exposure does it undertake the first moult and commence feeding. Hence, breeding in riverine and marine *Macrobrachium* involves migration to brackish water. The most critical first-instar duration is understandably the longest (> eight days) in the riverine shrimps such as *M. nobilii* and *M. malcolmsonii* (Table 49), whose larvae are passively carried to the estuarine waters; the duration is about five days in *M. rosenbergii*, the mature females of which migrate to the estuarine waters to hatch the larvae; it is the shortest in *M. carcinus* (three days), *M. acanthurus* and *M. idella* (two days), which inhabit fresh and brackish waters, and in *M. novaehollandiae* (four days), which inhabits brackish water throughout life. *Macrobrachium intermedium* is the only marine species which no longer visits the brackish water for spawning; likewise, land-locked freshwater species such as *M. choprai, M. lanchesteri*, and *M. idae* have completely cut off their connection to the brackish water; neither larvae nor adult females are migratory. Reserve yolk energy available to the first-instar zoea of *M. nobilii* is equivalent to 36 per cent of its body-energy content (Balasundaram, 1980); the corresponding value is only 8 per cent for *Homarus americanus* (see Pandian, 1970b). The extended first-instar zoeal duration and the greater reserve yolk energy go hand in hand to mitigate the ill effects of the possible interception, when larvae of *M. nobilii* and *M. malcolmsonii* are passively carried by the river to the estuarine waters. Indeed, their passage is intercepted by several man-made dams, reservoirs, and barrages. Pandian (1980) has summarized some of the ill effects suffered by migratory fishes and prawns at the intercepting dams.

Table 49

Duration of first zoeal stage (day) and migratory habit of some *Macrobrachium* spp.

Species	Duration of the first zoeal stage (day)	Remarks	Reference
M. choprai	?	Completes entire life cycle in fresh water	Raman (1976)
M. lanchesteri	?	Completes entire life cycle in fresh water	Raman (1976)
M. idae	3	Land-locked species	Pandian and Katre (1972)
M. intermedium	3	Completes entire life cycle in sea water	Williamson (1972)
M. idella	2	Brackish and freshwater species	Pillai and Mohammed (1973)
M. novaehollandiae	4	Brackishwater species	Fielder (1976)
M. carcinus	3	Adult in fresh and brackish waters	Choudhury (1971)
M. acanthurus	5	Adult in fresh and brackish waters	Choudhury (1970)
M. rosenbergii	5	Freshwater species; adults migrate to brackish water	George (1969)
M. nobilii	5	Freshwater species; larvae migrate to brackish water	Balasundaram (1980)
M. malcolmsonii	8	Freshwater species; larvae migrate to brackish water	Anonymous (1973), Kewalramani (1973), Ibrahim (1962)

?, Duration not recorded.

ACKNOWLEDGEMENTS

It is with great pleasure I acknowledge Dr. H.P. Bulnheim's (Hamburg) and Dr. T. Subramoniam's (Madras) critical reading of the manuscript, and the excellent cooperation of my collaborators Dr. C. Balasundaram (Trichy) and Dr. S.S.S. Sarma (Madurai) in preparation of the manuscript. I thank Dr. D.J. Wildish for his useful comments on an early draft of this chapter. Travel assistance to Davis (U.S.A.) and Bangkok (Thailand), extended by International Foundation for Science (Stockholm), and the National Fellowship awarded by the University Grants Commission (New Delhi) are also gratefully acknowledged.

REFERENCES

Adiyodi, K.G., and Adiyodi, R.G. (1970). 'Endocrine control of reproduction in decapod Crustacea', *Biol. Rev.*, **45**, 121–165.

Adiyodi, K.G., and Adiyodi, R.G. (1974). 'Comparative physiology of reproduction in arthropods', *Adv. comp. Physiol. Biochem.*, **5**, 37–107.

Adiyodi, K.G., and Adiyodi, R.G. (1985). 'Reproduction vs growth: Endocrine programming in the Brachyura', in *Current Trends in Comparative Endocrinology* (Eds. B. Loffer and W.N. Holmes), Hong Kong Univ. Press, Hong Kong, pp. 313–315.

Adiyodi, R.G. (1985). 'Reproduction and its control', in *Biology of Crustacea* (Eds. L.H. Mantel and D.E. Bliss), Vol. 9, Academic Press, New York, pp. 147–215.

Adiyodi, R.G. (1988). 'Reproduction and development', in *Biology of the Land Crabs* (Eds. W.W. Burggren and B.R. McMahon), Cambridge University Press, New York, pp. 139–185.

Ahl, J.S.B. (1991). 'Factor affecting contributions of the tadpole shrimp *Lepidurus packardi* to its over summering egg reserves', *Hydrobiologia*, **212**, 137–143.

Ahmed, M., and Mustaquim, J. (1974). 'Population structure of four species of porcellanid crabs (Decapoda: Anomura) occurring on the coast of Karachi', *Mar. Biol.*, **26**, 173–182.

Al-Adhub, A.H.Y., and Bowers, A.B. (1977). 'Growth and breeding of *Dichelopandalus bonnieri* in Isle of Man waters', *J. mar. biol. Ass. U.K.*, **57**, 229–238.

Allen, M.B. (1960). 'Utilization of thermal energy by living organisms', in *Comparative Biochemistry* (Eds. M. Florkin and R.S. Mason), Academic Press, New York, pp. 487–514.

Allen, W.V. (1972). 'Lipid transport in the Dungeness crab *Cancer magister* Dana', *Comp. Biochem. Physiol.*, **43**, 193–207.

Altes, J. (1962). 'Sûr quelques parasites et hyperparasites de *Clibanarius erythropus* (Latreille) en Corse', *Bull. Soc. Zool. Fr.*, **87**, 88–97.

Ambler, J.W. (1985). 'Seasonal factors affecting egg production and viability of eggs of *Acartia tonsa* Dana from East Lagoon, Galveston, Texas', *Est. Coast. Shelf. Sci.*, **20**, 743–760.

Ameyaw-Akumfi, C. (1975). 'The breeding biology of two sympatric species of tropical intertidal hermit crabs *Clibanarius chapini* and *C. senegalensis*', *Mar. Biol.*, **29**, 15–28.

Anderson, B.G., Lochhead, M.A., and Huebner, E. (1970). 'The origin and structure of the tertiary envelope in thick-shelled eggs of the brine shrimp, *Artemia*', *J. Ultrastr. Res.*, **32**, 497–525.

Anderson, J.F. (1978). 'Energy content of spider eggs', *Oecologia*, **37**, 41–57.

Anderson, G. (1974). 'Quantitative analysis of the major components of energy flow for the grass shrimp *Palaeomonetes pugio* with special reference to the effects of the parasite *Probopyrus pandalicola*', Ph.D. thesis, University of South Carolina, Columbia, 124 pp.

Anderson, G. (1977). 'The effects of parasitism on energy flow through laboratory shrimp populations', *Mar. Biol.*, **42**, 235–252.

Anonymous (1973). 'Studies on the breeding, rearing and culture of the freshwater prawn *Macrobrachium malcolmsonii* H. Milne-Edwards', *16th Meeting of the State Fish. Res. Council, Dept. of Fish., Govt. of Tamil Nadu*, **1**, 8–9.

Ansell, A.B., and Robb, L. (1977). 'The spiny lobster *Panulirus elephas* in Scottish waters', *Mar. Biol.*, **43**, 63–70.

Artom, C. (1931). 'L'origine eel evoluzione della partenogenesi attraversci differenti biotipi di una specie collecttiva (*Artemia salina* L.)', *Mem. Acad. Italia*, **2**, 215–238.

Bacci, C. (1965). *Sex Determination*, Pergamon Press, Oxford.

Badawi, H.K. (1975). 'On maturation and spawning in some penaeid prawns of the Arabian Gulf', *Mar. Biol.*, **52**, 1–6.

Balasundaram, C. (1980). 'Ecophysiological studies in prawn culture (*Macrobrachium nobilii*)', Ph.D. thesis, Madurai Kamaraj University, Madurai.

Balasundaram, C., and Pandian, T.J. (1981). '*In vitro* culture of *Macrobrachium* eggs', *Hydrobiologia*, **77**, 203–208.

Balasundaram, C., and Pandian, T.J. (1982a). 'Egg loss during incubation in *Macrobrachium nobilii* (Henderson & Mathai)', *J. exp. mar. Biol. Ecol.*, **59**, 299–331.

Balasundaram, C., and Pandian, T.J. (1982b). 'Yolk energy utilization in *Macrobrachium nobilii* (Hensen and Mathai)' *J. exp. mar. Biol. Ecol.*, **61**, 125–131.

Baldwin, E. (1964). *An Introduction to Comparative Biochemistry*, University Press, Cambridge.

Balss, H. (1955). 'Decapoda. Okologie', *Bronn's Kl. Ordn. Tierreichs*, Bd. 5, Abt. 1, Bch. 7, Akademische Verlagsgesellschaft, Leipzig, pp. 1285–1476.

Bamstedt, U. (1979). 'Reproductive bioenergetics within the summer and winter generations of *Euchaeta norvegica* (Copepoda)', *Mar. Biol.*, **54**, 135–142.

Banta, A.M. (1939). 'Studies on the physiology, genetics and evolution of some Cladocera', *Publ. Carnegie Inst., Washington*, **513**, 1–285.

Barange, M., and Stuart, V. (1991). 'Distribution patterns, abundance and population dynamics of the euphausiids *Nyctiphanes capensis* and *Euphausia hanseni* in the northern Benguela upwelling system', *Mar. Biol.*, **109**, 93–101.

Barigozzi, C. (1958). 'Différenciation des génotypes et distribution géographique d'*Artemia salina* Leach', *Ann. Biol.*, **33**, 241–250.

Barnes, H. (1962). 'The composition of the seminal plasma of *Balanus balanus*', *J. Exp. Biol.*, **39**, 345–351.

Barnes, H. (1963). 'Organic constituents of the seminal plasma of *Balanus balanoides*', *J. Exp. Biol.*, **40**, 587–594.

Barnes, H. (1965). 'Studies in the biochemistry of cirripede eggs. 1. Changes in the general biochemical composition during development of *Balanus balanoides* (L.) and *B. balanus* da Costa', *J. mar. biol. Ass. U.K.*, **45**, 321–339.

Barnes, H., and Barnes, M. (1956). 'The formation of the egg mass in *Balanus balanoides* (L.)', *Arch. Soc. Vanamo.*, **11**, 11–16.

Barnes, H., and Barnes, M. (1967). 'The effect of starvation and feeding on the time of production of egg masses in the Boreo-Arctic cirripede *Balanus balanoides* (L.)', *J. exp. mar. Biol. Ecol.*, **1**, 1–121.

Barnes, H., and Barnes, M. (1968). 'Egg numbers, metabolic efficiency of egg production and fecundity; local and regional variations in a number of common cirripedes', *J. exp. mar. Biol. Ecol.*, **2**, 135–153.

Barnes, H., Barnes, M., and Finlayson, D.M. (1963). 'The seasonal changes in body weight, biochemical composition, and oxygen uptake of two common Boreo-Arctic cirrepedes *Balanus balanoides* and *B. balanus*', *J. mar. biol. Ass. U.K.*, **43**, 185–211.

Barnes, H., and Evens, H. (1967). 'Studies in the biochemistry of cirripede eggs. 3. Changes in the aminoacid composition during development of *Balanus balanoides* and *B. balanus*', *J. mar. biol. Ass. U.K.*, **47**, 171–180.

Barnes, H., and Klepal, W. (1971). 'The structure of the pedicel of the penis in cirripedes and its relation to other taxonomic characters', *J. exp. mar. Biol. Ecol.*, **7**, 71–94.

Barnes, H., Klepal, W., and Munn, E.A. (1971). 'Observations on the form and changes in the accessory droplet and motility of the spermatozoa of some cirripedes', *J. exp. mar. Biol. Ecol.*, **17**, 173–196.

Barnes, H., and Powel, H.P. (1953). 'The growth of *Balanus balanoides* and *B. crenatus* under varying conditions by submersion', *J. mar. biol. Ass. U.K.*, **32**, 107–128.

Barnes, R.D. (1974). *Invertebrate Zoology*, W.B. Saunders Company, Philadelphia.

Barnes, W.B., and Wenner, A.M. (1968). 'Seasonal variation in the sand crab *Emerita analoga* (Decapoda: Hippidae) in the Santa Barbara area of California', *Limnol. Oceanogr.*, **13**, 465–475.

Battaglia, B. (1957). 'Recherche sul ofolo biologico di *Tisbe gracilis* (T. Scott). (Copepoda, Harpacticoida), studiato in condizioni di laboratorie', *Arch. Oceanogr. Limnol.*, **11**, 29–46.

Battaglia, B., and Volkmann-Rocco, B. (1973). 'Geographic and reproductive isolation in the marine harpacticoid copepod *Tisbe*', *Mar. Biol.*, **19**, 156–160.

Bauer, R.T. (1975). 'Grooming behaviour and morphology of the caridean shrimp *Pandalus danae* Stimpson (Decapoda: Natantia: Pandalidae)', *J. Linn. Soc. Zool.*, **56**, 45–71.

Bauer, R.T. (1977). 'Antifouling adaptations of marine shrimp (Crustacea, Decapoda, Caridea): Functional morphology and adaptive significance of antennular preening by the third maxillipeds', *Mar. Biol.*, **40**, 261–276.

Bauer, R.T. (1978). 'Antifouling adaptations of caridean shrimps: Cleaning of the antennal flagellum and general body grooming', *Mar. Biol.*, **49**, 69–82.

Bauer, R.T. (1979). 'Antifouling adaptations of marine shrimp (Decapoda: Caridea): Gill cleaning mechanisms and grooming of brooded embryos', *J. Linn. Soc. Zool.*, **65**, 281–303.

Bauer, R.T. (1981). 'Grooming behaviour and morphology in the decapod Crustacea', *J. Crust. Biol.*, **1**, 153–173.

Baylis, H.A. (1949). '*Fecampia spiralis*, a cocoon-forming parasite of the Antarctic isopod *Serolis schytiei*', *Proc. Linn. Soc. Lond.*, **161**, 64–71.

Beck, J.T. (1977). 'Reproduction of the estuarine mysid *Taphromysis bowmani* (Crustacea: Malacostraca) in fresh water', *Mar. Biol.*, **42**, 253–257.

Benesch, R. (1969). 'Auf Ontogenie und Morphologie von *Artemia salina* L.', *Zool. Jb. (Abt) Anat.*, **86**, 307–458.

Bennett, D.B. (1974). 'Growth of the edible crab (*Cancer pagurus*) off south-west England', *J. mar. biol. Ass. U.K.*, **54**, 803–823.

Bernice, R. (1970). 'Studies on the biology of *Streptocephalus*', Ph.D. thesis, University of Madras.

Berry, P.F. (1969). 'The biology of *Nephrops andamanicus* Woodmason (Decapoda: Reptantia)', *Invest. Rep. Oceanogr. Res. Inst. Durban*, **22**, 1–55.

Berry, P.F. (1971). 'The biology of spiny lobster *Panulirus homarus* (Linnaeus) off the coast of southern Africa', *Invest. Rep. Oceanogr. Res. Inst. Durban*, **28**, 1–75.

Berry, P.F., and Heydorn, A.F.F. (1970). 'A comparison of the spermatophoric masses and mechanisms of fertilization in southern African spiny lobsters (Panuliridae)', *Invest. Rep. Oceanogr, Res. Inst. Durban*, **25**, 1–18.

Birge, E.A., and Juday, C. (1908). 'A summer resting stage in the development of *Cyclops bicuspidatus* Claus', *Trans. Wis. Acad. Sci. Arts Lett.*, **23**, 1–9.

Blades, P.I. (1977). 'Mating behaviour of *Centropages typicus* (Copepoda: Calanoida)', *Mar. Biol.*, **40**, 57–64.

Bliss, D.E. (1968). 'Transition from water to land in decapod crustaceans', *Am. Zool.*, **8**, 355–392.

Bocquet-Vedrine, J., and Parent, J. (1973). 'Le parasitisms multiple du Cirripede operculé *Balanus improvisus* Darwin par le Rhizocephale *Boschmaella balani* (J. Bocquet-Vedrine)', *Arch. Zool. exp. gén.*, **113**, 239–244.

Born, J.W. (1967). '*Palaemonetes vulgaris* (Crustacea: Decapoda) as host for the juvenile stage of *Nectonema gracile* (Nematomorpha)', *J. Parasitol.*, **53**, 793–794.

Borowsky, B. (1983). 'Placement of eggs in their brood pouches by females of the amphipod Crustacea *Gammarus palustris* and *G. mucronatus*', *Mar. Behav. Physiol.*, **9**, 319–325.

Bosch, H.F., and Taylor, R.W. (1973). 'Distribution of the cladoceran *Podon polyphemoides* in the Chesaspeake Bay', *Mar. Biol.*, **19**, 161–171.

Bottrell, R.H. (1975). 'Generation time, length of life, instar duration and frequency of moulting and their relationship to temperature in eight species of Cladocera from the river Thames, Reading', *Oecologia*, **19**, 129–140.

Bourdon, R. (1960). 'Rhizocephales et Isopodes parasites des Décapodes marcheurs de la Baie de Quiberon', *Bull. Soc. Sci. Nancy*, **19**, 134–153.

Bourdon, R. (1963). 'Epicarides et Rhizocephales de Roscoff', *Cah. Biol. mar.*, **4**, 415–434.

Bourdon, R. (1964). 'Epicarides et Rhizocephales der Basin d'Aroacohn', *Proc. Verb. Soc. Linn. Bordeaux*, **101**, 51–65.

Branford, J.H. (1978). 'The influence of day-length, temperature and season on the hatching rhythm of *Homarus gammarus*', *J. mar. biol. Ass. U.K.*, **58**, 639–658.

Brewer, R.H. (1964). 'The phenology of *Diaptomus stagnalis* (Copepoda: Calanoida): The development and hatching of the resting stage', *Physiol. Zool.*, **37**, 1–20.

Brody, S. (1945). *Bioenergetics and Growth*, Rafner Publishing, New York.

Broekhuysen, G.J., Jr. (1936). 'On development, growth and distribution of *Carcinides maenas* (L.)', *Arch. néerl. Zool.*, **2**, 257–399.

Browdy, C.L. (1988). 'Aspects of the reproduction biology of *Penaeus semisulcatus* de Haan (Crustacea; Decapoda: Penaeidae)', Ph.D. thesis, University of Tel Aviv.

Browdy, C.L., and Samocha, T.M. (1985). 'The effect of eyestalk ablation on spawning, molting and mating of *Penaeus semisulcatus* de Haan', *Aquaculture*, **49**, 19–29.

Browne, R.A. (1980a). 'Reproductive pattern and mode in the brine shrimp', *Ecology*, **61**, 466–470.

Browne, R.A. (1980b). 'Competition experiments between parthenogenetic and sexual strains of the brine shrimp *Artemia salina*', *Ecology*, **61**, 471–474.

Browne, R.A., Davis, L.E., and Sallee, S.E. (1988). 'Effects of temperature and relative fitness of sexual and asexual brine shrimp *Artemia*', *J. exp. mar. Biol. Ecol.*, **124**, 1–20.

Buikema, A.L. (1973). 'Some effect of light on growth, moulting, reproduction and survival of the cladocerans *Daphnia pulex*', *Hydrobiologia*, **41**, 391–418.

Buikema, A.L. (1975). 'Some effects of light on the energetics of *Daphnia pulex* and implications for the significance of vertical migration', *Hydrobiologia*, **47**, 43–58.

Bulnheim, H.P. (1967). 'Mikrosporidieninfektion und Geschlechtbestimmung bei *Gammarus duebeni*', *Zool. Anz. Suppl.*, **30**, 432–442.

Bulnheim, H.P. (1969). 'Zur Analyse geschlechtsbestimmender Faktoren bei *Gammarus duebeni* (Crustacea, Amphipoda)', *Zool. Anz. Suppl.*, **32**, 244–260.

Bulnheim, H.P. (1970). 'Einfluss von microsporidien-Aufbestimmung und Vererbung des Geschlechts', *Umchau, wiss. Tec.*, **70**, 782–783.

Bulnheim, H.P. (1975). 'Microsporidian infections of amphipods with special reference to host-parasite relationships — A review', *Mar. Fish. Rev.*, **37**, 39–45.

Butler, T.H. (1964). 'Growth, reproduction and distribution of pandalid shrimps in British Columbia', *J. Fish. Res. Board, Can.*, **21**, 1403–1452.

Callan, R.C. (1940). 'The effects of castration by parasite and X-rays on the secondary sex character of prawns (*Leander* spp.)', *J. Exp. Biol.*, **17**, 168–179.

Carefoot, T.H. (1973). 'Studies on the growth, reproduction and life cycle of the supralittoral isopod *Ligia pallasi*', *Mar. Biol.*, **18**, 302–311.

Caroli, J. (1946). 'Un bopyride parasita di altro bopyride', *Pubbl. Staz. zool. Napoli*, **20**, 61–65.

Cattalano, N., and Restivo, F. (1965). 'Ulteriori notizic sulla *Pseudione luxinica* parasita di *Upogebia littoralis*, a Eapoli', *Pubbl. Staz. zool. Napoli*, **34**, 203–210.

Ceccaldi, H.J. (1968). 'Recherche sur la biologie des associations entre protéines et carotenoides chez les Crustacés Decapodes; aspects metaboliques et molécularies', Thése de doctorates, Science naturelles Université d'Ai-Marseille.

Ceccherlli, V.U., and Mistri, M. (1991). 'Production of the meiobenthic harpacticoid copepod *Canuella pepplexa*', *Mar. Ecol. Prog. Ser.*, **78**, 225–234.

Chace, F.A., and Dumont, W.H. (1949). 'Spiny lobsters — Identification, world distribution and U.S. trade', *Com. Fish. Rev.*, **11**, 1–11.

Charniaux-Cotton, H. (1960). 'Sex determination', in *The Physiology of Crustacea* (Ed. T.H. Waterman), Vol. 1, Academic Press, New York, pp. 411–447.

Charnov, E.L. (1979). 'Natural selection and sex change in pandalid shrimp: Test of a life history theory', *Amer. Natur.*, **113**, 715–734.

Checkley, D.M. (1980). 'The egg production of a marine planktonic copepod in relation to its food supply: Laboratory studies', *Limnol. Oceanogr.*, **25**, 430–446.

Cheung, T.S. (1966). 'The development of egg membranes and egg attachment in the shore crab *Carcinus maenas* and some related decapods', *J. mar. biol. Ass. U.K.*, **46**, 373–400.

Cheung, T.S. (1968). 'Transmolt retention of sperm in the adult female stone crab *Menippe mercenaria* (Say)', *Crustaceana*, **15**, 117–120.

Childress, J.J., and Price, M.H. (1978). 'Growth rate of the bathypelagic crustacean *Gnathophausia ingens* (Mysidacea: Lophogastridae). 1. Dimensional growth and population structure', *Mar. Biol.*, **50**, 47–62.

Chittleborough, R.G. (1976). 'Breeding of *Panulirus longipes* George under natural and controlled conditions', *Aust. J. mar. freshw. Res.*, **27**, 499–516.

Choudhury, P.C. (1970). 'Complete larval development of the palaemonid shrimp *Macrobrachium acanthurus* (Wiegmann, 1856)', *Crustaceana*, **18**, 113–132.

Choudhury, P.C. (1971). 'Complete larval development of the palaemonid shrimp *Macrobrachium carcinus* (L.) reared in the laboratory (Decapoda: Palaemonidae)', *Crustaceana*, **20**, 51–69.

Choy, S.C. (1987). 'Growth and reproduction of eyestalk ablated *Penaeus canaliculatus* (Olivier, 1811) (Crustacea: Penaeidae)', *J. exp. mar. Biol. Ecol.*, **112**, 93–107.

Clark, W.H., Jr., Yudin, A.I., Griffin, F.J., and Shigekawa, K. (1984). 'The control of gamete activation and fertilization in the marine penaeida *Sicyonia igentis*', in *Advances in Invertebrate Reproduction* (Eds. W. Engels, W.H. Clark Jr., A. Fischer, P.J.W. Olive and D.F. Went), Elsevier Science Publishers, New York, pp. 459–472.

Clarke, A. (1979). 'On living in cold water: K-strategies in Antarctic benthos', *Mar. Biol.*, **55**, 111–120.

Clarke, C.L. (1967). *Elements of Ecology*, Wiley, New York.

Clegg, J.S. (1962). 'Free glycerol in dormant cysts of the brine shrimp, *Artemia salina*, and its disappearance during development', *Biol. Bull.*, **123**, 295–301.

Clegg, J.S. (1964). 'The control of emergence and metabolism by external osmotic pressure and the role of free glycerol in developing cysts of *Artemia salina*', *J. Exp. Biol.*, **41**, 879–892.

Clegg, J.S. (1965). 'The origin of trehalose and its significance during formation of encysted dormant embryo of *Artemia salina*', *Comp. Biochem. Physiol.*, **14**, 135–143.

Clegg, J.S. (1967). 'Metabolic studies of cryptobiosis in encysted embryos of *Artemia salina*', *Comp. Biochem. Physiol.*, **20**, 801–809.

Clegg, J.S. (1974). 'Biochemical adaptations associated with the embryonic dormancy of *Artemia salina*', *Trans. Am. Microsc. Soc.* **93**, 481–490.

Clegg, J.S. (1978a). 'Hydration-dependent metabolic transitions and the state of cellular water in *Artemia* cysts, in *Dry Biological Systems* (Eds. J. Crowe and J.S. Clegg), Academic Press, New York, pp. 117–153.

Clegg, J.S. (1978b). 'Interrelationships between water and cellular metabolism in *Artemia* cysts. 8. Sorption isotherms and derived thermodynamic quantities', *J. Cell Physiol.*, **94**, 123–138.

Clegg, J.S., Zettlemeyer, A.C., and Hsing, R.H. (1978). 'On the residual water content of dried but viable cells, *Separatus Experientia*', **34**, 734–735.

Clutter, R.I. (1969). 'The microdistribution and social behaviour of some pelagic mysid shrimps', *J. exp. mar. Biol. Ecol.*, **3**, 125–155.

Clutter, R.I., and Theilacker, G.H. (1971). 'Ecological efficiency of pelagic mysid shrimp. Estimates from growth, energy budget, and mortality studies', *Biol. Bull.*, **6**, 93–115.

Coe W.R. (1902). 'The nemertean parasites of crabs', *Amer. Natur.*, **36**, 431–450.

Cole, G.A. (1953a). 'Notes on the vertical distribution of organisms in the profundal sediments of Douglas Lake, Mich.', *Am. Midland Naturalist*, **49**, 252–256.

Cole, G.A. (1953b). 'Notes on copepod encystment', *Ecology*, **34**, 208–211.

Collier, A., Ray, S., and Wilson, W.B. (1956). 'Some effects of specific organic compounds on marine organisms', *Science*, **124**, 220.

Comita, G.W. (1956). 'A study of a calanoid copepod population in an arctic lake', *Ecology*, **37**, 576–591.

Comita, G.W. (1964). 'The energy budget of *Diaptomus siciloides*, Lillijeberg', *Verh. int. Ver. Limnol.*, **15**, 646–653.

Comita, G.W., and Comita, J.T. (1966). 'Egg production in *Tigriopus brevicornis*, in *Some Contemporary Studies in Marine Science* (Ed. H. Barnes), George Allen and Unwin Ltd., London, pp. 171–185.

Comita, G.W., Marshall, S.M., and Orr, A.P. (1966). 'On the biology of *Calanus finmarchicus*. 13. Seasonal change in weight, calorific value and organic matter', *J. mar. biol. Ass. U.K.*, **46**, 1–17.

Conde, J.E., and Diaz, H. (1989). 'The mangrove tree crab *Aratus pisonii* in a tropical estuarine coastal lagoon', *Est. Coast. Shelf. Sci.*, **28**, 639–650.

Conklin, D.E., and Provasoli, L. (1977). 'Nutritional requirements of the water flea *Moina macrocopa*', *Biol. Bull.*, **152**, 337–350.

Conover, R.J. (1967). 'Reproductive cycle, early development and fecundity in laboratory populations of the copepod *Calanus hyperboreus*', *Crustaceana*, **13**, 61–72.

Cooley, J.M. (1971). 'The effect of temperature on the development of rearing eggs of *Diaptomus oregonensis* Lillij. (Copepoda: Calanoida)', *Limnol. Oceanogr.*, **16**, 921–926.

Cooley, J.M., and Minns, C.K. (1978). 'Prediction of egg development times of freshwater copepods', *J. Fish. Res. Board, Can.*, **35**, 1322–1329.

Corbet, P.S. (1962). *A Biology of Dragonflies*, Witherby, London.

Corey, S. (1981). 'Comparative fecundity and reproductive strategies in seventeen species of Cumacea (Crustacea: Peracaridae)'. *Mar. Biol.*, **62**, 65–72.

Corkett, C.J., and McLaren, I.A. (1969). 'Egg production and oil storage by the copepod *Pseudocalanus* in the laboratory', *J. exp. mar. Biol. Ecol.*, **3**, 90–105.

Corkett, C.J., and Zillioux, E.J. (1975). 'Studies on the effect of temperature on the egg laying of three species of calanoid copepods in the laboratory (*Acartia tonsa, Temora longicornia, Pseudocalanus elongatus*)', *Bull. Plankton Soc. Jap.*, **21**, 77–85.

Corner, E.D.S., Cowey, C.B., and Marshall, S.M. (1967). 'On the nutrition and metabolism of zooplankton, 5. Feeding efficiency of *Calanus finmarchicus*', *J. mar. biol. Ass. U.K.*, **47**, 259–270.

Costlow, J.D., Jr., and Bookhout, C.G. (1968). 'A method for developing the brachyuran eggs *in vitro*', *Limnol. Oceanogr.*, **5**, 212–225.

Cottareli, N.V. (1966). 'Notise sulla biologia di un Crustacea anosstraco: *Chirocephalus stagnalis* — Estratto dall', *Arch. Zool. Ital.*, **51**, 1031–1052.

Couch, J.N. (1942). 'A new fungus on crab eggs', *J. Elisha Mitchell Sci. Soc.*, **58**, 158–164.

Crisp, D.J. (1954). 'The rate of development of *Balanus balanoides* (L) embryo *in vitro*', *J. Anim. Ecol.*, **28**, 119–132.

Crisp, D.J., and Davies, P.A. (1955). 'Observations *in vitro* on the breeding of *Elminius modestus* grown on glass slides', *J. mar. biol. Ass. U.K.*, **34**, 357–380.

Crisp, D.J., and Spencer, C.P. (1958). 'The control of the hatching process in barnacles', *Proc. roy. Soc. London*, **148B**, 278–299.

Crocos, P.J., and Kerr, J.D. (1983). 'Maturation and spawning of the banana prawn *Penaeus merguiensis* de Man (Crustacea: Penaeidae) in the Gulf of Carpentaria, Australia', *J. exp. mar. Biol. Ecol.*, **69**, 37–59.

Crowe, J.H. (1971). 'Anhydrobiosis: An unsolved problem', *Amer. Natur.*, **105**, 563–574.

Crowe, J.H., and Madin, K.S. (1974). 'Anhydrobiosis in tardigrades and nematodes', *Trans. Am. Microsc. Soc.*, **92**, 513–524.

Cuellar, O. (1977). 'Animal parthenogenesis', *Science*, **197**, 837–843.

Daborn, G.R. (1973). 'Community structure and energetics in an argillotrophic lake, with special reference to the giant fairy shrimp, *Branchinecta gigas* Lynch', Ph.D. thesis, University of Alberta, Edmonton.

Daborn, G.R. (1975). 'Life history and energy relations of the giant fairy shrimp *Branchinecta gigas* Lynch 1937 (Crustacea: Anostraca)', *Ecology*, **56**, 1025–1039.

Dagg, M.J. (1976). 'Complete carbon and nitrogen budgets for carnivorous amphipod *Calliopius laeviusculus* (Kroyer)', *Int. Rev. ges. Hydrobiol.*, **61**, 297–357.

D'Agostino, A.S., and Provasoli, L. (1970). 'Dixenic culture of *Daphnia magna* Straus', *Biol. Bull.*, **139**, 485–494.

D'Appolito, L.M., and Stancyk, S.E. (1979). 'Population dynamics of *Euterpina acutifrons* (Copepoda: Harpacticoida) from North Inlet, South Carolina, with reference to dimorphic males', *Mar. Biol.*, **54**, 251–260.

Darnell, R.M. (1962). 'Sex ratios: Aquatic animals', in *Growth* (Eds. P.L. Altman and D.S. Dittmer), Federation for American Societies for Experimental Biology, Washington, pp. 439–442.

Dauvin, J.C. (1989). 'Life cycle, dynamics and productivity of Crustacea-Amphipoda from western English Channel — 5. *Ampelisca sarsi* Chevreux', *J. exp. mar. Biol. Ecol.*, **128**, 31–56.

Davis, C.C. (1959). 'Osmotic hatching in the eggs of some freshwater copepods'. *Biol. Bull.*, **116**, 15–29.

Davis, C.C. (1964). 'A study of hatching process in aquatic invertebrates. 10. Hatching in the freshwater shrimp *Potimirim glabra* (Kingsley) (Macrura: Atyidae)', *Pac. Sci.*, **18**, 378–384.

Davis, C.C. (1965). 'A study of hatching process in aquatic invertebrates. 16. An examination of hatching in *Palaemonetes vulgaris* (Say)', *Crustaceana*, **8**, 223–238.

Davis, C.C. (1966). 'A study of hatching process in aquatic invertebrates. 23. Eclosion in *Petrolisthes armatus* (Cibbes) (Anomura: Porcellanidae)', *Int. Rev. ges. Hydrobiol.*, **51**, 791-796.

Davis, C.C. (1968). 'Mechanisms of hatching in aquatic invertebrate eggs', *Oceanogr. mar. Biol. Ann. Rev.*, **6**, 325–376.

Davis, C.C. (1969). 'Hatching within the brood sac of the oviparous Isopod *Cirolana* sp. (Isopoda: Cirolanidae), 9. Hatching in the freshwater shrimp, *Potimirim glabra* (Kingsley) (Macrura: Atyidae)', *Pac. Sci.*, **18**, 378–384.

Dawson, R.M.C., and Barnes, H. (1966). 'Studies in the biochemistry of cirripede eggs. 2. Changes in lipid composition during development of *Balanus balanoides* and *B. balanus*', *J. mar. biol. Ass. U.K.*, **46**, 249–261.

DeCoursey, P.J. (1979). 'Egg hatching rhythms in three species of fiddler crabs', in *Cyclic Phenomena in Marine Plants and Animals* (Eds. E. Maylor and R.C. Hartnoll), Pergamon Press, Oxford, pp. 399–406.

Deecaraman, M. (1980). 'Some aspects of reproduction in a stomatopod crustacean with special reference to accessory sex glands', Ph.D. thesis, Madras University, Madras.

Delvi, M.R. (1972). 'Ecophysiological studies on the grasshopper *Poecilocerus pictus*', Ph.D. thesis, Bangalore University.

Delvi, M.R., and Pandian, T.J. (1971). 'Ecophysiological studies on the utilization of food in the paddy field grasshopper *Oxya velox*', *Oecologia*, **8**, 267–275.

DeVries, M.C., and Forward, R.B., Jr. (1991). 'Mechanisms of crustacean egg hatching: Evidence for enzyme release by crab embryos', *Mar. Biol.*, **110**, 281-291.

Diaz, H., and Conde, J.E. (1989). 'Population dynamics and life history of the mangrove crab *Aratus pisonii* (Brachyura, Grapsidae) in a marine environment', *Bull. mar. Sci.*, **45**, 148–163.

Dingle, H., and Caldwell, E.L. (1972). 'Reproductive and maternal behaviour of the mantis shrimp *Gonodactylus bredini* Manning (Crustacea: Stomatpoda)', *Biol. Bull.*, **142**, 417–426.

Do, T.D., and Kajihara, T. (1986). 'Sex determination and a typical male development in a poecilostomatid copepod, *Pseudomysicola spinosus* (Raffaele and Monticelli, 1985)', *Syllogeus*, **58**, 283-287.

Dobkin, S. (1969). 'Abbreviated larval development in caridean shrimps and its significance in the artificial culture of these animals', *F.A.O. Fish. Rep.*, **57**, 935–946.

Dolgopolskaya, M.A. (1958). 'Cladocera Tsernogo Morya (Cladocera of the Black Sea)', *Tr. Sevastopol. Biol. Sta.*, **10**, 27-75 (in Russian).

Downer, D.F., and Steele, D.H. (1979). 'Some aspects of the biology of *Amphiporsia lawrenciana* Shoemaker (Crustacea, Amphipoda) in Newfoundland waters', *Can. J. Zool.*, **57**, 257-263.

Drach, P., and Tchernigovitzeff, T. (1967). 'Sur le methode de determination des stades d' intermue et sous application generale aux crustaces', *Vie Milieu*, **18**, 595–607.

Dresel, D.I.B. (1948). 'Passage of haemoglobin from blood into eggs of *Daphnia*', *Nature, Lond.*, **162**, 736.

Dresel, D.I.B., and Hoyle, V. (1950). 'Nitrogenous excretion of amphipods and isopods', *J. Exp. Biol.*, **27**, 210–225.

Dugan, C.C., Hagood, W.R., and Frakes, A.T. (1975). 'Development of spawning and mass larval rearing techniques for brackishwater shrimps of the genus *Macrobrachium* (Decapoda: Palaemonidae)', *Fla. Dep. Nat. Resour. Publ.*, **12**, 1-28.

Dutrieu, J. (1960). 'Observations biochimiques et physiologiques sur le développement d'*Artemia salina* Leach', *Arch. Zool. exp. gén.*, **99**, 1-133.

Dutrieu, J., and Christia-Blanchine, D. (1966). 'Résistance des oeufs durablen hydratés d'*Artemia salina* a l'anoxia', *C. R. Acad. Sci. Paris*, **263D**, 998–1000.

Edmondson, W.T. (1955). 'The seasonal life history of *Daphnia* in an Arctic lake', *Ecology*, **36**, 439–455.

Edmondson, W.T., Comita, G.W., and Anderson, G.C. (1962). 'Reproductive rate of copepods in nature and its relation to phytoplankton population', *Ecology*, **43**, 625–634.

Efford, I.E. (1969). 'Egg size in the sand crab, *Emerita analoga* (Decapoda, Hippidae)', *Crustaceana*, **16**, 15–26.

Efford, I.E. (1970). 'Recruitment to sedentary marine populations, exemplified by sand crab *Emerita analoga* (Decapoda, Hippidae)', *Crustaceana*, **18**, 293–308.

Einsle, V. (1967). 'Die ausseren Bedingungen der Diapause planktischlebender Cyclopsarten', *Arch. Hydrobiol.*, **63**, 387–403.

Elgmork, K. (1959). 'Seasonal occurrence of *Cyclops strenuus strenuus*', *Folia Limnol. Scand.*, **11**, 1–196.

Elgmork, K. (1964). 'Dynamics of zooplankton communities in some small innunated ponds', *Folia Limnol. Scand.*, **12**, 1–83.

Ellis, H.J. (1961). 'A life history of *Asellus intermedius* Forbes', *Trans. Am. Microsc. Soc.*, **80**, 80–102.

Elner, R.W., and Elner, J.K. (1980). 'Observations on a simultaneous mating embrace between a male and two female rock crabs *Cancer irroratus*, Say, 1817 (Decapoda: Brachyura)', *Crustaceana*, **38**, 96–98.

El-Nockrashy, A.S., and Frampton, V.L. (1967). 'Destruction of lysine by non-reducing sugars', *Biochem. Biophys. Res. Commun.*, **28**, 675–681.

Emerson, D.W. (1967). 'Some aspects of free amino acid metabolism in developing encysted embryos of *Artemia salina*', *Biol. Bull.*, **132**, 156–160.

Emlen, J.M. (1973). *Ecology, an Evolutionary Approach*, Addison Wesley Publishing Company, Massachusetts.

Emmerson, W.D. (1980). 'Induced maturation of prawn *Penaeus indicus*', *Mar. Ecol.*, **2**, 121–131.

Enders, F. (1976). 'Clutch size related to hunting manner of spider species', *Ann. Ent. Soc. Amer.*, **69**, 991–998.

Engel, W.A. van (1958). 'The blue crab and its fishery in the Chesaspeake Bay', *U.S. Fish. Wildl. Serv. Comm. Fish. Rev.*, **20**, 6–17.

Ennis, G.P. (1973). 'Endogenous rhythmicity associated with larval hatching in the lobster *Homarus americanus*', *J. mar. biol. Ass. U.K.*, **53**, 531–538.

Ennis, G.P. (1975). 'Observations on hatching and larval release in the lobster *Homarus americanus*', *J. Fish. Res. Board, Can.*, **32**, 2210–2213.

Ewing, R.D. (1968). 'An analysis of lactate dehydrogenase and anaerobiosis in *Artemia salina*', Ph.D. thesis, University of Miami, Coral Gables, Florida.

Ewing, R.D., and Clegg, J.S. (1969). 'Lactate dehydrogenase activity and anaerobic metabolism during embryonic development of *Artemia salina*', *Comp. Biochem. Physiol.*, **31**, 297–307.

Fairbrain, D. (1958). 'Trehalose and glucose in helminths and other invertebrates', *Can. J. Zool.*, **36**, 787–795.

Fenwick, G.D. (1984). 'Life history tactics of brooding Curstacea', *J. exp. mar. Biol. Ecol.*, **84**, 247–264.

Fieber, L.A., and Lutz, P.L. (1984). 'Magnesium and calcium metabolism during moulting in the freshwater prawn *Macrobrachium rosenbergii*', *Can. J. Zool.*, **63**, 1120–1124.

Fielder, D.R. (1976). 'The larval life history of *Macrobrachium novaeholandiae* (deMan) (Decapoda: Palaemonidae) reared in the laboratory', *Crustaceana*, **30**, 252–286.

Finamore, F.J., and Clegg, J.S. (1969). 'Biochemical aspects of morphogenesis in the brine shrimp, *Artemia salina*', in *The Cell Cycle, Gene-Enzyme Interactions* (Eds. G.L. Padilla, G.L. Whitson, and I. Cameron), Academic Press, New York, p. 249.

Finamore, F.J., and Warner, A.H. (1963). 'The occurrence of p^1, p^4-diguanosine 5-tetraphosphate in brine shrimp eggs', *J. Biol. Chem.*, **238**, 344–348.

Fish, J.D., and Mills, A. (1979). 'The reproductive biology of *Corophium volutator* and *C. arenarium* (Crustacea: Amphipoda)', *J. mar. biol. Ass. U.K.*, **59**, 355–368.

Fisher, W.S. (1976). 'Relationships of epibiotic fouling and mortalities of the eggs of the Dungeness crab (*Cancer magister*)', *J. Fish. Res. Board, Can.*, **33**, 2849–2853.

Fisher, W.S., Nilson, E.H., and Shleser, R.A. (1975). 'Effect of the fungus *Halipthoros milfordensis* on the juvenile stages of the American lobster *Homarus americanus*', *J. Inv. Pathol.*, **26**, 41–45.

Fisher, W.S., Nilson, E.H., Steenbergen, J.F., and Lightner, D.V. (1978). 'Microbial diseases of cultured lobsters: A review', *Aquaculture*, **14**, 115–140.

Fleminger, A. (1967). 'Taxonomy, distribution and polymorphism in the *Labidocera jollae* group, with remarks on evolution within the group (Copepoda: Calanoida)', *Proc. U.S. natl Mus.*, **120**, 1–61.

Fleminger, A. (1985). 'Dimorphism and possible sex change in copepods of the family Calanoidae', *Mar. Biol.*, **88**, 273–294.

Fluchter, J., and Pandian, T.J. (1968). 'Rate and efficiency of yolk utilization in developing eggs of the sole *Solea solea*', *Helgoländer wiss. Meeresunters.*, **16**, 216–224.

Fox, H.M. (1948). 'The haemoglobin of *Daphnia*', *Proc. roy. Soc. London*, **135**, 195–212.

Frost, B., and Fleminger, A. (1968). 'A revision of the genus *Clausocalanus* (Copepoda: Calanoida) with remarks on distribution and patterns in diagnostic characters', *Bull. Scripps Inst. Oceanogr. Univ. California*, **12**, 1–235.

Fukui, Y., and Wada, K. (1986). 'Distribution and reproduction of four intertidal crabs (Crustacea, Brachyura) in the Tonda river estuary, Japan', *Mar. Ecol. Prog. Ser.*, **30**, 229–241.

Gaudy, R. (1974). 'Feeding four species of pelagic Copepoda under experimental conditions', *Mar. Biol.*, **25**, 125–142.

George, M.J. (1969). 'Genus *Macrobrachium* Bate 1968', in *Prawn Fisheries of India*, No. 14, 179–216; *Bull. Cent. Mar. Fish. Res. Inst.*, Mandapam Camp, **44**, 189–208.

Ghiselin, M.T. (1969). 'The evolution of hermaphroditism among animals', *Q. Rev. Biol.*, **44**, 189–208.

Gibbons, S.G. (1936). '*Calanus finmarchicus* and other copepods in Scottish waters in 1933', *Sci. Invest. Fish. Div., Scottish Home Dept.*, **1**, 1–37.

Giesbrecht, W. (1910). 'Stomatopoden', *Fauna Flora Golf. Neapel* (*Monogr.*), **33**, 1–239.

Giesel, J.T. (1976). 'Reproductive strategies as adaptations to life in temporally heterogenous environments', *Ann. Rev. Ecol. Syst.*, **7**, 57–79.

Gieskes, W.W.C. (1970). 'The Cladocera of the north Atlantic and North Sea: Biological and ecological studies', Ph.D. thesis, McGill University.

Gifford, J. (1934). 'Life history of *Argeia pauperata* from *Cragofraciscorum*', M.A. thesis, Leland Stanford Jr. University.

Gilchrist, B.M., and Green, J. (1960). 'The pigments of *Artemia*', *Proc. roy. Soc. London*, **152**, 118–136.

Gilchrist, B.M., and Lee, W.L. (1972). 'Carotenoid pigments and their possible role in reproduction in the sand crab *Emerita analoga* (Stimpson, 1857)', *Comp. Biochem. Physiol.*, **42**, 263–294.

Glosener, R.R., and Tilman, D. (1978). 'Sexuality and the components of environmental uncertainty: Clues from geographic parthenogenesis in terrestrial animals', *Amer. Natur.*, **112**, 659–673.

Goodbody, I. (1965). 'Continuous breeding in population of two tropical crustaceans, *Mysidium columbiae* (Zimmer) and *Emerita portoricensis* Schmidt', *Ecology*, **46**, 195–197.

Goodwin, T.W. (1951). 'Carotenoid metabolism during development of lobster eggs', *Nature, Lond.*, **167**, 559.

Goodwin, T.W. (1960). 'Biochemistry of pigments', in *Physiology of Crustacea*, Vol. 1, *Metabolism and Growth* (Ed. T.H. Waterman), Academic Press, New York, pp. 101–140.

Gophen, M. (1976). 'Temperature effect on lifespan, metabolism and development time of *Mesocyclops leukarti* (Claus)', *Oecologia*, **25**, 271–277.

Gore, P.S. (1971). 'A note on the successive egg-laying without mating of some Indian crabs', *Curr. Sci.*, **40**, 48.

Gravier, C. (1931). 'La ponte et l'incubation chez les Crustacés', *Ann. Sci. nat.* (*Zool.*), **14**, 303–419.

Green, J. (1956). 'Variation in the haemoglobin content of *Daphnia*', *Proc. roy. Soc. London*, **145**, 214–232.

Green, J. (1965). 'Chemical embryology of the Crustacea', *Biol. Rev.*, **40**, 580–600.

Green, J. (1966). 'Seasonal variation in egg production by Cladocera', *J. Anim. Ecol.*, **35**, 77–104.

Green, J. (1971). 'Crustaceans', *in Experimental Embryology of Marine and Freshwater Invertebrates* (Ed. G. Reverberi), North-Holland Publishing Co., Amsterdam, pp. 312–362.

Greenwood, J.C. (1965). 'The larval development of *Petrolisthes elongatus* (H. Milne Edwards) and *P. novazelandiae* Filhol (Anomura: Porcellanidae) with notes on breeding', *Crustaceana*, **8**, 285–307.

Grice, G.D., and Gibson, V.R. (1975). 'Occurrence, viability and significance of resting eggs of the calanoid copepod *Labidocera aestiva*', *Mar. Biol.*, **31**, 335–337.
Grice, G.D., and Gibson, V.R. (1977). 'Resting eggs in *Pontella maedi* (Copepoda: Calanoida)', *J. Fish. Res. Board, Can.*, **34**, 410–412.
Gupta, N.V.S., Kurup, K.N.P., Adiyodi, R.G., and Adiyodi, K.G. (1987). 'The antagonism between somatic growth and ovarian growth during different phases in intermoult (stage C4) in sexually mature freshwater crab *Paratelphusa hydrodromous*', *Int. J. Inv. Reprod.*, **12**, 307–318.
Gupta, N.V.S., Kurup, K.N.P., Adiyodi, R.G., and Adiyodi, K.G. (1989). 'The antagonism between somatic and testicular activity during different phases in intermoult (stage C4) in sexually mature, freshwater crab *Paratelphusa hydrodromous*', *Inv. Reprod. Dev.*, **16**, 195–204.
Hacher, V. (1902). 'Über die Fortpflanzung der limnetischen Copepoden des Titisees', *Ber. Naturf. Ges., Freiburg*, **12**, 1–133.
Hall, R.E., and MacDonald, L.J. (1975). 'Hatching of the anostracan Branchiopod *Chirocephalus diaphanus* Prevost. 1. Osmotic processes and the possible role of glycerol', *Hydrobiologia*, **46**, 369–375.
Hamer, M.L., and Appleton, C.C. (1991). 'Life history adaptations of phyllopods in response to predators, vegetation and habitat duration in north-east Natal', *Hydrobiologia*, **212**, 105–116.
Haq, S.M. (1965). 'Development of the copepod *Euterpina acutifrons* with special reference to dimorphism in the male', *Proc. Zool. Soc.* (*Lond.*), **144**, 174–201.
Haq, S.M. (1972). 'Breeding of *Euterpina acutifrons*, a harpacticoid copepod, with special reference to dimorphic males', *Mar. Biol.*, **15**, 221–235.
Haq, S.M. (1973). 'Factors affecting production of dimorphic males of *Euterpina acutifrons*', *Mar. Biol.*, **19**, 23–26.
Harrington, A., and Ikeda, T. (1986). 'Laboratory observations on spawning, brood size and egg hatchability of Antarctic Krill *Euphausia superba* from Prydz Bay, Antarctica', *Mar. Biol.*, **92**, 231–235.
Harris, R.P. (1972). 'Reproductive activity of the interstitial copepods of a sandy beach', *J. mar. biol. Ass. U.K.*, **52**, 507–524.
Harris, R.P. (1973). 'Feeding, growth, reproduction and nitrogen utilization by the harpacticoid copepod *Tigriopus brevicornis*', *J. mar. biol. Ass. U.K.*, **53**, 785–800.
Harris, R.P., and Paffenhofer, G.A. (1976). 'The effect of food concentration, cumulative ingestion and growth efficiency of two small marine planktonic copepods', *J. mar. biol. Ass. U.K.*, **56**, 875–888.
Hart, R.C., and McLaren, I.A. (1978). 'Temperature acclimation and other influence on embryonic duration in the copepod *Pseudocalanus* sp.', *Mar. Biol.*, **45**, 23–30.
Hartnoll, R.G. (1963). 'The biology of Mans spider crabs', *Proc. Zool. Soc.* (*Lond.*), **141**, 423–496.
Hartnoll, R.G. (1967). 'The effects of sacculinid parasites on two Jamaican crabs', *J. Linn. Soc.*, **46**, 275–296.
Hartnoll, R.G. (1972a). 'The biology of the burrowing crab, *Corystes aunus*', *Bijdr. Dierk.*, **41**, 423–496.
Hartnoll, R.G. (1972b). 'The biology of the burrowing crab, *Corystes cassivelaunus*', *Bijdr. Dierk.*, **42**, 139–155.
Hatchett, D.P. (1947). 'Biology of the Isopoda of Michigan', *Ecol. Monogr.*, **17**, 47–79.
Heeley, W. (1941). 'Observations on the life histories of some terrestrial isopods', *Proc. Zool. Soc. (Lond.)*, **111**, 79–149.
Heinle, D.R., and Flemer, D.A. (1975). 'Carbon requirements of a population of the estuarine copepod *Eurytemora affinis*', *Mar. Biol.*, **31**, 235–248.
Heinle, D.R., Harris, R.P., Ustach, J.F., and Flemer, D.A. (1977). 'Detritus as food for estuarine copepods', *Mar Biol.*, **40**, 341–354.
Heip, C. (1972). 'The reproductive potential of copepods in brackish water', *Mar. Biol.*, **12**, 219–221.
Heip, C. (1974). 'A comparison between models describing the influence of temperature on the development rate of copepods', *Biol. Jb. Dodonaea*, **42**, 121–125.
Heip, C. (1977). 'On the reproductive potentials in a brackish water meiobenthic community, *Mikrofauna d. Meeresbodens*, **61**, 105–112.

Heip, C., and Smol, N. (1976). 'Influence of temperature on the reproductive potential of two brackish water harpacticoids (Crustacae: Copepoda)', *Mar. Biol.*, **35**, 327-334.

Heip, C., Smol, N., and Absillis, V. (1978). 'Influence of temperature on the reproductive potential of *Oncholaimus oxyuris* (Nematoda: Oncholaimidae)', *Mar. Biol.*, **45**, 255–260.

Hentig, R. von (1971). 'Einfluss von Salzgehalt und Temperatur auf Entwicklung, Wachstum, Fortpflanzung und Energiebilanz von *Artemia salina*', *Mar. Biol.*, **9**, 145–182.

Herring, P.J. (1973). 'Depth distribution of pigment and lipid in some oceanic animals. 2. Decapod crustaceans', *J. mar. biol. Ass. U.K.*, **53**, 539–562.

Herring, P.J. (1974a). 'Observations on the embryonic development of some deep-living decapod crustaceans with particular reference to species of *Acanthephyra*', *Mar. Biol.*, **25**, 25–34.

Herring, P.J. (1974b). 'Size, density and lipid content of some decapod eggs', *Deep-sea Res.*, **21**, 91–94.

Hiatt, R.W. (1948). 'The biology of the lined shore crab *Pachygrapsus crassipes* Randall', *Pac. Sci.*, **2**, 135–213.

Hill, L.L., and Coker, R.E. (1930). 'Observations on mating habits of *Cyclops*', *J. Elisha Mitchell Scient. Soc.*, **45**, 206–220.

Hines, A.H. (1982). 'Allometric constraints and variables of reproductive effort in Brachyuran crabs', *Mar. Biol.*, **69**, 309–320.

Hinton, H.E. (1954). 'Resistance of the dry eggs of *Artemia salina* L. to high temperatures', *Ann. Mag. Nat. Hist.* (Ser. 7), 158–160.

Hippeau-Jacquotte, R. (1984). 'A new concept in the evolution of Copepoda: *Pachygypus gibber* (Notodelphyidae), a species with breeding males', *Crustaceana* Suppl., **7**, 60–67.

Hiratsuka, E. (1920). 'Researches on the nutrition of the silk worm', *Bull. Ser. Exp. Sta. Japan*, **1**, 257-315.

Hirche, H.J. (1989). 'Egg production of the Arctic copepod *Calanus glacialis*: Laboratory experiments', *Mar. Biol.*, **103**, 311-318.

Holdich, D.M. (1968). 'Reproduction, growth and bionomics of *Dynamene bidentata* (Crustacea: Isopoda)', *J. Zool. London*, **156**, 137–153.

Hopkins, C.C.E. (1977). 'The relationship between maternal body size and clutch size, development time and egg mortality in *Euchaeta norvegica* (Copepoda: Calanoida) from Loch Etive, Scotland', *J. mar. biol. Ass. U.K.*, **57**, 723–735.

Hopkins, C.C.E. (1978). 'The male genital system, and spermatophore production and function in *Euchaeta norvegica* Bosok (Copepoda: Calanoida)', *J. exp. mar. Biol. Ecol.*, **35**, 197–231.

Hopkins, C.C.E., and Machin, D. (1977). 'Patterns of spermatophore distribution and placement in *Euchaeta norvegica* (Copepoda: Calanoida)', *J. mar. biol. Ass. U.K.*, **57**, 113–131.

Hopkins, C.C.E., Mauchline, J., and Melusky, D.S. (1978). 'Structure and function of the fifth pair of pleopods of male *Euchaeta norvegica* (Copepoda: Calanoida)', *J. mar. biol. Ass. U.K.*, **58**, 631–637.

Hoshi, T. (1950). 'Studies on physiology and ecology of plankton. 3. Changes in respiratory quotient during embryonic development of daphnid, *Simocephalus vetulus* (O.F. Muller)', *Sci. Rep. Tôhoku Univ.*, **18**, 316–323.

Hoshi, T. (1957). 'Studies on physiology and ecology of plankton. 13. Haemoglobin and its role in the respiration of the daphnid, *Simocephalus vetulus*', *Sci. Rep. Tôhoku Univ.*, **23**, 35–58.

Hubschmann, J.H., and Broad, A.C. (1974). 'The larval development of *Palaemonetes intermedius* Holthuis, 1949 (Decapoda: Palaemonidae) reared in the laboratory', *Crustaceana*, **26**, 89–103.

Hurley, D.E. (1968). 'Transition from water to land in amphipod crustaceans', *Am. Zool.*, **8**, 327–353.

Hutchinson, G.E. (1967). *A Treatise on Limnology*, Vol. 2, *Introduction to Lake Biology and the Limnoplankton*, Wiley, London, 1115 pp.

Ianora, A., Scotto di Carlo, B., and Mascellaro, P. (1989). 'Reproductive biology of the planktonic copepod *Temora stylifera*', *Mar. Biol.*, **101**, 187–194.

Ibrahim, K.H. (1962). 'Observations on the fishery and biology of the freshwater prawn *Macrobrachium malcolmsonii* Milne-Edwards of river Godavari', *Indian J. Fish.*, **9**, 433–467.

Iwasaki, T. (1965). 'Sensitivity of *Artemia* eggs to r-irradiation. 7. Relationship between the biological damage and the decay of free radicals in irradiated eggs', *Int. J. Radiation Biol.*, **9**, 573–580.

Iwasaki, T. (1973). 'The differential radio sensitivity of oogonia and oocytes at different developmental stages of the brine shrimp *Artemia salina*', *Biol. Bull.*, **144**, 151–161.

Iwasaki, T., Maruyama, T., Kumamoto, Y., and Kato, Y. (1971). 'Effect of fast neutrons and 60 Co γ-rays on *Artemia*', *Radiation Res.*, **45**, 288–298.

Izawa, K. (1973). 'On the development of parasite Copepoda 1. *Arcotaces pacificus* Komai (Cyclopoida: Philichthyidae)', *Publ. Seto mar. Biol. Lab.*, **21**, 77–86.

Jacobs, J. (1961). 'Laboratory cultivation of the marine copepod *Pseudodiaptomus coronatus* Williams', *Limnol. Oceanogr.*, **6**, 443–446.

James, P.T. (1973). 'Distribution of dimorphic males of three species of *Nematoscelis* (Euphausiacea)', *Mar. Biol.*, **19**, 341–347.

Jamieson, B.G.M. (1991). 'Ultrastructure and phylogeny of crustacean spermatozoa', *Mem. Queensland Mus.*, **31**, 109–142.

Jensen, J.P. (1958). 'The relation between body size and number of eggs in marine malacostracans', *Meddr. Danm. Fisk. -Og Havunders*, II, **19**, 1–25.

Jerde, C.W., and Lasker, R. (1966). 'Moulting of euphausiid shrimps: shipboard observations', *Limnol. Oceanogr.*, **11**, 120–124.

Jewelt, S.C., Sloan, N.A., and Somerton, D.A. (1985). 'Size at sexual maturity and fecundity of fjord-dwelling golden king crab *Lithodes aequispina* Benedict, from northern British Columbia', *J. Crust. Biol.*, **5**, 377–388.

Jeyalectumie, C., and Subramoniam, T. (1989). 'Cryopreservation of spermatophores and seminal plasma of the edible crab *Scylla serrata*', *Biol. Bull.*, **177**, 247–253.

Jeyalectumie, C., and Subramoniam, T. (1991). 'Biochemistry of seminal secretions of the crab *Scylla serrata* with reference to sperm metabolism and storage in the female', *Mol. Reprod. Develop.*, **30**, 44–55.

Johnson, M. (1967). 'Some observations on the hatching of *Tortunus discaudatus* eggs subjected to low temperatures', *Limnol. Oceanogr.*, **12**, 405–410.

Johnson, M.W. (1940). 'The correlation of water movements and dispersal of pelagic larval stages of certain littoral animals, especially the sand crab *Emerita*', *J. Mar. Res.*, **2**, 236–245.

Johnson, M.W., and Lewis, W.M. (1942). 'Pelagic larval stages of the sand crabs *Emerita analoga* (Stimpson), *Blepharipoda occidentalis* Randall, and *Lepidopa myops* Stimpson', *Biol. Bull.*, **83**, 67–87.

Johnson, P.W., Sieburth, Mc.N.J., Sastry, A., Arnold, C.R., and Doty, M.S. (1971). '*Leucothrix mucor* infestation of benthic Crustacea, fish eggs and tropical algae', *Limnol. Oceanogr.*, **16**, 962–969.

Johnson, S.K. (1976). 'Chemical control of peritrichome ciliates on young penaeid shrimp, Texas A&M University', *Texas Agric. Ext. Serv. Publ.*, **1**, 1–2.

Johnson, T.W. (1958). 'A fungus parasite in the ova of the barnacle *Chthamalus fragilis denticulata*', *Biol. Bull.*, **114**, 205–214.

Johnson, W.S. (1976a). 'Biology and population dynamics of the intertidal isopod *Cirolana harfordi*', *Mar. Biol.*, **36**, 343–350.

Johnson, W.S. (1976b). 'Population energetics of the intertidal isopod *Cirolana harfordi*', *Mar. Biol.*, **36**, 351–358.

Jones, M.B. (1974). 'Breeding biology and seasonal population changes of *Jaera nordmanni nordica* Lemercier (Isopoda: Asellota)', *J. Zool. London*, **165**, 183–189.

Jones, M.B., and Naylor, E. (1971). 'Breeding and bionomics of the British members of the *Jaera albifrons* group of species (Isopoda: Asellota)', *J. Zool. London*, **165**, 183–199.

Juchault, P., and Legrand, J.J. (1968). Rôle des hormones sexuelles, des neurohormones et d'un facteur épigénètique dans la physiologie sexuelle d'individus intersexues d' *Armadillidium vulgare* Latr. (Isopoda: Oniscoide)', *C. R. Acad. Sci., Paris*, **267**, 2014–2016.

Juchault, P., and Legrand, J.J. (1975). 'Modalites de la transmission héréditaire du facteur epigénètique responsable de l'intersexualite masculinisante chez le Crustacé Oniscoide *Porcellio dilatatus* Brandt', *Bull. Soc. Zool. Fr.*, **100**, 467–476.

Kasahara, S., Uye, S., and Onbe, T. (1974). 'Calanoid copepod eggs in sea bottom muds', *Mar. Biol.*, **26**, 167–171.

Kasahara, S., Uye, S., and Onbe, T. (1975a). '2. Seasonal cycles of abundance in the populations of several species of copepods and their eggs in the Inland sea of Japan', *Mar. Biol.*, **31**, 16–25.

Kasahara, S., Uye, S., and Onbe, T. (1975b). '3. Effects of temperature, salinity and other factors on the hatching of resting eggs of *Tortanus forcipatus*', *Mar. Biol.*, **31**, 26–31.

Katona, S.K. (1973). 'Evidence for sex pheromones in planktonic copepods', *Limnol. Oceanogr.*, **18**, 574–583.

Katre, S. (1977a). 'The relation between body size and number of eggs in the freshwater prawn *Macrobrachium lamarrei* (H. Milne Edwards) (Decapoda: Caridea)', *Crustaceana*, **33**, 18–22.

Katre, S. (1977b). 'Yolk utilization in the freshwater prawn *Macrobrachium lamarrei*', *J. Anim. Morphol. Physiol.*, **24**, 13–20.

Katre, S., and Pandian, T.J. (1972). 'On the hatching mechanism of a freshwater prawn *Macrobrachium idae*', *Hydrobiologia*, **40**, 1–17.

Kensler, C.B. (1967). 'Fecundity in the marine spiny lobster *Jasus verreauxi* (H. Milne-Edwards) (Crustacea: Decapoda: Palinuridae)', *New Zealand J. mar. freshwat. Res.*, **1**, 143–155.

Kewalramani, H.C. (1973). 'Salinity requirements in the larval life history of freshwater prawn *Macrobrachium malcolmsonii* H. Milne-Edwards', *Spl. Publ. mar. biol. Ass. India*, **1**, 362–365.

Khalaf, A.N., and Hall, R.E..(1975a). 'Hatching of the anostracan branchiopod *Chirocephalus diaphanus* Prevost 2. The role of embryonic movements', *Hydrobiologia*, **46**, 377–390.

Khalaf, A.N., and Hall, R.E. (1975b). 'Embryonic development and hatching of *Chirocephalus diaphanus* Prevost (Crustacea: Anostraca) in nature', *Hydrobiologia*, **47**, 1–11.

Khmeleva, N.N. (1967). 'Transformatsiya energii u *Artemia salina* (L.)', *Dokl. Akad. Sci. U.S.S.R. (Biol. Sci.)*, **207**, 633–636.

Kinne, O. (1954). 'Zur Biologie und Physiologie von *Gammarus duebeni* Lillj., I', *Z. wiss. Zool.*, **157**, 427–491.

Kinne, O. (1955). '*Neomysis vulgaris* Thompson, Eine autokologisch biologische Studie', *Biol. Zentralbl.*, **74**, 160–202.

Kinne, O. (1963). 'The Geschlechtsbestimmung des Flohkrebses *Gammarus duebeni* Lillj. (Amphipoda). Ist Temperaturabhangig: Eine Entgegenung', *Crustaceana*, **3**, 56–69.

Kinne, O. (1977). *Marine Ecology: A Comprehensive, Integrated Treatise on Life in Oceans and Coastal Waters*. 3. *Cultivation*, Part 2, John Wiley & Sons, Chichester.

Kiorboe, T., Mohlenberg, F., and Hamburger, K. (1985). 'Bioenergetics of the planktonic copepod *Acartia tonsa*: Relation between feeding, egg production and respiration and composition of specific dynamic action', *Mar. Ecol. Prog. Ser.*, **26**, 85–97.

Klepal, W., Barnes, H., and Mann, E.A. (1972). 'The morphology and histology of the cirripede penis', *J. exp. mar. Biol. Ecol.*, **10**, 243–265.

Kolman, W.A. (1960). 'The mechanism of natural selection for the sex ratio', *Amer. Natur.*, **95**, 375–377.

Krishnakumar, R. (1985). 'Studies on spermatheca of some decapod crustaceans', Ph.D. thesis, Calicut University, India.

Krishnan, L., and John, P.A. (1974). 'Observations on the breeding biology of *Melita zeylanica* Stebbing, a brackish water amphipod', *Hydrobiologia*, **44**, 413–430.

Kuris, A.M. (1971). 'Population interactions between a shore crab and two symbionts', Ph.D. thesis, University of California, Berkeley.

Kuris, A.M. (1974). 'Trophic interactions: Similarity of parasitic castrators to parasitoids', *Q. Rev. Biol.*, **49**, 129–148.

Kuris, A.M. (1978). 'Life-cycle distribution and abundance of *Carcinonemertes epialti*, a nemertean egg predator of the shore crab, *Hemigrapsus oregonensis* in relation to host size, reproduction and molt cycle', *Biol. Bull.*, **154**, 121–137.

Kurup, K.N.P., and Adiyodi, R.G. (1981). 'The programming of somatic growth and reproduction in the crab *Paratelphusa hydrodromous* (Herbst)', *Int. J. Inv. Reprod.*, **3**, 27–39.

Kurup, K.N.P., and Adiyodi, R.G. (1984). 'Multiple limb autotomy can trigger either ovarian growth or somatic growth in the freshwater crab *Paratelphusa hydrodromus* (Herbst)', *Gen. Comp. Endocr.*, **56**, 433–443.

Kurup, K.N.P., and Adiyodi, R.G. (1987). 'Inhibitory effect of brood on somatic growth is mediated through eyestalks in the freshwater crab *Paratelphusa hydrodromous* (Herbst)', *Indian J. exp. Biol.*, **25**, 510–513.

Kuttner, O. (1911). 'Mitteilungen über marine Cladoceran', *Sber. Ges. naturf. Freunde Berl.*, **2**, 84–93.

Kyomo, J. (1988). 'Analysis of relationship between gonads and hepatopancreas in males and females of the crab *Sesarma intermedia* with reference to resource use and reproduction', *Mar. Biol.*, **97**, 87–93.

Lasker, R., Wells, J.B.J., and McIntyre, A.D. (1970). 'Growth, reproduction, respiration and carbon utilization of the sand dwelling harpacticoid copepod, *Asellopsis intermedia*', *J. mar. biol. Ass. U.K.*, **50**, 147–160.

Lawrence, A.L. (1973). 'Are bacteria nutritionally important to shrimp larvae and postlarvae?', *World Maricult. Soc. Mtgs, Mexico*, **1**, 3.

Lawrence, A.L., Mountain, J., Persyn, H.O., and Laramore, C.R. (1977). 'The nutritional importance of dissolved organic matter and bacteria to juvenile shrimp (*Penaeus vannamei*) in a controlled water exchange system', *8th Ann. World Maricult. Soc. Mtgs*, **1**, 15.

Lee, C.M. (1972). 'Structure and function of the spermatophore and its coupling device in the Centrophagidae (Copepoda: Calanoida)', *Bull. mar. Ecol.*, **8**, 1–20.

Leigh, E.G., Charnov, E.L., and Warner, R.R. (1976). 'Sex ratio, sex change and natural selection', *Proc. natl Acad. Sci., U.S.A.*, **73**, 3656–3660.

Le Tourneau, M. (1961). 'Contribution a l'etude des Cladoceres de plancton du Golfe de Marseille, *Recl. Trav. Stn mar. Endoume*, **22**, 123–151.

Leuken, W. (1963). 'Zur Spermienspeicherung bei Armadillidien (Isopoda terrestria)', *Crustaceana*, **5**, 27–34.

Leung-Trujillo, J.R., and Lawrence, A.L. (1987). 'Observations on the decline in sperm quality of *Penaeus setiferus* under laboratory conditions', *Aquaculture*, **65**, 363–370.

Lewis, R.W. (1970). 'The densities of three classes of marine lipid in relation to their possible role as hydrostatic agents', *Lipids*, **5**, 151–153.

Lloyd, A.J., and Yonge, C.M. (1947). 'The biology of *Crangon vulgaris* L. in Briston channel and Severn estuary', *J. mar. biol. Ass. U.K.*, **26**, 157–176.

Lochead, J.H. (1961). 'Oviparity versus ovoviviparity in the brine shrimp, *Artemia*', *Biol. Bull.*, **121**, 396.

Losse, G.F., and Merrett, N.R. (1971). 'The occurrence of *Oratosquilla investigatoris* (Crustacea: Stomatopoda) in the pelagic zone of the Gulf of Aden and the equatorial western Indian Ocean', *Mar. Biol.*, **10**, 244–253.

Lowndes, A.C. (1929). 'Results of breeding experiments and other observations on *Cyclops vernalis* Fischer and *Cyclops rebustus* Sara', *Int. Rev. ges. Hydrobiol. Hydrogr.*, **21**, 171–188.

Lowndes, A.G. (1935). 'The sperms of freshwater ostracods', *Proc. Zool. Soc. (Lond.)*, **1**, 35–48.

Lucas, J.S. (1970). 'Breeding experiments to distinguish two sibling species of *Halicarcinus* (Crustacea, Brachyura)', *J. Zool. London*, **160**, 267–278.

Lucas, J.S. (1971). 'The larval stages of some Australian species of *Halicarcinus* (Crustacea Brachyura, Hymenosomatidae), I. Morphology', *Bull. mar. Sci.*, **21**, 471–490.

Lucas, J.S. (1975). 'The larval stages of some Australian species of *Halicarcinus* (Crustacea, Brachyura, Hymenosomatidae). 3. Dispersal', *Bull. mar. Sci.*, **25**, 94–100.

Lumare, F. (1981). 'Artificial reproduction of *Penaeus japonicus* Bate as basis for the mass production of eggs and larvae', *J. World Maricult. Soc.*, **12**, 335–344.

Lynch, M. (1980). 'The evolution of cladoceran life histories', *Q. Rev. Biol.*, **55**, 23–42.

MacArthur, R.H., and Wilson, E.O. (1967). *Theory of Island Biogeography*, Princeton University Press, Princeton, 203 pp.

MacDiarmid, A.B. (1989). 'Size at onset of sexual maturity and size-dependent reproduction of female and male spiny lobster *Jasus edwardsii* (Hutton) (Decapoda: Palinuridae) in northern New Zealand', *J. exp. mar. Biol. Ecol.*, **127**, 229–243.

Main, R.A. (1962). 'The life history and food relations of *Epischura lacustris* Forbes (Copepoda: Calanoida)', Ph.D. thesis, University of Michigan, Ann Arbor.

Maly, E. (1973). 'Density, size and clutch of two high altitude diaptomid copepods', *Limnol Oceanogr.*, **18**, 840–848.

Maly, E.J. (1970). 'The influence of predation on the adult sex ratios of two copepod species', *Limnol. Oceanogr.*, **15**, 568–573.

Marcus, N.H. (1990). 'Calanoid copepod, cladoceran, and rotifer eggs in sea-bottom sediments of northern Californian coastal waters: Identification, occurrence and hatching', *Mar. Biol.*, **105**, 413–418.

Markus, H.C. (1930). 'Studies on the morphology and life history of the isopod *Mancasellus macrourus*', *Trans. Am. Microsc. Soc.*, **49**, 220–237.

Marshall, S.M., and Orr, A.P. (1952). 'On the biology of *Calanus finmarchicus*. 8. Factors affecting egg production', *J. mar. biol. Ass. U.K.*, **30**, 527-547.

Marshall, S.M., and Orr, A.P. (1955). *The Biology of a Marine Copepod,* Calanus finmarchicus *(Gunnerus)*, Oliver and Boyd, Edinburgh.

Martin, G.G., Herzig, C., and Narimatsu, G. (1987). 'Fine structure and histochemistry of the partly extruded and hardened spermatophore of the spiny lobster, *Panulirus interruptus*', *J. Morphol.*, **192**, 237-246.

Mathad, S.G., and Adiyodi, K.G. (1990). 'Changes in biochemical composition of the testis and semen during different phases of the intermoult in the freshwater crab, *Paratelphusa hydrodromous* (Herbst)', *Proc. Ind. nat. Sci. Acad.*, **56B**, 259-264.

Mathavan, S. (1985). 'Yolk utilization in the developing eggs of *Brachythermis contaminata* and *Orthetrum sabina*', *Proc. First Indian Symp. Odonatol.*, **1**, 107–114.

Mathews, D.C. (1954). 'The development of spermatophoric mass of the rock lobster *Parribacus antaroticus* (Lond.)', *Pac. Sci.*, **8**, 28–34.

Mathias, J.A. (1971). 'Energy flow and secondary production of the amphipods *Hyalella azteca* and *Crangonyx richmondensis occidentalis* in Marion lake, British Columbia', *J. Fish. Res. Board, Can.*, **28**, 711–726.

Mathias, P. (1937). 'Biologie des Crustacés Phyllopodes', *Actualité scient. ind.*, **47**, 1–107.

Mauchline, J. (1960). 'The biology of the euphausiid crustacean *Meganyctiphanes norvegica* (M. Sars)', *Proc. Roy. Soc. Edinb. (Biol.)*, **67B**, 141–179.

Mauchline, J. (1966). '*Thalassomyces fagei*, an ellobiopsid parasite of the euphausiid crustacean *Thysanoessa raschi*', *J. mar. biol. Ass. U.K.*, **46**, 531–539.

Mauchline, J. (1971a). 'The biology of *Praunus flexuosus* and *P. neglectus* (Crustacea: Mysidacea)', *J. mar. biol. Ass. U.K.*, **51**, 641–652.

Mauchline, J. (1971b). 'Seasonal occurrence of mysids (Crustacea) and evidence of social behaviour', *J. mar. biol. Ass. U.K.*, **51**, 809–825.

Mauchline, J. (1973). 'The broods of the British Mysidacea (Crustacea)', *J. mar. biol. Ass. U.K.*, **53**, 801–817.

Mauchline, J. (1988). 'Egg and brood sizes of oceanic pelagic crustaceans', *Mar. Ecol. Prog. Ser.*, **43**, 251–258.

Mauchline, J., and Fisher, L.R. (1969). 'The biology of euphausiids', *Adv. Mar. Biol.*, **7**, 1–454.

McClean, D.K., and Warner, A.H. (1971). 'Aspects of nucleic acid metabolism during development of the brine shrimp *Artemia salina*', *Develop. Biol.*, **24**, 88–105.

McIntyre, R.J. (1954). 'A common sand-hopper', M.Sc. thesis, Canterbury University.

McLaren, I.A. (1965). 'Some relationships between temperature and egg size, development rate and fecundity of the copepod, *Pseudocalanus*', *Limnol. Oceanogr.*, **10**, 529–538.

McLaren, I.A. (1966). 'Predicting development rate of copepod eggs', *Biol. Bull.*, **131**, 457–469.

McLaren, I.A. (1978). 'Generation lengths of some temperate marine copepods: Estimation, prediction and implications', *J. Fish. Res. Board, Can.*, **35**, 1330–1342.

McLaren, I.A., Corkett, C.J., and Zilloux, E.J. (1969). 'Temperature and adaptation of copepod eggs from the Arctic to the tropics', *Biol. Bull.*, **137**, 486–493.

McNaughton, S.J., and Wolf, L.L. (1974). *General Ecology*, Holt, Reinhart and Winston, New York.

Menge, B.A. (1974). 'Effect of wave action and competition on brooding and reproductive effort in the seastar *Leptasterias hexactis*', *Ecology*, **55**, 84–93.

Menzies, R.J. (1972). 'Experimental interbreeding between geographically separated populations of the marine wood-boring isopod *Limnoria tripunctata* with preliminary indications of hybrid vigor', *Mar. Biol.*, **17**, 149–157.

Mercier, L., and Poisson, R. (1929). 'Alteration des certains caractères sexuals secondaires du mâle de *Pinnotheres pisum* parasite par un Entoiscien', *Bull. Soc. Zool. Fr.*, **54**, 301–304.

Middleditch, B.S., Missler, S.R., and Hines, H.B. (1980). 'Metabolic profiles of penaeid shrimp: Dietary lipids and ovarian maturation', in *Advances in Chromatography* (Ed. A. Zlatkiss), University of Houston, Texas, pp. 713–721.

Moore, W.G. (1967). 'Factors affecting egg hatching in *Streptocephalus seali* (Branchiopoda: Anostraca)', *Proc. Symp. Crust.*, **2**, 724–735.

Moreira, G.S., and Vernberg, W.B. (1968). 'Comparative thermal matabolic patterns in *Euterpina acutifrons* dimorphic males', *Mar. Biol.*, **1**, 282–284.

Morgen, P.R. (1972). 'The influence of prey availability on the distribution and predatory behaviour of *Nucella lapillus* (L.)', *J. Anim. Ecol.*, **41**, 257–274.

Morizur, Y., Conam, G., Gruenole, A., and Omres, M.A. (1981). 'Fécundité de *Nephrops norvegicus* dans le golfe de Gascogne', *Mar. Biol.*, **63**, 319–324.

Morris, J.E. (1968). 'Dehydrated cysts of *Artemia salina* prepared for electron microscopy by totally anhydrous techniques', *J. Ultrastr. Res.*, **25**, 64–72.

Morris, J.E. (1971). 'Hydration, its reversibility, and the beginning of development in the brine shrimp *Artemia salina*', *Comp. Biochem. Physiol.*, **39A**, 843–857.

Morris, J.E., and Afzelius, B.A. (1967). 'The structure of the shell and outer membranes in encysted *Artemia salina* during cryptobiosis and development', *J. Ultrastr. Res.*, **20**, 244–259.

Morris, R.J. (1973). 'Relationships between the sex and degree of maturity of marine crusataceans and their lipid compositions', *J. mar. biol. Ass. U.K.*, **53**, 27–37.

Mouchet, S. (1931). 'Spermatophores des Crustacés décapodes, Anomoures et Brachyoures et castration parasitaire chez quelques *Pagurus*', *Annls Stn océanogr. Salammbô*, **6**, 1–203.

Munn, E.A., and Barnes, H. (1970). 'The structure of the axial filament complex of the spermatozoa of *Balanus balanus*', *Exp. Cell Res.*, **60**, 277–284.

Murphy, G.I. (1968). 'Patterns of life history and the environment', *Amer. Natur.*, **102**, 391–403.

Murugan, N. (1975a). 'The biology of *Ceriodaphnia cornuta* Sars (Cladocera, Daphnidae)', *J. Inland Fish. Soc. India*, **7**, 81–87.

Murugan, N. (1975b). 'Egg production, development and growth in *Moina micrura* Kurz (1874) (Cladocera: Moinidae)', *Freshwat. Biol.*, **5**, 245–250.

Murugan, N., and Sivaramakrishnan, K.G. (1973). 'The biology of *Simocephalus acutirostratus* King (Cladocera: Daphnidae). Laboratory studies of life span, instar duration, egg production, growth and stages in embryonic development', *Freshwat. Biol.*, **3**, 77–87.

Murugan, N., and Sivaramakrishnan, K.G. (1976). 'Laboratory studies on the longevity, instar duration, growth, reproduction and embryonic development in *Scapholeberis kingi* Sars (1903) (Cladocera: Daphnidae)', *Hydrobiologia*, **50**, 75–80.

Muthukrishnan, J., and Pandian, T.J. (1987). 'Insecta', in *Animal Energetics* (Eds. T.J. Pandian and F.J. Vernberg), Academic Press, New York, Vol. 1, pp. 373–511.

Naess, T. (1991). 'Marine calanoid resting eggs in Norway: Abundance and distribution of two copepod species in the sediment of an enclosed marine basin', *Mar. Biol.*, **110**, 261–266.

Nair, K.K.G., and Anger, K. (1979a). 'Life cycle of *Corophium insidiosum* (Crustacea: Amphipoda) in laboratory culture', *Helgoländer wiss. Meeresunters.*, **32**, 279–294.

Nair, K.K.G., and Anger, K. (1979b). 'Experimental studies on the life cycle of *Jassa falcata* (Crustacea: Amphipoda)', *Helgoländer wiss. Meeresunters.*, **32**, 444–452.

Nakanishi, Y., Iwasaki, T., Okigaki, T., and Kato, H. (1962). 'Cytological studies of *Artemia salina*. 1. Embryonic development without cell multiplication after the blastula stage', *Ann. Zool. Japon.*, **35**, 223–228.

Narayanan, S. (1990). 'Reproductive physiology of female decaped Crustacea', Ph.D. thesis, Calicut University, India.

Navaneethakrishnan, P., and Michael, R.G. (1971). 'Egg production and growth in *Daphnia carinata* King', *Proc. Indian Acad. Sci.*, **73**, 117–123.

Naylor, E. (1955). 'The life cycle of the isopod *Idotea emarginata* (Fabricus)', *J. Anim. Ecol.*, **24**, 270–281.

Needham, J. (1931). *Chemical Embryology*, Cambridge University Press, Cambridge.

Needham, J. (1950). *Biochemistry and Morphogenesis*, Cambridge University Press, Cambridge.

Needham, J., and Needham, D.M. (1930). 'On phosphorus metabolism in embryonic life. 1. Invertebrate eggs', *J. Exp. Biol.*, **7**, 317–348.

Nelson, W.G. (1980). 'Reproductive patterns in gammaridean amphipods', *Sarsia*, **65**, 61–71.

Nemoto, J., Kamada, K., and Hara, K. (1972). 'Fecundity of a euphausiid crustacean *Nematoscelis difficilis* in north Pacific Ocean', *Mar. Biol.*, **14**, 41–47.

Nemoto, T., Mauchline, J., and Kamada, K. (1976). 'Brood size and chemical composition of *Pareuchaeta norvegica* (Crustacea: Copepoda) in Loch Etiv, Scotland', *Mar. Biol.*, **36**, 151–158.

Neunes, H.W., and Pongolini, G.F. (1965). 'Breeding of a pelagic copepod, *Euterpina acutifrons* (DANA) in the laboratory', *Nature, Lond.*, **208**, 571–573.

Nicholls, A.G. (1931). 'Studies on *Ligia oceanica*. 1. a. Habitat and effect of change of environment on respiration. b. Observations on moulting and breeding', *J. mar. biol. Ass. U.K.*, **17**, 655–673.

Nilson, E.H., Fisher, W.S., and Shleser, R.A. (1976). 'A new mycosis of larval lobster (*Homarus americanus*)', *J. Inv. Pathol.*, **27**, 117–183.

Norse, E.A., and Fox-Norse, V. (1977). 'Studies on portunid crabs from the eastern Pacific. 2. Significance of the unusual distribution of *Euphylax dovii*', *Mar. Biol.*, **40**, 374–376.

Oikawa, P.G., and Smith, M. (1966). 'Nucleotides in the encysted embryos of *Daphnia magna*', *Biochemistry*, **5**, 1517–1521.

Olson, C.S., and Clegg, J.S. (1976). 'Nuclear numbers in the encysted dormant embryos of different *Artemia salina* populations', *Experientia*, **32**, 864–865.

Olson, C.S., and Clegg, J.S. (1978). 'Cell division during development of *Artemia salina*', *Roux's Arch. Dev. Biol.*, **184**, 1–13.

Onbe, T. (1974). 'Studies on the ecology of marine cladoceran', *J. Fac. Fish. Anim. Husb. Hiroshima Univ.*, **13**, 83–179.

Onbe, T. (1977). 'The biology of marine cladocerans in a warm temperature water', *Proc. Symp. Warm Water Zooplankton*, National Institute of Oceanography, Goa, **1**, 383–398.

Paffenhofer, G.A. (1970). 'Cultivation of *Calanus helgolandicus* under controlled conditions', *Helgoländer wiss. Meeresunters.*, **20**, 346–359.

Paffenhofer, G.A., and Harris, R.P. (1976). 'Feeding, growth and reproduction of the marine planktonic copepod *Pseudocalanus elongatus* Boeck', *J. mar. biol. Ass. U.K.*, **56**, 327–344.

Palanichamy, S. (1980). 'Ecophysiological studies on a chosen spider *(Cyrtophora cicatrosa)*', Ph.D. thesis, Madurai Kamaraj University.

Pandian, T.J. (1967). 'Changes in chemical composition and caloric content of developing eggs of the shrimp *Crangon crangon*', *Helgoländer wiss. Meeresunters.*, **16**, 216–224.

Pandian, T.J. (1970a). 'Ecophysiological studies on developing eggs and embryos of the European lobster *Homarus gammarus*', *Mar. Biol.*, **5**, 153–157.

Pandian, T.J. (1970b). 'Yolk utilization and hatching time in the Canadian lobster *Homarus americanus*' *Mar. Biol.*, **7**, 249–254.

Pandian, T.J. (1972). 'Egg incubation and yolk utilization in the isopod *Ligia oceanica*', *Proc. Indian Nat. Sci. Acad.*, **38**, 430–441.

Pandian, T.J. (1975). 'Mechanisms of heterotrophy', in *Marine Ecology*, Vol. 2, Part I (Ed. O. Kinne), Wiley, London, pp. 61–249.

Pandian, T.J. (1980). 'Impact of dam-building on marine life', *Helgoländer wiss. Meeresunters.*, **33**, 415–421.

Pandian, T.J. (1987) 'Fish', in *Animal Energetics* (Eds. T.J. Pandian and F.J. Vernberg), Academic Press, New York, Vol. 2, pp. 358–467.

Pandian, T.J., and Balasundaram, C. (1980). 'Contribution to the reproductive biology and aquaculture of *Macrobrachium nobilii*', *Proc. Symp. Inv. Reprod., Madras University, Madras*, **1**, 183–193.

Pandian, T.J., and Balasundaram, C. (1982), 'Moulting and breeding cycle in *Macrobrachium nobilii*, *Int. J. Inv. Reprod.*, **5**, 21–30.

Pandian, T.J., and Katre, S. (1972). 'Effect of hatching time on larval mortality and survival of the prawn *Macrobrachium idae*', *Mar. Biol.*, **13**, 330–337.

Pandian, T.J., and Schumann, K.H. (1967). 'Chemical composition and caloric content of egg and zoea of the hermit crab *Eupagurus bernhardus*', *Helgoländer wiss. Meeresunters.*, **16**, 225–230.

Panouse, J.B. (1947). 'La glands du sinus et la maturation des produits génitaux chez les crevetts', *Bull. Biol. Fr. Belg.* (Suppl.), **33**, 160–163.

Paris, O.H., and Pitelka, F.A. (1962). 'Population characteristics of the terrestrial isopod *Armadillidium vulgare* in California grassland', *Ecology*, **43**, 229–248.

Parrish, K.K., and Wilson, D.F. (1978). 'Fecundity studies on *Acartia tonsa* (Copepoda: Calanoida) in standardized culture', *Mar. Biol.*, **46**, 65–81.

Patel, B., and Crisp, D.J. (1960). 'Rates of development of the embryos of several species of barnacles', *Physiol. Zool.*, **33**, 104–119.

Pennak, R.W. (1953). *Freshwater Invertebrates of the United States*, The Ronald Press, New York.

Perez, O.S. (1990). 'Reproductive biology of the sandy shore crab *Mutata lunaris* (Brachyura: Calappidae)', *Mar. Ecol. Prog. Ser.*, **59**, 83–89.

Perkins, H.C. (1971). 'Egg loss during incubation from offshore northern lobsters (Decapoda: Homaridae)', *Fish. Bull.*, **69**, 451–453.

Pertzova, N.N. (1968). 'Life cycle and ecology of a thermophilous copepod *Centropages hamatus* in the White Sea', *Zool. Zh.*, **53**, 1013–1022.

Perysn, H.O. (1977). *Artificial Insemination of Shrimp*, United States Patent, **4**, 031, 855, 4 pp.

Peterson, W.T. (1988). 'Rates of egg production by the copepod *Calanus marshallae* in the laboratory and in the sea off Oregon, USA', *Mar. Ecol. Prog. Ser.*, **47**, 229–237.

Petipa, T.S. (1967). 'On the efficiency of utilization of energy in pelagic ecosystems of the Black Sea', *J. Fish. Res. Board, Can.* (Transl.), **973**, 44–65.

Phillips, G. (1971). 'Incubation of the eggs of the English prawn *Palemon serratus*', *J. mar. biol. Ass. U.K.*, **51**, 43–48.

Pianka, E.R. (1970). 'On r- and K-selection', *Amer. Natur.*, **104**, 592–597.

Pianka, E.R. (1974). *Evolutionary Ecology*, Harper and Row, New York.

Pike, R.B. (1960). 'The biology and post-larval development of the biopyrid parasites *Pseudions affinis* G.O. Sars and *Hemiarthrus abdominalis* (Krøyer) (*Phryxus abdominalis* (Krøyer)', *J. Linn. Soc.*, **44**, 239–251.

Pike, R.B. (1961). 'Observations on the epicaridea obtained from hermit crabs in British waters, with notes on the longevity of the host species', *Ann. Mag. Nat. Hist.*, **13**, 225–240.

Pillai, N.N., and Mohammed, K.H. (1973). 'Larval history of *Macrobrachium idella* (Higendorf) reared in the laboratory', *J. mar. biol. Ass. India*, **15**, 359–385.

Pillay, K.K., and Nair, N.B. (1971). 'The annual reproductive cycles of *Uca annulipes*, *Portunus pelagicus* and *Metapenaeus affinis* (Decapoda: Crustacea) from the south-west coast of India', *Mar. Biol.*, **11**, 152–166.

Pillay, K.K., and Nair, N.B. (1972). 'Reproductive biology of the sessile barnacle *Balanus amphitrite communis* (Darwin), of the south-west coast of India', *Indian J. mar. Sci.*, **1**, 8–16.

Ponnuchamy, R., Ayyappan, A., Reddy, S.R., and Katre, S. (1979). 'Yolk and copper utilization during embryogenesis of the freshwater prawn *Caridina nilotica*', *Proc. Indian Acad. Sci.*, **88**, 353–362.

Poulet, S.A. (1976). 'Feeding of *Pseudocalanus minutus* on living and non-living particles', *Mar. Biol.*, **34**, 117–126.

Prakash, R.N., and Pandian, T.J. (1978). 'Energy flow from spider eggs through dipteran parasite and hymenopteran hyperparasite populations', *Oecologia* **33**, 209–219.

Prophet, C. (1962). 'Ecology and reproduction of five species of Anostraca in Oklahoma', Ph.D. thesis, University of Oklahoma.

Purasjoki, K.J. (1958). 'Zur Biologie der Brachwasser Kladozera *Bosmina coregoni maritima* (P.E. Muller)', *Annls zool. Soc., Vanamo*, **19**, 1–117.

Quackenbush, L.S., and Herrnkind, W.F. (1981). 'Regulation of molt and gonadal development with spiny lobster *Panulirus argus* (Crustacea: Palinuridae): Effect of eyestalk ablation', *Comp. Biochem. Physiol.*, **69A**, 523–552.

Radha, T., and Subramoniam, T. (1985). 'Origin and nature of the spermatophoric mass of the spiny lobster *Panulirus homarus*', *Mar. Biol.*, **86**, 13–19.

Rahman, Md. K., and Subramoniam, T. (1989). 'Molting and its control in the female sand lobster *Thenus orientalis* (Lund)', *J. exp. mar. Biol. Ecol.*, **128**, 105–115.

Raman, K. (1976). *Report on the International Conference on Prawn Farming*, Vung Tan, South Vietnam, Misc. Contribution No. 12. Central Inland Fisheries Research Institute, Barrackpur, India.

Rammer, W. (1933). 'Wird der cladoceren-Embryo von Muttertier ernahrt?', *Arch. Hydrobiol.*, **25**, 692–698.

Rao, N., Ponnuchamy, R., Katre, S., and Reddy, S.R. (1980). 'Fecundity and energetics of embryonic metabolism of *Caridina weberi* (de Man) (Decapoda: Atyidae)', *Int. J. Inv. Reprod.*, **3**, 75–85.

Rasmussen, E. (1959). 'Behaviour of sacculanized shore crabs (*Carcinus maenas* Pennant)', *Nature, Lond.*, **183**, 479–480.

Reaka, M.L. (1979). 'The evolutionary ecology of life history patterns in stomatopod Crustacea', in *Reproductive Ecology of Marine Invertebrates* (Ed. S.E. Stancyk), Univ. South Carolina Press, Columbia, pp. 235–260.

Reeve, M.R. (1969). 'The laboratory culture of the prawn *Palaemon serratus*', *Fish. Invest. London*, **2**, 26–38.

Reid, G.K. (1961). *Ecology of Inland Waters and Estuaries*, Reinhold Publishing Corporation, New York.

Reinhard, R.G. (1956). 'Parasitic castration of Crustacea', *Exp. Parasitol.*, **5**, 79–107.

Reiper, M. (1978). 'Bacteria as food for marine harpacticoid copepods', *Mar. Biol.*, **45**, 337–346.

Reverberi, C., and Pittoti, M. (1942). 'Il ciclo biologico e la determinazione fenotipica del sesso di *Ione thoracica* Montagu, Bopiride parassito di *Callianassa laticaudata* Otto', *Pubbl. Staz. zool. Napoli*, **19**, 111–184.

Rice, A.L., and Chapman, C.J. (1971). 'Observations on the burrows and burrowing behaviour of two mud-dwelling decapod crustaceans, *Nephrops norvegicus* and *Coneplax rhomboides*', *Mar. Biol.*, **10**, 330–342.

Richman, S. (1958). 'The transformation of energy by *Daphnia pulex*', *Ecol. Monogr.*, **28**, 273–291.

Ricklefs, R.E. (1973). *Ecology*, Nelson, London.

Riley, J.P., and Skirrow, C. (1975). *Chemical Oceanography*, Vol. IV, Academic Press, London.

Roen, U. (1957). 'Contributions to the biology of some Danish free-living freshwater Copepoda', *Biol. Skr.*, **9**, 1–100.

Roff, J.C. (1972). 'Aspects of the reproductive biology of the planktonic copepod *Limnocalanus macrurus* Sars. 1863', *Crustaceana*, **22**, 155–160.

Runge, J.A. (1985). 'Relationship of egg production of *Calanus pacificus* to seasonal changes in phytoplankton availability in Puget Sound, Washington', *Limnol. Oceanogr.*, **30**, 382–396.

Ryan, B.P. (1965). 'Structure and function of the reproductive system of the crab *Portunus sanguinolentus* (Herbst) (Brachyura: Portunidae) 1. The male system', *Proc. Symp. Crust., Mar. biol. Ass. India*, Abstr. pap. 22.

Saito, S. (1965). 'Structure and energetics of the population of *Ligidium japonica* (Isopoda) in a warm temperate forest ecosystem', *Jap. J. Ecol.*, **15**, 47–55.

Sam, S.T., and Krishnaswamy, S. (1979). 'Effect of osmomolarity of the medium upon hatching of undried eggs of *Streptocephalus dichotomous* Baird (Anostraca-Crustacea)', *Arch. Hydrobiol.*, **86**, 125–130.

Sandifer, P.A. (1975). 'The role of pelagic larvae in recruitment to populations of adult decapod crustaceans in the York river estuary and adjacent lower Chesapeake Bay, Virginia', *Est. Coast. Mar. Sci.*, **3**, 269–279.

Sandifer, P.A., and Lynn, J.W. (1980). 'Artificial insemination of caridean shrimp', in *Advances in Invertebrate Reproduction* (Eds. W.H. Clark, Jr. and T.S. Adams), Elsevier/North-Holland, New York, pp. 271–288.

Sandifer, P.A., and Smith, T.I.J. (1979). 'A method for artificial insemination of *Macrobrachium* prawns and its potential use in inheritance and hybridisation studies', *Proc. World Mariculture Soc.*, **10**, 403–418.

Santharam, K.R. (1979). 'Biology of *Daphnia carinata* King (Cladocera: Daphnidae)', Ph.D. thesis, Madurai Kamaraj University.

Sarda, P., Cros, M.L., and Sese, B. (1989). 'Ca balance during moulting in the prawn *Aristeus antennatus* (Risso, 1816): The role of cuticle calcification in the life cycle of decapod crustaceans', *J. exp. mar. Biol. Ecol.*, **129**, 161–171.

Sasakawa, Y. (1975). 'Studies on blue king crab resources in the western Berring Sea III', *Bull. Jap. Soc. Sci. Fish.*, **41**, 941–944.

Sasikala, S.L., and Subramoniam, T. (1987). 'On the occurrence of acid mucopolysaccharides in the spermatophores of two marine prawns, *Penaeus indicus* (Milne-Edwards) and *Metapenaeus monoceros* (Fabricius) (Crustacea: Macrura)', *J. exp. mar. Biol. Ecol.*, **113**, 145–153.

Sassaman, C. (1991). 'Sex ratio variations in female-biased populations of Notostracans, *Hydrobiologia*', **212**, 169–179.

Saudray, Y., and Lemercier, A. (1960). 'Observations sur le développement des oeufs de *Ligia oceanica* Fab. Crustacé, Isopode, Oniscoide', *Bull. Inst. océanogr. Monaco*, **57**, 1–11.

Sazhina, L.I. (1968). 'On hibernating eggs of marine Calanoida', *Zool. Zh.*, **47**, 1554–1556.

Sazhina L.I. (1971). 'Fertility of the mass pelagic copepoda in the Black Sea', *Zool. Zh.*, **50**, 586–588.

Schmitt, W.L. (1965). *Crustaceans*, University of Michigan Press, Ann Arbor.

Schöne, H. (1961). 'Complex behaviour', in *The Physiology of Crustacea, 2. Sense Organs, Integration, and Behaviour* (Ed. T.H. Waterman), Academic Press, New York, pp. 465–515.

Scudamore, H.H. (1948). 'Factors influencing moulting and the sexual cycles in the crayfish', *Biol. Bull.*, **95**, 229–237.

Seale, D.B. (1987). 'Amphibia' in *Animal Energetics* (Eds. T.J. Pandian and F.J. Vernberg), Academic Press, New York, Vol. 2, pp. 467–552.

Sharma, S., Sharma, B.K., and Michael, R.G. (1984). 'The biology of *Daphnia lumholtzi* Sars (Cladocera: Daphniidae) — Laboratory studies on longevity, instar duration, egg production and growth', *Proc. natl Acad. Sci. India*, **54B**, 306–314.

Sheader, M. (1977a). 'The breeding biology of *Iodotea pelagica* (Isopoda; Valvifera) with notes on the occurrence and biology of its parasite *Clypeoniscus hanseni* (Isopoda: Epicaridea)', *J. mar. biol. Ass. U.K.*, **57**, 659–674.

Sheader, M. (1977b). 'Breeding and marsupial development in laboratory maintained *Parathemisto gaudichaudi* (Amphipoda)', *J. mar. biol. Ass. U.K.*, **57**, 943–954.

Sheader, M. (1978). 'Distribution and reproductive biology of *Corophium insidiosum* (Amphipoda) on the northeast coast of England', *J. mar. biol. Ass. U.K.*, **58**, 586–596.

Shillaker, R.O., and Moore, P.G. (1987). 'The biology of brooding of the amphipods *Lembos websteri* Bate and *Corophium bonnellii* Milne Edwards', *J. exp. mar. Biol. Ecol.*, **110**, 113–132.

Shirugur, A.G., and Naik, A.A. (1977). 'Observation on morphology, taxonomy, epihippial hatching and laboratory culture of a new species of *Alona* (*Alona taraporevalae* Shirgur and Naik), a chydorid cladoceran from back bay, Bombay'. *Proc. Symp. Warm Water Zooplankton*, National Institute of Oceanography, Goa, pp. 48–69.

Shlagman, A., Lewinsohn, C., and Tom, M. (1986). 'Aspects of the reproductive activity of *Penaeus semisulcatus* de Haan along the southeastern coast of the Mediterranean', *P.S.Z.N.I. Mar. Ecol.*, **7**, 15–22.

Silverbauer, B.I. (1971). 'The biology of South African rock lobster *Jasus lalandi* (H. Milne-Edwards). 1. Development', *Invest. Rep. Div. Fish. S. Afr.*, **92**, 1–70.

Sindhukumari, S., and Pandian, T.J. (1987). 'Effects of unilateral eyestalk ablation on moulting, growth, reproduction and energy budget of *Macrobrachium nobilii*', *Asian Fish. Sci.*, **1**, 1–17.

Sindhukumari, S., and Pandian, T.J. (1991). 'Interaction of ration and unilateral eyestalk ablation on energetics of female *Macrobrachium nobilii*', *Asian Fish. Sci.*, **4**, 227–244.

Skoultchi, A.I., and Morowitz, H.J. (1964). 'Information, storage and survival of biological systems at temperatures near absolute zero', *Yale J. Biol. Med.*, **37**, 158–163.

Sloan, N.A. (1984). 'Incidence and effects of parasitism by the rhizocephalan barnacle *Briarosaccus* Boschma in the golden king crab *Lithodes aequispina* Benedict from deep waters in northern British Columbia, Canada', *J. exp. mar. Biol. Ecol.*, **84**, 111–131.

Sloan, N.A. (1985) 'Life-history characteristics of fjord-dwelling golden king crab *Lithodes aequispina*', *Mar. Ecol. Prog. Ser.*, **22**, 219–228.

Slobodkin, L.B. (1960). 'Energy in animal ecology', *Adv. Ecol. Res.*, **1**, 69–101.

Slobodkin, L.B., and Richman, S. (1961). 'Calories/gm in species of animals', *Nature, Lond.*, **191**, 299.

Smith, C.W. (1905). 'Note on a gregarine (*Aggregata inachi* n.sp.) which may cause the parasitic castration of its host (*Inachus dorsettensis*)', *Mitt. Zool. Sta. Neapel*, **17**, 406–410.

Smyly, W.J. (1962). 'Laboratory experiments with stage V copepodids of the freshwater copepod, *Cyclops leuckarti* Claus, from Windermere and Esthwaite water', *Crustaceana*, **4**, 273–280.

Smyly, W.J.P. (1970). 'Observations on rate of development, longevity and fecundity of *Acanthocyclops viridis* (Jurine) (Copepoda: Cyclopoida) in relation to type of prey', *Crustaceana*, **18**, 21–36.

Snipes, W.C., and Gordy, W. (1963). 'Radiation damage to *Artemia* cysts: Effects of water vapour', *Science*, **142**, 503–504.

Snodgrass, R.E. (1956). 'Crustacean metamorphoses', *Smithsonian Inst. Publs Misc. Coll.*, **131**, 1–78.

Sorgeloos, P., Bossuyt, E., Lavina, E., Baeza-mesa, M., and Persoone, G. (1977). 'Decapsulation of *Artemia* cysts: A simple technique for the improvement of the use of brine shrimp in aquaculture', *Aquaculture*, **12**, 311–315.

Spaargaren, D.H. (1979). 'Hydrodynamic properties of benthic marine Crustacea. 1. Specific gravity and drag coefficients', *Mar. Ecol. Prog. Ser.*, **1**, 351–359.

Stancyk, S.E., and Moreira, G.S. (1988). 'Inheritance of male dimorphism in Brazilian populations of *Euterpina acutifrons* (Dana) (Copepoda: Harpacticoida)', *J. exp. mar. Biol. Ecol.*, **120**, 125–144.

Stearns, S.C. (1976). 'Life history tactics: A review of ideas', *Q. Rev. Biol.*, **51**, 3–47.

Steele, D.H. (1972). 'Some aspects of the biology of the *Gammarellus homari* (Crustacea: Amphipoda) in the northwestern Atlantic', *J. Fish. Res. Board, Can.*, **29**, 1340–1343.

Steele, D.H., and Steele, V.J. (1969). 'The biology of *Gammarus* (Crustacea, Amphipoda) in the northwestern Atlantic. *Gammarus duebeni* Lillij.', *Can. J. Zool.*, **47**, 235–244.

Steele, D.H., and Steele, V.J. (1970a). 'The biology of *Gammarus* (Crustacea: Amphipoda) on the northwestern Atlantic. 3. *Gammarus obtusatus* Dahl', *Can. J. Zool.*, **48**, 989–995.

Steele, D.H., and Steele, V.J. (1970b). 'The biology of *Gammarus* (Crustacea: Amphipoda) 4. *Gammarus lawrencianus* Bousefield', *Can. J. Zool.*, **48**, 1261–1267.

Steele, D.H., and Steele, V.J. (1972a). 'The biology of *Gammarus* (Crustacea: Amphipoda) in the northwestern Atlantic. 6. *Gammarus tigrinus* Sexton', *Can. J. Zool.*, **50**, 1063–1068.

Steele, D.H., and Steele, V.J. (1972b). 'The biology of *Jaera* spp. (Crustacea: Isopoda) in the northwestern Atlantic 1. *Jaera ischisetosa*', *Can. J. Zool.*, **50**, 205–211.

Steele, D.H., and Steele, V.J. (1975a). 'The biology of *Gammarus* (Crustacea: Amphipoda) in the northwestern Atlantic. 9. *Gammarus wilkitzkii* Birula, *Gammarus stoerensis* Reid, and *Gammarus* Say', *Can. J. Zool.*, **53**, 1105–1109.

Steele, D.H., and Steele, V.J. (1975b). 'The biology of *Gammarus* (Crustacea: Amphipoda) in the northwestern Atlantic. 10. *Gammarus finmarchicus* Dahl', *Can. J. Zool.*, **53**, 1110–1115.

Steele, D.H., and Steele, V.J. (1975c). 'The biology of *Gammarus* (Crustacea: Amphipoda) in the northwestern Atlantic. 11. Comparison and discussion', *Can. J. Zool.*, **53**, 1116–1126.

Steele, D.H., and Steele, V.J. (1975d). 'Egg size and duration of embryonic development in Crustacea', *Int. Rev. ges. Hydrobiol.*, **60**, 711–715.

Steele, E.A. (1961). 'Some observations on the life history of *Asellus aquaticus* (L.) and *Asellus meridianus* Racovitza (Crustacea: Isopoda)', *Proc. Zool. Soc. (Lond.)*, **137**, 71–87.

Steele, V.J., and Steele, D.H. (1970). 'The biology of *Gammarus* (Crustacea: Amphipoda) in the northwestern Atlantic. 2. *Gammarus setosus* Dementieva', *Can. J. Zool.*, **48**, 659–671.

Steele, V.J., and Steele, D.H. (1972). 'The biology of *Gammarus* (Crustacea: Amphipoda) in the northwestern Atlantic. 5. *Gammarus oceanicus* Segestrale', *Can. J. Zool.*, **50**, 801–813.

Steele, V.J., Steele, D.H., and MacPherson, B.R. (1977). 'The effect of photoperiod on the reproductive cycle of *Gammarus setosus* Dementieva, 1931', *Crustaceana* Suppl., **4**, 58–63.

Stocco, D.M., Beer, P.C., and Warner, A.H. (1972). 'Effect of anoxia on nucleotide metabolism in the encysted embryos of the brine shrimp', *Develop. Biol.*, **27**, 479–493.

Stross, R.G. (1969). 'Photoperiod control of diapause in *Daphnia*. 2. Induction of winter diapause in the Arctic', *Biol. Bull.*, **136**, 264–273.

Stross, R.G. (1971). 'Photoperiod control of diapause in *Daphnia*. 4. Light and Co_2-sensitive phase within the cycle of activation', *Biol. Bull.*, **140**, 137–155.

Stross, R.G., and Hill, J.C. (1968). 'Photoperiod control of winter diapause in the freshwater crustacean, *Daphnia*', *Biol. Bull.*, **134**, 176–198.

Subrahmanyam, C.B. (1963). 'A note on the annual reproductive cycle of the prawn *Penaeus indicus* (M-Edw.) of Madras coast', *Curr. Sci.*, **32**, 165–166.

Subramoniam, T. (1977). 'Aspect of sexual biology of the anomuran crab *Emerita asiatica*', *Mar. Biol.*, **43**, 369–377.

Subramoniam, T. (1979). 'Some aspects of reproductive ecology of a mole crab *Emerita asiatica* Milne-Edwards', *J. exp. mar. Biol. Ecol.*, **36**, 259–268.

Subramoniam, T. (1981). 'Protandric hermaphroditism in *Emerita*', *Biol. Bull.*, **160**, 161–174.

Subramoniam, T. (1991a). 'Sperm transfer and storage in decapod crustaceans', in *Aquaculture Productivity* (Eds. V.R.P. Sinha and H.C. Srivastava), Oxford & IBH Pub. Co. Pvt. Ltd., New Delhi, pp. 241–256.

Subramoniam, T. (1991b). 'Chemical composition of spermatophores in decapod crustaceans', in *Crustacean Sexual Biology* (Eds. R.T. Bauer and J.W. Martin), Columbia University Press, New York, pp. 308–321.

Sussman, A.S., and Halvorson, H.O. (1966). *Spores: Their Dormancy and Germination*, Harper and Row, New York.

Suyama, M. (1959). 'Biochemical studies on the eggs of aquatic animals', *Bull. Jap. Soc. Sci. Fish.*, **25**, 48–51.

Swartz, R.C. (1976). 'Sex ratio as a function of size in the Xanthid crab, *Neopanope sayi*', *Amer. Natur.*, **110**, 898–900.

Tagatz, M.E. (1968). 'Growth of juvenile blue crabs *Callinectes sapidus* Rathbun, in the St. Johns river, Florida', *Fish. Bull.*, **67**, 281–288.

Tchernigovitzeff, C. (1960). 'Nouvelles observations sur la mue de *Bopyrus fougerouxi*, Isopode parasite de *Leander serratus* (Pennant)', *C.R. Acad. Sci., Paris*, **250**, 185–189.

Teissier, C. (1960). 'Relative growth', in *The Physiology of Crustacea*, Vol. 1, *Metabolism and Growth* (Ed. T.H. Waterman), Academic Press, New York, pp. 537–556.

Telford, M. (1968). 'The identification and measurement of sugars in the blood of three species of Atlantic crabs', *Biol. Bull.*, **135**, 574–584.

Tharp, T.P., and Bland, C.E. (1977). 'Biology and host range of *Halipthoros milfordensis* Vishniac', *Can. J. Bot.*, **55**, 2936–2944.

Theisen, B.F. (1966). 'The life history of seven species of ostracods from a Danish brackish-water locality', *Meddr Danm. Fisk. -og Havunders.*, **4**, 215–270.

Thorson, G. (1950). 'Reproductive and larval ecology of marine bottom invertebrates', *Biol. Rev.*, **25**, 1–45.

Thorson, G. (1961). 'Length of pelagic larval life in marine bottom invertebrates as related to larval transport by ocean currents', in *Oceanography* (Ed. M. Sears), No. 47, American Association for Advancement of Science, pp. 455–474.

Thorson, G. (1966). 'Some factors influencing the recruitment and establishment of marine benthic communities', *Neth. J. Sea Res.*, **3**, 267–293.

Tinkle, D.W. (1969). 'The concept of reproductive effort and its relation to the evolution of life history of lizards', *Amer. Natur.*, **103**, 501–516.

Tinkle, D.W., and Hadley, H.F. (1975). 'Lizard reproductive effort: Caloric estimates and comments on its evolution', *Ecology*, **56**, 427–434.

Todd, C.D., and Doyle, R.W. (1981). 'Reproductive strategies of marine benthic invertebrates: A settlement-timing hypothesis', *Mar. Ecol. Prog. Ser.*, **4**, 75–83.

Treillard, M. (1925). '*Daphnia magna* en culture pure. Perennite de la parthenogénèse: nécesaité de facteurs bactérienes pour l'apparition des formes sexués', *C.R. Soc. Biol.*, **93**, 16–24.

Uchima, M., and Murano, M. (1988). 'Mating behaviour of the marine copepod *Oithona davisae*', *Mar. Biol.*, **99**, 39–45.

Uye, S., Kasahara, S., and Onbe, T. (1979). 'Effects of some environmental factors on the hatching of resting eggs', *Mar. Biol.*, **51**, 151–156.

Uye, S., and Fleminger, A. (1976). 'Effects of various environmental factors on egg development of several species of *Acartia* in southern California', *Mar. Biol.*, **38**, 253–262.

Vandel, A. (1940). 'La parthénogénèse géographique. Polyploidie et distribution géographique', *Bull. Biol. Fr. Belg.*, **74**, 94–100.

Van den Bosch, R., Schlinger, E.J., Lagace, C.F., and Hall, J.C. (1966). 'Parasitism of *Acyrthosiphon pisum* by *Aphidius smithi*, a density-dependent process in nature (Homoptera: Aphidae) (Hymenoptera: Aphididae)', *Ecology* **47**, 1049–1055.

Van Dolah, R.F., and Bird, E. (1980). 'A comparison of reproductive patterns in epifaunal and infaunal gammaridean amphipods', *Est. Coast. Mar. Sci.*, **11**, 593–604.

Varley, G.C. (1947). 'The natural control of population balance in the knapweed gall-fly (*Urophora jacaena*)', *J. Anim. Ecol.*, **16**, 138–187.

Veillet, A. (1945). 'Recherches sur le parasitisme des crabs et des Galathées par les Rhizocephales et les Epicarides', *Ann. Inst. Oceanogr. Monaco*, **22**, 193–341.

Veillet, A., and Graf, F. (1959). 'Dégénérescence de la glande androgene des Crustacés Décapodes parasites par les Rhizocephales', *Bull. Soc. Sci. Nancy*, **18**, 123–127.

Vernberg, F.J. (1987). 'Crustacea', in *Animal Energetics* (Eds. T.J. Pandian and F.J. Vernberg), Academic Press, New York, Vol. 1, pp. 301–372.

Vernberg, W.B., and Moreira, G.S. (1974). 'Metabolic-temperature responses of the copepod *Euterpina acutifrons* (Dana) from Brazil', *Comp. Biochem. Physiol.*, **49**, 757–761.

Vijayaraghavan, S., and Easterson, D.C.V. (1974). 'Biochemical changes and energy utilization in developing stages of the estuarine prawn *Macrobrachium idella* (Hilgendorf)', *J. mar. biol. Ass. India*, **16**, 275–279.

Vijayaraghavan, S., Wafar, M.V.M., and Royan, J.P. (1976). 'Changes in biochemical composition and energy utilization in developmental stages of the male crab *Emerita holthusi* Sankolli', *Mahasagar, Bull. Nat. Inst. Oceanogr.*, **8**, 165–170.

Vishniac, H. (1958). 'A new marine phycomycete', *Mycologia*, **50**, 66–79.

Vlasblom, A.G. (1969). 'A study of a population of *Marinogammarus marinus* (Leach) in the Oosterschelde', *Neth. J. Sea Res.*, **4**, 317–338.

Volkmann-Rocco, B., and Battaglia, B. (1972). 'A new case of sibling species in the genus *Tisbe* (Copepoda: Harpacticoida)', in *Fifth European Mar. Biol. Symp.* (Ed. B. Battaglia), Piccin Editore, Padova, pp. 67–80.

Volkmann-Rocco, B., and Fava, G. (1969). 'Two sibling species *Tisbe reluctans* (Copepoda: Harpacticoida) and *T. persimilis* n.sp. Research on their morphology and population dynamics', *Mar. Biol.*, **3**, 159–164.

Von Dehn, T. (1955). 'Die Geschlechtsbestimmung der Daphniden. Die Bedeutung der Fettstoffe untersucht an *Moina rectirostris*', *Zool. Jb. Physiol.*, **65**, 334–356.

Waldschmidt, S.R., Jones, S.M., and Porter, W.P. (1987). 'Reptilia', in *Animal Energetics* (Eds. T.J. Pandian and F.J. Vernberg), Academic Press, New York, Vol. 2, 553–619.

Walker, I. (1979). 'Mechanisms of density dependent population regulation in the marine copepod *Amphiascoides* sp. (Harpacticoida)', *Mar. Ecol. Prog. Ser.*, **1**, 209–221.

Walley, L.J. (1965). 'The development and function of the oviducal gland in *Balanus balanoides*', *J. mar. biol. Ass. U.K.*, **45**, 115–128.

Walley, L.J., White, F., and Brander, K.M. (1971). 'Sperm activation and fertilization in *Balanus balanoides*', *J. mar. biol. Ass. U.K.*, **15**, 489–494.

Ward, D.G., Middleditch, B.S., Missler, S.R., and Lawrence, A.L. (1980). Fatty acid changes during larval development of *Peneaus setiferus*, *Proc. World Maricult. Soc.*, **10**, 464–471.

Warner, A.H., and Finamore, F.J. (1965). 'Isolation, purification and characterization of P^1, P^3-diguanosine 5′-triphosphate from brine shrimp eggs', *Biochim. Biophys. Acta*, **108**, 525–530.

Warner, A.H., and Finamore, F.J. (1967). 'Nucleotide metabolism during brine shrimp embryogenesis', *J. Biol. Chem.*, **242**, 1933–1937.

Warner, A.H., and McClean, D.K. (1968). 'Studies on the biosynthesis and role of diguanosine tetraphosphate during growth and development of *Artemia salina*', *Develop. Biol.*, **18**, 278–293.

Warner, A.H., Puodziukas, J.G., and Finamore, F.J. (1972). 'Yolk platelets in brine shrimp embryos', *Exp. Cell Res.*, **70**, 365–375.

Warner, D.T. (1969). 'Structured water in biological systems', *Ann. Reports Med. Chem.*, **1969**, 256–265.

Warner, R.R. (1975). 'The adaptive significance of sequential hermaphroditism in animals', *Amer. Natur.*, **109**, 61–82.

Waterman, T.H., and Chace, F.A., Jr. (1960). 'General crustacean biology', in *The Physiology of Crustacea*, Vol. 1, *Metabolism and Growth* (Ed. T.H. Waterman), Academic Press, New York, pp. 1–30.

Watson, N.H.F., and Smallman, B.N. (1971a). 'The role of photoperiod and temperature in the induction and termination of an arrested development in two species of freshwater cyclopoid copepods', *Can. J. Zool.*, **49**, 855–862.

Watson, N.H.F., and Smallman, B.N. (1971b). 'The physiology of diapause in *Diacyclops navus* Herrick (Crustacea: Copepoda)', *Can. J. Zool.*, **49**, 1449–1454.

Wear, R.G. (1974). 'Incubation in British decapod Crustacea and the effects of temperature on the rate and success of embryonic development', *J. mar. biol. Ass. U.K.*, **54**, 745–762.

Weinstein, M.P., and Heck, K.L., Jr. (1977). 'Biology and host-parasite relationships of *Cymothoa excиza* (Isopoda: Cymothoida) with three species of snappers (Lutjanidae) on the Caribbean coast of Panama', *Fish. Bull.*, **75**, 875–877.

Weismann, A. (1886). 'Richtungskorper bei parthenogenetischen Eier', *Zool. Anz.*, **9**.

Wenner, A.M. (1972). 'Sex ratio as a function of size in marine Crustacea', *Amer. Natur.*, **106**, 321–350.

Wenner, A.M., Fusaro, O., and Oaten, A. (1974). 'Size at onset of sexual maturity and growth rate in crustacean population', *Can. J. Zool.*, **52**, 1095–1106.

Wharton, G.W. (1942). 'A typical sand beach animal, the mole crab, *Emerita talpoida* (Say), in ecology of sand beaches at Beaufort N.C.', *Ecol. Monogr.*, **12**, 137–181.

Whitaker, D.M. (1940). 'The tolerance of *Artemia* cysts for cold and high vacuum', *J. Exp. Zool.*, **83**, 391–399.

Wickham, D.E. (1979). 'Predation by the nemertean *Carcinonemertes errans* on eggs of the Dungeness crab *Cancer magister*', *Mar. Biol.*, **55**, 45–54.

Wickins, J.F. (1976). 'Prawn biology and culture', *Oceanogr. mar. Biol. Ann. Rev.*, **14**, 435–457.

Wickins, J.F., and Beard, T.W. (1974). 'Observations on the breeding and growth of the giant prawn *Macrobrachium rosenbergii* (de Man) reared in the laboratory', *Aquaculture*, **3**, 159–174.

Wierzbicka, M. (1962). 'On the resting stage and mode of life of some species of Cyclopoida', *Pol. Arch. Hydrobiol.*, **10**, 216–229.

Wilber, D.H. (1989). 'Reproductive biology and distribution of stone crabs (Xanthidae, *Menippe*) in the hybrid zone in the northeastern Gulf of Mexico', *Mar. Ecol. Prog. Ser.*, **52**, 235–244.

Wildish, D.J. (1977). 'Biased sex ratios in invertebrates', in *Advances in Invertebrate Reproduction* (Eds. K.G. Adiyodi and R.G. Adiyodi), Peralam-Kenoth, Karivellur, Kerala, India, pp. 8–23.

Williams, A.B. (1965). 'Marine decapod crustaceans of the Carolinas', *U.S. Fish Wildl. Serv. Fish. Bull.*, **65**, 1–298.

Williams, G.C. (1975). *Sex and Evolution*, Princeton University Press, Princeton.

Williams, R., and Lindley, J.A. (1982). 'Variability in abundance, vertical distribution and ontogenetic migration of *Thysanoessa longicaudata* (Crustacea: Euphausiacea) in the northeastern Atlantic Ocean', *Mar. Biol.*, **69**, 321–330.

Williams, R., Conway, D.V.P., and Collins, N.R. (1987). 'Vertical distributions of eggs, nauplii and copepodites of *Calanus helgolandicus* (Copepoda: Crustacea) in the Celtic Sea', *Mar. Biol.*, **96**, 247–252.

Williamson, D.I. (1951). 'On the mating and breeding of some semiterrestrial amphipods', *Rep. Dove Mar. Lab.*, **12**, 49–62.

Williamson, D.I. (1972). 'Larval development in a marine and freshwater species of *Macrobrachium* (Decapoda: Palaemonidae)', *Crustaceana*, **23**, 282–298.

Wilson, M.F., and Pianka, E.R. (1963). 'Sexual selection, sex ratio and mating system', *Amer. Natur.*, **97**, 405–407.

Wittmann, K.J. (1978). 'Adoption, replacement and identification of young in marine Mysidacea (Crustacea)', *J. exp. mar. Biol. Ecol.*, **32**, 259–274.

Wolf, E. (1905). 'Die fortapilanzunge Verhaltnisse unserer linheimischen Copepoden', *Zool. Jb. Syst.*, **22**, 101–280.

Wu, R.S.S., and Levings, C.D. (1979). 'Energy flow and population dynamics of the barnacle *Balanus glandula*', *Mar. Biol.*, **54**, 83–90.

Zagalsky, P.F., Cheesma, D.F., and Ceccaldi, H.J. (1967). 'Studies on carotenoid containing lipoproteins isolated from the eggs and ovaries of certain marine invertebrates', *Comp. Biochem. Physiol.*, **22**, 851–871.

Zamenhof, S., Eichhorn, H.H., and Rosenbaum, D. Oliver (1968). 'Mutability of stored spores of *Bacillus subtilis*', *Nature, Lond.*, **220**, 818–819.

Zillioux, E.J., and Gonzalez, J.G. (1972). 'Egg dormancy in a neretic calanoid copepod and its implications to overwintering in boreal waters', in *Proceedings of the Fifth European Marine Biology Symposium* (Ed. B. Battaglia), Piccin Editore, Padua and London, pp. 217–230.

Zurlini, G., Ferrari, I., and Nassogne, A. (1978). 'Reproduction and growth of *Euterpina acutifrons* (Copepoda: Harpacticoida) under experimental conditions', *Mar. Biol.*, **46**, 59–64.

4. ARTHROPODA — INSECTA

J. MUTHUKRISHNAN
School of Biological Sciences, Madurai Kamaraj University,
Madurai 625 021, India

I. INTRODUCTION

In terms of number, diversity, and distribution, insects constitute one of the most successful groups of animals (for classification of insects, see Systematic Résumé). The diversity of reproductive strategies adopted by them and their versatility to adapt to even hostile environmental conditions have contributed to their success. Sexual reproduction in insects is a complex phenomenon, involving such functional components as mate seeking, courtship, sperm transfer, oviposition, and post-copulatory mate guarding. To maximize progeny production and survival, insects seem to have experimented with a variety of tactics at each stage of the reproductive process. The most suitable ones have been maintained as evolutionarily stable strategies (Maynard Smith and Parker, 1976).

The reproductive success of a male depends on the number of matings he can achieve with different females within his short life span (Trivers, 1972; Parker, 1978a, b). To maximize reproductive success, a male adopts suitable strategies to seek and locate virgin females, thwart the attempts of the competing males, and successfully copulate and guard his mate till his sperm are used for fertilization or even till oviposition. Polygamy is widespread among different species of insects. Males of polygamous species adopt a variety of strategies to prevent competition between their sperm and those of subsequent males (Parker, 1970a).

The reproductive success of a female depends on her efficiency in transforming ingested food energy into eggs and her ability to produce fertile offspring (Engelmann, 1970). The success is usually assessed by potential and realized fecundity. Several extrinsic and intrinsic factors interact to modulate fecundity. Progeny survival depends mostly on the oviposition strategy of the mother. The ultimate aim of any oviposition strategy is to provide protection and adequate food for the freshly hatched young ones.

The pattern of reproduction (breeding once or several times in a generation) and allocation of the available energy to a few eggs or to a larger number of eggs varies vastly even among closely related insect species in different habitats. Some species breed prolifically and produce several less efficient offspring most of which succumb to the vagaries of the environment and are beaten by the few surviving better-adapted cohorts. They are called 'r-strategists'. Others, which produce fewer but highly competitive offspring per generation, are called 'K-strategists'. The pattern of partitioning of the available energy into a few or more offspring is mostly determined in insects, as in other animals, by the stability of the habitat.

Environmental conditions suitable for growth, development, and reproduction prevail only during definite seasons. With their ability to sense environmental cues such as photoperiod, temperature, population density, and availability of food, some of the insects enhance their survival during unfavourable conditions by entering into an energetically inexpensive state of dormancy called 'diapause' or by undertaking energetically expensive migrations. By accurately timing the termination of diapause or emergence and onset of reproduction, they realize rapid population growth during favourable conditions.

Biased sex ratio constitutes a predominant force in the selection of reproductive strategies. Differences in the maternal investment, mortality, and longevity between the male and female progeny as well as resource availability are some of the factors which cause biased operational sex ratios in populations. The altered sex ratios modulate the overall reproductive strategy of the species at the population level (see Adiyodi and Adiyodi, 1974). Unusual modes of reproduction such as parthenogenesis, paedogenesis, and polyembryony (see Gillott *et al.*, Volume V of this series) also alter the sex ratio of insect populations. Parthenogenesis is a widespread phenomenon among insects except Odonata and Hemiptera. In extreme cases, as in some gall wasps, males have never been observed and there is little doubt about their non-existence (Richards and Davies, 1977). Parthenogenesis involving thelygeny accelerates reproduction by restricting the activities of females to feeding and reproduction. Reduction in genetic variability of the population is one of the disadvantages of parthenogenesis.

Insects provide by far the best example of eusocial life. With a perfect chemical coordination and sharing of the different tasks of the colony, the social insects have been successful. They regulate the proportion of different castes as well as the size of the colony as a whole and found new populations.

Following the principle of allocation, Harper and Ogden (1970), Baltzi (1974), Price (1974a), and Giesel (1976) have shown that reproductive strategies can be better understood with pertinent data on energy budget. Each reproductive strategy levies a reproduction cost or an investment by a parent in an individual offspring that increases the offspring's chances of survival (Trivers, 1972). Besides investing in the production of gametes and associated structures (e.g. spermatophore: see Davies and Dadour, 1989, or mating plug in males and oötheca in females), parents have to expend considerable energy on various reproductive functions such as mate seeking, mating, and mate guarding (by males) and locating oviposition site, oviposition, and progeny maintenance and defence (by females) (Fig. 1). Therefore, Baltzi (1974) emphasized that the energy budget should be more detailed and should include fractions of assimilated energy, allocated to somatic growth as well as to reproductive structures and functions (see also Muthukrishnan and Pandian, 1987a). Growth, metabolism, and reproduction are the major energy-demanding processes in an insect's life history and they compete for the limited energy resources and time at the disposal of an insect. The success of a reproductive strategy depends on efficient partitioning of these resources for various somatic and reproductive processes. A detailed energy budget, furnishing data on various reproductive structures and functions, is still wanting, and workers on reproductive strategies are frustrated by lack of pertinent data (Price, 1974a). Allocation of time for various reproductive activities is equally important, and this aspect of insect ecology has been totally neglected.

II. COMPETITIVE MATE-SEEKING AND MATING STRATEGIES

As in other animals, internal fertilization in insects resulted in anisogamy leading to severe intraspecific sexual selection. Consequently, the females evolved the strategy of maximizing their gametic and other investments (e.g. progeny protection) to increase survival and fitness of their offspring, and the males evolved the strategy of minimizing their gametic investment in order to spare relatively more energy for mate seeking and maximizing the number of copulations with different females (Trivers, 1972; Thornhill, 1976; Parker, 1978a, b). An inescapable consequence of these contrasting strategies of parental investment is the production of a large number of sexually active males (i.e. male-biased operational sex ratio: Emlen and Oring, 1977) and the resulting intra-male competition for copulating with the limited number of fertile females available in the population. The severity of the competition is intensified by distribution of females in patches of foraging, nesting, oviposition, and emergence sites. Therefore, males are compelled to reallocate their initial gain from gametic investment on alternate strategies such as enhanced mobility in search of mates, territoriality and intra-male aggression to gain a receptive female, courtship to increase the receptivity of the female, and presentation of gifts to the mate and supplementing her investment to ensure their paternity (see Thornhill, 1976).

A. Mating-site Search Strategy

Mate-seeking males, often at the danger of encountering predators, avoid random searching and generally look for females in specific, conventional sites, where the probability of locating them is high. Such places also provide suitable oviposition and/or larval or adult feeding sites; hence, they are called 'resource-based leks' (see Alexander, 1975). A delay in mating often results in resorption of oocytes

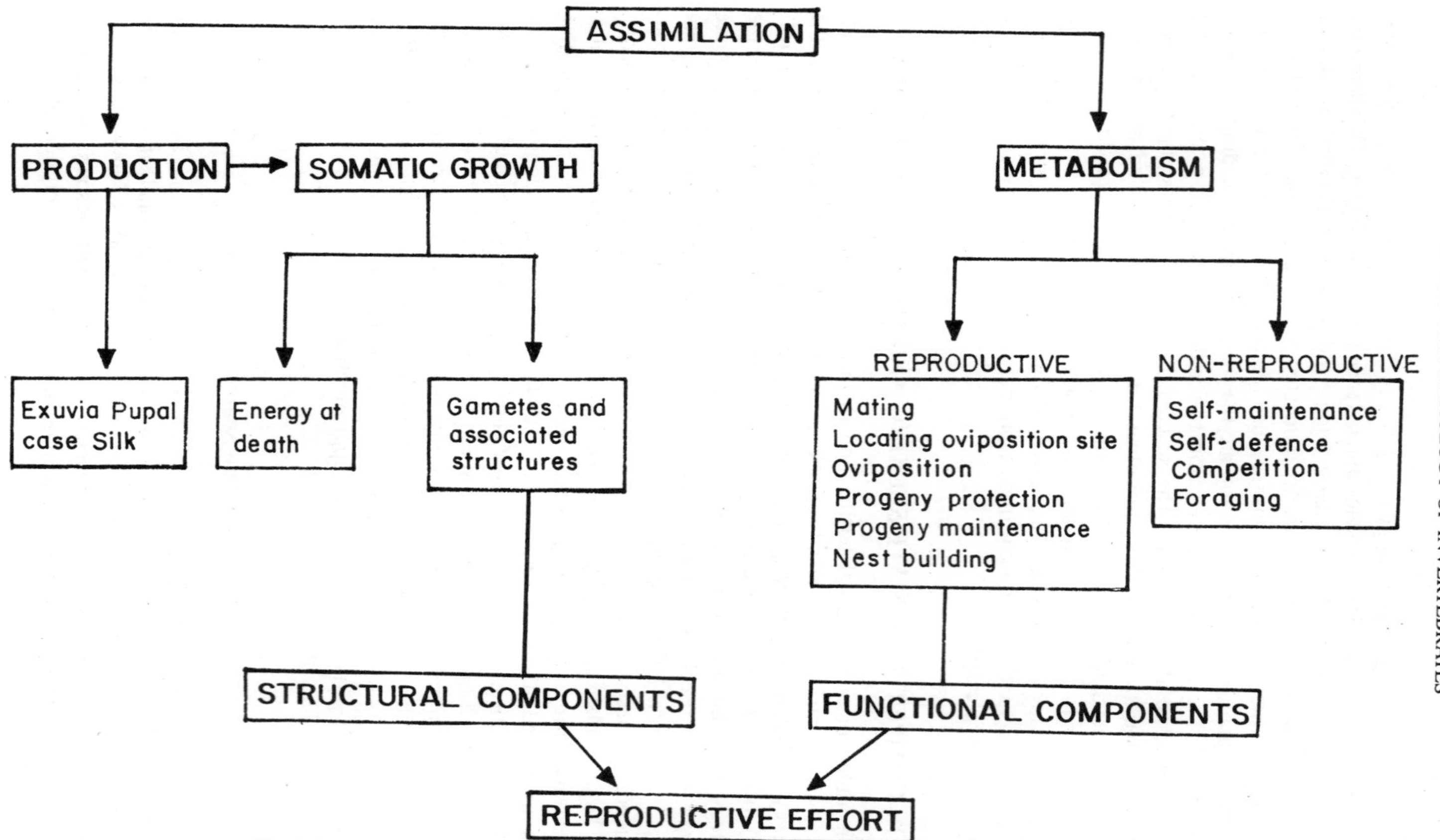

Fig. 1. Partitioning of assimilated energy between structural and functional components of reproduction.

or deposition of infertile eggs. Females may also move into a favourable locality, where the density of searching males is high (Parker, 1978a). Males of several species (e.g. bees) haunt female-emergence sites, foraging sites, and/or nesting sites. Males of hymenopterous parasites are promiscuous and passively wait near host species for emerging females and mate with them (Table 1). Males of several orders of insects guard territories in the vicinity of oviposition sites and wait for the arrival of virgin or ovipositing females. Non-resource-based hill tops or sun-lit spots are also occupied by searching males of Lepidoptera and Diptera (Table 1). Larvae developing at low temperatures prevailing at high altitudes grow and metamorphose slowly into larger adults with high fecundity; on the other hand, development is faster and fecundity of the adult lower at low altitudes (Baker, 1966).

The reward gained by way of suitable oviposition sites and larval and adult feeding sites seems to have prompted most insects to adopt resource-based lekking strategy. However, according to Alexander (1975), non-resource-based lekking strategy decreases the temporal cost of waiting and enables the females to make a wider choice of potential mates and improve the genetic make-up of the offspring.

B. Male-male Interactions and Mate-seeking Strategies

Having arrived at a mating site, driven by the goal of maximizing the number of copulations, males display a variety of intrasexual competitive interactions in order to secure a potential female. The severity of the competition is intensified for reasons given below:

(1) A highly male-biased operational sex ratio becomes effective due to the relatively greater maternal investment (Trivers, 1972). For instance, a ratio as high as 8 : 1 in favour of males has been reported for the butterfly *Pieris protodice* (see Shapiro, 1970) as well as the dungfly *Scatophaga sterocoraria* (see Parker, 1970a). Over 75 per cent of the individuals of *Argia plana* (Odonata) at mating sites (water) are males (Bick and Bick, 1968).

(2) In most insects, particularly bees and wasps, males emerge prior to females and wait together for the arrival of females (see Alcock *et al.*, 1978; Van den Assem *et al.*, 1980).

(3) Several species of Diptera, Lepidoptera, and chalcidoid and pteromalid wasps mate only once in their lifetime. Even among other insects, females with stores of active sperm in spermatheca are non-receptive and stay away from competing males (see Jones, 1974; Engelmann, 1970; Van den Assem *et al.*, 1980).

(4) Patchy distribution of mating sites in relation to the high density of searching males (see Alcock *et al.*, 1978).

Competing males are never too similar. Even males of the same brood vary in size, especially among bees and wasps (e.g. *Anthidium maculosum* and *Centris pallida*: Alock *et al.*, 1977a, b; *Nasonia* sp.: Van den Assem *et al.*, 1980; several species of odonates: Corbet, 1980). The males are usually drawn from various

Table 1

Mate-seeking strategies of some insects

Species	Remarks	Reference
	A. Resource-based strategies 1. Oviposition site	
Ephemeroptera		
Several species	Female enters a swarm of males and leaves with male; mid-air copulation near oviposition site	Brinck (1957)
Odonata		
Onychogomphus forcipatus	Simple aggression without territoriality	Kaiser (1974a)
Anax imperator	Patrolling littoral area of pond, but often showing preference to certain sites	Corbet (1980)
Enallagma aspersum	Perching and cruising specific sites	Bick and Hornuff (1966)
Calopteryx maculata	Strongly territorial and aggressive	Waage (1973)
Brachythemis contaminata	Strongly territorial and aggressive	Mathavan (1975)
Plecoptera		
Pteronarcys proteus and *Perla marginata*	Exchange of drumming vibrations brings males and females together; mating and oviposition near water margin	Hynes (1976)
Lepidoptera		
Aglais urticae and *Inachus ie*	Guard territories in the vicinity of oviposition sites	Baker (1972)
Prionoxystus robiniae	Females attract males using pheromones; usually females mate only once	Solomon and Neel (1974)
Acrolepiopsis assectella	Host plant chemical brings both partners together, induces mating and oviposition	Labeyrie (1978b)
Diptera		
Scatophaga steroceraria	Mating and oviposition and larval and adult feeding on cow pat; males are aggressive and territorial	Parker (1978b)
Drosophila spp.	The site serves the adults for feeding, oviposition, and mating	Parker (1978b)
Coleoptera		
Aphodius rufipes	Males and females are attracted by fresh cow dung on which they mate, oviposit, and feed	Holter (1979)

Contd.

Table 1 Contd.

Species	Remarks	Reference
	2. Nest sites and female-emergence sites	
Hymenoptera		
Centris pallida	Dig soil (1 cm depth) and locate pre-emergent females. Weak males hover at the emergence site and mate with females	Alcock *et al.* (1978)
Cephalononia spp.	Chew their way into female cocoons and copulate prior to emergence	Gordh (1976)
Goniozus spp. and *Parasierola* spp.	Chew their way into female cocoons and copulate prior to emergence	Gordh and Evans (1976)
Andrena mojavensis	Search for females near foraging sites after the emergence of females	Linsley *et al.* (1963a, b)
Tachytes distinctus	Search nest site when the emergence of females in a site is over	Lin and Michener (1972)
Philanthus multimaculatus	Mate with females entering or leaving nests	Alcock (1975a)
Andrena flavipes	Mate with females entering or leaving nests	Butler (1965)
Chalcidoid and pteromalid wasps	Wait on cocoons of host species and mate with the emerging females	Van den Assem (1976)
Note: Several species of ectoparasitic Mallophaga and Siphunculata mate on the host. Endoparasitic females of Strepsiptera are mated by free-living males attached to the host.		Richards and Davies (1977)
	3. Adult or larval foraging sites	
Several species of grasshoppers, locusts, and phytophagous Lepidoptera mate near the larval or adult foraging sites and oviposit in the same area.		Parker (1978a)
	B. Non-resource-based strategies 1. Hill topping	
Lepidoptera		
Papilio zelicaon	Males and females are attracted by topographic prominences like hilltops	Schields (1967)
Pieris protodice and 21 other species of butterflies	Males highly territorial on hilltops; when hilltops provide the larval host-plant, the strategy is resource-based	Shapiro (1970)
Diptera		
Several species of botflies	Swarms of males are attracted by distinct areas on hilltops. Females are attracted by the swarms	Grunin (1959)
	2. Sunspotting	
Males of several species of butterflies and nematoceran flies passively wait on sun-lit spots on rows of tree or edge of wood and mate with passing females. Swarms of fireflies, attracted by light-flashes emitted by females, mate in non-resource-based leks		Downes (1969) Baker (1972)

populations and hence are genotypically different. Males of gregarious parasitic wasps, emerging from the same host, belong to different parents (e.g. *Pteromalus puparum*: Van den Assem, 1976). Intrasexual selection acting on such a group of phenotypically and genotypically different males has often resulted in dual mate-seeking strategies (see Gadgil, 1972). Strong and larger males follow territorial strategies; the weak and less competitive ones adopt non-territorial strategies, avoid combat and direct competition, and resort to alternate means of securing females.

1. Behavioural and size dimorphism

The odonates provide several examples of the behavioural dimorphism of mate-seeking males. Territorial and non-territorial or wandering behavioural manifestations have been noticed among them. Four different degrees of territoriality have been recognized: (1) simple aggression without preference for any specific territory, as in *Onychogomphus forcipatus* (see Kaiser, 1974a); (2) patrolling and occasionally showing preference for certain territories, as in *Anax imperator* and several species of *Aeshna* (see Kaiser, 1974b); (3) perching and cruising around a territory as in *Lestes disjunctus, Enallagma aspersum*, and *Ischnura verticalis* (see Bick *et al.*, 1976); and (4) intense territoriality and aggressiveness as in *Plathemis lydia* (see Campanella and Wolf, 1974) and *Calopteryx maculata* (see Waage, 1973), which guard specific territories for one to three weeks without any interruption.

Based on their investment in territorial behaviour, the males are rewarded with one to several matings a day. For instance, the territorial *Sympetrum parvulum* mates 1.4 times a day compared to 0.5 mating a day by the non-territorial male (Ueda, 1979). Besides this, the territorial male is assured of his sperm being used in fertilization, as he guards his mate till she has completed oviposition. The non-territorial male waits passively till the territory owner gives chase to an intruder off his territory and earns a sneak copulation. A high proportion of the sperm of the last-mated male is used for fertilization and hence the non-territorial wanderers are occasionally successful (see Corbet, 1980).

The behavioural dimorphism of males of several insect species is the consequence of size variation between them. For instance, two distinct size classes exist among males of the megachilid bee *Anthidium maculosum* (see Alcock *et al.*, 1977a). The larger ones are the territory owners, defending a territory of 1–10 m^2. The smaller ones, called 'satellite males', occupy corners of guarded territories passively waiting to cheat the territory owners and take off their females. Between these two groups is a group of wanderers, who visit several territories and try to disturb the territory owners and attempt sneak copulations with their females. The larger territorial males spend over 75 per cent of their time budget guarding their territory and are duly rewarded with 82 per cent of the total number of copulations (see Alcock *et al.*, 1977a). Satellite males and wandering intruders exploit the remaining 25 per cent time of the territorial males when they are away from their territory. For a considerably smaller amount of energy expended, non-territorial males are benefited by about 18 per cent of the total number of copulations.

Similarly, males of the anthophorid bee *Centris pallida* constitute two behavioural groups—the 'diggers' and 'hovers'. The diggers are significantly larger (head width 5.14 mm) than the hovers (head width 4.8 mm) (Alcock *et al.*, 1977b). The diggers locate emergence site of females, dig 1–2 cm through the soil to reach them, and copulate before the females emerge from the soil. The hovers remain perched 5–8 m above the ground or near flowering shrubs or trees, which the females visit for collecting nectar. They bank on mainly uncaptured, airborne females which escape the diggers. The diggers are often interrupted by hovers. As the proportion of buried pre-emergent females is larger than that of the uncaptured airborne females and as the diggers successfully intimidate the hovers by chasing them away, the diggers enjoy a greater success in securing females.

Non-territorial males, who are not capable of challenging their strong rivals, withdraw from competition and resort to one of the following alternate strategies to exploit the efforts of their rivals: (1) displacing a territorial rival or occupying his territory when he is away (e.g. *Philanthus multimaculatus*: Alcock, 1975a, b; Alcock *et al.*, 1977a); (2) stealing a digging site (e.g. *Centris pallida* hover: Alcock *et al.*, 1977b); and (3) stealing a captured female from the rival males (e.g. *Emphoropsis pallida*: Boharat *et al.*, 1972).

Notwithstanding their success, the larger males are relatively rare in bee and wasp populations because of the high energy cost of provisioning larger cells accommodating larger males, and the high probability of parasitic and predatory attack on the larger males (Alcock *et al.*, 1977b, 1978). Thus, the mother determines the mate-seeking strategy of her sons by varying her investment in them (Fig. 2).

Parasitic hymenopterans oviposit male-producing eggs in smaller hosts. A male developing in a smaller host emerges as a smaller individual relatively earlier than the one emerging from a larger host. Among the non-host-specific parasitic hymenopterans, size of the offspring varies widely depending on host size. For instance, Salt (1937) demonstrated that males of *Trichogramma semblidis* emerging from eggs of the neuropteran *Sialis* sp. are smaller and apterous, compared to those emerging from lepidopterous eggs which are larger and macropterous. Females of many parasitic wasps are monogamous and are ready to mate immediately after their emergence. They do not show any size preference in accepting males for copulation. The first male to come into contact with an emerging female is the most successful (Van den Assem, 1976). Therefore, males of parasitic wasps adopt the strategy of shortening their larval duration and emerge into smaller adults so as to keep themselves ready for mating near the host cocoon, from which emerge the females. Males of chalcidoid wasps, such as *Nasonia vitripennis* and *Melittobia* sp., are brachypterous and emerge from their hosts just before the emergence of females. Their brachypterous nature and the consequent restricted radius of activity do not hamper their mate finding as the females emerge very close to the waiting males.

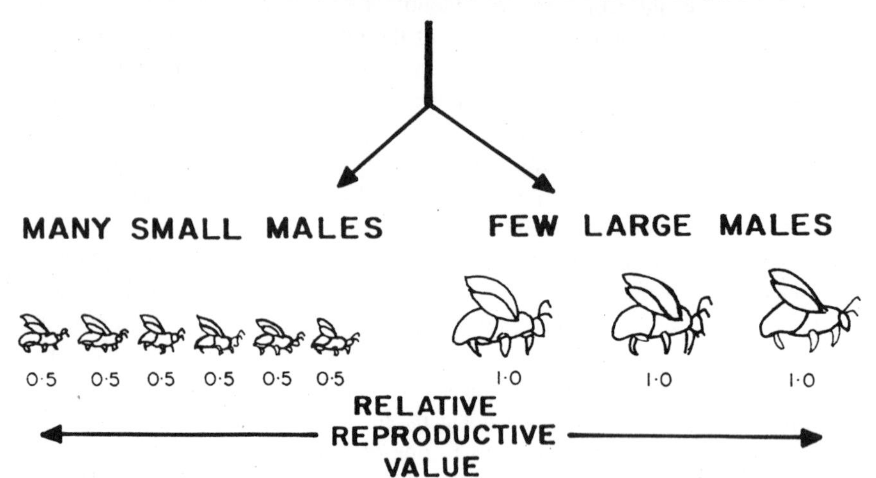

Fig. 2.. Patterns of maternal investment in the bee *Centris pallida* resulting in size dimorphism among the offspring (After Alcock *et al.*, 1977b).

However, smaller males suffer from decreased longevity, increased vulnerability to attack by larger males, and limited capacity to inseminate. These disadvantages of smaller males are outweighed by the advantage of emerging considerably earlier and being ready for mating well ahead of larger males. By emerging earlier, the smaller males spare more host material for the sib-sisters to feed on and increase their fecundity; subsequently, they mate with their sib-sisters and sire more offspring.

2. Success of territoriality

In dungflies (e.g. *Scatophaga sterocoraria* and *Sepsis* sp.: Parker, 1972, 1978b), a greater fraction of the sperm of the last-mated male is used for fertilization. Therefore, males of these species display territoriality and continue to guard their mate till she has completed oviposition. To avoid interruptions or take-over by the rival non-territorial males, the territorial males intercept the virgin female even before she has arrived at the dung pat and copulate with her, or remove her to the peripheral grass (if she lands on the pat) and mate with her. Occasionally, the intruder enters into combat with the territory owner and wins his female. However, the majority of the copulations are achieved by the territory owners themselves (Parker, 1978b). Among syrphid flies, a spatial and temporal partitioning of the mating resources has been observed. A few species (e.g. *Chrysogaster nitida*) mate with females visiting the blossoms for collecting nectar, while a few others (e.g. *Brachyplus oarus*) wait for females near the oviposition sites and mate with them.

Males of the former type are called 'patrollers' and are active most of the time; males of the latter type are 'waiters', active only during the afternoon (Maier and Waldbauer, 1979).

Territorial mate-seeking behaviour has been poorly documented among lepidopterans. Males of *Aglais urticae* and *Inachus ie* locate their mating site adjacent to a hedge or rows of trees and wait for females migrating in search of oviposition sites. Success of these males mostly depends on the area of the territory and the territory-male ratio. With increasing area and decrease in intensity of competition within a territory, a male's success increases. Territorial males of desert grasshoppers, such as *Lighrotettix* sp. and *Bootettix* sp., defend bushes harbouring females and repel rival males by stridulating and by fighting. When the number of challenging males increases, the territorial males abandon their territory and thus save time and energy (Otte and Joern, 1975).

The success of a territorial male depends on (1) his diligent choice of a territory — one which is not too big nor too small so that it does not require increased expenditure of time and energy; (2) the time spent on guarding the territory — males intermittently leaving their territory are liable to lose their mate (see Parker, 1978b); (3) use of female-attracting signals, such as stridulating sound as in gryllids and desert grasshoppers, or pheromones as in euglosine bees and philanthine wasps (Alcock, 1975a); and (4) spatial and/or temporal partitioning of the resource patches of females, as in syrphid flies (Maier and Waldbauer, 1979). Maintenance of dual strategies (territorial and non-territorial) helps the non-territorial males also earn a few matings (see Maynard Smith and Price, 1973; Maynard Smith and Parker, 1976).

C. Courtship and Copulation

Before attempting any physical contact with the female, it is essential for the male to impress his specific identity on her, relieve her of her 'coyness', induce receptivity, align her to a suitable position to facilitate genital contact, and display his supremacy over other males. Such pre-copulatory behavioural activities of males constitute courtship. Besides employing a variety of mechanisms and diverse cues, males of some insects lure conspecific females by providing nutritious 'food'; this sort of nuptial feeding, especially among carnivorous insects, may represent a form of appeasement of the aggressive females by males. Briefly, subjected to intense competition for securing conspecific females for mating, males have evolved courtship as a strategy for attracting females and inducing receptivity in them (for details of sexual behaviour, see Gillott *et al.*, Volume V of this series).

1. Receptivity induction

Several insects, e.g. grasshoppers, fruitflies, beetles, and parasitic wasps, exchange a series of responses in a definite order, such as antennal manipulation, abdominal vibration, and attempted genital insertion. Each step increases excitement and receptivity of the females. The sequence may be repeated several times before successful copulation as in the meloid beetle *Lytta nuttali* (see Gerber and Church, 1973). Depending on the receptivity of the female, the courtship may last for different durations. The megachilid bee *Anthidium maculosum* courts the female in a simple way. It lands on the back of the female and strokes the sides of her head and thorax by quick and vigorous movements of its fore- and mid-legs, and within a couple of seconds begins copulation (Alcock *et al.*, 1977a). *Locusta migratoria* repeats a complex sequence of courtship behaviour in the pre-copulatory passive phase. Even among closely related species, mechanisms of courtship differ drastically. Such variations in courtship behaviour are important to maintain sexual isolation among sympatric species.

2. Courtship cycles

The courting male always tries to occupy a strategic position which enables him to exchange signals with the females and establish genital contact with a female immediately after receiving a receptivity signal from her. For instance, the receptivity signal of a female pteromalid wasp consists of a primary component displayed by raising of her abdomen and a secondary component including specific antennal movements (Van den Assem, 1974, 1976). Males of some pteromalid wasps, which are smaller, have shifted their courtship position from the rear of the female to the front so as to receive the receptivity signal without any hindrance (Van den Assem, 1974). However, occupying a frontal position is disadvantageous in that it leaves the posterior side of the female unguarded and provides chances for competitive rival males. To compensate for this disadvantage, the males, as in *Pteromalus puparum*, after completion of each cycle of courtship sequence, move behind, attempt copulation, and test the receptivity of the female (Van den Assem, 1976). A delay in backing up by a male amounts to missing a copulation. Having induced receptivity in a female, a courting male *Nasonia vitripennis* which occupies frontal position in courtship misses copulation (especially at high male densities near emergence site) as a non-courting male from behind takes up the chance.

A courting male which cannot induce receptivity in a female even after repeating a few cycles of courtship sequence should decide whether to continue courtship with the same female or to give it up and seek another mate. Males who give up courtship too soon without arousing receptivity in females may not successfully copulate and those who court unreceptive females too long are liable to miss copulation with other females. Sexual selection acting on males seems to have determined the optimum duration of courtship (see Parker, 1974a, b). Therefore, it is essential for a male to recognize the unreceptivity signals of the female as well as the receptivity signals.

3. Unreceptivity signals

Females of *Heliconius erato* repel courting males with the help of an 'antiaphrodisiac' substance received from males during the previous mating (Gilbert, 1976; Weaver, 1978). Female *Drosophila melanogaster* reject courting males with the help of paragonial gland secretion of males received during previous copulation (Burnet *et al.*, 1973). Such unreceptivity signals from females not only help the males avoid losing time in wasteful courtship but also help the females maintain monogamy.

4. Nuptial feeding

Males of a few species lure conspecific females with nutritious food. They offer their own protein-rich glandular secretions (dorsal gland, salivary gland, or accessory gland secretions), or the prey organisms collected by them, or their own bodies after copulation (Thornhill, 1976). Male *Gryllus assimilis* and *Blatta* sp. feed the female with their dorsal gland secretion to overcome female 'coyness' and to protect their spermatophores from being consumed by females before the sperm have been used for fertilization. Males of some dipterans (e.g. *Rivellia boscii*) and mecopterans (e.g. *Panorpa* sp.) offer protein-rich secretions of their salivary gland to females before copulation (Pritchard, 1967; Thornbill, 1974). Spermatophores, which were primarily evolved for protection of sperm against desiccation in land environment, are used for nourishing copulating females in several orders of insects (Table 2). Spermatophores provide a rich source of protein. Whereas females of some orthopterans and neuropterans remove the spermatophores deposited into their bursa and consume them, lepidopterous and trichopterous females digest and absorb them in their genital tract (see Engelmann, 1970). Males producing larger spermatophores are preferred by females not only for the numerous sperm in them but also for their nutritive value. Female *Heliconius erato, H. hecale*, and *Danaus plexippus* utilize spermatophore proteins for egg production (Boggs and Gilbert, 1979). The mating plug of several lepidopterans and dipterans, intended for prevention of further matings, is presumably digested and absorbed by females in their genital tract (within 6–24 hours after copulation in mosquitoes and honey bees).

To avoid being eaten by females during copulation, dipterans of the family Empididae wrap their prey in their anal secretion of silk and offer them to their mates during courtship (Svensson and Peterson, 1987). Mecopterans such as *Panorpa* sp. also offer prey to females. In all these cases, duration of copulation is likely to be determined by the quantity of the prey offered. Males of tiphiid wasps and the bug *Stilbocoris natalensis* offer nectar and fig seed as gifts respectively. Mantids, the carabid beetle *Carasus auratus* (see Thornhill, 1976), and two species of midges have been reported to cannibalize the copulating males. Incidence of such cannibalism by females in nature is likely to be rare and in most cases caused by their mistaking the males for prey. Males which guard reproductive territories also provide food for females indirectly, by permitting them to feed in their guarded

Table 2

Provision of nourishment as a part of courtship by males of some insects (from Thornhill, 1976)

I. Glandular products

A. From dorsal glands

1. Some Orthoptera
2. Some Dictyoptera

B. From salivary secretions

1. A few Diptera, e.g. *Rivellia boscii* (Otitidae, *Cardiacephala myrmex* (Calobatidae), and *Rioxa pornia* (Tephritidae)
2. A few Mecoptera, e.g. *Panorpa* sp.

C. Spermatphores

1. Collembola
2. Orthoptera, Dictyoptera, Dermaptera, Psocoptera, Neuroptera, Lepidoptera, Hemiptera, Trichoptera, Coleoptera, Hymenoptera, Diptera

D. Mating plug

1. Diptera: Culicidae; Drosophilidae
2. Lepidoptera
3. Some Coleoptera, Orthoptera, Isoptera

II. Other food

A. Nuptial prey

1. Mecoptera, e.g. Bittacidae
2. Diptera, e.g. Empididae

B. Nuptial feed

1. Nectar: Hymenoptera, e.g. Tiphiidae
2. Seeds: Hemiptera, e.g. Lygaeidae

III. Their own body

1. Dictyoptera, e.g. Mantidae
2. Coleoptera, e.g. Carabidae
3. Diptera, e.g. Ceratopogenidae

territories. The female's choice for increased male investment and intrasexual selection seems to have resulted in courtship feeding behaviour.

D. Sperm Transfer

Successful courtship leads to mating and sperm transfer. Associated with internal fertilization, suitable organs for the transfer of sperm in males and sperm storage in females have been developed. In several orders of insects sperm transfer is direct and through the spermatophore. However, several species belonging to the higher orders of Insecta such as the Diptera, Coleoptera, and Hymenoptera have dispensed with spermatophore production independently. Depending on the adult longevity and duration of gametogenic cycles, the frequency of mating varies in different insects.

1. Mating frequency

After successful completion of courtship and copulation, males tend to guard their mates against reinsemination by other males and ensure that their sperm is used for fertilization, or they may seek new mates, court them, and copulate with them. On the contrary, the tendency of just-mated females (e.g. mosquitoes and butterflies) is to pass through a sexually refractive period and resist the mating attempts of fresh males at least till the sperm stored in the spermatheca have been exhausted. Solitary bees such as *Anthidium maculosum* (Alcock *et al.*, 1977a) find it less economical to resist the mating attempts of males and yield to repeated matings. Although most insects mate several times in their lifetime, a few with ephemeral adult stage, such as chironomids and mayflies, mate hardly more than once in their lifetime (Parker, 1978a).

Among other factors such as availability of potential female and intensity of competition among mate-seeking males, mating frequency of a male is determined by the quantity of sperm and accessory gland substances transferred to a female during the previous mating. In spermatophore-producing insects, the size of the spermatophore and the duration required to replenish sperm storage and accessory gland secretion set a physiological limit to mating frequency of males. For instance, *Ephippiger* sp. (Orthoptera) loses as much as 25 per cent of its body weight in the production of spermatophore and requires three to five days to remate. Sperm transfer in the katydids is through an expensive spermatophyllax, which drains most of the materials stored in the accessory gland. The accessory gland of an unmated male *Requena verticalis* weighs 26.5 per cent of the whole body weight. Even after 120 hours of mating the accessory gland weighs only 17 per cent of the body weight. *Requena verticalis* does not mate till its accessory gland recoups its normal weight (Davies and Dadour, 1989). *Galleria* sp. (Lepidoptera) requires 12 hours to produce a normal spermatophore and hence cannot remate within that duration. The leek moth, *Acrolepiopsis assectella*, can mate and form a normal spermatophore only once in 24 hours (Thibout, 1979).

However, the mating frequency of odonate males is high. From his field data on the damselfly *Calopteryx maculata*, Waage (1979a, b) reported that a male can mate once in about 10 minutes without affecting its potency. Male *Glossina* mates about six times without any deterioration of its inseminating ability (Jordan, 1972). About two days following emergence, *Aedes aegypti* has a large store of sperm in its seminal vesicles and is able to inseminate five to six females in quick succession; it is capable of mating with 30 females within 30 minutes, though most of the matings do not result in ejaculation. It refills its seminal vesicles and accessory glands in about two to three days after mating, but most such 'recovered' males do not copulate, and even if they do, fail to ejaculate (Jones, 1973a). However, depletion of sperm or accessory gland secretion does not prevent immediate remating in several butterfly males; for instance, male *Heliconius erato* remates even after it has exhausted its spermatophore material in one or two matings, but prolongs the copulation to more than 48 hours, within which period the stores are replenished (Thornhill, 1976).

Availability of females influences the mating frequency of several insects. In an adult span of about 14 days, the male spiny bollworm, *Earias insulana* (Lepidoptera), mates more than four times, when presented with two or more females (Kehat and Gordon, 1977). Mating frequency of some of the male wasps is extremely high. In a *Melittobia* (Chalcidoidea) population, with a sex ratio skewed in favour of females (1 M : 19 F), a male mates and inseminates over 100 females in quick succession (Van den Assem *et al.*, 1980). Larger males of *Nasonia vitripennis* and *Lariophagus* sp. inseminate more than 25 females within a short time (Van den Assem, 1976).

Although males have evolved such devices as the mating plug (e.g. Lepidoptera), spermatophores (e.g. Orthoptera and Dictyoptera), and post-copulatory behavioural tactics (e.g. Odonata, dungflies, etc.) to discourage immediate reinsemination of females, females of several species are known to mate repeatedly in their lifetime. Mating frequency of a female depends on her ability to store sperm, quantity of sperm received in previous mating, and longevity of sperm in the female genital tract. However, queens of several species of ants store in their spermatheca just enough sperm to fertilize the eggs present in the ovariole. A significant linear relationship between sperm count in the spermatheca and ovariole number has been reported by Tschinkel (1987). As the maintenance of the sperm in the genital tract is energetically expensive, storing more sperm than necessary is considered less advantageous. On the other hand, *Glossina* females receive a substantial quantity of sperm in the first mating and store them in a viable condition in the spermatheca till the end of their life; yet, most of them are reinseminated several times. Likewise, female mosquitoes receive more than sufficient sperm to fertilize all their eggs; nevertheless, they too mate frequently in their lifetime. For instance, Jones (1973b) has shown that an *Aedes aegypti* female mates repeatedly with the same male usually, but gets inseminated only once. However, in a subsequent paper, Jones (1974) demonstrated that *A. aegypti* females can be reinseminated, if the duration of copulation in the first mating was not sufficiently long to transfer an adequate quantity of sperm. Although sperm in the female genital tract of *Locusta migratoria* survive for more than 10 weeks, a female copulates more than once with different males within an oviposition cycle (Parker and Smith, 1975). Female syrphid flies (Diptera) are receptive for most of their adult span and mate repeatedly (Maier, 1978). Of the 59 females of *Earias insulana* (Lepidoptera) tested for their mating ability, about 9 and 15 per cent mated thrice and twice in a night, respectively (Kehat and Gordon, 1977). Incidence of remating in females of Lepidoptera has been found to be high (Leopold, 1976). Counting the number of spermatophores in the genital tract of plugged and unplugged female butterflies, Ehrlich and Ehrlich (1978) showed that plugging does not preclude the possibility of a second mating in them. Several species of butterflies are forced to resort to second mating after a short refractory period, because of ineffective plugging and insufficiency of sperm to fertilize a large number of eggs (e.g. 1,100 eggs in *Euphdryas editha*: Labine, 1966). Female *Pieris brassicae* accept second mating six to nine days after the first one. However, sperm of *Heliconius* spp. are long-lived and in most species, the females do not mate more than once in their lifetime (Gilbert, 1972).

Males of several species of Hymenoptera with their dogged persistence in mating with unreceptive females interfere with their nesting activities (Barrows, 1976; see also Alcock *et al.*, 1978). Instead of losing time and energy in resisting these persistent males, females of *Dianthidium, Anthidiellum*, and *Callanthidium* accept brief copulations (see Alcock *et al.*, 1977a).

Females of *Drosophila paulistorum* (see Richmond and Ehrman, 1974) *D. pseudoobscura* (see Richmond, 1976), and *D. melanogaster* (see Pyle and Gromko, 1978) mate repeatedly, though viable sperm are still available in their spermatheca. Over 50 per cent of *D. melanogaster* females remate five days after their first mating (Gromko and Pyle, 1978). Repeated mating in *D. melanogaster* is under the control of a gene in the X chromosome; selection for this character for over 20 generations resulted in the production of flies, over 50 per cent of which remated just 24 hours after the first mating.

Females resort to multiple matings in order to (1) build up a sufficient store of sperm, (2) minimize the cost of maintaining a large spermatheca and several sperm over a long period, (3) acquire nutrients from males through semen, (4) produce genetic variations in the offspring by mating with different males, and (5) conserve time and energy by accepting short copulations. Multiple matings help the males promote their reproductive success by siring a large number of offspring. Sexual selection seems to have varied the timing of remating in males and females. Whereas males are ready to remate for most of their adult life, females, in general, are receptive for remating only when their sperm store is depleted.

E. Sperm Competition and Utilization

Whereas the females of parasitic wasps, chironomids, and mayflies mate only once, others such as butterflies and moths mate intermittently, i.e. whenever their sperm store is exhausted; still others (e.g. fruit flies and solitary bees) mate repeatedly. Based on mating systems, they can be grouped into monandrous forms which mate only with one male, or polyandrous forms which mate and receive sperm from different males. The pteromalid hymenopteran *Nasonia vitripennis* mates only once; mated females do not respond to courting males and form a good example of monandry among insects (Van den Assem and De Bruzin, 1977). Short-lived adults, such as chironomids and the antarctic midge *Beigica antarctica*, hardly mate more than once (MacLean, 1975). The moth *Byrdia* of Devon Island mates immediately after emergence and oviposits inside the cocoon itself (MacLean, 1975). Physiological influence of the first-mated males on females and the ephemeral existence of some adults perhaps account for the monandry in insects.

In view of the fact that the males are sexually motivated for most of their adult life, and often interfere with the routine activities of females by their persistent, desperate copulatory attempts, and that multiple matings confer certain advantages to females, sexual selection seems to have favoured polyandry of females of several species of insects. Consequently, there is competition not only among males to acquire a potential mate but also among the ejaculates of different males within the genital tract of the female to fertilize as many ova as possible. Such a competition between the sperm of different males in a polyandrous female profoundly influences sexual selection, acting on males, and determines the male mating strategy (see Parker, 1970a, 1974a). In monandrous species such as *Pectinophora gossypiella*, the first male to copulate is successful (Bartlett and Lewis, 1985), whereas in polyandrous species such as bees and wasps, the male that copulates last fertilizes a major proportion of the eggs laid (see Alcock *et al.*, 1978).

1. Male mating strategies

The following male mating strategies are some of the manifestations of sexual selection: (1) To remain near the emergence site of a female and gain priority as the first male to mate with a virgin female (e.g. the bee *Centris pallida*: Alcock *et al.*, 1977b; the parasitic wasps *Nasonia vitripennis* and *Pteromalus puparum*: Van den Assem, 1976). (2) To mate with a female and prevent reinsemination by other males by such devices as mating plug or spermatophore (e.g. Orthoptera, Dictyoptera, Lepidoptera, and Diptera: Parker, 1970a; Parker and Smith, 1975). (3) To mate with a female and guard her till oviposition has been completed (e.g. Odonata: Waage, 1979b; dungflies: Parker, 1978b). (4) To mate with a female just before she begins to oviposit (e.g. syrphid flies: Maier and Waldbauer, 1979; the free-living wasps *Oxybelus* spp.: Alcock *et al.*, 1978).

2. Avoidance of sperm competition

Sperm competition is a common phenomenon among Orthoptera, Lepidoptera, Coleoptera, Diptera, and Hymenoptera. To reduce sperm competition, these insects seem to have evolved different structural and behavioural mechanisms independently. During copulation, *Locusta* male transfers to the female a long (35–45 mm) spermatophore with a tube 0.3 mm in diameter. The mating pair separates following the breaking of the spermatophoral tube. The distal part of the tube remaining in the female disappears within a day; the proximal part persists in a turgid condition and prevents sperm displacement for a few days and ensures the use of sperm of the first-mated male for fertilization up to one or two ovipositions. Subsequently, the second male transfers his spermatophore and removes the sperm of the first male from the spermatheca simultaneously. Parker and Smith (1975) proposed that the predominancy of the last male in removing sperm from the spermatheca is a sexually selected adaptation for the first male to ensure that most of his sperm are utilized in fertilization, curtailing precedence of the sperm that might be deposited by a subsequent male.

In several species of odonates, males resort to behavioural tactics to prevent reinsemination of their mates and sperm displacement. After the completion of copulation, a male guides his mate to the oviposition site and defends her till she has laid all the eggs (Corbet, 1980). Some males remain in tandem position with the female till she completes oviposition (e.g. *Chromagrion conditum*: Bick *et al.*, 1976; *Tramae* sp. : Jurzitza, 1974). A few others establish tandem position with the mate repeatedly, till she completes egg-laying (e.g. *Pantala flavescens* and *Hadrothemis* sp.: Corbet, 1980). Such post-copulatory behavioural tactics of the male are mainly intended to prevent the take-over of the female by other males. However, Corbet (1980) considers that the intensity of the male's association with his mate depends on the density of rival males in the territory and the frequency of interference with the ovipositing female. For instance, a male *Sympetrum parvulum* remains in tandem with his mate at high male densities in the territory, or guards her by hovering over her, or leaves her alone, when rival males are too few (Ueda, 1979).

3. Sperm displacement

Investment of a considerable amount of time and energy by males in affording protection to their mates gains a high premium against mechanisms of sperm displacement evolved in a few species of odonates. Waage (1979a) attributes a dual function to the penis of the damselfly *Calopteryx maculata*. Besides transferring sperm into the female, it is capable of removing 88–100 per cent of the sperm already present in the bursa copulatorix and spermathecal tube of the female. Using scanning electron microscopy, Waage (1979a) showed that the head of the penis, extended by internal fluid pressure, scoops the sperm from the bursa copulatorix; the backwardly directed hairs at the ventrolateral side of the base of the penis remove the sperm mass; and the backwardly directed hairs on the horns of the penis head remove the sperm from the spermathecal tube. Comparing the index of sperm volume in the bursa and spermatheca of a once-mated female (4.2 ± 0.6; $N = 14$) with that of a twice-mated female (4.4 ± 0.7; $N = 16$), Waage (1979a) also showed that the sperm volume in the bursa and the spermatheca, before and after the second mating, was not statistically significant. He confirmed his observation by examining the seminal vesicle of the male used for the second mating, which was found to be empty. Hence, undoubtedly, the second-mated male replaces the sperm of the first-mated male with his own sperm.

Mating a female boll weevil *Anthonomus grandis* with a P^{32} tagged male and a normal male subsequently, Villavaso (1975) showed that the second-mated male displaces as much as 66 per cent of its predecessor's sperm. From a female with spermathecal muscle surgically removed, the second male could displace only 22 per cent of the sperm. Obviously, the spermathecal muscle of the female plays an important role in sperm displacement. The second-mated male gains precedence of his sperm over those of his predecessor, and sires a major proportion of the offspring. The meal moth *Tribolium castaneum* uses the sperm received from previous matings only when those received in the mating just prior to oviposition have been exhausted. Apparently, the sperm received from the first male are forced to the rear end of the spermatheca resulting in a 'last-in-first-out' pattern. Elongate spermatheca usually facilitates sperm precedence following this principle (see Brower, 1975; Riemann and Thorson, 1974; Ridley, 1988).

Incidence of remating in lepidopteran females has been reported to be quite high, despite the use of mating plug such as spermatophragma to prevent reinsemination (Holt and North, 1970). Females of most species have only one genital duct, which accomplishes both insemination and oviposition. The mating plug formed by the male during previous mating is discarded at each oviposition by the female, enabling remating and sperm competition (Leopold, 1976). Most females make use of the sperm contributed by the male which copulated just prior to oviposition, for fertilizing a major proportion of the eggs laid. But nothing is known about the mechanism employed by the last male to achieve sperm precedence (see Ehrlich and Ehrlich, 1978). Rutowski *et al.* (1989) and Dickinson and Rutowski (1989) proposed that variation in the size of the mating plugs deposited into the females accounts for the differences in the reproductive success of male butterflies. In the checkerspot butterfly, *Euphdryas chalcedona*, males occasionally circumvent the mating plugs deposited by the first-mated male (Rutowski *et al.*, 1989). Therefore, a larger plug may protect the male's investment and increase his reproductive success. In some cases, sperm of the first-mated male are physically displaced as in *Calopteryx maculata* (see Waage, 1979a) and *Anthonomus grandis* (see Villavaso, 1975), or pushed to the rear end of the spermatheca as in *Plodia interpunctella* (see Brower, 1975). In the bruchid *Callasobruchus maculatus*, the second-mated male sires 77 to 87 per cent of the offspring (Eady, 1991). However, sperm of the first-mated male get priority in some cases and about 76 per cent of them are used for fertilization (e.g. *Bombyx mori*).

Among Diptera, the mating plug is less effective in preventing reinsemination; it is rather useful in preventing sperm loss (see Leopold, 1976). Parker (1970b) demonstrated sperm competition in the dungfly, *Scatophaga sterocoraria*. In this species males are not likely to discriminate between virgin females and mated ones. Although mated females are non-receptive, high male densities on the dung surface increase the probability of females being remated. A territorial male prevents take-over and reinsemination of his mate by other males by emigrating with her to a nearby grass patch where male density is less and by remaining mounted over her in a 'post-copulatory passive phase' till she completes oviposition. Such passive-phase-guarding also decreases sperm competition. These strategies repay him richly, by enabling him to fertilize almost all the eggs laid by his mate. Occasionally a non-territorial intruder, after a fierce struggle, manages to take over a copulating female and mate with her. Mating a female with a normal male first and an irradiated male subsequently, Parker (1970b) showed that the second male manages to achieve sperm precedence and fertilizes as many as 80 per cent of the eggs laid by his mate. The number of eggs fertilized by the second male increases with increasing duration of copulation (Fig. 3) and about 50 minutes is sufficient to fertilize over 90 per cent of the eggs (Parker, 1978b; also see Saul *et al.*, 1988). Nothing is known about the mechanism of physical displacement of the sperm from the female. *Drosophila melanogaster* male, by mating last with a female about to lay eggs, displaces the sperm of his predecessor(s), achieves sperm precedence,

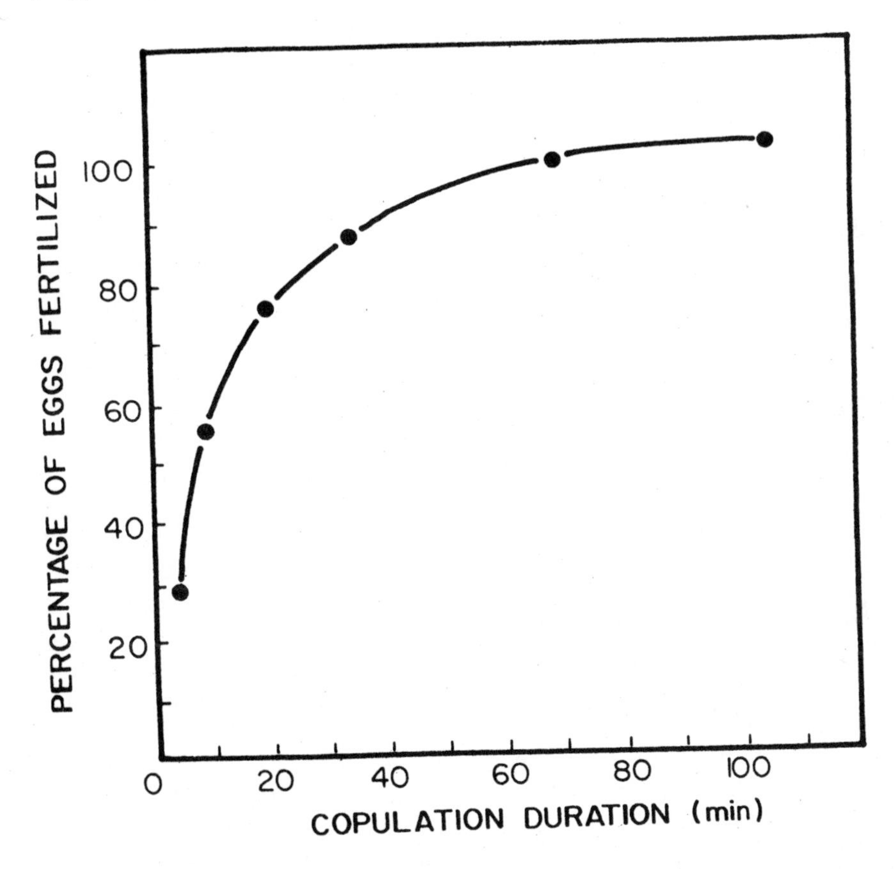

Fig. 3. Percentage of eggs fertilized by the sperm of the second-mated male as a function of copulation duration in *Scatophaga sterocoraria* (from Parker, 1978b).

and gains a distinct selection advantage over those who mated first (Boorman and Parker, 1976). In the tephritid fly *Rhagoletis pomonella* too, the second-mated male gains sperm precedence and sires 79–93 per cent of the offspring (Opp *et al.*, 1990).

However, the first-mated male gains sperm precedence in a few dipterans such as *Culicoides melleus.* Males mostly depend on the unreceptive behaviour of their mates for preventing reinsemination and sperm displacement. For instance, a just-mated *C. melleus* female, by kicking the males attempting to copulate with her, decreases the amount of sperm ejaculated by the second- and third-mating males by about 15 and 21 per cent respectively (Linsley and Hind, 1975). Besides losing about 15 per cent of sperm during sperm transfer, the second-mating male contributes his sperm to only 35 per cent of the offspring.

Among bees and wasps, timing of emergence of males in relation to that of conspecific females seems to have determined the mating strategy of males. For species in which males emerge only after females, the last male to mate is favoured by selection (Alcock *et al.*, 1978). For instance, in wasps such as *Oxybelus* spp., *Trypoxylon* spp., and *Sphex tepanecus*, in which males emerge after their females, the males remain near the nest and mate with the female every time she returns to the nest after foraging (Alcock *et al.*, 1978). These males are reported to be capable of displacing the sperm of their predecessors with or without the assistance of their mate. Females of the social Hymenoptera are polyandrous and some of them, such as *Apis cerana* and *A. mellifera*, mate as many as 17 to 30 times. In such species, forceful ejaculation by the male and abdominal contractions of the queen result in the mixing of sperm of different males. The queen seems to control the insemination success of the males (see Page, 1986).

4. Sperm utilization

Males which emerge prior to females always prefer to mate first. In the wasps *Cephalonomia* spp., *Goniozus* spp., and *Parasierola* spp., the males are capable of locating females concealed in cocoons and of chewing their way into the cocoon to mate with the female, which is about to emerge (Gordh, 1976; Gordh and Evans, 1976). The 'digger' male of *Centris pallida* digs through soil and copulates with the concealed female (Alcock *et al.*, 1977b). In all these cases as well as several species of chalcidoid wasps, females do not mate for a second time and occasional second matings do not result in insemination (Van den Assem, 1976). Therefore, there is no scope for sperm displacement or precedence in these cases.

The honey bee, *Apis*, is reported to mate several times with different males and use their sperm for fertilization. Sperm from different males do not mix in the spermatheca and there is no definite role for sperm precedence. Of 23 females studied, the sperm of the first male predominated in 11 females, sperm of the last male predominated in five cases, and in others an equal volume of sperm from each male was found to be used (see Parker, 1970a).

In summary, the tendency of females of several species to mate repeatedly with different males, the drive of males to copulate with as many females as possible within the short span of their adult life, the structural design of the reproductive system of females enabling storage of sperm received from different males, and the prolonged survival of sperm in the female genital tract all seem to have contributed to a high level of sperm competition in insects. Some of the adaptations that resulted from sexual selection are physiological (through accessory gland secretion: Gillott, Volume III of this series for details) and behavioural (prolonging copulation duration) mechanisms evolved to inhibit receptivity of females for subsequent mating, structural innovations such as spermatophore and mating plug to prevent reinsemination of females, structural modification of male genitalia (as in damselfly) for removal of previous stock of sperm from females, and post-copulatory mate-guarding behaviour to prevent take-over of females by other males.

III. FECUNDITY AND OVIPOSITION

The reproductive success of a female depends on her ability to produce eggs (i.e. fecundity) as well as to promote the survival of her offspring by providing adequate energy reserve for embryonic development and by placing the eggs on suitable habitats containing a rich source of larval food. Ultimately, fecundity is an aspect of food conversion into eggs by the female. Hence, environmental factors which alter the rate and efficiency of food utilization of a female also decisively affect her fecundity. Fecundity of insects ranges from one or two larvae in the viviparous tsetse fly *Glossina austeni* (see Langley, 1977) to 600,000 eggs in the domestic honey bee (Hinton, 1981). Even within a species, fecundity varies widely from time to time or from generation to generation. For instance, fecundity of the pine looper moth, *Bupalus piniarius*, ranged from 161 to 299 eggs in the 1951-52 generation or from 136 to 210 in the 1957-58 generation (Kolmp, 1966). Apparently, each species has to strike a compromise between egg number and size in relation to the resources available in the environment. Natural selection seems to have optimized the size and rate of egg production of each species for a given set of environmental conditions.

Fecundity estimation is of great value, as it reflects the reproductive potential of an individual or the intrinsic growth rate of a population. The cumbersome method of counting the number of eggs laid by a female in a generation has been replaced by several indirect methods such as the use of ovariole number, body weight or length, and quantity of food consumed during larval or adult life. However, before attempting to estimate the fecundity of a species either directly or indirectly, it should be remembered that fecundity is related to the environmental conditions of the population as a whole (Labeyrie, 1978a). Yet, in view of the difficulties involved in monitoring the multifarious factors, most of the available data on fecundity relate to a few isolated factors only.

A. Influence of Nutrition

As stated already, fecundity is an aspect of conversion of ingested food. Food intake profoundly influences oocyte growth and fecundity through mechanical, neural, and endocrine mechanisms (Engelmann, 1970). Ametabolous and hemimetabolous insects feed throughout life and utilize mostly the nutrients accumulated during the adult stage for egg production. But, the holometabolous insects have set apart their larval stages for feeding and growth, and the adult stage for reproduction and dispersal. Therefore, food consumed during the larval stages plays a dominant role in determining the fecundity of most holometabolous insects. The ephemeropterans and a few dipterans (e.g. chironomids) with transient adult stage depend entirely on the nutrients accumulated during the larval stage. The herbivores are exposed to scarcity and abundance of food like the carnivores. Therefore, quality and quantity of food available, life style such as growing larger over a long duration or smaller in a short period, and feeding pattern significantly influence fecundity. In view of the growing importance of predators and parasitoids in biocontrol in agriculture and of prey organisms like the chironomid larvae in aquaculture, insect culture is a promising area. Therefore, the role of nutrients such as proteins in fecundity has been highlighted in this section.

1. Dietary requirements and food-quality effects

Several reviews on dietary requirements of insects have been published during the last 15 years (e.g. Dadd, 1985; Slansky and Rodriguez, 1986). Most of them deal with the requirements for growth and maintenance and provide scattered information on the requirements for reproduction. Basically, the dietary components which promote growth and longevity also help improve reproductive performance. This is essentially true of several holometabolous insects which do not subsist on adult feeding. Supplementation of adult diet with nutrients such as amino acids, unsaturated fatty acids, cholesterol, and minerals increases the fecundity of the insects which feed throughout their lifetime. Therefore, variations of dietary requirements between different species are apparently due to differences in the quality and quantity of larval metabolites transferred to the adult stage at the time of metamorphosis and the degree of dependence on adult feeding.

As fecundity depends on availability of protein for yolk synthesis, a protein-rich diet increases fecundity. Differences in the nitrogen (or protein) and water contents of various foods or of the same host plant due to phenological changes significantly influence the fecundity of insects. Data presented in Table 3 demonstrate that in most cases fecundity increased proportionately with nitrogen (or protein) content or water/nitrogen ratio. Fecundity of *Spodoptera litura*, fed on *Brassica oleracea* leaf with 3.2 per cent N during the larval period, was 50 per cent less than that fed on the flower of the same host plant. A decrease in the water/nitrogen ratio of the host plant from 10.67 to 6.97 decreased the fecundity of *Aiolopus*

Table 3

Effect of host-plant quality on the fecundity of some insects

Insect	Food	Remarks	Fecundity	Reference
Aiolopus thalassinus (Orthoptera)	*Panicum maximum*	Water/N : 8.32	61	Sanjayan and Murugan (1987)
	Cyperus rotundus	10.67	72	Sanjayan and Murugan (1987)
	Cynadon dactylon	6.97	34	Sanjayan and Murugan (1987)
Atractomorpha crenulata (Orthoptera)	*Ricinus communis* leaf	Water/N : 41.6	257	Senthamizhselvan and Murugan (1988)
	P. maximum leaf	Water/N : 8.37	79	Senthamizhselvan and Murugan (1988)
Oxycarenus hyalinipennis (Hemiptera)	*Gossypium hirsutum* seed	8.7 per cent N	25	Ananthakrishnan *et al.* (1982)
	Abutilon indicum seed	4.9 per cent N	15	Ananthakrishnan *et al.* (1982)
	Abutilon crispum seed	4.7 per cent N	14	Ananthakrishnan *et al.* (1982)
	Sida rhomboidea seed	3.1 per cent N	7	Ananthakrishnan *et al.* (1982)
O. laetus (Hemiptera)	*G. hirsutum* seed	8.7 per cent N	31	Raman (1987)
	A. crispum seed	4.9 per cent N	21	Raman (1987)
Megacopta cribraria (Hemiptera)	*Sesbania grandiflora* seed	1.1 per cent protein	73	Srinivasaperumal *et al.* (1992)
	Crossandra undulifolia seed	0.9 per cent protein	60	Srinivasaperumal *et al.* (1992)
	G. hirsutum seed	0.91 per cent protein	49	Srinivasaperumal *et al.* (1992)
Dialeurodes vulgaris (Homoptera)	*Jasminum multiflorum* leaf	0.47 per cent phenol	82	Sundararaj and David (1990)
	J. grandiflora leaf	0.9 per cent phenol	47	Sundararaj and David (1990)
	J. pubescens leaf	2.1 per cent phenol	26	Sundararaj and David (1990)
Spodoptera litura (Lepidoptera)	*Brassica oleracea* flower	6.4 per cent N	453	Senthamizhselvan and Muthukrishnan (1989)
	B. oleracea leaf	3.2 per cent N	211	Senthamizhselvan and Muthukrishnan (1989)
Diatraea saccharalis (Lepidoptera)	Wild corn	Rich in nitrogen	714	Bessin and Reagan (1990)
	Jonson's grass	Poor in nitrogen	427	Bessin and Reagan (1990)
Heliothis armigera (Lepidoptera)	*Cicer arietinum* pod	3.6 per cent N	405	Tripathi and Singh (1989)
	Vigna radiata pod	2.8 per cent N	306	Tripathi and Singh (1989)

(Contd.)

Table 3 (Contd.)

Insect	Food	Remarks	Fecundity	Reference
	Lens culinaris pod	2.2 per cent N	257	Tripathi and Singh (1989)
Porthesia scintillans (Lepidoptera)	*Ricinus communis*	1.2 per cent N; 72.1 per cent W	47	Muthukrishnan and Sen Selvan (1993)
	Applied with	1.8 per cent N; 74.8 per cent W	62	Muthukrishnan and Sen Selvan (1993)
	different doses	2.3 per cent N; 76.4 per cent W	98	Muthukrishnan and Sen Selvan (1993)
	of nitrogen fertilizer	2.7 per cent N; 80.7 per cent W	114	Muthukrishnan and Sen Selvan (1993)
Cloeon sp. (Ephemeroptera)	Alga	43.5 per cent protein	540	Sivaramakrishnan and Venkataraman (1987)
	Detritus	17.8 per cent protein	102	Sivaramakrishnan and Venkataraman (1987)
Baetis sp. (Ephemeroptera)	Alga	43.5 per cent protein	151	Sivaramakrishnan and Venkataraman (1987)
	Detritus	17.8 per cent protein	200	Sivaramakrishnan and Venkataraman (1987)
Kiefferulus barbitarsis (Diptera)	Mixed algal powder	7.1 per cent N	1,218	Palavesam and Muthukrishnan (1992)
	Chlorella	7.0 per cent N	1,078	Palavesam and Muthukrishnan (1992)
	Vegetable waste	5.8 per cent N	1,060	Palavesam and Muthukrishnan (1992)
	Silt from pond	5.1 per cent N	800	Palavesam and Muthukrishnan (1992)
Graphognathus peregrinus (Coleoptera)	*Pisum sativum*		1,031	Ottens and Todd (1979)
	V. unguiculata		470	Ottens and Todd (1979)
	Corn		25	Ottens and Todd (1979)
	Okra		325	Ottens and Todd (1979)
	Cotton		134	Ottens and Todd (1979)
Raphidopalpa atripennis (Coleoptera)	*Luffa acutaugula*	Mature leaf: 2.9 per cent N; 73.8 per cent W	205	Annadurai (1987)
		Senescent leaf: 2.0 per cent N; 66.0 per cent W	70	Annadurai (1987)
	L. cylindrica	Mature leaf: 6.44 per cent N; 77.6 per cent W	170	Annadurai (1987)
		Senescent leaf: 4.79 per cent N; 62.7 per cent W	70	Annadurai (1987)
Glischrochilus quadrisigratus	Apple	More nitrogen	271	Peng and Williams (1991)
	Tomato	Less nitrogen	116	Peng and Williams (1991)

Acrosternum graminea (Hemiptera)	*Cleome viscosa* leaf	6.2 per cent N: Total lipids 2.2 per cent	126	Velayudhan (1987)
	Gynandropsis pentaphylla leaf	5.0 per cent N: Total lipids 2.1 per cent	81	Velayudhan (1987)
	Croton sparciflorus leaf	3.9 per cent N: Total lipids 1.1 per cent	46	Velayudhan (1987)
Leptinotarsa decemlineata (Coleoptera)	*Solanum tuberosum*	Increases soluble proteins of haemolymph to 55 mg/ml	3,350	Brown *et al.* (1980)
	S. sarrachoides	Soluble protein of haemolymph 44 mg/ml	2,100	Brown *et al.* (1980)

thalassinus from 72 to 34 eggs (Table 3). A significant linear relationship between the number of egg rafts, eggs per raft, and total number of eggs oviposited by the detritivorous *Kiefferulus barbitarsis*, on the one hand, and nitrogen content of the larval rearing media on the other (Table 4) has been reported by Palavesam and Muthukrishnan (1992). However, the fecundity of the detritivore *Baetis* feeding on nitrogen-poor detritus was (200 eggs/female) more than that (157 eggs/female) feeding on algal diet (Sivaramakrishnan and Venkataraman, 1987). The authors attribute this inconsistency to the non-suitability of the mandibles of the larva to deal with algal diet.

Increased nitrogen fertilization of the soil augments leaf nitrogen and water contents of the host plants leading to increased fecundity of the pests. For instance, fertilization of the soil at the rate of 400 g urea/m^2 increased the nitrogen content of *Ricinus communis* leaf to 2.7 per cent from 1.2 per cent in soil fertilized with 75 g/m^2. Consequently, fecundity of the pest *Porthesia scintillans* increased three-fold (see Table 3). Baylis and Pierce (1991) reported that fecundity and fertility of the lycaenid butterfly *Jalmenus evagoras* colonizing fertilized *Acacia decurrens* plants were significantly higher than of those colonizing unfertilized plants. Increase in nitrogen content of the fertilized plants helps the adults produce a secretion rich in amino acids and sugars and tends a large population of ants which protect the eggs and larvae of the butterfly from predators. A high nitrogen-potassium ratio in the soil helps the plants increase the soluble nitrogen content of their sap which ultimately helps the aphids *(Brevicoryne brassicae* and *Myzus persicae*) increase their fecundity (Van Emden, 1966). The hymenopteran parasitoid *Trichogramma* sp. plays an important role in biological pest control. Eggs of the bug *Corcyra cephalonica* are used for the mass culture of the parasitoid. The host seed on which *C. cephalonica* is cultured seems to determine the fecundity of *Trichogramma* sp. Fecundity of *T. japonicum*, raised on *C. cephalonica* eggs obtained from cultures provided with green gram rich in both nitrogen (3.1 per cent) and free amino acids (29.6 mg/g), is almost twice (24 eggs/female) that raised on hosts reared on groundnut.

Table 4

Effect of nutrition on fecundity of *Kiefferulus barbitarsis* (Palavesam and Muthukrishnan, 1992)

Nutrient	Nitrogen content (per cent dry wt.)	No. egg rafts	No. eggs/raft	Egg production efficiency (per cent)	Hatchability (per cent)
Mixed algal powder	7.1	5.25 ± 1.48	232.0 ± 18.12	67.2 ± 3.1	99.0 ± 0.00
Chlorella	7.0	4.75 ± 0.43	227.0 ± 11.0	64.5 ± 2.7	99.0 ± 0.00
Sheep manure	6.5	3.67 ± 0.47	213.0 ± 8.62	58.8 ± 2.8	97.33 ± 0.94
Vegetable wastes	5.8	5.33 ± 0.94	199.0 ± 1.63	32.4 ± 1.4	96.67 ± 0.94
Silt from pond	5.1	4.67 ± 0.47	171.0 ± 8.04	37.0 ± 1.8	95.0 ± 1.41

High-protein diets are detrimental to survival and reproduction of the female cockroach *Blattella germanica*. As the fat body of the female fed on high-protein diet gets filled with uric acid, she prefers to take a low protein diet and compensate it by increasing the food consumption. Consequently, the number of eggs per oötheca was 41 in the female fed on 5 per cent protein diet compared with 25 in that fed on 65 per cent protein diet (Hamilton and Schal, 1988). Switching over from 5 to 25 per cent protein diet after metamorphosis doubled the number of oöthecae produced by the brown-banded cockroach *Supella longipalpa*. Females which were continued on 5 per cent protein diet even after metamorphosis mated late and their reproductive rate decreased (Hamilton *et al.*, 1990).

Most insects do not synthesize sterol and hence require it in the diet. Supplementation of the diet with sterol and unsaturated fatty acids increases the fecundity of both herbivorous and carnivorous insects. For instance, a relatively high level of total lipids (2.2 per cent), especially sterol and unsaturated fatty acids, in the host plant *Cleome viscosa* enabled the pentatomid bug *Acrosternum graminea* to realize the maximum fecundity of 126 eggs/female. Decrease in the total lipids to 1.1 per cent in the host plant *Croton sparciflorus* resulted in about 65 per cent decrease in the fecundity (46 eggs/female) (see Table 3). The colarado potato beetle, *Leptinotarsa decemlineata*, fed on senescent *Solanum tuberosum* leaves regained normal fecundity when fed with leaves treated with 2 per cent lecithin (Grison, 1952). Hymenopteran parasitoids which do not feed during the adult stage seek hosts rich in unsaturated fatty acids such as stearic acid. Thus, *Gryon* sp. and *Telenomus lucellus* realize maximum fecundity of 38 and 30 eggs per female respectively when host egg rich in stearic acid (e.g. *Acrosternum graminea*) is accessible to them (see Table 5). Deprivation of cholestrol in the diet did not decrease the fecundity of the fly *Dacus olea* but significantly affected the hatchability. Supplementation of the diet with vitamins, minerals, and amino acids increased the hatchability (Tsiropoulos, 1980). Stored-product pests, such as the flour beetle *Tribolium castaneum*, developing on exclusively carbohydrate diets, depend on the symbiotic bacteria in the gut for their sterol requirements. Predators and parasitoids in general depend solely on the prey organisms for their lipid and protein requirement. However, predators such as *Orius insidiosus* subsist on plant products (e.g. pollen) but fail to realize their normal fecundity because of low concentration of lipids in the plant food. For instance, fed on *Heliothis* eggs, *O. insidiosus* produced 103 eggs as against 40 eggs produced by females fed on pollen alone (Table 5). Similarly, the tiger beetle *Cicindela cancellata* fed on lipid and protein-rich *Corcyra* eggs produced eight times more eggs than that fed on ants (Shivashankar and Veeresh, 1987).

Carbohydrates are useful for the maintenance of adult insects which do not generally feed. The importance of carbohydrates for egg production in insects which depend on food ingested during the adult stage for egg production is far less than that of proteins and lipids. However, carbohydrate deficiency in the adult diets of these insects decreases general vitality and activity which ultimately results in

Table 5

Fecundity of some predators and parasites as a function of prey quality

Predator/parasite	Host species/diet	Remarks	Fecundity (eggs/female)	Reference
Orius insidiosus (Hemiptera)	*Heliothis* eggs + pollen		106	Kibman and Yeargan (1985)
	Pollen alone		40	Kibman and Yeargan (1985)
	Heliothis eggs alone		103	Kibman and Yeargan (1985)
	Bean + pollen		27	Kibman and Yeargan (1985)
	Mites + pollen		48	Kibman and Yeargan (1985)
	Mites + bean		19	Kibman and Yeargan (1985)
	Thrips + pollen		38	Kibman and Yeargan (1985)
	Thrips alone		20	Kibman and Yeargan (1985)
Gryon sp. (Hymenoptera)	*Acrosternum graminea* egg	Stearic acid content high	38	Velayudhan (1987)
	Agonoscelis nubila egg	Stearic acid content low	25	Velayudhan (1987)
Telenomus lucellus (Hymenoptera)	*Acrosternum graminea* egg	Stearic acid content high	30	Velayudhan (1987)
	Agonoscelis nubila egg	Stearic acid content low	22	Velayudhan (1987)
Trichogramma chilonis (Hymenoptera)	*Corcyra cephalonica* egg			Velayudhan (1987)
	C. cephalonica reared on:	Free amino acid	176	Velayudhan (1987)
	Green gram seed	29.6 mg/g, 3.1 per cent N		
	Sorghum seed	27.4 mg/g, 2.8 per cent N	172	Velayudhan (1987)
	Groundnut seed	31.8 mg/g, 2.8 per cent N	172	Velayudhan (1987)

T. japonicum	*C. cephalonica* egg			
(Hymenoptera)	host reared on:	Free amino acid		
	Green gram	29.6 mg/g, 3.1 per cent N	23	Velayudhan (1987)
	Sorghum	27.4 mg/g, 2.8 per cent N	14	Velayudhan (1987)
	Groundnut	31.8 mg/g, 2.8 per cent N	12	Velayudhan (1987)
Cicindela cancellata	3 *Corcyra* eggs/day		40	Shivashankar and Veeresh (1987)
(Coleoptera)	1 *Corcyra* egg/day		30	Shivashankar and Veeresh (1987)
	Ants		5	Shivashankar and Veeresh (1987)
	Mixed prey		15	Shivashankar and Veeresh (1987)

shorter life span and decreased fecundity (Lang, 1978). Insects which depend on larval nutrition for egg production enhance their fecundity when provided with sugar solution. Sugars are used by the lepidopteran adults for the mobilization of proteins stored in the fat body. An increase in concentration of sucrose fed to the moth *Trichoplusia ni* from about 3 to 12 per cent increased the fecundity from 300 to 550 eggs (Shorey, 1963). *Earias* spp. prefer the trisaccharide raffinose to other forms of sugar. Fecundity of *E. vitella* fed on 15 per cent raffinose solution was significantly higher (605 eggs) than that fed on distilled water (165 eggs) (Table 6). Female *Trichogramma chilonis* provided with fructose solution produced three times more eggs (106 eggs) than those provided with water (35 eggs) (Table 6). Natural foods such as honey, aphid honey dew, yeast, pollen, and raisins improve the fecundity of parasitoids. However, the fecundity of *Exeristes comstocki* reared on a mixture of amino acids, minerals, vitamins, fatty acids, RNA, and cholesterol was not significantly different from that reared on its natural host *Galleria mellonella* larva (Thompson, 1986). Besides the nutritive value of the components of the adult diet, the flavour of certain substances in the diet stimulates oviposition and increases fecundity. The response seems to vary with the substance (see Table 6). For instance, treatment of the oviposition bottle with orange juice enhanced the fecundity of the citrus fly *Dacus dorsalis* to 20.8×10^4 eggs/female compared with 7.4×10^4 eggs on a bottle treated with water (Vargas and Chang, 1991).

Apart from sugars, the amino acids, vitamins, nucleic acids, and minerals added to the adult diet significantly increase fecundity. The flour beetle, *Tribolium castaneum*, requires wheat flour supplemented with yeast or vitamin B complex mixture with high concentration of folic acid to realize the maximum fecundity of 43–57 eggs. Sterile starch, glucose, or cholesterol decreased the fecundity significantly (Table 6). Apparently, supplementation of wheat flour with yeast helps the beetle derive its vitamin requirements from yeast. For several insects, a protein-rich diet is required at the early stages of egg production. Most housefly (*Musca domestica*) females provided with only sugars as adults do not develop their ovarian follicles beyond stage 4; pulse feeding of protein to these flies stimulates ovarian maturation and increases the levels of ecdysteroid and vitellogenin in the haemolymph (Adams and Gerst, 1991). Lack of essential amino acids, such as histidine, leucine, and lysine, in the diet (e.g. orchard grass) resulted in vitellogenic arrest in the grasshopper *Melanoplus sanguinipes* (see Krishna and Thorsteinson, 1972). Among the essential amino acids, tryptophan is more important than methionine. A tryptophan-plus and methionine-minus diet helped the walnut huskfly, *Rhagoletis competa*, produce 484 eggs compared with 178 eggs produced by the fly reared on tryptophan-minus and methionine-plus diet (Tsiropoulos, 1978). Dietary nucleic acid is required for maintaining egg production in *M. domestica* (see Reinecko, 1985). RNA supplementation increased the longevity and fecundity of the fruitfly *Dacus olea* (see Tsiropoulos, 1980).

Fecundity of insects reared on artificial diet under laboratory conditions may significantly vary from that of insects reared on natural hosts under field conditions.

Table 6

Effect of adult nutrition on fecundity of selected insects

Insect	Adult food	Fecundity (eggs/female)	Reference
Earias vitella (Lepidoptera)	15 per cent raffinose	605	Krishna (1987)
	15 per cent sucrose	445	Krishna (1987)
	15 per cent glucose	317	Krishna (1987)
	Distilled water	165	Krishna (1987)
E. fabia	15 per cent raffinose	121	Krishna *et al.* (1971)
	0.1 per cent raffinose	51	Krishna (1987)
Tribolium castaneum (Coleoptera)	Wheat flour enriched with:		Krishna (1987)
	Yeast	57	Krishna (1987)
	Starch	19	Krishna (1987)
	Sucrose	5	Krishna (1987)
	Glucose	9	Krishna (1987)
	Cholesterol	2	Krishna (1987)
	Vitamin B complex mixture with 63 μg folic acid/g diet	43	Krishna (1987)
	Vitamin B complex mixture with 0.63 μg folic acid/g diet	2	Krishna (1987)
Trichogramma chilonis (Hymenoptera)	No feed	46	David *et al.* (1987)
	Fructose	106	David *et al.* (1987)
	Yeast extract + glucose	69	David *et al.* (1987)
	Corcyra egg extract + glucose	69	David *et al.* (1987)
	Tap water	35	David *et al.* (1987)
Dacus dorsalis (Diptera)	Oviposition bottle treated with:		
	Orange juice	20.8×10^4	Vargas and Chang (1991)
	Papaya juice	14.0×10^4	Vargas and Chang (1991)
	Guava juice	13.4×10^4	Vargas and Chang (1991)
	Coffee	12.8×10^4	Vargas and Chang (1991)
	Water	7.4×10^4	Vargas and Chang (1991)
D. cucurbitae	Pumpkin juice	26.8×10^4	Vargas and Chang (1991)
	Cucumber juice	22.9×10^4	Vargas and Chang (1991)
	Papaya juice	17.5×10^4	Vargas and Chang (1991)
	Water	2.8×10^4	Vargas and Chang (1991)
Kiefferulus barbitarsis (Diptera)	Glucose 2 per cent	2,362	Palavesam and Muthukrishnan (1992)
	Amino acid mixture, 2 per cent	2,402	Palavesam and Muthukrishnan (1992)

Contd.

Table 6 Contd.

Insect	Adult food	Fecundity (eggs/female)	Reference
	Amino acid mixture, 5 per cent	2,085	Palavesam and Muthukrishnan (1992)
	Amino acid mixture, 10 per cent	2,645	Palavesam and Muthukrishnan (1992)
	Glucose + amino acid, 2 per cent mixture	2,068	Palavesam and Muthukrishnan (1992)
Oxycarenus hyalinipennis (Hemiptera)	Continuous feeding on cotton seed soaked in water	28	Raman and Sanjayan (1983)
	Cotton seeds on first three days and water subsequently	12	Raman and Sanjayan (1983)
	Water on first three days and subsequently:		
	Gossypium seed	7	Raman and Sanjayan (1983)
	Abelmoschus esculentes	5	Raman and Sanjayan (1983)
	Abutilon indicus	3	Raman and Sanjayan (1983)
	Water alone	0	Raman and Sanjayan (1983)
Culex pipiens (Diptera)	Cannery blood	187	Worke (1937)
	Human blood	58	Worke (1937)

Although the different components of the food influence fecundity independently, the realized fecundity is the result of the interaction of the different components between themselves and with the environment. Maturation of leaf increases the nitrogen content and tends to increase food consumption, growth, and fecundity; simultaneous increase in secondary metabolites, such as phenols, decreases the overall fecundity. In their natural habitat, insects time their adult eclosion to synchronize with the availability of preferred food in order to maintain the optimum population growth. For instance, the western corn rootworm, *Diabrotica virgifera*, synchronizes adult eclosion with corn flowering in the habitat to realize the maximum fecundity. The growth stage of the corn, in which shedding of pollen and emergence of silks occur, maximizes the fecundity to 441 eggs/female (Elliott *et al.*, 1990).

2. Effect of food availability and consumption

In most hemimetabolous insects, the quantity of food consumed during the adult stage plays a major role in reproduction and hence, restriction of adult feeding is likely to interfere with egg production, unless there is a carry-over of larval reserves at least to meet the metabolic demands of the adult.

The work done by Slansky (1980) on the milkweed bug *Oncopeltus fasciatus* illustrates the effects of restricted availability of food on fecundity of an insect

Table 7

Effect of restriction of feeding on food consumption and egg production of *Oncopeltus fasciatus* fed on *Asclepias syriaca* seeds and *Bombyx mori* fed on *Morus alba* leaf

Parameter	*O. fasciatus* (data from Slansky, 1980) Ration (mg dry seed/week)		
	100	50	25
Total food consumption (mg dry)/ pair of M and F	422.4	268.2	189.8
Pre-oviposition period (day)	3.4	3.2	9.4
Pre-oviposition consumption (mg/pair)	113.5	88.7	141.1
Pre-oviposition body wt. (mg fresh)	93.8	91.7	93.1
Biomass of egg production (mg dry/female)	119.0	59.4	14.6
*Egg production efficiency (per cent)	27.9	21.1	7.7
Fecundity (eggs/female)	1,217.0	576.0	131.0
Interoviposition period (day)	2.3	4.7	10.0

Parameter	*B. mori* (data from Haniffa *et al.*, 1988) Feeding duration (times/day)			
	2	4	6	8
Food consumption (mg dry leaf/female)	3,084	4,022	4,849	5,331
Pre-oviposition period (hr after eclosion)	8.7	8.4	7.9	7.0
Fecundity (eggs/female)	217	389	494	563
**Egg production efficiency (per cent)	2.8	3.1	3.3	3.5
Postoviposition period (hr upto death)	123	152	194	223

* $\frac{\text{Biomass of egg}}{\text{Food consumption}} \times 100$

** $\frac{\text{Number of eggs}}{\text{Food consumption}} \times 100$

which feeds throughout its lifetime. A critical analysis of his data, presented in Table 7, points out that the drain of energy on maintenance metabolism during the extended interoviposition period in order to recoup the required nutrients for a fresh bout of oviposition by the insect, exposed to restricted food availability, considerably decreases the fecundity and egg-production efficiency. In females fed on glucose solution from eclosion and receiving a transplantation of corpus allatum from mature, well-fed females, the ingestion of restricted quantities of food may not sufficiently activate the corpora allata and thereby decrease the fecundity (Walker, 1976; also see Ralph, 1976).

For the seed-eating bug *Oxycarenus hyalinipennis*, ingestion of food immediately after metamorphosis is obligatory to realize its normal fecundity. Delay in availability of the cotton seed by three days significantly decreased the egg output to 7/female as against 28/female (Raman and Sanjayan, 1983). Haniffa *et al.* (1986) demonstrated that *Bombyx mori* fed eight times a day during the larval period produced more eggs (563/female) than females fed only twice a day (217/female). Besides, hatchability of the eggs produced by females which were fed frequently was greater than that of females fed less frequently. Therefore, feeding directly influences egg production by regulating the energy available for egg production and indirectly through the neuroendocrines. Fertility of males does not, however, seem to decrease significantly due to partial or total starvation.

For several predators, acquisition of adequate energy to meet the metabolic demand is essential to begin egg production. Once this demand has been met, fecundity of the predators holds a definite relation to the prey ingested (Hodek, 1973; Lawton *et al.*, 1975; Beddington *et al.*, 1976a, b). For instance, a *Coccinella undecimpunctata aegyptiaca* female (Coleoptera) consumes about 25 aphids a day before commencing oviposition; thereafter, with increasing prey consumption to 40 and 70 aphids a day, total egg production increases to 60 and 200 per female (Fig. 4A). With increasing prey density from 2 to 4 and 8 prey per female, total egg production by *Anthocoris confusus* (Hemiptera) increases from 40 to 90 and 130 eggs/female, respectively (Fig. 4B). Evans (1973) has shown that egg hatchability of *A. confusus* also increases to 95 per cent at the prey density of 4 per female. Similar relations obtained for the beetle *Adalia decempunctata* by Dixon (1959) and for the bug *Notonecta undulata* by Teth and Chew (1972) are shown in Fig. 4C and D. The levelling off of fecundity beyond a particular limit of prey density is apparently due to the fact that food consumption does not appreciably increase at higher prey density and that ovarian production is always limited to a particular level.

The pattern of feeding varies widely among insects: collembolans, orthopterans, dictyopterans, and hemipterans feed throughout their lifetime while others such as most lepidopterans, parasitic hymenopterans, and ephemeropterans restrict feeding mostly to the larval period. Although uptake of honey or sugar during the adult period helps them initiate oviposition, the nutrients accumulated during the larval period alone are converted into eggs. Therefore, fecundity in these insects holds a significant correlation to food consumption and weight of the terminal larva or pupa. For instance, fecundity of the moth *Pseudoletia unipuncta* increases from 425 eggs/female to about 600 eggs for an increase in the food supply to the final instar larva from 90 mg to 250 mg (Mukerji and Guppy, 1970). Dependence of fecundity on larval food consumption as well as weight of terminal larva or pupa has been demonstrated in the army worm *Mamestra configurata* (Lepidoptera) by Bailey (1976) (Fig. 5). The pattern of daily oviposition in these insects is charac-

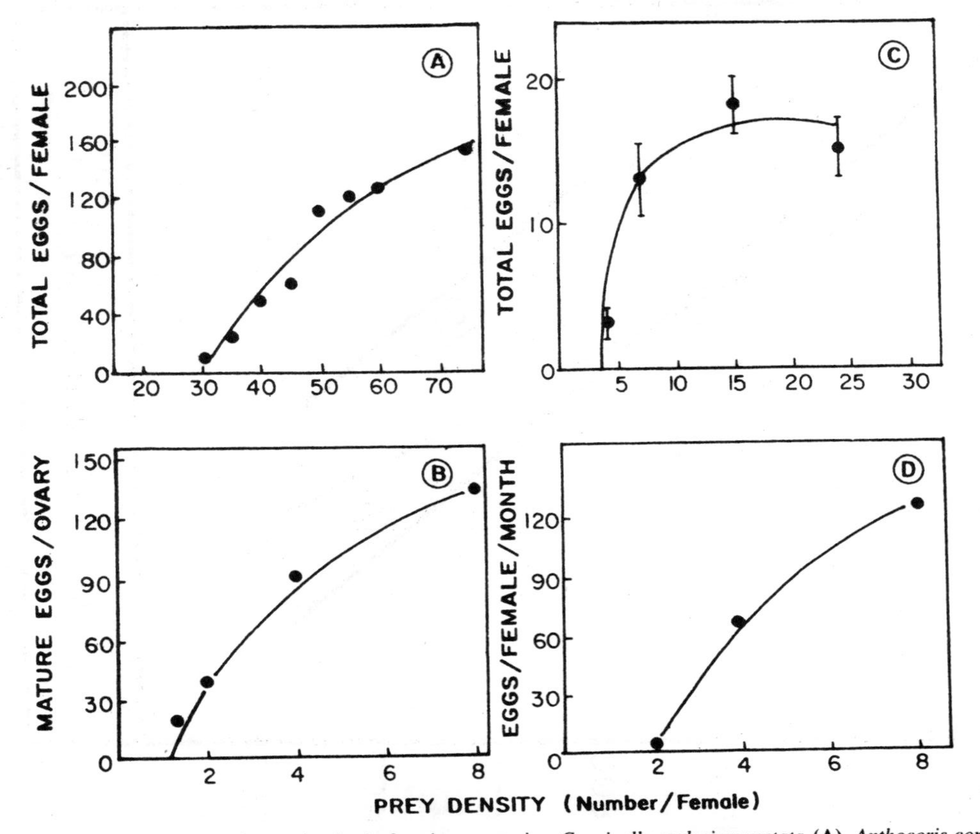

Fig. 4. Fecundity of predators as a function of prey density in four insect species, *Coccinella undecimpunctata* (**A**), *Anthocoris confusus* (**B**), *Adalia decempunctata* (**C**), and *Notonecta undulata* (**D**). (**A**, after Beddington *et al.*, 1976a; **B**, **C**, and **D**, after Beddington *et al.*, 1976b.)

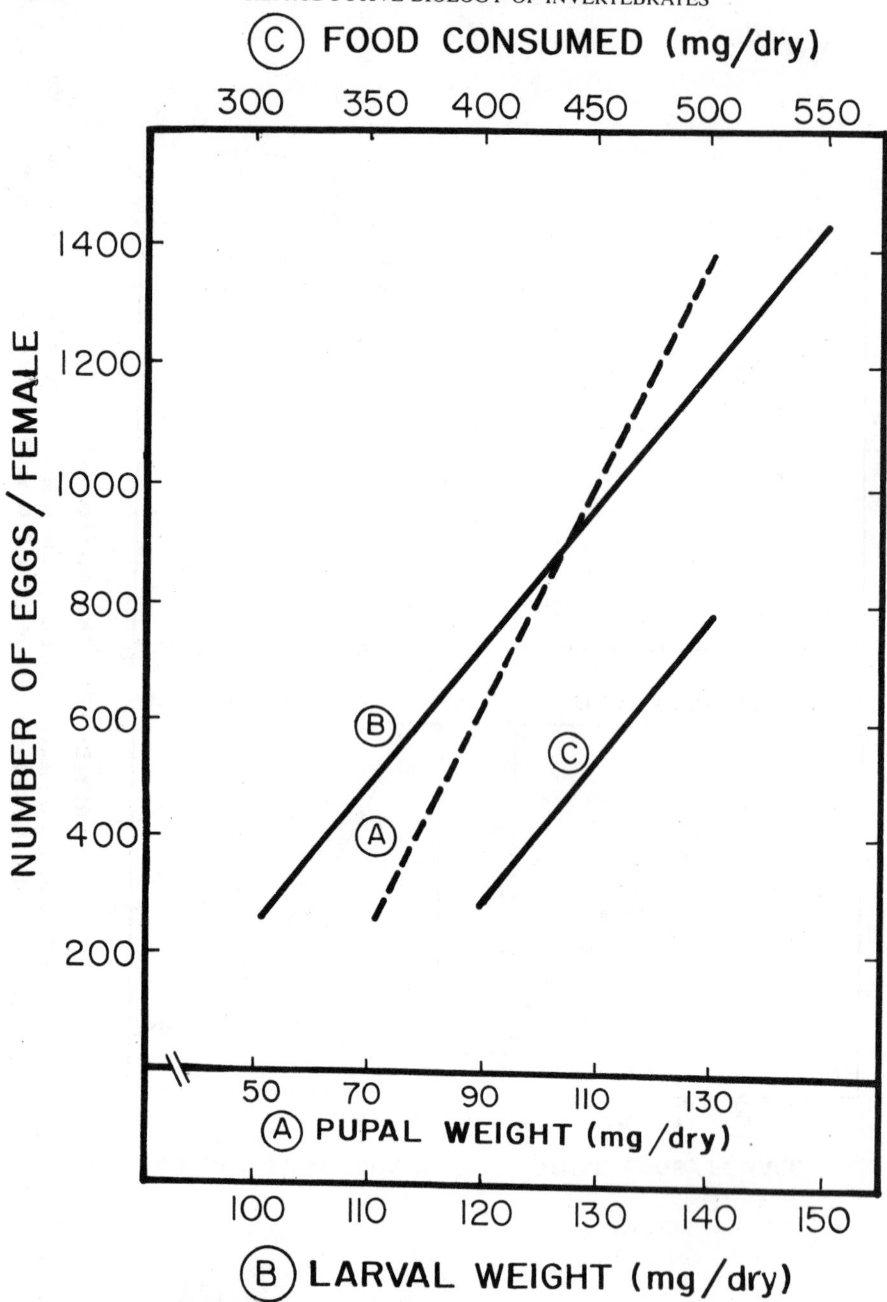

Fig. 5. Fecundity of *Mamestra configurata* as functions of (**A**) pupal weight, (**B**) terminal larval weight, and (**C**) food consumption during final instar. The lines were drawn using the following simple regression equations: (A) $Y = -1055.9 + 18.688 X$; $r = 0.997$; (B) $Y = -2107.9 + 23.535 X$; $r = 0.997$; (C) $Y = -1690.5 + 4.948 X$; $r = 0.945$. (Basic data from Bailey, 1976.)

terized by a peak just after maturity followed by a gradual decrease to zero prior to death. Exceptional among the Lepidoptera are a few species of *Heliconius* butterflies, whose fecundity is not limited by larval food reserves. They are capable of extracting amino acids from pollen collected in their imaginal life and of incorporating them rapidly into eggs (Dunlap-Pianka *et al.*, 1977). Procurement of amino acids during adult stage seems to have changed the egg-production pattern of these butterflies (Gilbert, 1972). *Heliconius charitonius* is able to lay eggs, even when it is deprived of pollen. A pollen-provided female oviposits around 12 eggs/day from the 10th day of its imaginal life for a period of two months. Its total fecundity approaches, 1,000 eggs. But in a pollen-deprived *H. charitonius*, the number of eggs produced per day decreases significantly by the 15th day of its imaginal life and drops to zero after 22 days; its total fecundity (330) is just one third of that provided with pollen. However, oviposition of *H. hecale* is erratic and varies from 7 to 15 eggs/day and seems to be determined by the availability of pollen (Dunlap-Pianka, 1979).

3. Autogeny

Importance of protein for egg production in blood-feeding insects has been demonstrated by mixing trypsin inhibitor with the blood meal. The stable fly, *Stomoxys calcitrans*, fed on a blood meal mixed with 3 mg of soya bean trypsin inhibitor fails to oviposit eggs, whereas that receiving blood with phospholipase A inhibitor oviposits more eggs than the one on normal diet (Spates, 1979). Surprisingly, a few insects are autogenous and are capable of laying a batch of eggs without ingesting a proteinaceous diet in the adult stage (see Engelmann, 1970). Arctic mosquitoes, such as *Aedes impiger* and *A. nigriceps* as well as *A. detritus* inhabiting the deserts of Tunisia and the Sahara, adopt the strategy of autogenous egg production, as their chances of finding a mammalian host in the habitat are remote (Corbet, 1964). Autogeny seems to have evolved as an adaptation to adverse climatic conditions; eight out of nine species of blackflies (Simulidae) in arctic Canada and all species of *Culicides* in the tundra are autogenous (Downes, 1965). Adult life span of most tropical chironomids is not longer than a day: soon after their emergence, mostly before dawn, they undertake copulation in the air and die subsequent to oviposition on the following night (Palavesam and Muthukrishnan, 1992). Most of them are autogenous and do not depend on adult feeding for egg production. Providing exclusive glucose or amino acid solution ranging from 2 to 10 per cent as well as a mixture of them to *Kiefferulus barbitarsis* adults, Palavesam and Muthukrishnan (1992) found that adult nutrition does not influence either the number of egg rafts deposited or the number of eggs per raft (see Table 6).

Larval nutrition is one of the major factors controlling autogeny in haematophagous insects (Lea, 1964; Friend *et al.*, 1965). For instance, the bug *Rhodnius prolixus* fed on full blood meal at each nymphal instar is larger, has larger fat bodies, and retains considerable quantities of blood (equivalent to 50 per cent of an adult blood meal) in the midgut, compared to that fed on partial blood meal in the larval stage. The store of blood in the midgut and the reserves in the fat body enable the well-fed bugs to lay a few eggs autogenously prior to an adult meal. Bugs of partly fed larvae are anautogenous. After the first adult blood meal, a bug fed full larval meal lays 51 eggs as against 18 eggs by that fed on partial larval meal (Patterson, 1979). In another allied species, *Triatoma infestans*, blood consumption of a female fed eight times after eclosion ranges from 930 to 1,510 mg; correspondingly, the number of eggs produced also varies from 50 to 95. On an average, about 16.6 mg of blood is required for the production of an egg (Regis, 1979). The ratio of blood ingested to eggs produced remains relatively constant.

Among the Diptera, the salt-marsh deer fly, *Chrysops atlanticus*, is autogenous in the first gonotrophic cycle but requires vertebrate blood for egg production in subsequent cycles. Developmental rates of autogenous and anautogenous cycles do not vary significantly. On an average, an egg mass contains 168 eggs (Magnarelli and Anderson, 1979). In the stable fly, *Stomoxys calcitrans*, proteins derived from larval nutrition are sufficient for the maturation of oocytes only to stage II (resting stage); for further maturation, the fly must consume a proteinaceous diet (Moobola and Cupp, 1978). In the anautogenous Australian sheep blowfly, *Lucilia cuprina*, proteins consumed during the adult stage play an important role in ovarian development; a sucrose-supplemented protein diet is essential for the maturation of oocytes (Williams *et al.*, 1979). Feeding *Tabanus nigrovittatus* with different quantities of blood, Magnarelli and Stoffolano (1980) arrived at a significant correlation between egg production and blood consumed. The regression equation ($Y = 46.2 + 3.972X$; $r = 0.85$) given by them can be used for predicting fecundity (Y) of the fly from the weight of blood meal consumed (X). Blood consumption ranged from 0 to 60 mg. These authors have also shown that the autogenous and anautogenous fecundity rates of the fly do not differ significantly.

The mosquito *Aedes aegypti* requires a minimum of 0.82 mg of blood to begin egg-laying, and a significant correlation exists between blood consumed and the number of eggs produced. The minimum quantity of blood meal for maturation of an *A. aegypti* egg is 0.4 mg (Roy, 1936). Lea *et al.* (1978) reported that an *A. aegypti* female produces 107 eggs, 48 hours after administration of 4 μl of heparinized rat blood. Over 45 per cent of the females receiving 1 μl blood each gradually resorbed their oocytes and the remaining females produced only 48 eggs each in as long as four days. Figure 6 shows the relation between blood ingested and eggs produced by *Rhodnius prolixus, Triatoma infestans, A. aegypti*, and *Tabanus nigrovittatus*. The negative intercept on Y of the regression lines for the hemipteran bugs *R. prolixus* and *Tr. infestans* clearly indicates that blood consumption after eclosion is obligatory for the initiation of oviposition; on the other hand, the lines for the two dipterans, *A. aegypti* and *T. nigrovittatus*, intercept Y at positive values showing the relative independence of adult blood meal for ovipositing at least a few eggs. It is tempting to conclude that such sanguivorous insects displaying positive intercept on Y may fall under autogeny and others are likely to be anautogenous (see also Muthukrishnan and Pandian, 1987b).

4. Feeding pattern and life style

Table 8 provides data on food consumption, growth, and egg-production efficiency of insects. The efficiency ranges from 0.7 per cent for *Poecilocerus pictus* to 34.4 and 41.9 per cent in the granivorous *Oryzaephilus surinamensis* and the juice-feeding *Aphis fabae*, respectively. The wide range in the efficiency may partly be attributed to the life style (living for longer or shorter durations, and growing larger or smaller) and feeding pattern (feeding throughout or for part of the life span). The two primitive collembolans *Orchesella cincata* and *Tomocerus minor* moult throughout life comprising intermoult periods of active feeding and reproduction, and short non-feeding moult periods. About 5 and 7 per cent of the ingested energy (ca. 38 J) is allocated by them to egg production. *Tomocerus minor* is metabolically less active than *O. cincata* and hence can afford to allocate a higher percentage of ingested energy to production of eggs (Testerink, 1982). Of the two species of grasshoppers considered, *Oxya velox* displays egg-production efficiency of 4.1 per cent as against 0.7 per cent by *P. pictus*. Although their life span is around 250 days, the energy cost of growing larger (6.9 g in *P. pictus* compared with 0.3 g in *O. velox*) and of maintenance of a larger adult biomass for longer duration (> 105 days in *P. pictus* compared with < 75 days in *O. velox*) has restricted the egg-production efficiency of *P. pictus* to 0.7 per cent (also see Calow, 1977). The scarabaeid beetle *Rhopaea verreauxi* lives longer (924 days) than *P. pictus* and attains a biomass of 3.3 g, but allocates 4.8 per cent of the ingested energy to egg production. It does not feed as adult but converts most of the energy accumulated during the larval period into eggs. On the other hand, *P. pictus* allocates a part of the ingested energy during nymphal period to somatic growth and depends mostly on the energy ingested during the adult period for maintenance and egg production.

The strategy of the moth *Cyclophragma leucosticta* is similar to that of *Rhopaea verreauxi*. Despite growing as large as 5.1 g and passing through a larval period of 107 days, *C. leucosticta* allocates 3.8 per cent of the ingested energy to egg production. *Bombyx mori* spins an expensive pupal case and yet manages to spare 5.8 per cent of the ingested energy for egg production. Granivorous beetles such as *Sitophilus granarius*, *Oryzaephilus surinamensis*, and *Cryptolestes ferrugineus* adopt a different strategy. They shorten the vulnerable larval period and prolong the adult period to acquire sufficient energy and nutrients and maximize egg production. Shortening the larval period to four days and decreasing the adult biomass to 13.9 mg have helped the aphidophagous ladybird beetles *Coccinella transversalis* and *Menochilus sexmaculatus* to allocate as much as 29.0 and 22.7 per cent of the ingested energy to egg production. Ingestion of amino acid-rich phloem sap enables *Aphis fabae* to enhance its egg-production efficiency to 41.9 per cent.

Table 8

Food consumption and energy allocation to egg production in some insects

Insect	Food consumption (J/insect)			Growth (J/insect)	Egg production (J/insect)	Egg production efficiency (per cent)	Remarks	Reference
	Larva	Adult	Total					
Collembola								
Orchesella cincata	—	—	18.3	3.334	1.87	4.9	Moults and feeds throughout life	Testerink (1982)
Tomocerus minor	—	—	38.8	4.467	2.80	7.2	Moults and feeds throughout life	Testerink (1982)
Orthoptera								
Oxya velox	10,208	15,967	26,175	2,180	1,077	4.1	Feeds throughout life; grows to about 300 mg in 240 days	Delvi and Pandian (1971)
Poecilocerus pictus	717,587	211,704	1,929,291	48,704	13,053	0.7	Feeds throughout life; grows to about 6.9 g in 267 days	Delvi (1972)
Atractomorpha crenulata	8,263	27,978	36,241	2,979	1,668	4.6	Feeds throughout life; grows to 0.4 g in 131 days	Senthamizhselvan and Murugan (1988)
Dicyoptera								
Mantis religiosa	39,678	55,327	95,005	14,630	5,413	5.7	Feeds throughout life; grows to about 2.2 g in 131 days	Muthukrishnan (1987)
Homoptera								
Aphis fabae	—	—	62	32.3	26	41.9	Feeds throughout life; grows to about 1.0 mg in 39 days	Llewllyn and Qureshi (1978)
Lepidoptera								
Bombyx mori	59,469	0	59,469	18,245	3,467	5.8	Feeds only during larval period (25 days); spins an expensive pupal case	Hiratsuka (1920)

Cyclophragma leucosticta	189,083	0	189,083	33,773	7,160	3.8	Feeds only during larval period (107 days) and attains a maximum body weight of 5.1 g	Mackey (1978)
Porthesia scintillans	12,128	0	12,128	1,754	94.7	0.8	Feeds throughout life; 8 days larval period; 3 days adult period	Senthamizhselvan and Muthukrishnan (1988)
Spodoptera litura	13,446	0	13,446	1,868	175.1	1.3	Feeds throughout life; 8 days larval period, 3 days adult period	Senthamizhselvan and Muthukrishnan (1989)
Coleoptera								
Coccinella transversalis	226	964	1,190	—	345	29.0	Feeds throughout life; 4 days larval period; 28 days adult period	Senthmizhselvan (1987)
Menochilus sexmaculatus	292	1,424	1,716	126.4	389	22.7	Feeds throughout life: 4 days larval period; 32 days adult period	Senthamizhselvan (1987)
Cicindela cautena	—	5,547	5,547	385	296	5.3	Larval consumption not estimated; adult life span 25 days	Senthamizhselvan (1987)
Rhopaea verreauxi	124,880	0	124,880	11,140	5,414	4.3	Feeds only during larval period (924 days); attains maximum growth of 3.3 g	Cairns (1982)
Oryzaephilus surinamensis	38	273	311	120	107	34.4	Feeds during larval (13 days) and adult period (60 days); attains maximum growth of 25 mg	White and Sinha (1981)
Sitophilus granarius	331	1,189	1,520	87	39.4	2.6	Feeds during larval (19 days) and adult period (30 days); attains maximum growth of 7 mg	Campbell *et al.* (1976)

Contd.

Table 8 Contd.

Insect	Food consumption (J/insect)			Growth (J/insect)	Egg production (J/insect)	Egg production efficiency (per cent)	Remarks	Reference
	Larva	Adult	Total					
Cryptolestes ferrugineus	25.5	310	336	3.2	21	6.0	Feeds during larval (20 days) and adult periods (123 days); attains maximum growth of about 0.7 mg	Campbell and Sinha (1978)
Callasobruchus maculatus	386	0	386	87	21.0	5.4	Feeds only during larval period (16 days); oviposits for 8 days; attains maximum growth of 4 mg	Chandrakantha (1985)
Cynaeus angustus	598	8,569	9,167	178	333	3.5	Feeds both during larval and adult life	White and Sinha (1987)

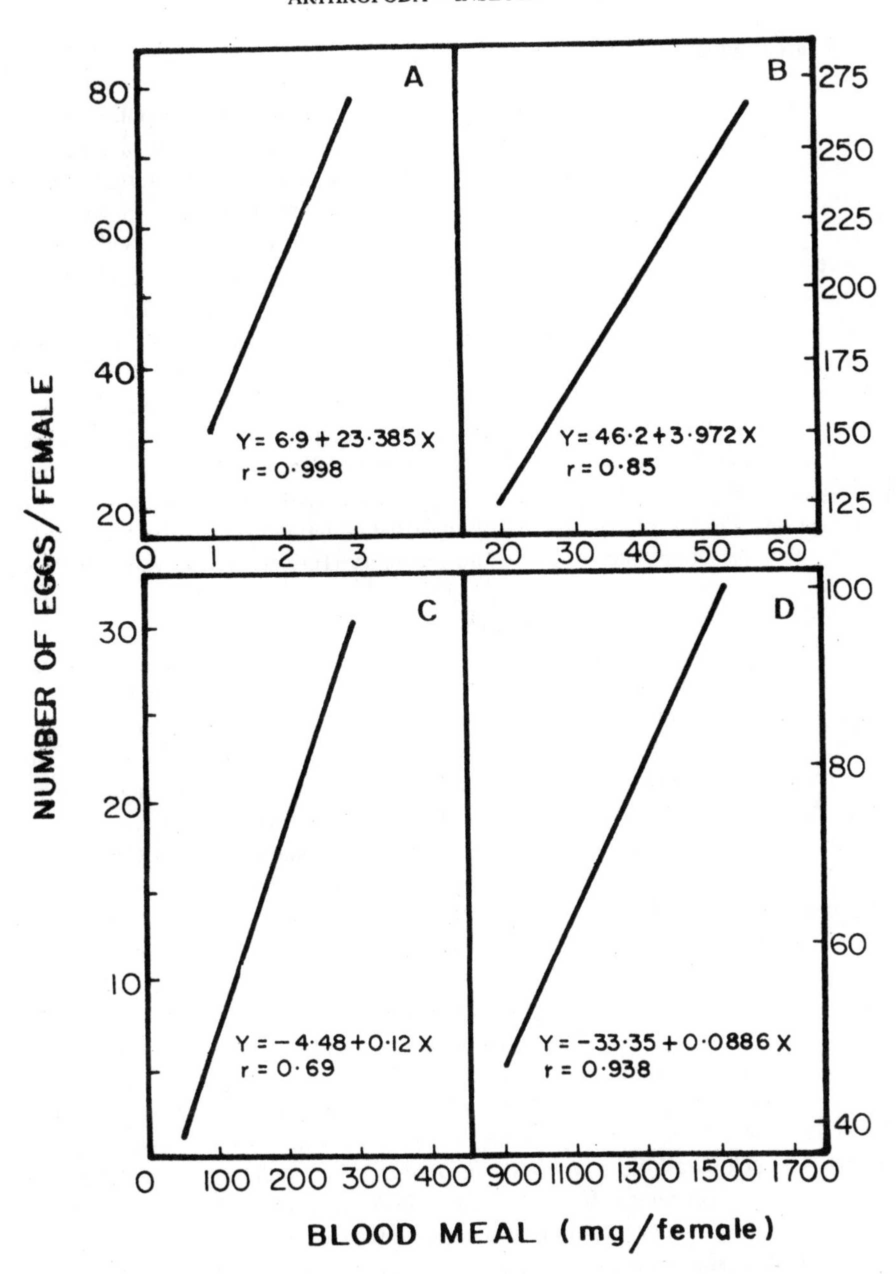

Fig. 6. Fecundity of some haematophagous insects as a function of blood meal consumed (Muthukrishnan and Pandian, 1987b). **A**: *Aedes aegypti* (basic data from Roy, 1936). **B**: *Tabanus nigrovittatus* (basic data from Magnarelli and Stoffolano, 1980). **C**: *Rhodnius prolixus* (basic data from Patterson, 1979). **D**: *Triatoma infestans* (basic data from Regis, 1979). Note the positive intercept of the regression lines for the flies and the negative intercept for bugs.

5. Strategy of parasitoids

Although holometabolous parasitoids require an adult feeding on pollen or nectar prior to oviposition (e.g. *Cyzenis albicans*), the surest strategy of several of them is to carry over sufficient nutrients from larval stages for maturation of a full complement of oocytes. As parasitoids derive nourishment (or share it with other larvae in the case of gregarious parasitoids) from a single host individual, their success depends mostly on the survival of the host. Fecundity of closely related parasitoids varies widely. Although *Trissolcus euschisti* and *Telenomus podisi* parasitize eggs of the bug *Podisus maculiventris*, the former produces significantly more offspring (64 ± 7 per female) than the latter (40 ± 5 per female). A longer period of potential reproduction and better adaptablity to patches with low host density are responsible for the higher reproductive potential of *Tr. euschisti* (see Yeargan, 1982). Among the Ichneumonidae (Hymenoptera), fecundity ranges from 30 eggs/female in *Pleolophus indistinctus* to over 1,000 eggs/female in *Euceros frigidus* (see Price, 1973a). Probability of host-finding and survival in the host explain such vast differences in parasitoid fecundity (Hassell and May, 1973; Price, 1973a).

Probability of host finding is in direct correlation with the abundance and availability of the host. Host abundance decreases as the host grows old. Parasitoids infecting eggs and earlier larval stages of host have higher probability of host finding than those infecting cocoons, some of which may be concealed in the leaf litter or soil (Price, 1972, 1975a, b). Parasitoids with short ovipositors lay eggs superficially on the host and their fecundity is higher than those impregnating the eggs deep into the host with the help of long ovipositors. Among ichneumonid wasps, members of the subfamilies Diplazontinae, Metaopiinae, and Cermastinae infecting eggs and earlier larvae of syrphid flies and sawflies possess around 20 ovarioles per ovary (Fig. 7A); Tryphoninae, Ichneumoninae, and Gelinae, which infect terminal larvae and pupae of sawflies, possess only about five ovarioles per ovary (Price, 1973a, 1975b). Exceptionally, *Euceros frigidus* has as many as 50 ovarioles per ovary and produces over 1,000 eggs (Price, 1973a). Its high fecundity is correlated with the mode of deposition of eggs on the foliage very near the host eggs as in several tachinids (Fig. 7A).

Fecundity of *Coccygomimus turionella* emerging from nutitionally rich larger hosts is greater than fecundity of those emerging from smaller hosts (Sandlan, 1979). The ectoparasitoid *Epiricania melanoleuca* lays more eggs (1,002) on the larger and sluggish female host *Pyrilla perpusilla* than on the active and smaller adult male (556 eggs) or fourth instar nymph (766 eggs) (Krishna and Misra, 1992). Increased host size increases the size of the parasitoid as well as its longevity; extended longevity, incidentally, enhances fecundity by providing longer duration for the ovarioles to produce more eggs.

Price (1975b) obtained a significant correlation ($r = 0.8$; $p < 0.001$) between ovariole number and mode of oviposition (near host eggs; outside or inside of egg,

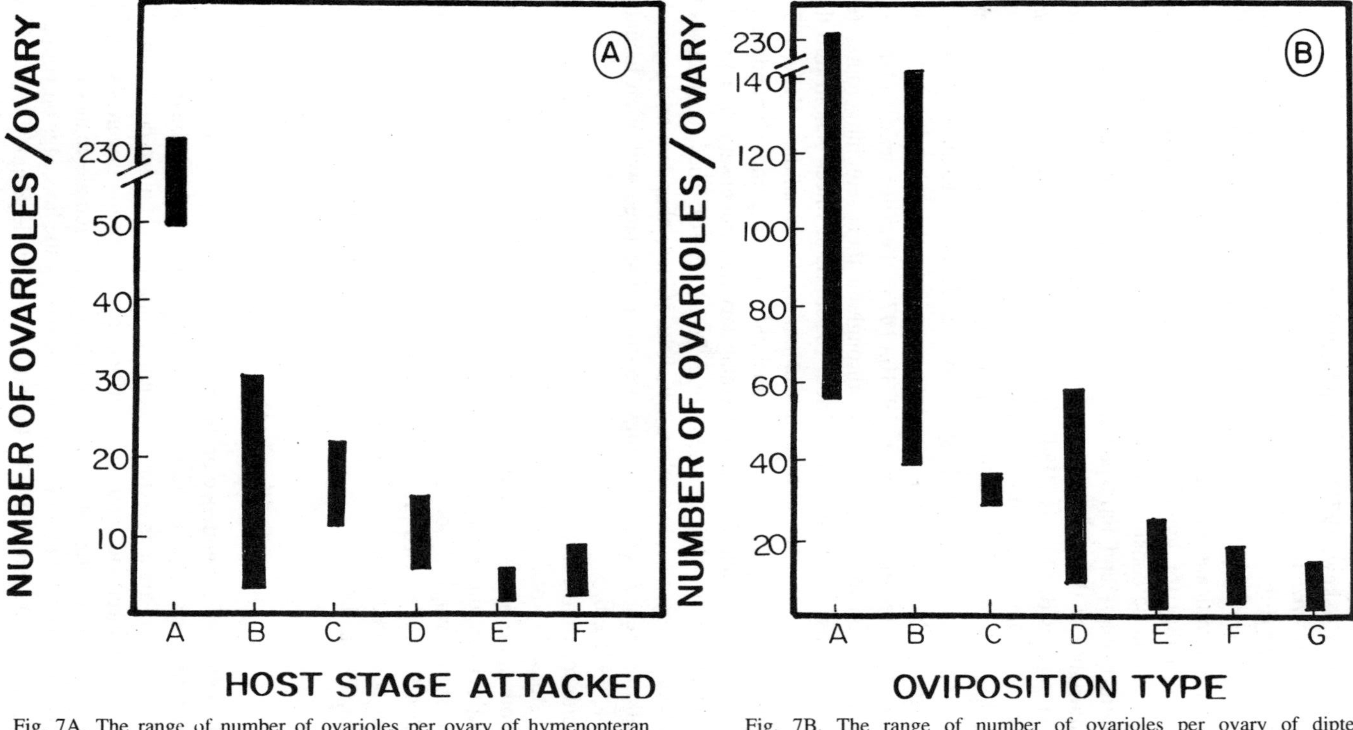

Fig. 7A. The range of number of ovarioles per ovary of hymenopteran (Ichneumonidae) parasites ovipositing on food (**A**), host egg (**B**), young larva (**C**), mid-larva (**D**), old larva (**E**), and pupa (**F**). Host survival increases from **B** to **F**.

Fig. 7B. The range of number of ovarioles per ovary of dipteran (Tachinidae) parasites oviposting on food (**A**), on food or soil (**B**), at the tunnel entrance (**C**), on host larva (**D**), in host larva (**E**), on host adult (**F**), and in host adult (**G**). Host abundance decreases from **A** to **G** (data collected from Price, 1975b).

young, middle, or old larva, and pupa) for the tachinid and conopid parasitoids of Diptera (see also Askew, 1971). Tachinids such as *Leschenaultia exul, Zenillia libatrix*, and *Gonia atra* lay eggs on host's food and possess 230, 85, and 80 ovarioles per ovary (Fig. 7B). Others infecting young, middle, and old larva or pupa possess approximately 14, 11, 5, and 4 ovarioles, respectively (Price, 1975b). Price obtained a significant correlation ($r = 0.97$; $p < 0.0001$) between ovariole number and fecundity for hymenopterous and dipterous parasitoids, and hence ovariole number can be considered an index of fecundity of these insects.

To minimize inter- and intraspecific larval competition and to decrease the premature mortality of the host due to superparasitization, most parasitoids prefer to oviposit in unparasitized hosts (Wylie, 1971a, 1972, 1973). *Muscidifurax raptor* and *Nasonia vitripennis* discriminate between parasitized and unparasitized hosts with their ovipositor; when unparasitized hosts are extremely scarce, they prefer to lay unfertilized eggs on parasitized hosts; such eggs develop into males. Obviously, the mothers try to decrease larval competition among their daughters in a host and maximize their fecundity (Wylie, 1971b, 1973). Females leave allelopathic repellents not only on the hosts but also throughout their searching routes (e.g. *Pleolophus basizonus*, Ichneumonidae: Price, 1970a, b) or inject venom into their host (e.g. *Muscidifurax raptor* and *Nasonia vitripennis*, Pteromalidae: Wylie, 1971b, 1973) and prevent superparasitization. Parasitoids attacking more abundant and easily accessible hosts, such as eggs and young larvae, compensate their high rate of mortality by increasing their fecundity (e.g. Price, 1974b). They locate hosts by flight at a lower energy cost than others which locate less abundant and concealed hosts by walking. Hence, long forewings are often associated with high fecundity (Price, 1973a, 1975a, b).

Parasitoids maximize fecundity by increasing (1) ovariole number, e.g. *Euceros frigidus*; (2) storability of oocytes in ovarioles, e.g. wasps of subfamilies Cremastinae and Anomalinae store as many as three oocytes per ovariole; (3) storability of oocytes in oviduct, e.g. members of Prozontinae, Scolobatinae, Metopiinae, and Cremastinae store over 100 oocytes in their long oviduct; and (4) by decreasing egg size. Less fecund species produce a few larger eggs, which hatch into larger larvae, infest advanced stages of host, develop faster, and emerge within a short period (Price, 1973a, 1975b).

B. Biocenotic Influence

Based on the extent of dependence on host plants, the phytophagous insects range from 'broad generalists to pure specialists' (Orians *et al.*, 1974; Labeyrie, 1978b). The dependence is so great in a few specialists that they require host-plant cues not only for oviposition and maturation of oocytes and/or vitellogenesis but also for mating. For instance, *Bruchus pisorum* (Coleoptera) female mates only after feeding on the pollen of *Pisum sativum*; after a few days she lays eggs on young pea pods (Pajin and Sukeska Sood, 1974). Dependence of the leek moth *Acrolepiopsis assectella* on the volatile compounds of its larval host plant *Allium porrum* has made it 'a true drug addict' (Rahn, 1969). Host plants stimulate ovarian activity of a few generalists. Such insects exploit the periodic occurrence of a few 'indicator plant species' for synchronizing their reproductive activity with the onset of suitable environmental conditions for larval development.

1. Extraspecific influence

About 50 per cent of virgin *Acanthoscelides obtectus* from a high-altitude (2,000 m) Colombian population, reared in the absence of bean (*Phaseolus vulgaris*) seeds, failed to mature oocytes beyond the 'pre-vitellogenic' phase even up to 150 days of imaginal life. But over 85 per cent of the females reared in the presence of seeds began production of mature oocytes within 12 to 18 days following emergence. Ovarian production of a female in the presence of seeds is significantly higher (39 $\pm$ 3 oocytes) than production in the absence of seeds (17 $\pm$ 2) (Huignard and Biemont, 1979). In the presence of seeds, ovarian production of low-altitude (1,000–1,400 m) virgin females surpassed that of females in the absence of seeds as early as the eighth day of imaginal life (Huignard and Biemont, 1979). The seeds initiate oviposition in a few females of both the populations. In the absence of host-plant cues, production and retention of oocytes in the lateral oviducts for a long period decreases the fertility of the eggs. Further, expenditure of most of the energy on vitellogenesis restricts the mobility of the female in search of new oviposition sites (Biemont, 1975; Labeyrie, 1978b). Therefore, ovarian inactivity of the high-altitude females in the absence of host-plant cues is essentially an adaptation to the habitat, in which the host plant *P. vulgaris* occurs only at a particular period of the year.

Members of a population are often polymorphic and exhibit different strategies with regard to their dependence on host-plant cues for initiation of ovarian activity. Among the virgin females of the high-altitude population of the bean weevil, *Acanthoscelides obtectus*, the following strategies prevail: (1) about 50 per cent of the females are highly dependent on host-plant cues and their ovaries remain in a state of 'diapause', till they receive host-plant cues (Fig. 8); (2) some 5–10 per cent of the females are independent of host-plant cues and begin production of oocytes after passing through a latent period of different durations; and (3) the remaining females are not only independent of host-plant cues but also produce 10–20 oocytes each, within the first 15 days of imaginal life, without passing through long latent periods (Huignard and Biemont, 1979).

Under conditions of relative rarity of hosts, host-plant independent morphs are of high value in colonizing new niches or exploiting the coevolved mimetic hosts (Labeyrie, 1978a, b). Huignard and Biemont (1979) call such host-independent morphs 'pioneer individuals'. Further, a specialist with its various aspects of reproductive process under the control of host-plant stimulation (e.g. *Acrolepiopsis*

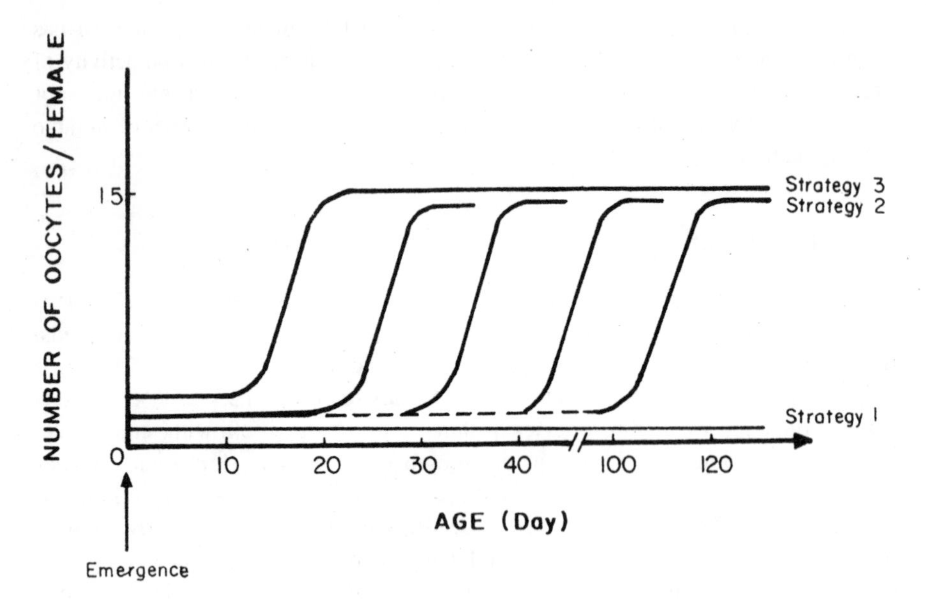

Fig. 8. Strategies of egg production in high-altitude virgin females of *Acanthoscelides obtectus* (from Huignard and Biemont, 1979).

assectella) is able to adjust its fecundity according to the carrying capacity of the biocenosis. With decreasing host population, a specialist decreases its fecundity and invests more in seeking patchy hosts (Labeyrie, 1978b).

Probably, host-dependent reproduction in insects is under genetic control (see Wiklund, 1974). Rearing the larvae of the swallowtail butterfly, *Papilio machon*, a generalist on four different host plants, and observing the oviposition preference of females, Wiklund (1974) found that memory of the larval host plant does not determine the oviposition preference of this species. Irrespective of the larval host plant, females oviposited mostly on *Peucedanum*. Wiklund, therefore, concluded that host-plant preferences for larval feeding and oviposition are determined by 'two independently recombining and mutating genetic systems'.

Biocenotic stimulation of vitellogenesis and oviposition has been reported for the hymenopteran parasite *Diadromus pulchellus* (see Labeyrie, 1964). Ovarian inactivity of *D. pulchellus*, till it finds its host (pupa of *Acrolepiopsis assectella*), helps it channel its time and energy into host seeking. Olfactory perception of the host pupa initiates rapid maturation of oocytes and oviposition ensues within a short period. A few plants are known to deter the ovipositing attempts of parasites and protect the eggs and larvae of hosts. For instance, *Telenomus* sp. (Hymenoptera) attempting to parasitize the eggs of *Manduca sexta* (Lepidoptera) on tobacco plants are in danger of being trapped in the exudate secreted by the plant and hence prefer to infect the eggs deposited on other plants (Rabb and Bradley, 1968).

2. Intraspecific influence

Besides intensifying intraspecific competition for food and oviposition site, co-existence of females with conspecific individuals in the biocenosis accelerates ovarian maturation and activity in several species. Locusts and termites undertake pre-reproductive mass migration. Among the locusts, crowding produces different effects on ovarian maturation. For instance, *Schistocerca gregaria* and *Nomadacris septemfasciata* females crowded with conspecific individuals reduce the pre-oviposition period from 34 and 58 days to 26 and 45 days (Norris, 1952, 1959), respectively. However, the response of *Locusta migratoria* to crowding is contrasting: pre-oviposition period of a crowded female is extended (Norris, 1950); females reared in crowded conditions possess a few larger ovarioles compared to many small ovarioles in those reared in isolation. However, hatchability of the eggs laid by crowded females is higher than that of females reared individually.

As pointed out by Okelo (1979) for *Schistocerca vaga*, pheromonal induction by males enhances the fecundity of females. A density of eight aphids per bean plant increased the fecundity of the aphid *Aphis fabae* (see Dempster, 1975). However, studies of Park (1933) on *Tribolium confusum* pointed out that increase in the number of males from one to two and to three per female decreased the lifetime fecundity from 1,010 to 922 and 644 per female (Table 9). Similarly, increase in the number of males has been found to reduce the fecundity of *Gryllodes sigillatus* and *Anthonomous grandis* (Table 9). The flour beetle *T. castaneum* has the ability to modulate its fecundity in relation to population density. Females from a low-density cohort exhibit a higher rate of oviposition (41 eggs/female/day) than high-density females (30 eggs/female/day) early in life (at the age of 66–88 days). With advancing age, high-density females maintain higher rate of oviposition (11 eggs/female/day) than low-density females (3 eggs/female/day), at the age of 266–268 days. Total fecundity for both groups averages to approximately 2,600

Table 9

Effect of the number of males associated with a female on fecundity in some insects

Insect	Number of male/female	Fecundity (eggs/female)	Reference
Anthonomous grandis	1	388	
(Coleoptera)	10	238	Nilakhe (1977)
	20	88	
	40	56	
Tribolium confusum	1	1,010	Park (1933)
(Coleoptera)	2	922	
	3	644	
Gryllodes sigillatus	3	2,979	Subramaniam
(Orthoptera)	8	2,300	*et al.* (1988)

eggs/female (Boyer, 1978). This process of adjusting the oviposition rate in relation to age and density has been called 'reproductive compensation' by Boyer (1978). Aiken and Gibo (1979) found that the presence of larvae in the oviposition medium decreases the fecundity of *Drosophila melanogaster* and *D. simulans* females. The decrease in fecundity of *D. melanogaster* females from 20 to 6 eggs/female at the end of 48 hours has been attributed to water-soluble byproduct of larval activity in the medium. Competition for food and space by the males as well as their frequent interferences with the ovipositing females is apparently responsible for the decrease in fecundity of monandrous females associated with more than the required numbers of males.

However, among social insects, such as the bees and termites, ovarian maturation of workers or secondary reproductives is inhibited in the presence of queen or primary reproductives. For instance, mandibular gland secretion, produced by a queen *Apis mellifera*, spreads rapidly among the workers of a colony by mutual feeding and inhibits ovarian development (Butler and Fairey, 1963). Similarly, a pheromone by the primary reproductive (queen) is picked up by workers through proctodaeal feeding and their ovarian maturation is arrested (Lauscher, 1961; see also Brian, 1965). In bumblebees and wasps (e.g. *Myschocyttarus drewseni*), mere harassing and physical domination of a female with well-developed ovary suppress ovarian development in other females (also see Butler, 1964).

C. Influence of Mating

Apart from the availability of basic materials for vitellogenesis and the release of environmental signals for oviposition, fecundity depends on the availability of sperm for fertilization and the mating stimulus received from the male. Despite the fact that males in general transfer adequate sperm for fertilization of almost all the eggs produced by the female and that the sperm survive in the female genital tract for a long time, males of several species mate repeatedly. Spermatophores or sperm constitute an additional source of nutrition for the female. Increase in fecundity with the number of matings undergone by a female has been reported in several cases (see Ridley, 1988). Production and release of oocytes are two distinct processes in insects, which may occur in quick succession or may temporally be separated until the next mating takes place. By releasing oviposition behaviour and clearing the eggs from the oviduct, mating accelerates maturation of oocytes and increases the clutch size in subsequent ovipositions. Influence of mating stimulus has been studied by implantation of male accessory glands or by injecting their extract into virgin females and by rearing females in single or multiple pairs. Reports on the effects of mating stimulus on fecundity and oviposition in insects are conflicting. This section highlights the effects of mating and repeated mating of a female with one or more males and of factor(s) transferred to the female by male(s) on fecundity.

1. Fecundity of virgin and mated females

A comparison of the performance of virgin and mated females of different species, presented in Table 10, clearly indicates that mating increases fecundity in several cases. Fecundity of mated *Earias insulana* (see Kehat and Gordon, 1977) and *Rhinocoris marginatus* (see Ambrose and Livingstone, 1985) is several times greater than that of virgin females (Table 10). The mated sawfly *Neodiprion sertifer* is 19 per cent more fecund than the virgin (Lyons, 1976). Gerber (1975) demonstrated that the stimuli associated with copulation and insemination are responsible for the increased fecundity seen in mated *Tenebrio molitor*. However, in several insects, association of a male at least for some time with the mated female is essential for the potential fecundity to be realized. For instance, daily egg production of mated *Drosophila melanogaster* females reared in isolation decreased rapidly, perhaps due to the depletion of sperm store. On day 12, the female reared with a male oviposited 45 eggs/day compared with 16 eggs/day by that reared alone after mating (Bouletreau, 1975). On the other hand, fecundity of *Corcyra cephalonica* mated and associated with a male throughout the adult life was less (227 eggs/female) than that (283 eggs/female) of a mated female associated with a male only for 16 hours a day (Krishna and Narain, 1976). Krishna and Narain proposed that mating and association with a male are two independent but significant biological factors which act synergestically and improve the fecundity of *C. cephalonica*. The mere association of a male failed to increase the fecundity of the virgin *Acrolepiopsis assectella* beyond 55 eggs; but the mated female associated with a male could increase its fecundity to 227 eggs (Thibout, 1979). Apparently, monandrous females, which mate once in their lifetime, do not require the stimulation associated with the presence of a male for realizing their potential fecundity. Presence of fertile males continuously with a polyandrous female may help replenish the stock of sperm and increase the fecundity and fertility of the mated females.

Similarly, mating status of the male significantly influences the fecundity of the females. For instance, fecundity of the mediterranean fruitfly *Ceratitis capitata*, mated with a virgin male, is significantly higher (110 eggs/female) than fecundity of that mated with a non-virgin male (73 eggs/female) (Whittier and Kaneshiro, 1991). In such species the females adopt a dual mating strategy. Some of them wait for a virgin male and others, who cannot afford to wait and run the risk of never getting a male, mate with small non-virgin males and produce a smaller number of offspring.

The age of the monandrous female at the time of mating plays a significant role in determining the fecundity. A delay in mating, after emergence, may significantly curtail the fecundity. For instance, the butterfly *Chilo partellus*, mated on the fourth day after emergence, realized the maximum fecundity of 469 eggs. Mating on the sixth day decreased the fecundity by 64 per cent and the fertility by 68 per cent (Unnithan and Pye, 1991). *Adoxophyes orana* mated on the sixth day after eclosion laid 269 eggs compared with 347 eggs by that mated on the same day of eclosion (Van der Kraan and Van der Straten, 1988). Mating after emergence

Table 10

Effect of mating on fecundity of some insects

Insect	Mating status	Fecundity (egg/female)	Reference
Plebeiogryllus guttiventris	Virgin	251	Bentur and Mathad (1973)
	Mated	363	
Gryllodes sigillatus	Virgin	539	Subramaniam *et al.*(1988)
	Mated	863	
Schistocerca vaga	Virgin	110	Okelo (1979)
	Mated	122	
Podisus maculiventris	Virgin	197	Baker and Lambdin (1985)
	Mated for 48 hours	335	
Rhinocoris marginatus	Virgin	9	Ambrose and Livingstone (1985)
	Mated	363	
Earias insulana	Virgin	31	Kehat and Gordon (1977)
	Mated	190	
Epiphyas postvittana	Virgin	234	Danthanarayana and Gu (1991)
	Mated	446	
Corcyra cephalonica	Virgin	196	Krishna and Narain (1976)
	Unmated but associated with male for 16 hr/day	171	
	Mated and associated with male throughout	227	
	Mated and associated with male for 16 hr/day	283	
Tenebrio molitor	Virgin	18	Gerber (1975)
	Mated	135	
Acrolepiopsis assectella	Virgin; not associated with male	82	Thibout (1979)
	Virgin; associated with male	55	
	Mated; not associated with male	227	

and association with a male for a while enable monandrous females to realize their maximum fecundity. Association with the male beyond a definite duration after mating does not help increase the fecundity any further. For instance, fecundity and fertility of the mated female *Podisus maculiventris* associated with a male for durations ranging from 48 to 192 hours and that of a female always paired with a male ranged from 257 to 367 eggs/female and did not significantly differ (Baker and Lambdin, 1985). These insects require mating stimulus for oviposition and the sperm transferred during a single mating are adequate to fertilize all the eggs matured in the ovary.

Interestingly, mating does not increase fecundity in some insects. For instance, the fecundity of the virgin and mated apple maggot fly, *Rhagoletis pomonella*, is not significantly different (Neilson, 1975). Further, Nilakhe (1976) has shown that the fecundity of the mated and virgin females of the rice stink bug, *Oebalus pugnax*, is more or less the same.

2. Fecundity-enhancing substances

Fecundity-enhancing substances have been extensively reviewed by Gillott and Friedel (1977) and by Gillott (Volume III of this series, pp. 423–437).

3. Role of eupyrene sperm

The importance of motile eupyrene sperm in the spermatheca of females in initiating maturation of oocytes, or oviposition, has been stressed by a number of workers. The moth *Zeiraphera diniana* mated with an irradiated male, which transfers spermatophores containing non-motile apyrene sperm, oviposited only 54 per cent of the eggs matured in the ovary (Benz, 1969). This observation suggests that, in addition to the accessory gland secretion, the presence of motile sperm in the female genital tract is essential for laying of all the eggs. By removing selectively different parts of the spermatheca from the female leek moth, *Acrolepiopsis assectella*, and mating it with a male or implanting male accessory glands in females, Thibout (1979) concluded that the presence of eupyrene sperm in the receptaculum of spermatheca is essential for stimulation of oocyte production and oviposition. With increasing density of sperm, the number of oocytes matured in the ovary and the eggs laid increased significantly.

Presence of motile sperm in the receptaculum seems to stimulate reproductive activity by means of a humoral mechanism. However, ovulation of the tsetse fly *Glossina morsitans* may be triggered mechanically by insertion of male genitalia into female and insemination is not a prerequisite. Rate of ovulation by *G. morsitans* increases with increasing copulation duration and the female adds up its sexual experience with the male. A female mated four times (each lasting for 45 minutes) ovulates as many as 79 per cent of the oocytes matured in the ovary as against 24 per cent in a female mated once. *Glossina morsitans* female mated with *G. austeni* also increases its rate of oviposition (Saunders and Dodd, 1972).

Corpora allata and pars intercerebralis are the two principal centres controlling maturation of oocytes in insects (Raabe, 1982 and Kunkel and Nordin, 1985 for reviews; also see Wyatt, 1991 for gene regulation in insect reproduction), though in Diptera vitellogenin production is regulated by both juvenile hormone and 20-hydroxyecdysone (see Adams and Gerst, 1991). These centres are activated directly or indirectly by humoral factors present in the male accessory gland secretion and/or neurally by the motile sperm in the spermatheca (see Bouletreau, 1975).

Considerable differences in the competency to mating stimuli are likely to exist among females of the same population (Leopold, 1976). Environmental factors, such as temperature and photoperiod, are likely to alter the competency of females to mating stimuli. Therefore, care should be exercised before comparing the oocyte production of females and their response to mating stimuli.

4. Effect of multiple mating

As stated in Section II D1, multiple mating is a common feature among several species of insects except the strictly monandrous parasitic Hymenoptera, which are capable of regulating sperm use for fertilization and laying unfertilized eggs. Data presented in Table 11 clearly point out that multiple mating enhances the fecundity of most insects with a few exceptions in which either the difference is not significant or there is a decrease due to multiple mating. Critically analysing 67 selected reports on multiple mating in insects, Ridley (1988) found a significant increase in fecundity at least in 54 of them. The extent of increase over the performance of the female mated once widely varies from species to species. For instance, a five-fold increase has been reported for the cerambycid beetle *Tetraopes tetraophthalmus* by Lawrence (1990) and a two-fold increase for *Drosophila melanogaster* by Pyle and Gromko (1978) (see Table 11). In *Gryllus sigillatus*, mating six times decreased the fecundity by a small margin compared to the four-times level, apparently because of the disturbance in the oviposition process owing to repeated matings (Subramaniam *et al.*, 1988). Multiple mating gives females the following advantages: (1) sperm replenishment, (2) nutrient acquisition, (3) oviposition stimulation, (4) increase in fertility, (5) increase in longevity, and (6) protection from predators.

Monandrous females, such as *Orius indictus, Glossina* spp., *Papilio glaucus*, and *Spodoptera* spp., receive enough sperm from a single mating and never run out of sperm for fertilizing the eggs (Ridley, 1988). However, polyandrous females, such as *Drosophila melanogaster, Locusta migratoria*, and *Rhodnius prolixus*, do not meet their sperm requirement from a single mating and hence have to mate repeatedly to fertilize all their eggs. Although sperm are viable for a long time in the genital tract of the female, maintenance of the sperm in a viable condition is energetically expensive and hence females of several species prefer to mate at frequent intervals (see Tschinkel, 1987).

Spermatophores and associated structures are highly nutritious. Apart from providing sperm for fertilization, they substantially increase the nutrient reserves of the multiple-mated females and help them allocate more materials for egg production. Fecundity of the katydid *Requena verticalis* increased with the number of spermatophores ingested by them (Gwynne, 1984). Spermatophore consumption and the consequent increase in fecundity have also been reported for several orthopterans. However, females of several species which do not receive any paternal investment also show an increase in fecundity due to multiple mating

Table 11

Effect of multiple mating on fecundity of insects

Insect	Mating status	Fecundity (Egg/female)	Remarks	Reference
Orthoptera				
Gryllus bimaculatus	Mated once	47	Associated with a male for 7 hours	Simmons (1988)
	Multiple-mated	67	Increase in fecundity due to oviposition stimulus from the male associated with it every day	Simmons (1988)
G. sigillatus	Mated once	1,451	Increase in fecundity due to spermatophore consumption by multiple-mated females; multiple mating also resulted in production of heavier eggs	Subramaniam *et al.*(1988)
	Mated twice	1,624		
	Mated four times	2,355		
	Mated six times	2,292		
Acheta domesticus	Mated once	121	Spermatophore consumption may be responsible	Sakaluk and Cade (1983)
	Mated twice	172		
Requena verticalis	—	32	Fed on one spermatophore	Gwynne (1984)
	—	71	Fed on seven spermatophores	
Hemiptera				
Triatoma brasiliensis	Mated once	138	Mating every day facilitates oviposition;	
	Multiple mated	293	Multiple mating required to realize the potential fecundity	Brassilerio (1982)
Rhinocoris marginatus	Mated once	116	Limitation of sperm available	
	Multiple mated with a single male of the same age	66	Depletion of sperm due to mating in quick succession or impotent successive matings	Ambrose and Livingstone (1985)
Nezara viridula	Mated once	632		

Contd.

Table 11 Contd.

Insect	Mating status	Fecundity (Egg/female)	Remarks	Reference
	Multiple mated	724	Transfer of more nutrients into the multiple-mated female	Mclain *et al.* (1990)
Plautia stali	Mated once	80	Mating at different intervals provides more sperm	Mau and Mitchell (1978)
	Multiple mated	220		
Dysdercus cardinalis	Male-associated upto the first oviposition	391	Single insemination enough to fertilize all eggs; hence no difference	Kasule (1986)
	Male-associated throughout	381		
Coleoptera				
Trogoderma granarium	Mated once	66	Females mate normally once	Karanavar (1972)
	Twice mated	109		
Henosepilachna pustulosa	Mated once	336	No significant increase in fecundity as one mating is enough to fill the spermatheca	Nakano (1985)
	Mated twice	377		
Epilachna varivestris	Mated once	1,808	Probably mates once in the lifetime	Webb and Smith (1968)
	Multiple mated	1,598		
Tetraopes tetraophthalmus	Mated once	8	Transfer of nutrients during successive matings; increased longevity of multiple-mated females	Lawrence (1990)
	Multiple mated	40		
Diptera				
Drosophila melanogaster	Mated once	528	Multiple mating required to store adequate sperm	Pyle and Gromko (1978)
	Multiple mated	1,053		
D. hydei	Mated once	56	Multiple mating required to store adequate sperm	Markow (1985)
	Mated twice	80		

Aedes aegypti	Male provided before blood meal	70	Some of the matings are impotent, hence repeated matings are required to realize maximum fecundity	Lang (1956)
	After blood meal	65		
	Before and after blood meal	86		
Lepidoptera				
Epiphyas postvittana	Mated with one male	493	Sperm replenishment in the multiple-mated female	Danthanarayana and Gu (1991)
	Mated with three males	623		
Bombyx mori	Associated for 30 min	301	Prolonged association helps repeated mating and acquisition of sperm	Thomas Punitham *et al.* (1987)
	for 1 hr	407		
	for 6 hr	546		
Heliothis virescens	Mated once	608		
	Mated twice	581	Lack of adequate time interval between two matings; frequency of impotent mating is also high	Pair *et al.* (1977)

(Ridley, 1988). Therefore, the nutritional benefit hypothesis, proposed to explain the increase in fecundity of the multiple-mated females, remains ambiguous. On the other hand, multiple mating may provide oviposition stimuli repeatedly to the females and enhance their fecundity as in *Acrolepiopsis assectella* (see Thibout, 1979), or help the females save expenditure of energy on maintenance of sperm in the genital tract and thereby increase their longevity and fecundity. Therefore, several factors independently and collectively influence the fecundity and fertility of the multiple-mated females. However, multiple mating by the same male at frequent intervals may result in a high percentage of impotent mating (without sperm transfer) with no increase in fecundity. For instance, the reduvid *Rhinocoris marginatus* female repeatedly mated with the same male produced fewer eggs (66 eggs) than the female mated only once (116 eggs) or the female mated with different males at different times (144 eggs) (Ambrose and Livingstone, 1985). Lack of an adequate time interval between successive matings resulted in insufficiency of sperm in the male and decreased fecundity in *Heliothis virescens* (see Pair *et al.*, 1977). Mating at different oviposition periods is essential for increasing the fecundity of orthopterans.

D. Environmental Influences

1. Photoperiod

Of all the fluctuating variables of the environment, the duration of photoperiod is the most consistent variable in nature (Danilevskii *et al.*, 1970). A major factor responsible for the reproductive success of insects is their ability to make use of photoperiodic cues to adjust the timing of their reproductive activity to favourable enironmental conditions. Photoperiods bring about remarkable changes in the fecundity of insects either directly (through the endocrine system) or indirectly (through feeding and metabolism) (Engelmann, 1970). Under favourable conditions for reproduction (short day as in *Locusta migratoria*: Cassier, 1965; or long day as in *Culex tarsalis*: Harwood and Halfhill, 1964; and *Anacridium aegyptium*: Geldiay, 1966), maturation of oocytes is accelerated and egg output is increased. Insects react to changing photoperiods so sharply that a shift to unfavourable photoperiod even after the onset of oviposition results in decreased output of eggs (Tauber and Tauber, 1969).

Daily egg production of the lacewing *Chrysopa carnea* reared at 16 hour photoperiod averages to 12 and 8 eggs/female for the first 15 days and a subsequent period of 45 days, respecitvely. A shift in the photoperiod from 16 to 12 hours on the sixth oviposition not only decreases the daily production to 3 eggs/female up to 35 days but also stops it thereafter; returning the female to 16 hour photoperiod on the 71st day results in a peak production of about 22 eggs on the 80th day (Tauber and Tauber, 1969). Obviously, resources accumulated during the reproductively inactive period (from day 35 to 70) are used to maximize egg production on return to the favourable photoperiodic regime (16 hours). A decrease in the photoperiod from 16 to 14, 13, and 12 hours prolongs the pre-oviposition period from 6 to 8, 22, and 46 days, respectively. Freshly emerged adults of *C. harrisii* shifted from 16 to 10 hour photoperiod do not oviposit at all (Tauber and Tauber, 1974).

However, a few lepidopterans seem to prefer a longer scotophase for reproduction. Rearing the flour moth *Ephestia kuhniella* under 24 hour DD, 12 hour LD, and 24 hour LL photoperiods, Cymborowski and Giebultowicz (1976) found that fecundity of the moths reared at 12 hour photoperiod is 141 eggs compared to 89 and 5 eggs produced by those reared at DD and LL conditions, respectively. By crossing females and males under different photoperiods, they concluded that males reared at LL condition are responsible for the decrease in fecundity of LL and DD females. Average fecundity of an LL female mated with a DD male is 117 eggs, as against 14 eggs for the LL female mated with an LL male. None of the eggs oviposited by females mated with LL males hatch. The spermatophores transferred by LL males contain fewer eupyrene sperm bundles than DD or LD males. As stated already (Section III C3), a decrease in the number of motile sperm in spermatheca decreases the fecundity and fertility of females. Leepla (1976) observed that the noctuid moth *Anticarsia gemmatalis* prefers to mate mostly in the scotophase, when the sperm are transferred into the spermatophore. Disturbance in the function of accessory sex glands of males due to unfavourable photoperiod is also likely to decrease the fecundity of females mated by them (Adlakha and Pillai, 1975).

The mosquito *Culiseta inornata* prefers short day for feeding and oviposition. Whereas 75 per cent of the females exposed to 8 hour photoperiod oviposit, most of them (over 55 per cent) exposed to 16 hour photoperiod retain eggs in the ovary and defer oviposition (Barnard and Mulla, 1977). However, *Culex pipiens pipiens* prefers long days (15 hours) for feeding and ovarian maturation (Sanburg and Larson, 1973). Females exposed to long days develop longer ovarian follicles (0.085 mm, suggesting high fecundity) than those reared under short-day conditions with smaller follicles (0.045 mm). Similarly, a decrease in the photoperiod at the onset of winter forces *Anopheles maculipennis* to stop egg production. Ecogeographical, climatic, and genetic factors have been held responsible for the variations in the photoperiodic response of different insects (see Danilevskii, 1965; Tauber and Tauber, 1972; Saunders, 1976; Dingle *et al.*, 1977). Arctic and polar species prefer shorter photoperiods for active reproduction. The optimum photoperiod for a hybrid of two geographical races lies between those of the parental populations (Danilevskii *et al.*, 1970).

2. Temperature and humidity

The biokinetic range of temperature permitting reproduction in insects is much narrower than that for other physiological processes. At extremes of this range, oocyte production is arrested and reproduction is suspended. Extreme temperatures induce sterility in males (e.g. *Drosophila* sp. at 32°C; *Musca domestica* at 34°C: see Bursell, 1964a). Whereas the aphid *Toxoptera graminum* can oviposit even at 5°C, the louse *Pediculus* sp. requires temperature above 25°C for oviposition (see Bursell, 1964a). Higher temperatures accelerate the rate of development and advance the beginning of oviposition. An inverse relation between pre-oviposition period and temperature has been reported for several insects (e.g. *Anopheles pharoensis*: Gaaboub *et al.*, 1971). Pre-oviposition period of the pentatomid bug *Euschistus conspersus* decreases from 37 days at 21°C to 24 and 23 days at 27°C and 32°C (Toscano and Stern, 1976). Likewise, the pre-oviposition period of the predatory coccinellid beetle *Scymnus frontalis* decreases from 20.5 days at 15°C to 8.3 days at 26.2°C. Correspondingly, the daily egg production increases from 0.5 to 7.3/female (Naranjo *et al.*, 1990). The temperature threshold for the reproductive activity of the beetle in Russia ranged from 10 to 13°C compared with 0.5 to 5°C of its aphid prey. Such differences in temperature preference between the predator and its prey render biological control less successful.

Within the permissible range of temperature for reproduction, egg production increases with temperature, and reaches the maximum at a temperature nearer to the highest in the range. On either side of the optimum temperature egg production decreases. For instance, fecundity of *Euschistus conspersus* is 229 eggs/female at 27°C; at 21 and 32°C it decreases to 57 and 158 eggs/female, respectively (Toscano and Stern, 1976). *Heliothis virescens*, reared on cotton, laid more eggs at 20°C (1,968 eggs/female) than at 30°C (1,361 eggs/female). Nadganda and Pitre (1983) attribute the decreased fecundity of this species at high temperature (30°C) to excess cost of maintenance. For the wax moth *Galleria mellonella*, 26°C is the optimum temperature for maximizing larval feeding and oviposition after emergence (Oldiges, 1959). Surprisingly, the cricket *Gryllus bimaculatus* prefers temperatures fluctuating during the day and night cycles around 27°C rather than a constant temperature for maximizing egg production. Whereas the production of a female at a constant temperature of 34°C averages to 1,000, that of a female reared at fluctuating temperatures (around 27°C) is 1,400 eggs. However, eggs produced at fluctuating temperatures suffer a decreased hatchability (50 per cent) over those produced at constant temperature (53 per cent) (Hoffmann, 1974).

The high-altitude (1,700–1,900 m) Colombian strain of the bean weevil, *Acanthoscelides obtectus*, experiences fluctuating temperatures of 21°C during daytime and 13°C during the night. In the presence of bean seeds and at the habitat temperatures (23/11°C), the egg production of a mated female averages to 46. At a higher temperature cycle (25/15°C), the production increases to 56 eggs. Fecundity of a low-altitude (1,200–1,400 m) female, enjoying a habitat temperature of 25/15°C, is 68 eggs. When the temperature cycle is lowered to 21/13°C, egg production significantly decreases from 68 to 43 (Huignard and Biemont, 1979). Fecundity of the migratory *Locusta migratoria migratorioides* in Highveld, South Africa, varies with season. During spring and summer, it averages to 316 eggs/female and in autumn to 125 eggs/female (Price and Brown, 1990). Thus, temperature effects on fecundity appear to be secondary, stemming from the primary impact on energy allocation and rate of development. Besides decreasing the rate and efficiency of food utilization, low temperatures slow down the developmental rate and prolong the pre-oviposition period; hence, insects reared at low temperatures have to drain considerable energy on maintenance metabolism. High metabolic rates of insects at above optimum temperatures also entail a considerable reduction in the allocation of energy to egg production.

Humidity influences oviposition rate in insects, which lay their eggs in soil. *Locusta* prefers a relative humidity of 70 per cent for maximum oviposition (Bursell, 1964b). A decrease in humidity affects the water retention capacity of dung pats and retards the rate of oviposition of the dung beetle *Aphodius rufipes* (see Bursell, 1964b).

3. Intrinsic factors

Fecundity and pattern of reproduction are species-specific characters, determined genetically (see Wyatt, 1991 for gene regulation of insect reproduction). However, environmental factors complement the genetic factors considerably. Heterosis has been known to promote fecundity. Fecundity of an inbred line of *Drosophila melanogaster* is about 900 eggs/female. In the hybrids, the fecundity is doubled. Likewise, the longevity and fecundity of hybrids of inbred lines of *D. subobscura* are greater than those of inbred parents. Synthesis of vitelline proteins is under genetic control in *D. melanogaster*. Mutants for the gene controlling vitelline protein synthesis suffer from deficiency of juvenile hormone and produce fewer eggs (Gavin and Williamson, 1976). Continuous inbreeding in different strains of the milkweed bug *Oncopeltus fasciatus* results in decreased fecundity and decreased survival of the offspring (Turner, 1960). The rate of production of progeny by the parasitoid *Muscidifurax raptor* decreases from 90.8/female/day after two generations of inbreeding to 26.1/female/day after 132 generations of inbreeding (Geden *et al.*, 1992). Inbreeding-depression and amplification of maternally transmitted pathogens are supposed to be responsible for the decrease in the rate of progeny production by the parasitoid.

Fecundity varies with age. In long-lived adults, it increases with age. For instance, fecundity of one-year-old carpenterworm, *Prionoxystus robiniae* fed on various host plants ranges from 331 to 737 eggs/female; two-year-old moths produce 514–570 eggs/female (Solomon and Neel, 1974). In the mole crickets, *Scapteriscus acletus* and *S. vicinus*, with advancing oviposition period, there is a decrease in clutch size as well as the number of eggs per clutch (Forrest, 1986). In short-lived forms, on the other hand, fecundity decreases with age. Fecundity of *Aedes aegypti* depends on the age at first meal. Fed on blood meal on the fifth day after emergence, *A. aegypti* produces 87 eggs. Providing blood meal only on the 30th day following emergence decreases the fecundity to 56 eggs/female. Similar trends have also been reported for *Musca domestica* and *Drosophila melanogaster* (see

Clark and Rockstein, 1964). The predaceous tree-hole mosquito *Toxorhynchites rutilus* displays a progressive drop in its egg-laying potential with advancing age; the number of eggs laid on days 1, 9, and 15 of adult life averages to 15, 6, and 2 per female, respectively. However, such a pattern of continuous oviposition helps decrease the chances of prey depletion and cannibalism among the larvae (Trimble, 1979).

4. Sterility

Parasitization renders several species of insects totally or partly sterile. Frequently cited examples belong to families Delphacidae (Hemiptera) and Vespidae, Specidae, and Andrinidae (Hymenoptera), which are parasitized by different species of Stylopoidea (Strepsiptera) (Askew, 1971). Both males and females are infected and their secondary sexual characters undergo several modifications; the process is called 'stylopization'. Ovaries of the bee *Andrena vaga*, parasitized by *Stylops*, are atrophied due to inadequate nutrition and the oocytes degenerate in their follicles; the females are never fertile. Although the testes of stylopized *A. vaga* are usually little affected, most males become sterile. The female develops male features such as reduced pollen-collecting apparatus and yellow colouration of the body; the copulatory apparatus of the male is reduced. Changes undergone by delphacids, parasitized by *Elenchus*, vary with the duration that the parasite spends in the host. Females of chironomids, parasitized by mermithid nematodes, display intersexuality (for more information on intersexuality in insects, see Gillott *et al.*, Volume V of this series, pp. 346–347). In extreme cases, the ovaries are replaced by testes, which begin to mature spermatocytes (Rempel, 1940). Nemestrinid, calliophorid, and muscid (Diptera) parasites usually castrate their acridoid hosts and kill them at the time of emergence (Kuris, 1978).

Toxic substances, released by the parasites acting through gonadotropic hormones of the corpora allata, render the females sterile (e.g. hymenopteran *Bombus* parasitized by the nematode *Sphaerularia bombi*: Palm, 1948). Tylenchoid and sphelenchoid nematode parasites infecting coleopteran, dipteran, and lepidopteran females decrease their fecundity considerably (Welch, 1965). Dilepidid cestode larval parasites not only castrate their ant hosts but also convert them into social parasites, demanding more and more food from unparasitized workers (Plateaux, 1972).

Use of gamma radiation to induce male sterility has been successful in tsetse control (see Curtis and Langley, 1972). Chemosterilants find wide application in sterile-male technique for mosquito control. Male *Aedes aegypti* have been successfully sterilized by hempa, a non-alkylating chemosterilant (George and Brown, 1967). However, the sterilizing effect of the chemical gradually decreases as the male grows. A male on the first day of treatment with hempa renders 97 per cent of the eggs deposited by a normal female sterile, compared to 77 per cent sterility induced by a male after 17 days of treatment (Grover and Pillai, 1970a). Apholate and tepa are known to induce a high degree of sterility in males of *Culex fatigans* (see Grover and Pillai, 1970b). Antijuvenile hormone agents, such as precocene 1 and 2, have been reported to inhibit female reproduction probably by interfering with vitellogenesis (Lubzense *et al.*, 1981; Virag and Darvas, 1983; see also Stall, 1986).

Use of synthetic pesticides for control of pests has been very effective in decreasing the pest population. At sublethal concentrations, pesticides significantly reduce fecundity. Treatment of males and females of the cricket *Gryllus sigillatus* with 0.003 per cent aqueous solution of an organophosphate (cythion) resulted in a decline in fecundity to 102 eggs/female compared with 458 eggs for an untreated female (Haniffa and Stephen Jose, 1987a). The organochlorine DDT also exerted a similar effect on the fecundity of the cricket (Haniffa and Stephen Jose, 1987b). Sublethal doses of permethrin and fenvalerate decreased the fecundity of the lacewing *Chrysopa carnea* (see Grafton-Cardwell and Hoy, 1985). On the other hand, Yokoyama and Pritchard (1984) reported an increase in fecundity of the lygaeid bug *Geocoris pallens* exposed to sublethal doses of a variety of pesticides (see also Jones and Parrella, 1984). Morse and Zareh (1991) also found a significant increase in fecundity of the citrus thrip *Scirtothripus citri*, fed on leaves containing residues of dicophol, fluvalinate, formetanate, and malathion. They attributed the increase in fecundity to hormoligotic effect of the pesticides at low doses. The effect of pesticides on fecundity is likely to vary with the dose and chemical nature of the pesticide and the response may differ with species (see Ibrahim and Knowles, 1986).

Topical application of chemicals such as caffeine, thiourea, and dimilin also significantly reduces the fecundity of insects (e.g. *Bombyx mori*: Haniffa *et al.*, 1986). Oviposition deterrents, present in wild plants, induce partial sterility in insects. For instance, treatment of the natural host plant, cabbage, with an ethanol extract of the wild mustard plant *Erysimum cheiranthoides* decreased the fecundity of *Pieris rapae* by 50 per cent (Dimock and Renwick, 1991). Exposure of the larvae or pupae of the tephritid flies *Dacus cucurbitae, D. dorsalis* and *Ceratitis capitata* to sand containing 1.85 ppm of azadirachtin decreased their fecundity from 398, 781, and 525 eggs/female to 179, 525, and 308 eggs/female, respectively (Stark *et al.*, 1990). The natural and synthetic compounds seem to interfere with endocrine secretions (see Bhatnagar *et al.*, 1982) as well as with the synthesis and incorporation of macromolecules in the ovary (see Hall *et al.*, 1976).

IV. PARTHENOGENESIS

Sexual reproduction, which introduces new gene combinations and increases the heterozygosity of populations, has been most successful in fluctuating environments. Yet, the heavy premium levied on energy and time budgets (especially in insects with a short life cycle), of the pre- and post-mating behavioural activities and the incredible waste of gametes in the process render sexual reproduction a luxury. Maintenance of males, whose main role in sexual reproduction is to contribute sperm for fertilization, is a costly ecological luxury (Ricklefs, 1973; also see Pandian, this volume), as the males deplete the resources which can profitably be utilized by the productive female, if her gametes can develop parthenogenically. Parthenogenesis, a reproductive strategy which circumvents copulation and fertilization, is widespread among insects (see Gillott *et al.*, Volume V of this series, pp. 351–353). Considering the number of offspring produced on an individual basis, the reproductive potential of a parthenogenic individual is theoreticlly twice that of its disexual relative (Glesener and Tilman, 1978). Yet, parthenogenesis has not replaced sexual reproduction mainly because of the genetic incompetency of parthenogenic forms to adapt themselves to dynamic environmental conditions.

A. Origin of Parthenogenesis in Insects

In insects, transition from disexual reproduction to parthenogenesis demands oviposition without mating, and development of unfertilized eggs. Gene mutations and their accumulation favoured by natural selection seem to have enabled some females to dispense with mating and to lay parthenogenic eggs. Unfertilized eggs of many insects begin to develop, but development comes to a standstill at various stages prior to hatching. In *Drosophila* spp. and several lepidopterans, development of such eggs proceeds further resulting in hatching (Suomalainen *et al.*, 1976). Astaurov (1940) demonstrated that certain strains of *Bombyx mori*, depending on their gene complements, display a strong parthenogenic tendency. Obviously, parthenogenesis has a strong genetic basis as shown, for instance, by the finding of Carson (1967) that the low rate (0.1 per cent) of parthenogenesis in the field populations of *D. mercatorum* can be increased 60-fold through selection in the laboratory.

Parthenogenesis has originated independently in different insects (Carson, 1967). Conditions such as periodic shortage of males in the habitat (e.g. Arctic), where populations are scattered or reduced and mate-finding is difficult, seem to be ideal for the appearance and spread of parthenogenesis (Tomlinson, 1966). Eight out of 14 species of insects which have invaded Japan since 1887 are parthenogenic (Kawada, 1990). Populations of small animals, such as insects, rely exclusively on parthenogenesis for their reconstruction from a few survivors after seasonal catastrophe or in disturbed areas (Ricklefs, 1973). Colonization of new favourable areas is easier for parthenogenic strains, as transport of one individual is sufficient to establish a colony. Parthenogenic weevils have spread from Europe to North America (Suomalainen, 1966).

B. Types of Parthenogenesis

Based on the behaviour of chromosomes during the maturation of oocytes, parthenogenesis has been classified into three types: (1) apomictic or ameiotic, (2) automictic or meiotic, and (3) generative or haploid (Suomalainen, 1962).

1. Apomictic parthenogenesis

In apomictic parthenogenesis, the primary oocyte undergoes single equational division forming a diploid egg; chromosomes do not undergo crossing over. The diploid ovum develops parthenogenically without amphimixis. However, two equational divisions occur in phasmids (e.g. *Carausius morosus* and *Sipyloidea sipylus*). Cognetti (1961) claims to have observed a transient pairing and apparent chiasma formation in the oogenesis of aphids (e.g. *Eriosoma lanigerum*) but failed to provide any convincing experimental proof. Although meiotic features are lacking in apomictic parthenogenesis, chromosomes contract, become stout, and look entirely different from mitotic chromosomes (see White, 1977). In the absence of meiotic features, apomictic parthenogenesis merely permits polyploidy and increases the heterozygosity of the population (Glesener and Tilman, 1978). With increasing heterozygosity and possibility for polyploidy, apomictic populations maintain a high degree of genetic polymorphism (Lokki *et al.*, 1976; Saura *et al.*, 1976a, b). In the absence of new genic combinations, rapid conversion to polyploidy enhances survival of apomictic forms, especially under harsh conditions. Triploidy and tetraploidy enabled a series of apomitic parthenogenetic races and species of curculionid weevils (Otiorrhynehinae, Brachyderianae, Eremminae, and Cylindrorrhininae) to colonize a vast area extending over Canada, Finland, Poland, Germany, Austria, and Switzerland (see Suomalainen, 1955; White, 1977). Apomictic eggs usually develop into females, hence called 'thelygenous'.

Several species of chironomids are apomictically thelygenous and seven out of 12 species studied are triploids (White, 1977). Among Orthoptera, *Saga pedo* has been extensively studied by Matthey (1941). A tetraploid (n = 68) race of *S. pedo* has, after a period of successful dispersal from South Russia to Central Spain, become extinct in different parts of its range, because of its inability to survive in fluctuating environments. Among phasmids, *Carausius morosus* and *Sipyloidea sipylus* are common examples of apomictic thelygeny. The latter is disexual in the Malay archipelago, its original home, and exclusively thelygenous in Madagascar. The sawfly *Strongylogaster macula* is apomictic; but two of its relatives, *Diprion polytomus* and *Pristophora pallipes*, retain meiosis and later restore diploidy by automictic mechanisms. For a taxonomic survey of apomictic parthenogenesis in insects, see reviews by Suomalainen (1962), Suomalainen *et al.* (1976), and White (1977).

2. Automictic parthenogenesis

In contrast to the apomictic type, automictic parthenogenesis retains the meiotic features of normal meiosis; the primary oocyte gives rise to four post-meiotic haploid nuclei, which lie in a row, the innermost one representing the egg nucleus. Subsequently, the egg nucleus restores diploidy by one of the mechanisms shown in Table 12.

Table 12

Automictic parthenogenesis and the mode of restoration of diploidy in the offspring (from different sources)

Type	Mode of restoration of diploidy	Remarks	Example
I	Fusion of the first two haploid cleavage nuclei which are homologous	Homozygosity of the mother is perpetuated; occasional gene mutations provide heterozygosity	Phasmida: *Clitummus extradentatus* Hemiptera: *Gueriniella serratula*
II	Fusion of egg nucleus and second polar nucleus which are homologous	Homozygosity of the mother is perpetuated; occasional gene mutations provide heterozygosity	Thysanoptera: *Heliothrips Haemorrhoidalis* Hymenoptera: *Diprion polytomus* *Pristophora pallipes*
III	Fusion of first and second polar nuclei which are non-homologous	*Heterozygosity of the mother is perpetuated	Lepidoptera: *Solenobia triquetrella*
IV	Fusion of secondary oocyte with first polar nucleus; both are non-homologous	*Heterozygosity of the mother is perpetuated	Lepidoptera: *Solenobia lichenella* *Apterona helix*
V	Unusual premeiotic doubling of chromosome (endomitosis) resulting in diploid egg	Heterozygosity of the mother is perpetuated	Orthoptera: *Warramba (= Moraba) virgo*
VI	Multiple conversions of haploid nuclei to diploid nuclei during or after embryonic blastoderm stage	Homozygosity of the population increases; males are rare	Orthoptera: *Euhadenoecus insolitus*

*As females of Lepidoptera are heterogametic for sex, the heterozygoid offspring is female (see Bacci, 1965).

In addition to these mechanisms, diploidy in *Drosophila parthenogenetica* and *D. mangabeiria* is restored by fusion of any two haploid polar nuclei; occasionally, the egg nucleus may fuse with two polar nuclei and form a triploid offspring (Stalker, 1954). In *D. mercatorum*, diploidy is restored either by nuclear fusion or by endomitotic pronuclear duplication (Carson, 1973). Similar endomitotic duplication occurs at the second anaphase in the parasitic wasp *Nemeritis canescens*.

As in the apomictic type, the offspring produced automictically are females, hence the name 'automictic thelygeny'. Genetic consequences of automictic parthenogenesis depend largely on the nuclei involved in fusion and restoration of diploidy. Fusion of genetically homologous nuclei, e.g. the first two cleavage nuclei (type I, see Table 12), or fusion of the egg nucleus with the second polar nucleus (type II) leads to a gradual decrease in heterozygosity. On the other hand, fusion of non-homologous nuclei (types III and IV) or premeiotic doubling of chromosomes (type V) helps retain maternal heterozygosity for several generations. However, the general trend in automictic thelygeny is toward a reduction in heterozygosity. As crossing over and segregation of chromosomes take place in the formation of automictic eggs, several new gene combinations are possible. But the chances for polyploidy and structural aberrations, which interfere with normal meiosis, are much less in automictic parthenogenesis. Members of a population which are homozygous for a few specific genes and are different from others in their gene combination gradually diverge from others, constitute a separate race, and occupy a niche (e.g. the scale insects *Lecanium hesperidium* and *L. hemesphaericum*).

3. Thelygeny

Several species of insects, such as weevils, *Solenobia*, stick insect (*Carausius*), some aphids, and gall wasps, adopt apomictic or automictic thelygeny as their only mode of reproduction; consequently, males of these species are absent, extremely rare, or at least non-functional when they exist. Among crickets (e.g. *Triglophilus cavicola* and *Euhadenoecus insolitus*) inhabiting the caves of Europe and Virginia, U.S.A., males are extremely rare. None of the 976 specimens of *E. insolitus* collected by Lamb and Willey (1989) from the caves of Virginia was a male. These crickets are peculiar in that haploidy persists in them till the embryonic blastoderm stage and diploidy is restored by multiple conversion of haploid nuclei; with a high degree of homozygosity, the parthenogenic populations rapidly colonize new habitats. Such a process of parthenogenic reproduction is called 'obligate thelygeny'. The most significant feature of this is that it permits a high reproductive potential, restricting the females to feeding and reproduction. However, in a few thelygenous insects, males are as common as females and parthenogenesis is only facultative (e.g. the hemipteran *Coccus hesperidium*). In such females, fertilized eggs give rise to both males and females, and unfertilized eggs (after restoring diploidy by automixis) produce females alone. Exceptionally, a few aphids such as *Teraneura ulmi* (see Schwartz, 1932), *Phylloxera* sp., and the gall wasp *Neuropterus baccarum* (see Doods, 1939) produce a sexual generation of males and females parthenogenically. Such forms are called 'amphitokous'females. Besides an increase in reproductive potential, a high degree of polymorphism and polyploidy gives thelygenous insects the advantage of dispersal over a vast area (e.g. curculionid weevils). Once a parthenogenic form manages to colonize a new favourable habitat, it is able to exploit the resources more rapidly than are the sexual individuals (see also Ghiselin, 1974). In addition to parthenogenesis, a rare phenomenon called 'gynogenesis', in which the sperm is required just to activate the egg without contributing its genetic material, occurs among a few insects (e.g. the lepidopteran *Luffia lapidella:* Narbel-Hofstetter, 1963).

4. Generative parthenogenesis

In generative parthenogenesis, as in automictic parthenogenesis, the oocytes undergo normal reduction division and produce haploid ova. Fertilized eggs give rise to diploid females. Parthenogenic production of males, arrhenogeny, is common among wasps, bees, and ants (Hymenoptera), scale insects (e.g. *Icerya purchasi*: Hemiptera), the paedogenous *Micromalthus* sp. (Coleoptera), and a few thrips (Thysanoptera). As recessive alleles express in haploid males, deleterious mutations are eliminated and favourable ones are retained. One of the important genetic consequences of generative parthenogenesis is the increase in homozygosity of females, as in automictic parthenogenesis. Males have given up mitotic division in spermatogenesis for self-regulation of population, through biased sex ratio. For instance, rarity of males leaves more eggs unfertilized leading to the restoration of male ratio.

C. Heterogony

Whereas weevils and most gall wasps have resorted exclusively to the strategy of parthenogenic mode of reproduction, others such as aphids, cynipids (Hymenoptera), and cecidomyids (Diptera) have combined the reproductive and dispersional advantages of parthenogenesis with the genetic advantages of sexual reproduction. Alternation of a series of parthenogenic generations with an occasional sexual generation in the life cycle of these insects has been called 'heterogony'.

The aphids have a complicated life history, involving different structural and functional morphs which appear in a definite sequence, controlled by genetic and environmental factors (Blackman, 1971; Dixon and Glen, 1971). Combination of parthenogenesis with viviparity in almost all the aphids and with paedogenesis (see Gillott *et al.*, Volume V, p. 351) in a few cases has enabled them to extensively telescope successive generations and accelerate the reproductive rate in the favourable season. The sequence presented in Fig. 9 pertains to migratory aphids such as *Aphis fabae* and *Rhopalosiphum padi*, which use stable and woody shrubs for overwintering (e.g. spindle tree for *A. fabae*) and unstable annuals (e.g. bean for *A. fabae*) for rapid multiplication in summer (Taylor, 1975). The sequence of morphs in the life cycle of a typical migratory aphid is summarized in Table 13. Non-migratory aphids, such as *Megoura viciae*, have a simple life history without involving migrantes and alienicolae morphs (see Richards and Davies, 1977).

Each morph has a definite function to perform; correspondingly, the reproductive strategy differs from morph to morph. For instance, the alate morph of *Hyperomyzus lactucae* passes through a longer duration of development, and realizes less lifetime fecundity and intrinsic rate of population increase than the apterous morph. The intrinsic rate of increase for the former is 0.227 female/female/day compared with 0.263 female/female/day of the latter (Liu Shu-Sheng and Hughes, 1987). Survival and fecundity of the morphs are size dependent (e.g. *Aphis fabae*:

Table 13

Sequence of morphs in the life cycle of the migratory aphid, *Aphis*

	Morph	Features
I	Fundatrice	Develop from overwintering fertilized eggs; apterous; high reproductive potential at the cost of reduced somatic growth; parthenogenic
II	Fundatrigeniae	Apterous, live on primary host; parthenogenic; produce two or three generations of fundatrigeniae
III	Migrantes	Alate (winged); migrate by flight to secondary host; parthenogenic
IV	Alienicolae	Apterous or alate; produce several parthenogenic generations on secondary host; parthenogenic
V	Sexuparae	Apterous or alate; migrate to primary host; parthenogenic; produce sexual males and females; dimorphic in some forms (male-producing androparae and female-producing gynoparae); e.g. *Aphis fabae* and *Phylloxera*
VI	Sexuales or oviparae	Appear only once in the life cycle; females apterous and oviparous; differ from parthenogenic fundatrigeniae in having a long body and thick hind tibia

Taylor, 1975; *Rhopalosiphum padi*: Dixon, 1976). Migrantes, whose function is to colonize new secondary hosts, are larger, have the highest fecundity, retain a high percentage of embryos in advanced stages of development, and give birth to as many young ones as possible during the first few days of adult life. These features enable the migrantes to rapidly build up large populations on secondary hosts.

Apterous and alate alienicolae face the problem of locating their annual, ephemeral secondary host plants; when overcrowded, they move from one host to another. Apterous alienicolae walk from plant to plant and decrease the population pressure. Alate alienicolae are the smallest in size, comparatively less fecund than migrantes, and reproduce only after finding a suitable host plant. Decrease in body size reduces the energy cost of host-plant-seeking and deferring production of offspring to later days helps them increase the survival of their offspring on the right host plant.

The strategy of sexuparae or gynoparae is to return to the primary host and to quickly produce as many larvae as possible before the leaves of the primary host plant begin to fall. To suit this strategy, the sexuparae have almost all the embryos in advanced stage of development and deliver most of them on the first day of adult life. The significantly low fecundity of the sexuparae, compared to migrantes and alienicolae, is probably intended to decrease the competition between sexuales. The sexuales develop and produce one or a few eggs, which overwinter and escape from the severe winter climate. The host-alternating aphid *Rhopalosiphum cerasifoliae* does not survive on its primary host plant *Prunus virginiana* beyond early summer. Colonies which survive till late summer or fall do not produce sex-

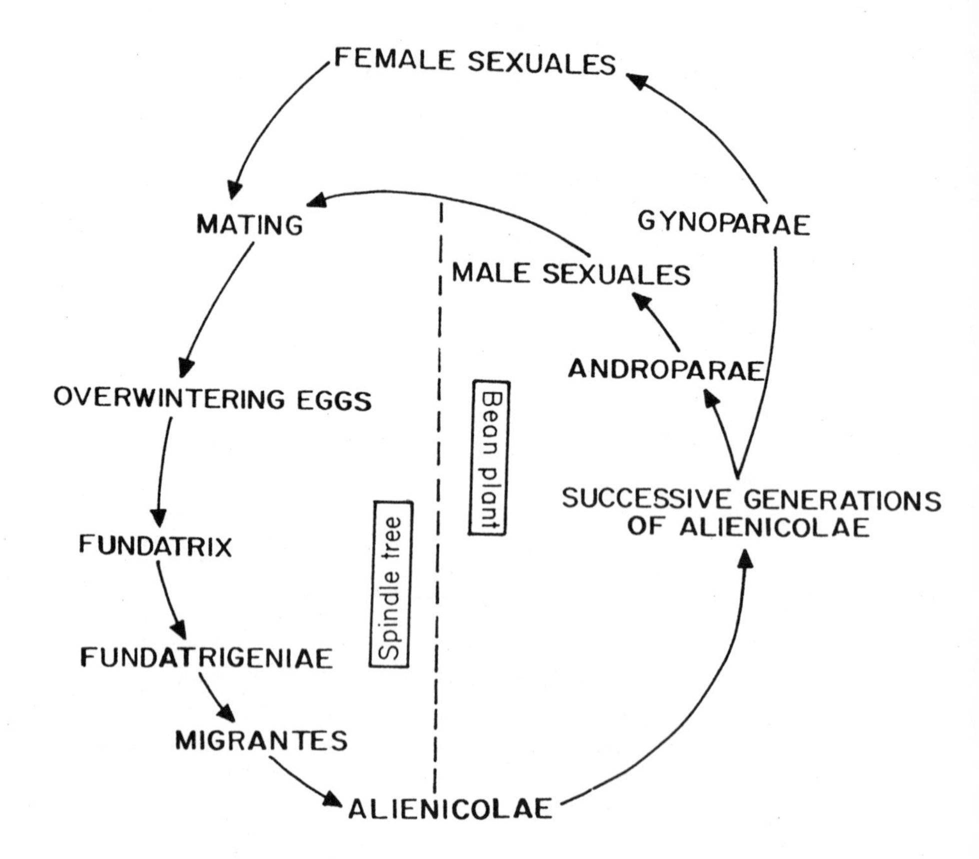

Fig. 9. Life cycle of the host-alternating aphid *Aphis* (redrawn from Chapman, 1971).

uales at all (Voegtlin and Halbert, 1990). As Dixon (1976) points out, "the wide range of environments in which an aphid lives is likely to favour the evolution of different reproductive strategies of each morph." Data presented in Table 14 for the cherry-oat aphid, *Rhopalosiphum padi*, throw light on the reproductive strategies of different morphs.

1. Factors influencing heterogony

The number of generations produced per year differs in different aphids. In adelgids, e.g. *Adelgee cooleyi*, the life cycle is extremely complicated and extends over two years, but most species have several summer generations; the phylloxerans produce three generations in the course of a year. In extreme cases, as in some members of the subfamily Pemphiginae, the sexual generation has been completely

Table 14

Reproductive strategies of the morphs of *Rhopalosiphum padi* (from Dixon, 1976; modified)

Parameter	Morphs		
	Migrantes	Alienicolae	Sexuparae
Total fecundity (larvae/female)	38	25	15
Weight (mg)	777	330	457
Reproductive rate on the first day (larvae/female)	6.8	0.23	9.2
Larvae retained in advanced stage	19	8	13
Adult life (day)	13	17	8

eliminated from the life cycle, which has an endless succession of thelygenous generations. Such a life cycle without sexual generation is called 'anholocyclic' (White, 1977). Tropical aphids fail to produce sexuales occasionally or sporadically. Several temperate species are entirely anholocyclic (e.g. *Myzus ascalonicus, Rhopalosiphum latysiphon*). Anholocyclic and holocyclic strains of *M. persicae* are found sometimes in the same locality. Blackman (1974) observed clonal differences between sexuparae of *M. persicae* in their readiness to produce sexuales. Cognetti and Pagliai (1963), who investigated *Brevicoryne brassicae*, think that intraclonal genetic variability to 'endomeiosis' results in a few offspring with the parthenogenesis-promoting gene combinations, and that others without them readily produce sexuales. According to Lees (1966), anholocyclic character arises from a gene mutation in the fertilized egg and the mutation is preserved in all the parthenogenic generations. Once the ability to produce sexual generations is lost, it is never regained. Anholocyclic clones of *Acyrthosiphon pisum* reared under short days for over three years failed to produce sexuales (Lees, 1989).

Environmental conditions, such as crowding, photoperiod, temperature, and a combination of these factors, are capable of modifying the expression of genes responsible for the production of sexuales. Triggered by low temperature and short photoperiod, sexuales cease producing further generations of sexuales and only produce alienicolae. Non-host-alternating monoecious aphids (e.g. *Aphis chloris, Brevicoryne brassicae, Megoura viciae* and heteroecious aphids such as *Aphis fabae* and *Myzus persicae* have been successfully induced to produce sexuales prematurely in summer, by decreasing the photoperiod (below 12 hours); prolongation of photoperiod forces them to continue parthenogenic generations (Lees, 1966). High temperature usually prevents the production of sexuales. Even at short photoperiod, 25°C defers production of sexuales in *B. brassicae*; at 22°C, males alone are produced. At low temperatures, photoperiod plays a major role. A long photoperiod, even at low temperatures, only induces production of parthenogenic generations. In the range of 10–20°C, the response of *Megoura viciae* to photoperiod is completely independent of temperature; at 15°C, a photoperiod of 14 hours and 55 minutes results in the production of sexuales. An increase in temperature by 5°C (below 20°C) decreases the critical photoperiod by 15 minutes (Lees, 1966).

Females of the pea aphid *Acyrthosiphon pisum*, maintained in continuous darkness, respond differently. Some of them produce only virginoparae, some produce only oviparae, and others virginoparae or oviparae followed by males (Lees, 1989). The production of sexuales is regulated by photoperiod through the endocrine system. The finding of Corbeitt and Hardie (1985) that *A. fabae, A. pisum* and *Myzus viciae*, previously programmed to produce oviparae produce virginoparae or intermediates on receiving a topical application of juvenile hormone or its analogue, kinoprene, at the developing stages confirms the role of the corpus allatum in regulation of polymorphism in the aphids (see also Hardie and Lees, 1985). Short photoperiods are associated with increased size of the corpus allatum, which activates sexual reproduction. Nutrition seldom interferes with production of sexuales. However, Schwartz (1932) reported that annual grasses such as oats convert the entire colony of *Tetraneura ulmi* into sexuparae, while perennial hosts such as *Lolium* induce continuation of parthenogenic generations. *Myzus persicae* reared under crowded conditions produces more gynoparae.

The life cycle of gall wasps is simpler than that of aphids, and does not involve functional and structural polymorphism. In most cases (e.g. *Cynipis*), there are two generations in a year—the summer thelygenous generation alternating with the winter disexual generation. In *Neurotenus lenticularis*, fertilized eggs produce different male-producing (androparae) and female-producing (gynoparae) females. Oocytes of androparae undergo normal meiosis and produce haploid eggs which develop parthenogenically into males; in gynoparae, meiosis does not occur in thelygenic morphs and polar bodies are not formed; the diploid eggs develop parthenogenically into females (White, 1977).

A combination of parthenogenesis and viviparity with paedogenesis is seen in some cecidomyid dipterans (e.g. *Heteropeza*: White, 1977; also see Chapman, 1971). Their larvae mature precociously; the eggs in the ovarioles of these larvae begin to develop without being fertilized. Normal males and females of *Heteropeza pygmaea* appear during the unfavourable dry season, under crowded conditions, or when the host plant begins to wilt. A female lays normal and parthenogenic eggs; a normal egg after fertilization produces paedogenic larva which gives rise to a generation of daughter larvae. The latter may produce (1) adult female, (2) purely arrhenogenic paedogenic larva, (3) paedogenic larva, which may produce adult male or another generation of paedogenic larva, or (4) paedogenic larva like their parent (Fig. 10). On no occasion do imagines directly develop from eggs through larvae and pupae (see Richards and Davies, 1977 for discussion). Food quality seems to favour the production of males in some cases. Winged females help dispersal and colonization of new hosts. Larviparous

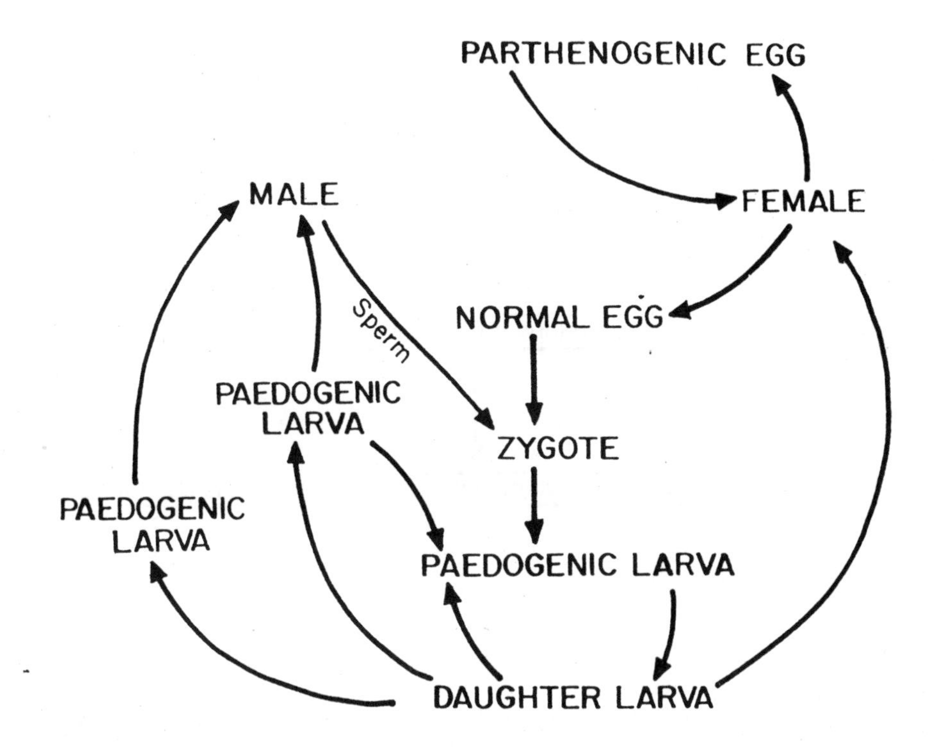

Fig. 10. Life cycle of *Heteropeza pygmaea* (based on information collected from different sources).

Tekomyia and *Henria* produce two types of pupae; some of them metamorphose into adults while others (hemipupae) with vestiges of wings and legs produce a brood of about 60 paedogenic larvae each; the larvae escape by rupturing the pupal wall (Wyatt, 1961).

Among the Coleoptera, *Micromalthus debili* has an exceptionally complex life history (Fig. 11). A high degree of parthenogenesis, paedogenesis, and viviparity combined with their extremely isolated life (in timbers present in mines) restricts the diversity of the species. *Micromalthus debili* is the only species of the genus Micromalthidae (Richards and Davies, 1977). Males are haploid and the females diploid (Scott, 1941). Males and some of the larvae are known to consume their parental larvae.

Facilitating a high rate of reproduction, parthenogenesis is quite common in disclimax communities where competitive interactions are less, and in habitats with unexploited resources (see Wright and Lawe, 1968). Parthenogenic populations tend to occur at higher latitudes and altitudes, in disturbed rather than undisturbed habitats, and in islands rather than mainlands (Ghiselin, 1974). Sexuality is favoured

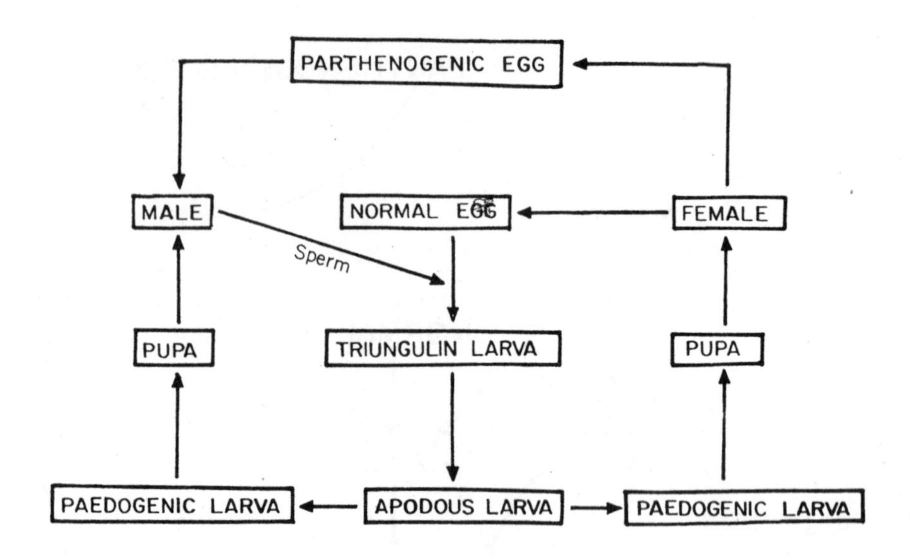

Fig. 11. Life cycle of *Micromalthus debili* (from Chapman, 1971; modified).

under greater biotic stress and in biologically uncertain environments (Glesener and Tilman, 1978). However, conditions at which sexual phase is timed by heterogonic insects such as aphids (short photoperiod and low temperature or high population density as in *Myzus persicae*) support the above conclusion to some extent (see Williams, 1966).

V. REPRODUCTIVE STRATEGIES

In their natural habitat, insects are exposed to an environment which is heterogeneous in both space and time. The short duration of ontogenesis in insects does not permit them to develop homeostatic mechanisms and hence environment plays a decisive role in determining their reproductive strategy (Labeyrie, 1978a). Environment limits the resources of energy and time for insects as for other organisms. When resource limitations are severe, animals, especially insects with a short generation time, should optimally allocate the resources for growth, maintenance, and reproduction (Gadgil and Bossert, 1970). Environmental factors such as temperature and availability of food not only affect the rates of intake and accumulation of energy but also its partitioning between somatic structures and functions (Muthukrishnan *et al.*, 1978; Muthukrishnan and Pandian, 1983, 1984) as well as reproduction. Therefore, several workers (e.g. Price, 1974a; Giesel, 1976) have proposed an energy-budget approach for understanding reproductive strategies.

A. Energy Allocation

The net metabolizable energy available to an individual is channeled into broad compartments devoted to somatic and reproductive structures as well as reproductive and non-reproductive functions (see Fig. 1). Increased expenditure of energy on non-reproductive functions, such as maintenance, defence, competition, and foraging, drastically decreases the reproductive effort. A major fraction of the assimilated energy is expended on maintenance metabolism. Several insects allocate about 14 per cent of the total assimilated energy to egg production (Table 15). Barring *Mamestra configurata*, for which egg production was estimated only for the first 12 days of adult life, the lowest value (10.6 per cent) is recorded for the lasiocampid moth *Cyclophragma leucosticta*, which spends as much as 76 per cent of its assimilated energy on metabolism during the larval and pupal period. To maximize energy allocation to egg production, *C. leucosticta* female adopts the following strategies: (1) decrease in the maintenance metabolism of adult by shortening the adult life span as well as by remaining sluggish and not tending to disperse from the place of eclosion, and (2) economizing energy expenditure on

Table 15

Partitioning of assimilated energy in some insects. Values are expressed in percentage of assimilated energy

Example	Metabolism	*Net growth	Exuvia and/or silk, pupal case	Egg production	Reference
Orthoptera					
Oxya velox	70.8	11.7	3.6	13.8	Delvi and Pandian (1971)
Poecilocerus pictus	48.8	35.1	2.4	13.7	Delvi (1972)
Homoptera					
Macrosiphum liriodendri	—	—	—	26.3	Van Hook *et al.* (1980)
Lepidoptera					
Cyclophragma leucosticta	75.7	5.9	7.8	10.6	Mackey (1978)
Mamestra configurata	43.4	40.5	8.2	7.8	Bailey and Singh (1977)
Bombyx mori	52.0	13.0	22.0	13.0	Hiratsuka (1920)
Lymantria dispar	—	—	—	44.8	Montgomery (1982)
Coleoptera					
Sitophilus oryzae	74.0	11.4	0.4	14.3	Singh *et al.* (1976)

*Energy content of spent adult.

exuvia and cocoon; a female spends only 7.8 per cent of the assimilated energy, whereas a male spends 15.3 per cent (Mackey, 1978). Similarly, the high energy cost of metabolism in the grasshopper *Oxya velox* (see also Delvi and Pandian, 1971) and the weevil *Sitophilus oryzae* (see Singh *et al.*, 1976) does not permit them to allocate more than 14 per cent of the assimilated energy to egg production. Despite a decrease in the expenditure on metabolism, the high cost of silk production permits *Bombyx mori* to spare only 13 per cent of its assimilated energy to egg production (Table 15).

With increasing energy demand for egg production, energy reserves accumulated in fat body and wing muscles are drawn to supplement the allocation for egg production. In extreme cases, females are prepared for phenotypic sacrifice to enhance egg production. For instance, some crane flies (e.g. *Tipula carinifrons* and *Pedicia hannai*) inhabiting the Arctics (where moist atmosphere and low temperature limit the power of flight) display a tendency to increase their investment on egg production at the cost of reduced wings, legs, eyes, antennae, and mouth parts

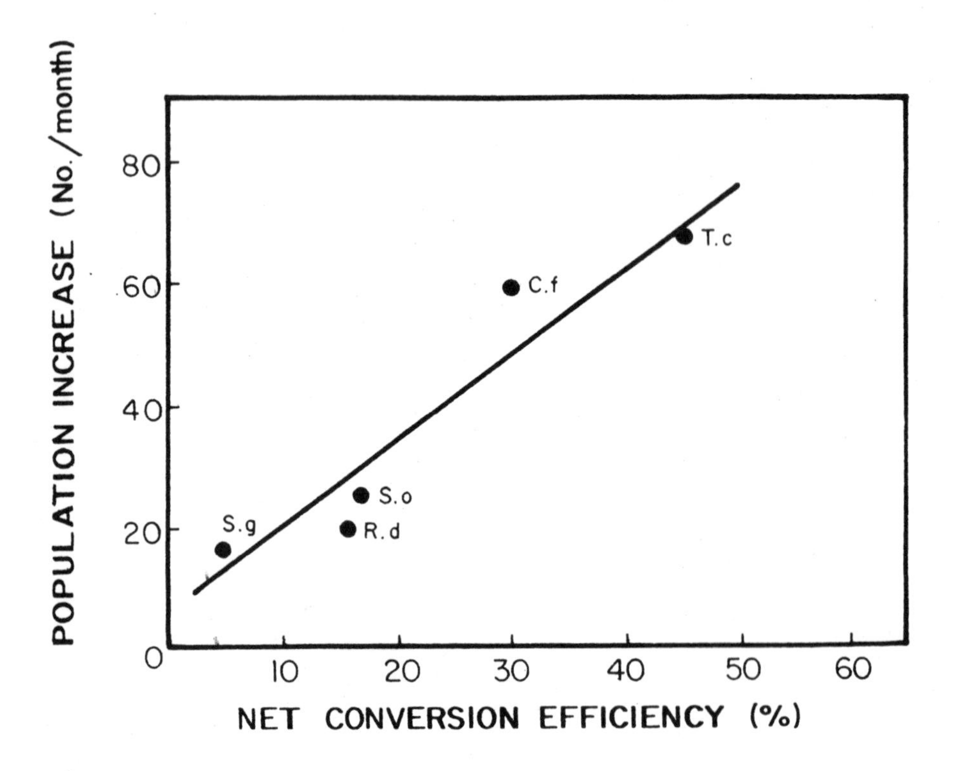

Fig. 12. Relationship between net conversion efficiency and rate of population increase in some stored-product pests. S.g, *Sitophilus granarius*; S.o, *S. oryzae*, R.d, *Rhyzopertha dominica*; C.f, *Cryptolestes ferrugineus*, and T.c, *Tribolium castaneum* (redrawn from Campbell and Sinha, 1978).

(MacLean, 1975). Winter crane flies (*Chinea* spp.) forego their flight muscles to gain a 4 per cent increase in egg production (Byers, 1969). A reduction in the investment on flight muscles in the brachypterous plant hoppers significantly increases their fecundity over that of the winged forms (Denno *et al.*, 1989). During tethered flight lasting for 5 hours, female *Nilaparvata lugens* lose 20 per cent of their lipid and carbohydrate reserves (Padghan, 1983). The brachypterous females which do not expend their energy on flight channel their reserves to increase their egg production.

In carabid beetles, excessive energy drain on egg production in one breeding season decreases the residual energy of females and affects their survival till the next breeding season (Murdoch, 1966). Most insects are semelparous (breeding once in lifetime) and hence reduction in the survival value of the post-breeding females is of less significance. Van Dijk (1979) rejected the hypothesis of Murdoch (1966) and proved that inverse correlation between number of eggs produced and post-breeding survival of the mother does not exist in *Pterostichus coerulescens* (Carabidae).

Insects with a high efficiency of conversion of assimilated energy into body tissue allocate more energy to egg production. Campbell and Sinha (1978) obtained a significant correlation between the net conversion efficiency and rate of population increase (reproductive consequence) for five species of stored-product pests (Fig. 12). Addition of more data is likely to transform the linear relation to an exponential one. The flour beetle *Tribolium castaneum* exhibits the highest net conversion efficiency (45 per cent) and registers a population growth rate of 72 individuals per month.

Among insects, the parasitic hymenopterans which feed on the energy-rich host tissues exhibit the highest net conversion efficiency, ranging from 60 to 75 per cent (e.g. *Phygadeuon dumetorum*, 60 per cent: Fisher, 1971; *Idechthis* (= *Nemeritis*) *canescens*, 66 per cent: Fisher, 1977; *Cidaphus alarius*, 75 per cent: Varley, 1961; *Sceliphron violaceum*, 75 per cent: Marian *et al.*, 1982; *Apanteles flavipes*, 75.3 per cent; *Microplitis ophiusae*, 90.6 per cent; and *Bracon brevicornis*, 78.8 per cent: Senthamizhselvan, 1987). Endoparasitic mode of life provides superabundant and easily obtained food supply and permits the parasites to develop high fecundity (Jennings and Calow, 1975). High net conversion efficiency combined with increased reproductive effort enables most parasites and stored-product pests (e.g. ichneumonids, *Tribolium castaneum*, and *Cryptolestes ferrugineus*) to build up larger populations within a short period.

B. *r*- and *K*-strategies

The pattern of parental investment varies with the environment. In highly unstable environments, characterized by seasonal catastrophes and density-independent mortality, as in temperate regions, the strategy of a female is to divide her total investment into as many offspring as possible and compensate the mortality (r-strategy); contrastingly, in stable environments such as the tropics, the strategy of a female is to increase her investment in individual offspring and produce a few highly competitive individuals (K-strategy) (MacArthur and Wilson, 1967; Gadgil and Solbrig, 1972). In line with the above principle of investment, the temperate *Oncopeltus fasciatus* has a higher reproductive rate than the tropical *O. unifasciatellus* (see Landhal and Root, 1969). The beetle *Chrysomelina quadrigemina* lays more eggs in California than in southern British Columbia (Peschken, 1972).

The marine collembolans inhabiting the littoral area adopt different strategies of reproductive allocation. *Isotoma viridis* and *Hypogastrura viatica* are exposed to a wide range of salinities ranging from freshwater to hyper-saline (200 per cent) sea water; they produce three or more generations per year at the rate of three to four egg batches per month. Their high rate of reproduction is related to the highly unstable nature of their habitat. Hence they may be described as r-strategists. On the other hand, *Anurida maritima* inhabits a more or less stable and predictable habitat and produces only one generation per year at a reproductive rate of about one egg batch per month. Its highly rhythmic behaviour helps it escape flooding of the habitat and avoid unfavourable conditions. It can be described as a K-strategist (Witteveen and Joosse, 1987).

Ichneumonid parasitoids belonging to different subfamilies provide suitable examples of the strategy of parental investment. The total reproductive effort of a female (calculated by multiplying egg volume in cubic mm by the number of eggs laid) increases from 0.16 for Anomalinae ovipositing in eggs to 0.98 for Ichneumoninae attacking old larvae and pupae (Table 16). An inverse relationship between egg volume and number exists among the different subfamilies (Smith and

Table 16

Reproductive effort of some ichneumonid females in relation to host stage attacked (from Price, 1974a; modified)

Subfamily	Host stage attacked	Number of eggs/female	Egg volume (mm^3)	Total reproductive effort
Ichneumoninae	Old larva and pupa	9.2	0.1065	0.98
Ephialtinae	Old larva and pupa	8.0	0.1104	0.88
Gelinae	Pupa	8.0	0.0940	0.75
Acaenitinae	Middle and old larva	26.2	0.0269	0.70
Ophioninae	Middle and old larva	29.3	0.0239	0.70
Metopiinae	Young larva	70.4	0.0092	0.65
Diplazontinae	Young larva	55.3	0.0090	0.49
Anomalinae	Egg	84.1	0.0019	0.16

Fretwell, 1974). Like the individuals exposed to seasonal catastrophes in temperate climate, parasites ovipositing in earlier stages of hosts have low survival value and hence tend to maximize their fecundity by decreasing their investment in each egg. For instance, an egg parasitoid of the family Anomalinae produces as many as 84 small eggs (0.0019 mm^3) compared to eight larger eggs (0.11 mm^3) produced by the cocoon parasitoid (e.g. Ephialtinae), whose strategy is to produce a few highly competitive offspring with high survival value (Price 1973a, b). The former may be called an 'r-strategist' and the latter a 'K-strategist'; others infecting young and middle larvae fall in between the r–K continuum (see Pianka, 1970, 1974; Gadgil and Solbrig, 1972; Demetrius, 1975).

Studying an endemic host-parasitoid community consisting of four species of hymenopteran parasitoids infecting the gall midge *Rhopalomyia californica*, Force (1972) concluded that less competitive and highly fecund parasitoids (r-strategists, e.g. *Tetrastichus* sp.) dominate the harsh, early stages of succession and that they are followed by highly competitive and less fecund species (K-strategists) late in the process of succession. Similar conclusions have also been drawn by Price (1973b) for the parasitoid guilds of the sawfly (*Neodiprion swainei*) host. Early colonizers, such as *Lamachus laphyri* and *Olesicampe lophyri*, are replaced after a short period by cocoon parasitoids such as *Gelis urbanus* and *Pleolophus basizonus*. In addition to the position in the sequence of succession, parasitoids can be distinguished by certain anatomical features (Price, 1973b). Early colonizers have a larger number of ovarioles and other ovarian adaptations for increased egg production and storage. Force (1972) proposed that r-strategist parasitoids (early colonizers) with high fecundity, rapid dispersability, and high tolerance to harsh conditions can effectively control pest populations invading new habitats. The key anatomical features provided by Price (1975b) are of immense value in choosing an r-strategist parasitoid for biological pest-management programmes.

Southwood *et al*. (1974) consider that the generation time of an individual determines its reproductive strategy. For instance, among lepidopterans, *Spodoptera exempta, Pieris rapae*, and *P. brassicae* with a generation time of about 25 days produce around 500 eggs/female; with a generation time of 365 days, *Apature iris* and *Genopterynx rhamni* produce about 80 eggs/female (Southwood *et al*., 1974). Members of the former group are r-strategists and those of the latter group are K-strategists. But insect life histories are so diverse and varied that the all-embracing concept of r- and K-selection is inadequate to explain the reproductive strategies of some insects. For instance, the hispine beetle *Chelobasis perplexa* undergoes a prolonged larval development (200 days) and lays a few large eggs like a K-strategist (see Strong and Wang, 1977). The view that the prolonged maturity of *C. perplexa* is a K-selected adaptation to divert more energy into competitive ability does not hold good; for *C. perplexa* lives in an uncrowded condition with little competition. The long generation time and low reproductive rate of *C. perplexa*, typical of a K-strategist, are not due to K-selection but due to the low nitrogen content and high concentration of phenoloxidase (which decreases protein assimilation) in the host plant, *Helioconia imbricata* (see Strong and Wang, 1977). Therefore, besides r- and K-selection, food quality is another factor which can determine the reproductive strategy of some insects.

C. Sex Ratio

Partitioning of the total reproductive effort between male and female offspring is an important component of the reproductive strategy. Differential investment between the offspring results in a biased sex ratio, which is one of the factors regulating intrinsic growth rate of populations. According to Emlen and Oring (1977), the 'operational sex ratio' (= 'quaternary sex ratio' of Pianka, 1974) of a population depends on the ratio between the energetic investments of the mother and the father on the offspring. An excess of maternal investment and asynchrony in the onset of sexual receptivity of females result in male-biased operational sex ratio (Emlen and Oring, 1977).

Fisher's (1930) theory of sex ratio predicts a 1 : 1 ratio for stable populations of randomly mating individuals, in which the cost of production of males is equal to the cost of production of females. This conclusion has been confirmed theoretically by several workers, but the prediction holds good only for populations that experience severe competition for mates. In the absence of a meiotic drive, genetic mechanisms also envisage a 1 : 1 primary sex ratio at birth; but during the slow stepwise process of sexual differentiation in insects, young ones are subjected to differential growth and mortality leading to a biased secondary sex ratio. More vulnerable to environmental stress than females, the males suffer higher mortality (Werren and Charnov, 1978). *Aphidius ervi*, which parasitizes the mortality-prone larval stages of its host, displays a male-biased operational sex ratio. This is attributed to the mortality of the progeny developing in unstable host stages (Wellings *et al*., 1986). Factors such as meiotic drive, mating system, host size and availability, and delayed fertilization or oviposition significantly change the sex ratio and alter the reproductive strategy of parasitoids. Extrinsic factors such as photoperiod, temperature, and availability of food also alter the sex ratio. The population-homeorheostat model of Wildish (1977) suggests density-dependent changes in sex ratios.

1. Genetic and cytoplasmic factors

Several endogenous and exogenous factors have been known to alter the primary sex ratio of insects. Presence of a sex-ratio (SR) gene in the Y chromosome of *Drosophila pseudoobscura* leads to degeneration of male-producing sperm in the testis; consequently, over 99 per cent of the offspring produced by SR males are females (Policansky and Ellison, 1970; Policansky, 1974). Reduced fertility of SR males decreases the advantage of meiotic drive in overproduction of females and maintains the SR trait in the population. A similar phenomenon has been reported for a few wood-boring scotylid beetles. Low-hatchability broods of *Dendroctonus jeffreyi* are all-female (Table 17). Lanier and Wood (1968) presume that a cytoplasmic factor, transmitted by females, selectively kills the eggs developing into males and decreases the hatchability. This in turn leads to a complete absence of males among the offspring. However, larger broods of *D. jeffreyi* contain a few males probably not susceptible to the lethal cytoplasmic factor of the mother (Lanier and Wood, 1968). A high proportion of females in the population of bark beetles enables them to exploit a sudden increase in the supply of wood, following drought, fire, or catastrophic wind; a single male fertilizes several females and compensates the unequal sex ratio (Reid, 1962). Among chrysomelid beetles, interpopulational hybrids suffer differential mortality during development due to genetic imbalance; in areas of population overlaps, females are represented in large numbers (e.g. *Calligrapha philadelphica*: Robertson, 1964).

2. Mating systems and parental characteristics

Influence of mating system on sex ratio has been well documented among some beetles and parasitic hymenopterans. Under conditions of panmixia or random mating, equal sex ratio is expected (Hamilton, 1967). Species of *Xyleborus* and *Ips* (Coleoptera) adopt a polygamous mating system (i.e. several females are fertilized by one male) and their populations display a female-biased sex ratio (Hopkins, 1909). Male *Nasonia vitripennis* are brachypterous with a restricted radius of activity; within this radius a male fertilizes as many females as he can; a high rate of fertilization decreases the proportion of males as males develop only from unfertilized eggs (Van den Assem *et al*., 1980). Gregarious parasitoids infecting small, well-dispersed hosts practise sib-mating (e.g. *Nesolynx albiclavus:* Van den Assem *et al*., 1980). Species exhibiting extreme sib-mating (e.g. the fig wasp *Blastophaga*) produce only very few sons who can ensure insemination of all their sisters (Hamilton, 1967). Most female parasitic Hymenoptera are monandrous. Occasional second mating and reinsemination of females within 24 hours of first mating cause an obstruction to the flow of sperm in the oviduct and result in the production of unfertilized eggs which subsequently develop into males. Thus, most of the offspring of *Nasonia vitripennis* females successively reinseminated within 24 hours are males (58 M : 17 F) (Van den Assem and De Bruzin, 1977).

In *Muscidifurax raptor*, a delay in oviposition results in decreased production of females (Wylie, 1979). A delay in mating results in male-biased sex ratios in *Drosophila melanogaster* and the butterfly *Talaeproi tubulosa* (see Werren and Charnov, 1978). Female mealy bug *Pseudococcus*, mated 0, 6, 8, and 10 weeks after emergence, produced males and females in the ratio of 1 : 1, 2 : 1, 3 : 1, and 10 : 1, respectively (see Werren and Charnov, 1978). With advancing age, the number of males produced by the eulophid parasitoid *Sympiesis marylandensis* increased; a

Table 17

Skewed sex ratios observed in some insects

Insect	Sex ratio M:F	Remarks	Reference
Thysanoptera			
Frankliniella occidentalis	0:1	Virgin females associated with 10 males for one week; more females are produced to build the population rapidly	Higgins and Myers (1992)
Orthoptera			
Gryllus rubens	1.0:1.4	Males mature prior to females and die. Mating call of males exposes them to predators	Veazey *et al.* (1976)
Lepidoptera			
Prionoxystus robiniae			
1-year moths	5.2:1.0	Dimorphic males; most males require two years to complete larval development	Solomon (1976)
2-year moths	1.0:1.9		
Diptera			
Glossina sp.	1:2	Males have shorter adult life	Jackson (1944)
Coleoptera			
Oxyleborus sp.	1:60	Polygamous mating system	Hopkins (1909)
Ips sp.	1:2	Polygamous mating system	Hopkins (1909)
Dendroctonus jeffreyi	0:3 to 0:26	Low hatchability broods; a maternally transmitted cytoplasmic factor selectively kills eggs developing into males	Lanier and Wood (1968)
Hymenoptera			
Mellittobia japanica	1:24	Gregarious parasite; oviposits only a few unfertilized eggs developing into males	Van den Assem *et al.* (1980)
Tetrastichus atriclavus	1:3	Gregarious parasite; oviposits only a few unfertilized eggs developing into males	Van den Assem *et al.* (1980)
Nasonia vitripennis	1:3	Brachypterous males with restricted radius of activity	Van den Assem *et al.* (1980); King (1961)
Nesolynx albiclavus	1:20	Sib-mating species; restricts male progeny	Van den Assem *et al.* (1980)
Coccygomimus turionella	1.0:2.3	Oviposition of more fertilized eggs on large host	Sandlan (1979)
C. turionella	1.0:0.25	Oviposition of more unfertilized eggs on small host	Sandlan (1979)
Coccophagus atratus	7:3	Hosts suitable for male and female eggs are equal; males are hyperparasitic over conspecific females	Donaldson and Walter (1991a)

Contd.

Table 17 Contd.

Insect	Sex ratio M : F	Remarks	Reference
C. atratus	4 : 6	70 per cent of the hosts available are suitable for female eggs	Donaldson and Walter (1991b)
Epidinocarsis lopezi	7 : 3	Smaller second instar hosts (mealy bug) are abundant	Van Dijken *et al.* (1991)
E. lopezi	2 : 8	Larger third instar hosts abundant	Van Dijken *et al.* (1991)
Sympiesis marylandensis	9.9 : 0.1	Reared on IV instar host	Ridgway and Mahr (1990)
S. marylanden-sis	1.3 : 1.0	Reared on V instar host	Ridgway and Mahr (1990)
S. marylanden-sis	0.9 : 1.0	Reared on host pupa	Ridgway and Mahr (1990)
Encarsia pergandiella	13 : 5	Male- and female-suitable hosts, 20 each	Hunter (1989)
E. pergandiella	16 : 3	Male-suitable hosts 30, female-suitable hosts 10	Hunter (1989)
E. pergandiella	7 : 6	Female-suitable hosts 30, male-suitable hosts 10	Hunter (1989)
Goniozus leg-neri	1 : 19	At a density of 25 host insects per female parasitoid	Legner and Warkentin (1988)
Muscidifurax raptor	25 : 65	After 2 generations in culture	Geden *et al.* (1992)
M. raptor	6.2 : 13.1	After 8 generations in culture	Geden *et al.* (1992)
M. raptor	14.5 : 12.8	After 45 generations in culture	Geden *et al.* (1992)
M. raptor	11.8 : 14.3	After 132 generations in culture	Geden *et al.* (1992)
Leptopilina boulardi	45.4 : 129.3	Exposed to host (*Drosophila*) larvae for 1 hr/day	Kopelman and Chabora (1992)
L. boulardi	48.4 : 60.9	Exposed to hosts for 1 day/3 days	Kopelman and Chabora (1992)
L. boulardi	116.1 : 60.0	Exposed to hosts continuously	Kopelman and Chabora (1992)

26-day-old mother produced 90 per cent females compared with none by a 44-day-old female (Ridgway and Mahr, 1990).

Multiple insemination of females before oviposition results in male-biased sex ratios in *Microcentrus ancylivones* (see King, 1987). Males mated repeatedly undergo sperm depletion and hence their mating results in male-biased sex ratios of the offspring (e.g. *Spalangia cameroni*: Hurlbutt, 1987). However, after

30 minutes of the previous mating, female *Pachycrepoideus vindemiae* replenish their sperm supplies and resume production of daughters in subsequent matings (Nadel and Luck, 1985). Small and medium-sized females of parasitoids such as *Trichogramma brevicapillum* with their potential fecundity less than that of larger females produce female-biased sex ratios (see Pak and Oatman, 1982). In several species of parasitoids, the genetic basis of sex ratio has been clearly demonstrated (e.g. *Mastrus carpocapssea, Dhalbominus fuscipennis*, and *Nasonia vitripennis*: see King, 1987). Apart from nuclear genes, four extrachromosomal factors inherited through the mothers regulate the offspring sex ratios of *N. vitripennis* (E.D. Parker and Orzack, 1985; also see Skinner, 1985).

3. Host size and availability

A host-discriminating machinery in the ovipositor helps the female parasitoids distinguish parasitized and unparasitized hosts as well as hosts of various sizes (Wylie, 1965, 1967). The ectoparasitic *Pnigalio flavipes* oviposits on early and late larval instars and pupae of its lepidopteran host. More females emerged from the larger leaf-mining and tissue-feeding larval stages and pupae of the host. The smaller sap-feeding stage of the host gave rise to more males of the parasitoid progeny (Barrett and Brunner, 1990). This is because the females are larger and need larger hosts in which to develop (see Waage, 1986). About 70 to 75 per cent of *Epidinocarsis lopezi* emerging from small nymphs of the cassava mealy bug, *Phenacoccus manihoti*, are males. The percentage of males emerging from larger hosts decreases to 13–25 per cent (Van Dijken *et al*., 1991). Exposed to five different lepidopterous pupae of various sizes, female *Coccygomimus turionella* (Ichneumonidae) follow a quantitative rather than qualitative cue and oviposit more fertilized (female-producing) eggs on larger hosts; consequently, with increasing host size from 100 to 200 mg, the proportion of females to males (Fig. 13) is doubled (Sandlan, 1979). In general, males emerge from smaller hosts, males and females from medium-sized hosts, and a high percentage of females from larger hosts (e.g. *Pimpala turionella*: Arthur and Wylie, 1959; *Itoplectis cristatae*: Nozato, 1969; *Nasonia vitripennis*: Smith and Pimental, 1969; see also Fisher, 1971). An increase in the size of the host results in a greater increase in the size of the gregarious female parasitoid emerging out of it and promotes its survival and reproductive success. Therefore, sex ratio of the progeny arising out of larger hosts is female biased.

Density of the host population also plays an important role in determination of the sex ratio of parasitoids. For *Epidinocarsis lopezi*, Van Dijken *et al*. (1991) observed a male-biased sex ratio at high densities of the host, *Phenacoccus manihoti*. At high host densities, the rate of encounter of the parasitoids with the male-producing smaller hosts is larger because of the following factors: (1) due to density-dependent factors acting on the host population, the proportion of smaller hosts at higher host densities is higher; (2) smaller hosts are more vulnerable to parasitization as they do not defend themselves efficiently against the parasitoids at-

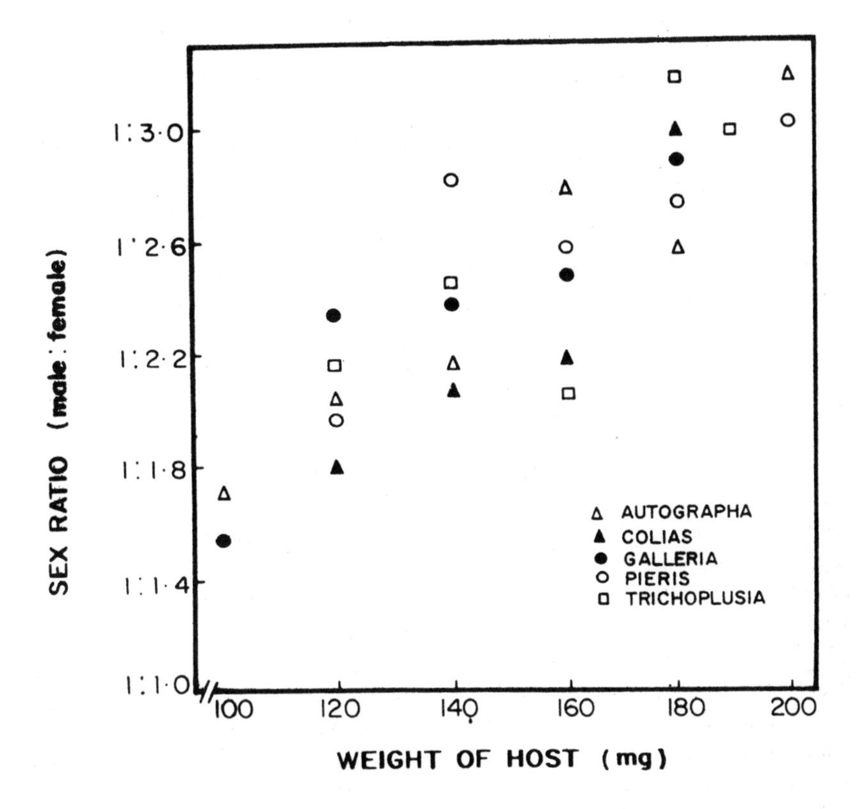

Fig. 13. Sex ratio of *Coccygomimus turionella* as a function of host size (redrawn from Sandlan, 1979).

tempting to oviposit on them; and (3) smaller hosts always remain in the periphery of the host population and are frequently exposed to the searching parasitoids.

On the other hand, Legner and Warkentin (1988) believe that at high host densities, parasitoids prefer to lay more female-producing eggs on the host in order to suppress the rising host population. For instance, over 95 per cent of the offspring of *Goniozus legneri* exposed to high host densities are females. *Epidinocarsis lopezi* oviposits 70–75 per cent of male-producing eggs in the small second instar larva of the cassava mealy bug and 75–87 per cent of female-producing eggs in the large third instar larva (Van Dijken *et al*., 1991). High host densities in the field result in male-biased sex ratio of the parasitoid. Such host-size-dependent variations in sex ratio may be attributed to: (1) the increase in the proportion of small hosts suitable for males with increasing host density; (2) decrease in host size at high host densities due to density-dependent factors; (3) preference of smaller hosts for occupying the periphery of the population leading to greater encounter with ovipositing parasitoid females; and (4) weak defence of smaller hosts against the ovipositing parasitoids and hence their greater vulnerability to parasitic attack. For several hymenopterous parasitoids, availability of suitable hosts for oviposition and development of male- and female-producing eggs varies spatially and temporally. In *Coccophagus atratus*, parasitic on the scale insect *Filippia gemina*, Donaldson and Walter (1991a, b) reported a male-biased brood sex ratio (0.7 M : 0.3 F) when male- and female-suitable hosts were available in equal numbers, and a slightly female-biased (0.4 M : 0.6 F) ratio when 70 per cent of the hosts were suitable for female progeny. Hunter (1989) has also reported similar results for the autoparasitoid *Encarsia pergandiella* (Table 17). In summary, female parasitic wasps adjust the sex ratio of their offspring in response to host conditions by controlling fertilization (Charnov, 1982).

4. Density of ovipositing females

Restricted availability and patchy distribution of hosts often result in a severe competition between ovipositing parasitoid females. Hamilton (1967) proposed the 'local-mate-competition theory' to explain the relation between changes in sex ratios of offspring and the density of ovipositing females of parasitoids (see also Frank, 1986). Accordingly, isolated females primarily produce daughters and just enough sons to inseminate all the daughters. The limited production of males helps avoid competition between sib-brothers for mating and maximize the production of daughters. When a female realizes the presence of other ovipositing females nearby, she lays more male-producing unfertilized eggs. This results in male-biased sex ratio and severe competition for mating between non-sibling males. Under conditions of extreme restriction of host availability, either the same female or several females repeatedly oviposit on one host. Such a phenomenon is called 'superparasitization'.

Hymenopterous parasitoids make use of chemical traces, left on the host by the female which parasitized previously, to discriminate between parasitized and unparasitized hosts and prefer to lay a greater percentage of male-producing unfertilized eggs on parasitized hosts (Suzuki and Iwasa, 1980; Werren, 1980). Therefore, with increasing density of parasitoid females, the number of parasitized hosts increases and a male-biased offspring ratio results. This has been found true in 11 out of 13 species reported by King (1987), *Bracon hebetor* and *Spalanzani cameroni* being the exceptions. Scelinoid parasitoids, such as *Trissolcus euschisti*, do not respond to chemical traces left by the previous female. But their tendency to lay male eggs early in an ovipositional sequence results in male-biased offspring ratio due to superparasitization (Braman and Yeargan, 1989). When a host is superparasitized by two females and the number of offspring of the second female is greater than that of the first, the offspring sex ratio of the second female is female biased (e.g. *Nasonia vitripennis*: Werren, 1980). This helps avoid competition for mates between sib-sons of the second female. Apart from regulated laying of fertilized or unfertilized eggs in previously parasitized hosts, severe intraspecific competition among the young ones developing in a host results in differential mortality of the sexes and biased offspring sex ratio (e.g. *Trichogramma chilonis*: Suzuki *et al*., 1984).

Sex-ratio regulation is a means of population control in parasitic wasps. At high population densities, more males are produced and the population thus decreases. At low densities, less interference between females results in overproduction of females and the population increases. With increasing proportion of ovipositing females to hosts, percentage of females in the progeny decreases (Table 18). At high female densities, interference of ovipositing females by conspecific females reduces the rate of egg fertilization and production of females (e.g. *Caraphractus cinctus*: Jackson, 1966; *Nasonia vitripennis*: Wylie, 1966). Following interruption by conspecific females, ovipositing *N. vitripennis* withholds sperm and temporarily reduces the level of fertilization (Wylie, 1976a). An uninterrupted female *N. vitripennis* produces offspring in the ratio of 1 : 3 in favour of females. Arrangement of ova in the ovary is responsible for the female-biased ratio. Approximately one-fourth of the eggs in the ovary move into the oviduct in such a way that the micropyle of one out of four eggs is turned away from the opening of the spermathecal duct and hence left unfertilized. The sequence of oviposition of female-producing fertilized eggs varies between species as well as within species. The rate of oviposition may change the oviposition sequence of male- and female-producing eggs (King, 1987). In *Trissolcus grandis*, the effect of interruption and the consequent decreased rate of fertilization persists for some time even after isolation of the female (Victorov, 1968).

5. Photoperiod and temperature

Sex ratio varies with photoperiod and temperature in some insects. For instance, the cricket *Nemobius vezoensis* exposed to long days produces over 75 per cent

Table 18

Sex ratio of two hymenopteran parasitoids as a function of parasite: host ratio. Pupa of the housefly *Musca domestica* served as host for both species of parasites (data from Wylie, 1976b, 1979; modified)

Parasite : Host	Number of progeny	Sex ratio M : F	Female percentage
	Eupteromalus dubius		
1 : 5	450	1.0 : 10.3	91.3
2 : 5	744	1.0 : 3.2	76.4
5 : 5	1,118	1.0 : 1.2	55.4
	Muscidifurax raptor		
1 : 10	477	1.0 : 2.4	70.8
2 : 10	500	1.0 : 1.3	57.0
2 : 20	974	1.0 : 1.7	62.5
5 : 25	618	1.0 : 1.2	53.7
5 : 100	862	1.0 : 3.2	76.3

females. At 12 hour photoperiod, sex ratio of the hymenopteran parasitoid *Campoletis perdistinctus* is skewed in favour of females (Hoelscher and Vinson, 1971). At 10 : 14 LD, another hymenopteran parasite *Pteromalus puparum* produces two to three times more females than at 14 : 10 LD (Bouletreau, 1976). Sex ratio of *Chrysopa harrisii* (Neuroptera) at 24°C is 7 : 9 in favour of females; an increase in temperature to 27°C decreases the proportion of males to females to 5 : 3; at 18°C males and females are in the ratio of 7 : 6 (Tauber and Tauber, 1974). Low temperature prevailing in temperate regions favours the survival of females of the flea *Xenopsylla cheopis* and distorts the ratio in favour of females; but at a higher temperature range (21–24°C) males outnumber females. High temperatures increase male production by one of the following mechanisms: sterilization of males, incapacitation of sperm, differential mortality, and interference with mating (see P.E. King, 1961; B.H. King, 1987). However, such temperature-induced changes in offspring sex ratios are rare under field conditions, as the parasitoids behaviourally avoid such extremes of temperature.

6. Temporal and spatial variations

Sex ratio varies temporally and spatially in some insects. For instance, the proportion of males in the monarch butterfly (*Danaus plexippus*) populations at Muir Beach, Santa Cruz, and Santa Barbara (California) increases from about 58 per cent in December to 75 per cent during March (Tuskes and Brower, 1978). The increase in the proportion of males is presumably due to the tendency of mated females to leave the population at a faster rate than males. Similarly, migration of mated *Pieris protodice* females from the population during peak reproductive season in July results in a high proportion (93 per cent) of males (Shapiro, 1970). In *P. protodice*, males make 57 per cent of the population in May; with advancing breeding period and increasing population density, the proportion of males increases to 93 per cent in July but returns subsequently to the original level in September (Table 19). Between August and September, populations of *Gryllus rubens* and *G. firmus* reach their maximum peak, when the percentage of males is higher than that at summer (Veazey *et al*., 1976). During the pre- and post-monsoon seasons, males dominate the *Pheropsophus hilaris* (Coleoptera) populations; male-female sex ratio becomes 1 : 1 in summer (Kalyanam *et al*., 1972). The African monarch butterfly, *Danaus chrysippus*, has a female-biased sex ratio at Kazi and Makerere, a male-biased one at Kangolomolo (see Wildish, 1977). Females of migrating species of leaf hoppers (e.g. *Circulifer tenellus*) migrate farther than males, causing a spatial difference in sex ratio (Cook, 1967). The biased sex ratio in relation to space and time is a strategy to regulate population to a small size when resources are inadequate, and to increase the size to exploit abundantly available resources.

In mass production of parasitoids, continuous inbreeding for several generations results in a decrease in the proportion of female progeny. For instance, in

Table 19

Sex ratio of *Pieris protodice* as a function of population density in New Jersey, U.S.A. (from Shapiro, 1970; modified)

Date	Density (number/hectare)	Sex ratio M : F	Male percentage
28-5-67	60	1.3 : 1.0	56.6
28-6-67	133	1.2 : 1.0	54.1
17-7-67	459	3.8 : 1.0	79.1
22-7-67	1,019	12.5 : 1.0	92.5
16-8-67	805	5.8 : 1.0	85.3
20-9-67	171	1.3 : 1.0	56.6
22-10-67	30	1.4 : 1.0	56.7

the culture of the hymenopterous parasitoid *Muscidifurax raptor*, the proportion of female progeny decreased from 73 per cent after two generations to 49 per cent after 132 generations of the culture (see Table 17). The decrease in the efficiency of utilization of resources available in the culture medium with increasing number of generations may be responsible for the decrease in the proportion of females.

7. Parental investment

Based on the measurements of parental investments in social Hymenoptera with the haplo-diploid type of sex determination, Trivers and Hare (1976) proposed that the queens are benefited by the 1 : 1 investment among their reproductive offspring, and the workers from the 1 : 3 (male : female) investment among reproductive siblings. Alexander and Sherman (1977) have shown that the predictions of Trivers and Hare (1976) hold good only when (1) there is no local mate competition, (2) workers do not lay eggs, and (3) females are strictly monogamous (see also Charnov, 1979). Differential parental investment in the offspring results in biased sex ratios. In the solitary bee *Ceratina calcarata*, for instance, provisioning of 1.3 times more prey in the daughter cells resulted in a male-biased sex ratio. Variations in the quality and quantity of the resources available and the mother's ability to provision the cells have been attributed to the male-biased sex ratios in *C. calcarata* (see Johnson, 1988).

Although considerable information is available on sex-ratio variation in natural populations, much remains to be known about their implication *per se* on population. In a strictly monogenous species, such as *Pseudococcus longispinus*, overproduction of males is a means of decreasing the size of the population. In the all-female populations of cynipid wasps which reproduce by parthenogenic viviparity, a linear relationship exists between number of females and population growth (White, 1977). Exposed to adverse conditions such as scarcity of food, a K-strategist (e.g. *Lucilia sericata*) tends to produce more males and maintains the population size close to the carrying capacity; the response is contrasting in an r-strategist such as the mosquito *Theobaldia incidens*, which produces more females, so that at least a few of the offspring will survive and repopulate when conditions improve (see Wildish, 1977).

D. Time Allocation

Besides energy allocation for reproduction and partitioning the total reproductive effort between male and female offspring, allocation of time for reproduction is an important component of the reproductive strategy. Insects synchronize their growth and reproduction with favourable climatic conditions. The duration of favourable climate varies with latitude. Consequently, the generation time, as well as the time allocated for reproduction, markedly varies in populations inhabiting different latitudes. Rapid development, allocation of a major proportion of generation time for reproduction, and turnover of several generations per year are characteristic features of tropical insects, which experience favourable climate for most of the year. On the other hand, slow development, dormancy for most of the life, and restriction of reproductive duration are features common to temperate and polar insects which are exposed to the vagaries of the environment.

Most temperate insects have a long generation time (Table 20) and require several years to complete a single generation. For instance, the beetle *Coccinella semipunctata*, which enjoys a cosmopolitan distribution, produces one or two generations per year in Euro-Siberia (70°N) (Hodek, 1973), as against 20 generations per year in India (15°N) (Puttarudriah and Basavanna, 1953). Within Europe, the dragonfly *Ischnura elegans* is semivoltine at 58°N, univoltine at 54°N, and trivoltine at 44°N (Johannson, 1978; D.J. Thompson, 1978). Indeed, the proportion of generation time allocated for development and reproduction forms the basis of latitudinal differences in the reproductive strategies of insects. In environments where reproduction is possible only for a short duration each year, reproductive rate (r_m) is maximized (r-strategy) in order to provide 'seeds' for repopulating the habitat when the environment is favourable (Lewontin, 1965).

Survival during the unfavourable dry summer in tropics and the freezing winter in temperate and polar regions, especially at a low energy cost, is of great value for reviving growth and reproduction during favourable environmental conditions. Diapause and migration are the two important life-history tactics which enable the insects to escape unfavourable conditions in time and space, respectively (e.g. *Oncopeltus fasciatus*: Dingle, 1972; *Coccinella septempunctata* and *Semiadalia undecimnotata*: Hodek, 1967; *Danaus plexippus*: Brower *et al*., 1977). An increase in diapause duration considerably decreases the time allocated for reproduction as well as the number of offspring produced. For instance, prolonged larval diapause (over eight months) decreases the fecundity of the noctuid stem borer *Busseola fusca* and the pyralid *Chilo partellus* by 50 per cent (Kfir, 1991).

Table 20

Life span of some diapausing insects

Species	Life span (year)	Reference
Gryllobattoidea		
Gryllobatta compodeiformis	6	Kamp (1970)
Odonata		
Tanypteryx pryeri	3	Taketo (1971)
Oplonaceschna armata	5	Johnson (1968)
Plecoptera		
Perla bipunctata	2	Hynes (1976)
Siphonoptera		
Xenopsylla cheopis (Pacific coast)	1	Richards and Davies(1977)
Lepidoptera		
Gynaephora rossi	5	Downes (1965)
Coleoptera		
Pterostichus brevicornis	2	Kaufmann (1971)
Upis ceramboides	2	Kaufmann (1969)
Diptera		
Tipula carinifrons	4	MacLean (1975)
Pedicia hannai	4	MacLean (1973)
Culicoides spp.	2	Downes (1962)
Chironomus hyperboreus	2	Rempel (1936)
Neuroptera		
Sialis	1	Kaiser (1961)

1. Types of diapause

Diapause is an 'actively induced' state of dormancy, involving the cessation of neuroendocrine activity usually triggered by environmental conditions (Lees, 1965). Insects make use of photoperiod and/or temperature — the most reliable cues of environment for timing growth, reproduction, and diapause. The temperate and polar species are summer-active (long-day insects) and undertake hibernation during the cold winter, while those in tropics are active in winter and spring (short-day insects) and evade the dry summer by undertaking aestivation. Subtropical insects undergo aestivo-hibernation during winter and summer and restrict growth and development to spring and autumn (e.g. *Nebria brevicollis* and *Patrobus atrorufus*: Thiele, 1969).

In certain univoltine insects with a long period of post-embryonic development, diapause is obligatory, intervening at a species-specific life stage; the onset of this type of diapause is genetically determined (e.g. egg diapause in the gryllid *Teleogryllus* sp.: Masaki and Ohmachi, 1967; nymphal diapause in the bug *Aleurolobus asari*: Bahrmann, 1972). In polyvoltine insects, diapause is facultative. The intervention occurs only in that generation which passes through the unfavourable period; the other generations are not interfered with (Müller, 1970).

The stages sensitive and susceptible to diapause vary in different insects (Fig. 14). A species' choice of the 'resting stage' is on good selective grounds. It could be in the embryonic or egg stage (e.g. *Bombyx mori*: Kogure, 1933; *Peripsocus quadrifasciatus*: Eertmoed, 1978); postembryonic stage (larval, nymphal, or pupal) (e.g. *Dendrolimus pini*: Geyspitz, 1965); or reproductive (imaginal) stage (e.g. *Leptinotarsa decemlineata*: De Wilde and De Boer, 1961; *Chrysopa carnea*: Tauber and Tauber, 1969). For a comprehensive list of examples for the different types of diapause, see Saunders (1976) and Tauber *et al*. (1986).

2. Embryonic diapuase

Embryonic diapause occurs at various stages of embryonic development in different species. For instance, the developing eggs of *Bombyx mori* begin diapause as early as the blastokinesis stage, those of *Locusta migratoria* at the time of differentiation of appendages, and those of *Lymantria dispar* after the completion of embryonic development but before hatching (see Lees, 1965). Embryonic diapause is determined solely by the photoperiod experienced by the mother during her last embryonic or early larval development as in *B. mori* (see Fukuda, 1951), or during late larval development as in the lymantrid moth *Orygia antigua* (see Kind, 1965). In *Peripsocus quadrifasciatus*, embryonic diapause is maternally induced, facultative, and a more temperature-dependent response to photoperiod; the length of the critical photoperiod increases by 15 minutes for an increase of 1° in latitude (Eertmoed, 1978). Short days induce embryonic diapause; exceptionally, long days are effective in *B. mori*.

Rotational cropping pattern of corn (the preferred host plant) with non-host plants such as soya bean induces egg diapause in the northern corn rootworm, *Diabortica barberi*. In South Dakota (U.S.A), the incidence of egg diapause of *D. barberi* holds a positive correlation with the percentage of corn grown in rotation (Levine *et al*., 1992).

Very little is known about embryonic diapause among aquatic insects. For the temperate intertidal collembolan *Anurida maritima*, during winter, the duration of photophase (6–8 hours) is inadequate to search for its food and escape the high tide. It therefore undertakes winter diapause at the egg stage (Witteveen *et al*., 1988). Most species of black flies produce non-diapausing eggs, but a few (e.g. *Cnephia dacotensis*) produce aestivating eggs and others such as *Simulium arcticum* and *S. decorum* produce overwintering eggs (Hynes, 1970). Among Ephemeroptera, the eggs of *Ephemerella ignita* hatch only after six months of diapause (Percival and Whitehead, 1929). The duration of diapause in *Baetis* sp. extends over 200 days (Illies, 1959). Eggs of the stonefly *Pteronarcys proteus* remain dormant for a period of 11 months before hatching (Holdsworth, 1941). Embryonic diapause among other stoneflies, such as *Nemoura cinerea* and *Brachyptera risi*, is apparently under thermal control. High temperature prevailing in the stream during summer forces

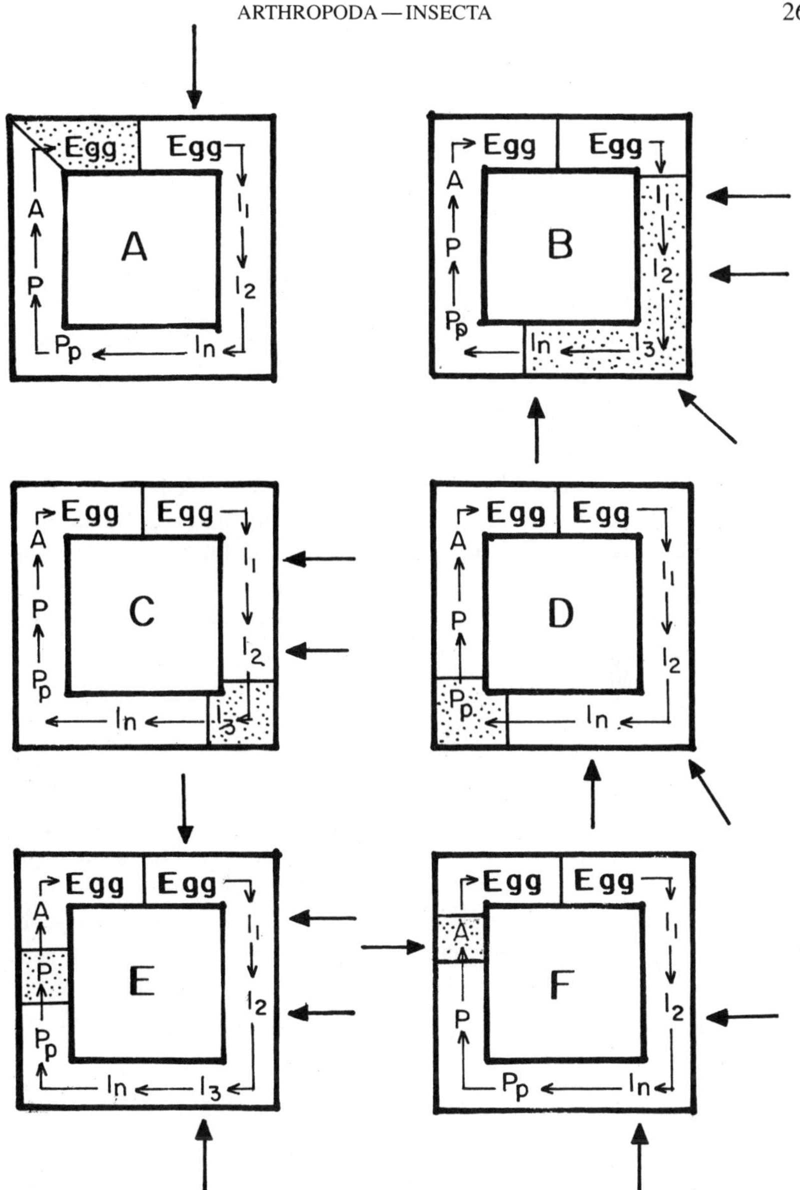

Fig. 14. Diapause-sensitive (indicated by discontinuous areas) and diapause-susceptible (stippled) stages of some insects. **A**: Embryonic diapause in *Bombyx mori*. **B–E**: Post-embryonic diapause in *Dendrolimus pini* (**B**), *Pandomis ribeana* (**C**), *Loxostego sticticalis* (**D**), and *Polychrosis botrana* (**E**). **F**: Reproductive diapause in *Leptinotarsa decemlineata* (after Highnam and Hill, 1979).

them to undertake prolonged diapause for over five months (Khoo, 1964). In these as well as some mayflies, diapause is broken irregularly, as a result of which the eggs hatch at different intervals (Hynes, 1970). Many stoneflies protect their eggs from drying with the help of a gelatinous coat. The eggs survive for long periods among stones in the dry stream bed (Khoo, 1964). Thus the eggs of many aquatic species serve as resting stage in the life cycle and enable them to survive unfavourable conditions by remaining dormant.

3. Post-embryonic diapause

Larval/nymphal and pupal diapause are induced by photoperiodic inactivation of the prothoracic glands by the brain and the consequent arrest in growth and development. In the corn borer *Diatraea grandiosella*, a reduced titre of juvenile hormone under the influence of the diapause-maintaining photoperiod (12 hour) produces stationary moults without growth. Long days decrease the juvenile hormone production and permit moulting into pupa (Yin and Chippendale, 1970).

Post-embryonic diapause intervenes at a particular life stage of the species. In some insects undergoing development over a long period, onset of diapause occurs at two different life stages during successive years. For instance, the dragonfly *Tetragoneuria cynosura*, which requires two years to complete nymphal development, diapauses as a young nymph during the first winter and as a late nymph during the next winter (Lutz, 1974). Likewise, diapause occurs at one or more life stages of the pine moths *Dendrolimus pini* and *D. sibiricus* (see Geyspitz, 1965). In such insects with prolonged generation time and occurrence of diapause at one or more stages, individuals belonging to different generations coexist; hence failure of reproduction during one year does not hamper the fitness of the population. Among the Trichoptera, the last larval instar is susceptible to diapause. Most of them overwinter in Europe (e.g. *Apatidea muliebris*) and aestivate in North America (e.g. *Pycnopsyche guttifer*); the overwintering last stage larvae coexist with active larvae of the same or earlier instar (Nielsen, 1942). The dormant larvae are of great ecological significance, as they ensure the existence of the population when the active larvae succumb to environmental vagaries.

Several arctic insects owe their extended life cycle to larval diapause. Most of them complete maturation of eggs at the pupal stage itself and the adult span is drastically reduced and devoted exclusively to reproduction. The moth *Gynaephora* mates immediately after emergence and oviposits on the wall of the cocoon (Downes, 1965). In an extreme case, the adult stage is completely absent in the black fly *Prosimulium ursinum* and the eggs are released by rupture of the pupal wall (Carlsson, 1962). However, a reduction in the total life span as an adaptation to arctic conditions is reported for the high-altitude grasshopper *Aeropedellus clavatus*; the reduction is accomplished by a decrease in the number of instars from five to four (Alexander and Hilliard, 1964). Diapause prevents unseasonal emergence and the consequent reproductive failure in arctic insects (Danks and Oliver, 1972).

Larval diapause in the culicid *Chaoborus americanus* is caused by different environmental factors. During winter, when there is a threat of freezing of water, short days induce diapause; but in spring and summer, lack of prey in the habitat triggers diapause. Using these environmental cues, *C. americanus* increases the survival of the larvae (Bradshaw, 1969). Although photoperiodically induced diapause is common among several aquatic insects (e.g. *Chitonophora krieghoffi*, *Capnia bifrens*, and *Taeniopteryx burksi*: Hynes, 1976), a thermally imposed quiescence occurs in a few species. For instance, high temperature directly induces quiescence in larvae of *Polypedium vanderplankei*, a chironomid, breeding in rocky pools in Africa and enables them to evade drought (Hinton, 1951). The larvae are capable of tolerating repeated hydration and dehydration, can survive a short exposure to a temperature as high as 104°C (Hinton, 1960), and can withstand complete desiccation for 20 months at 63°C (Hinton, 1952).

Pupal diapause, in most insects, serves as a strategy to synchronize emergence at the onset of the favourable season. Synchronous emergence considerably decreases the time and energy costs of mate-seeking and mating. In insects with short imaginal life, such reductions are of great significance. The tachinid parasite *Anthrycia cinerea* requires 24 weeks at 2°C for synchronizing adult emergence (Wylie, 1977). Besides temperature, age of the female and nature of the host also induce pupal diapause in *A. cinerea*. Old females and females emerging from diapause-programmed hosts produce eggs which give rise to diapausing pupae (Wylie, 1977). The ichneumonid parasitoid *Phytodietus vulgaris* undergoes a prepupal diapause which is photoperiodically determined. Short days (10 L : 14 D) induced 100 per cent diapause, while transfer to long days (14 L : 10 D) terminated the diapause (Coop and Croft, 1990). The critical photoperiod for the induction of prepupal diapause in the braconid parasitoid *Microplitis croceipes* increased from < 10 L : 14 D at 18–21°C to > 11 L : 13 D at < 18°C (Brown and Phillips, 1990). Pupal diapause in the temperate species of fleshflies, such as *Sarcophaga* and *Poecilometopa*, is triggered by short days of late summer. However, in the tropics, the importance of day length in inducing diapause is less and temperature plays a major role in inducing as well as terminating diapause (Denlinger, 1978). Furthermore, there is an inverse relationship between temperature and duration of pupal diapause (73 days at 18°C compared to 235 days at 12°C).

4. Reproductive diapause

Timing of reproductive effort to coincide with a favourable environment to promote the survival of offspring is an important component of the reproductive strategy (Giesel, 1976; Cohen, 1976). The timing of reproductive diapause is very precise in several insects. For instance, the Colorado potato beetle, *Leptinotarsa decemlineata*, in its habitat in western Massachusetts goes into reproductive diapause exactly on August 15 every year. Adults emerging 7 days prior to the onset of diapause do not oviposit whereas those emerging 21 days before, oviposit 120 eggs/female (Voss *et al.*, 1988). Reproductive diapause is characterized by cessation of ovarian activity in females and regression of accessory sex glands resulting in the arrest of mating behaviour in males (Riddiford and Truman, 1978). Inactivation of median neurosecretory cells controlling the corpora allata leads to the absence of juvenile hormone and induces reproductive or imaginal diapause. The diapause may intervene following the emergence and intense feeding activity as in *L. decemlineata* (see De Wilde, 1962) or following copulation but prior to ovarian maturation as in several syrphid (Diptera) females (Schneider, 1969). Ultimately, reproductive diapause renders the sexual state of the population homogeneous (e.g. the bug *Psylla pyri*: Thanh-Xuan, 1967). However, in many regions, reproductive diapause does not affect the entire population and consequently diapausing and non-diapausing adults occur in the same population (e.g. the beetle *Aelia acuminata* in southern Slovakia: Hodek, and Honek, 1970). Such populations are capable of adjusting their life cycle and extending their habitat to latitudes midway between those where reproductive diapause is indispensable and those where it is useless (Danilevskii *et al.*, 1970).

Reproductive diapause in males is less intense and less well defined than in females. Compared to females, males develop at a faster rate and almost complete gonadal maturation at the time of emergence (Beck, 1968). A few males are sexually active prior to the onset of diapause and even possess viable sperm during diapause (see Hodek, 1962). In general, diapausing males do not diaplay any reproductive behaviour (Orshan and Penner, 1979a, b). In the pre-diapause female, ovarioles are small and narrow and vitellogenesis is suppressed; active oogenesis and vitellogenesis are postponed to the post-diapause period. However, in some females, diapause is less intense and a few mature oocytes are present at the time of termination of diapause (e.g. the water-beetle *Dysticus marginalis*).

Like other types of diapause, reproductive diapause is regulated by photoperiod and temperature. For instance, with increasing photophase, incidence of reproductive diapause in the anthocorid predatory bug *Orius insidiosus* decreases. At 10 L : 14 D all the females enter diapause. Most of them avert diapause when they are reared in 13 to 15 L. At 20°C, the critical photoperiod falls between 12 L and 13 L (Ruberson *et al.*, 1991). The photoperiod experienced by the freshly emerged adult seems to be more critical than that experienced during the larval period. For instance, shifting of the squash bug *Anasa tristis* to 12 L : 12 D at 24°C immediately after emergence from the pre-eclosion photoperiod of 17 L : 7 D at 24°C results in all the females entering diapause. On the other hand, a switch to 12 L : 7 D at 24°C, after 9 days of eclosion, delayed the onset of diapause of the ovipositing females by 29 days (Fielding, 1988). The response of a species to diapause, including critical photoperiods, may be modified by temperature and the quality of food experienced by the species in the wide range of its distribution. A very sharp difference in the critical photoperiods has been observed for the populations of *Leptinotarsa decemlineata* inhabiting Riverhead, Long Island (latitude 42°58′ N) and Freeville (40°27′

N), New York, U.S.A. The former responds to 15 L : 9 D, the latter to 15.8 L : 8.2 D. At temperatures below 29.4°C, the response of the populations to their respective critical photoperiods is inversely related to temperature. The response of the population in the cooler Freeville to low temperatures is greater than that in the warmer Riverhead (Tauber *et al.*, 1986, 1988). In both habitats, incidence of diapause is increased by the presence of senescent host-plant foliage (see also Hare, 1980; De Kort, 1990).

Tauber and Tauber (1973) reported that nutritional factors modify the response of *Chrysopa mohave* to the photoperiod and temperature. Irrespective of the photoperiod experienced in the habitat, feeding on rape seeds induces reproductive diapause in the cabbage bug *Eurydema rugosa*. The presence of the seed in combination with short photoperiods increases the incidence of diapause to 100 per cent (Numata and Yamamoto, 1990). Non-availability of seeds of the sapindaceous host trees triggers diapause in the Oklahoma (U.S.A.) population of the rhopalid bug *Jadera haematoloma* during October (Carroll, 1988). Several publications stress the importance of quality and quantity of food in diapause regulation. In the pink bollworm *Pectinophora gossypiella* reared at 10 : 14 LD, a diet containing 0.25 per cent wheat-germ oil induces diapause in 15 per cent of the population compared to the incidence of 80 per cent induced by a diet containing 15 per cent cottonseed oil (Adkisson *et al.*, 1963). Absence of food restricts diapause to 2–8 per cent of the population of the culicid *Chaoborus americanus*, whereas food scarcity enhances the incidence to 50 per cent (Bradshaw, 1969). Besides the nutritional factors, hormones transferred from host to parasite (Schoonhoven, 1962) or vice versa (Wellso and Adkisson, 1966) are known to significantly influence induction and termination of diapause (see also Fisher, 1971; Askew, 1971).

Reproductive diapause is basically a strategy evolved by insects to tide over unfavourable environmental conditions, e.g. non-availability of quality food to the progeny. Apart from responding to environmental cues that forecast approach of adverse environmental conditions, many insects have also evolved polymorphism for the timing of diapause. For instance, some of the females in the North American potato beetle populations directly undergo reproductive diapause while several of them oviposit for a short period before entering into diapause (see Tauber *et al.*, 1988). Directly entering into diapause for an entire year without oviposition entails considerable reproductive cost. However, this loss is compensated by other groups of females which oviposit briefly prior to diapause. The survivors of the second generation coexist with their parental generation after the termination of diapause (Tauber *et al.*, 1986, 1988; see also De Kort, 1990). Istock (1981) has also reported such a phenomenon among mosquitoes. Environmental conditions seem to vary the strategy of temperate and tropical races of the same species. The temperate (Oklahoma) *Jadera haematoloma* population enters diapause in response to the total exhaustion of host seeds. Despite the host seeds' availability at different patches throughout the year, the tropical (Florida) population undertakes short migrations seeking fresh resources due to severe competition for the limited seed resources (see Carroll, 1988).

5. Biochemical aspects of diapause

Certain biochemical changes taking place during diapause enable the incumbent to tide over unfavourable conditions as well as to restore physiological vigour after the termination of diapause. Increasing concentrations of polyhydric alcohols and low-molecular-weight carbohydrates during diapause enhance supercooling in freezing-susceptible insects (e.g. the arctic carabid beetle *Pterostichus brevicornis*: Baust, 1972) and minimize the damage in freezing-tolerant insects (e.g. the gall-fly *Eurosta solidagensis*: Baust *et al.*, 1979). Glycerol, sorbitol, and trehalose are the major components of the antifreeze cryoprotective system (Morrisey and Baust, 1976). Tolerance of the overwintering pupae of the fleshfly *Sarcophaga bullata* to cold temperature increases with haemolymph glycerol concentration (Chen *et al.*, 1991). With a haemolymph concentration of 2.4 mM glycerol, *P. brevicornis* manages to survive even at −35°C (Baust, 1972). Utilizing sorbitol and theritol in haemolymph, the tenebrionid beetle *Upis ceramboides* tolerates temperatures as low as −50°C (Chippendale, 1978). Trehalose contributes to an increased cold-hardiness in insects which do not accumulate polyols (e.g. the sawfly *Trichocampus populi*: Chippendale, 1978).

6. Migration

Like diapause, migration forms an integral part of the life history of several insects and involves temporary suppression of feeding and reproduction. In many insects, both migration and diapause have evolved together as parts of the same physiological syndrome (e.g. *Oncopeltus* spp.). While a vast majority of insects migrate as adults (e.g. bugs, butterflies, moths, and beetles), a few locusts begin migration as nymphs. Insects such as the pentatomid bug *Eurygaster integriceps*, the milkweed butterfly *Danaus plexippus*, and many coccinellid beetles migrate to hibernation sites and return to the feeding and breeding sites in the spring. Host-alternating aphids migrate from summer- to winter-host plants.

Besides helping to escape from unfavourable environments, migration is useful for dispersal and colonization of new habitats. Dingle (1972) considers insect migrants colonizers rather than refugees. For instance, the intrinsic rate of population increase (r/day) of high-migratory *Dysdercus fasciatus* (0.094), *D. bimaculatus* (0.106), *Oncopeltus fasciatus* (0.044) and *Tribolium castanuem* (0.128) is greater than that of their corresponding low-migratory relatives: *D. superstitiosus* (0.062), *D. mimulus* (0.066), *O. unifasciatellus* (0.034), and *T. confusum* (0.100) (see Dingle, 1979). The high rate of population increase is realized by recourse to reproduction immediately after settling, or by increasing fecundity (Dingle, 1979). Migrants are adapted to unstable habitats and are referred to as r-strategists (Dingle, 1972; Southwood, 1977).

Although migration is genetically determined, environmental factors such as temperature, photoperiod, availability of food, and stimuli from other individuals regulate initiation and termination of migration (Dingle, 1978).

Low temperature (< 23°C), shorter photoperiods (< 12 hours), and lack of food increase not only the proportion of migrants in *Oncopeltus fasciatus* population but also the duration of migration and the distance travelled by them. These factors also delay the onset of reproduction. *Oncopeltus fasciatus* escapes from the winter by undertaking prolonged migration at the cost of colonization potential. Long days and higher temperature, experienced by the North American spring and summer *O. fasciatus* populations, promote rapid population growth and enable them to take full advantage of the previously unoccupied habitats (Dingle, 1972, 1979). Impact of availability of food on migration and reproductive strategies has been well documented in the African cotton stainer bug, *Dysdercus*. Whereas *D. fasciatus* feeds on fruits of a few selected species which are highly seasonal, *D. superstitiosus* feeds on fruits of a wider range of host plants. Consequently, the former faces a greater risk of scarcity of food than the latter. Both *D. fasciatus* and *D. superstitiosus* terminate migration as soon as they encounter a suitable host plant. In the presence of food, the opportunistic *D. fasciatus* begins to reproduce early and realizes a higher fecundity than *D. superstitiosus* (see Dingle and Arora, 1973). At higher population densities, locusts such as *Schistocerca gregaria* and *Locusta migratoria* enter the gregarious phase and undertake extensive migration.

Morphogenesis into migratory winged form or macropter of several species of plant hoppers is under polygenic control (Roff, 1986). However, in some of them environmental factors such as population density and host plant quality as well as availability interact with the genotype of the insects and determine the morphogenetic pathway. The critical population density that triggers macroptery varies from species to species. For instance, a density of 40 nymphs in a small cage triggered macroptery in 75 per cent of the population of *Prokelisia dolus*, whereas a density as low as 5 nymphs per cage produced the same result in *P. marginata* (see Denno *et al.*, 1985). The migratory macropters partly compensate the disadvantage due to energetically expensive migration and shorter adult life span by feeding intermittently during migration, choosing high quality host plants, sacrificing egg size to realize maximum fecundity, reproducing continuously after colonization, and supplementing nutrient requirements for egg production by breaking down the wing muscles after arriving in a new habitat (see Denno *et al.*, 1985).

Migration, especially the duration of flight, has been shown to be under polygenic control in milkweed bugs (e.g. *Lygaeus kalmii*: Caldwell and Hegmann, 1969; *Oncopeltus fasciatus*: Dingle, 1978). Likewise, reproductive potential, as determined by various life-table characteristics such as age at first reproduction, number of clutches, and clutch size, is apparently under the control of various genes. As migration leads to a mixing of populations of different genotypes, studies on genetic correlations between migration and life-history parameters would provide greater prospects for understanding reproductive strategies of migratory insects.

7. Breeding cycles

Physical and biological features of the environment regularly vary at definite time intervals and give rise to seasonal climatic cycles. The magnitude of temporal variations in climatic factors differs at different latitudes. For instance, day length does not vary as much in tropics, as it does in the temperate regions. With sensitive monitoring systems, insects have attuned the major features of their life cycles with seasonal climatic cycles of their habitat. Climatic cycles, which are apparent even at a latitude of 8°, are known to influence important processes such as ovarian maturation and oviposition (e.g. *Nomadacris septemfasciata*: Mordue *et al.*, 1970). Most insects synchronize their breeding activities with favourable seasons to enhance the survival of their offspring; duration of the favourable season decreases with increasing latitude. Consequently, tropical insects produce several generations (multivoltine) in a year, while several temperate and polar species require several years to complete a single generation. For instance, the fleshfly *Sarcophaga bullata* breeds throughout the year and produces seven generations in the tropics (1.8°N); but in the temperate region, due to a prolonged pupal diapause, the fly is able to produce only two to three generations (Denlinger, 1978). Likewise, the tachinid fly *Anthrycia cinerea* produces a generation once in seven weeks at latitudes where the habitat temperature exceeds 25°C; at higher latitudes, with temperature of 20°C or less, intervention of an extended pupal diapause permits the fly to produce only a single generation (univoltine) in a year (Wylie, 1977).

Photoperiod, temperature, precipitation, and availability of food are some of the exogenous factors which influence the breeding cycles of insects. These factors operate through the endocrine system and regulate reproductive activities (see Highnam and Hill, 1979). In several multivoltine summer-active insects, long days initiate ovarian maturation and reproductive activities (e.g. *Pectinophora gossypiella*: Pittendrigh and Minis, 1971; *Pieris brassicae*: Danilevskii, 1965). In the spring-breeding carabid beetles, long days following short days initiate ovarian maturation (e.g. *Pterostichus oblongopunctatus*: Thiele, 1971; Paarmann, 1977). On the other hand, short days following long days are effective in the autumn-breeding beetles (e.g. *Patrobus atrorufus* and *Nebria brevicollis*: Thiele, 1969).

In the tropics, where seasonal fluctuations in temperature are more marked than changes in day length, temperature plays a major role in controlling the breeding cycles (Tauber and Tauber, 1976). A daytime temperature of over 25°C breaks pupal diapause and initiates emergence and reproductive activities in the fleshflies of Africa (Denlinger, 1979). At Jerusalem, a thermoperiod of 26 : 13°C at 14 : 10 LD breaks reproductive diapause and initiates mating behaviour in the male grasshopper, *Oedipoda miniata* (see Orshan and Penner, 1979a, b). A greater daily temperature amplitude induces gonadal dormancy in beetles inhabiting dry habitats as well as river banks, and a smaller one prevailing during the rainy season initiates gonadal maturation and reproduction (e.g. *Abacetus* spp: Paarmann, 1979). In subtropical North African winter-breeding beetles, a combination of low temperature and short photoperiod facilitates completion of gonadal maturation and beginning of reproduction (e.g. *Brosucus laevigatus*: Paarmann, 1973).

Rain induces reproductive activities in several insects. Mating songs of crickets are heard following rains. Shortly after a rainstorm, winged termites and ants undertake swarming migrations resulting in mating and founding of new colonies (see Brian, 1965). In equatorial Africa, the red locust *Nomadacris septemfasciata* lays eggs after summer rains; individuals emerging in late summer enter reproductive diapause (Norris, 1959). *Schistocerca gregaria* begins to reproduce following heavy rain when availability of food as well as moist sand for oviposition is increased (Norris, 1959).

The impact of food on breeding cycles has been well documented in the beetle *Heppodamnia convergens*. In California, facing shortage of prey after emergence in May, the beetle had to migrate to adjacent mountains and valleys and undertake diapause. Following improvement of irrigation facilities, the prey *Therioaphis maculata* (aphid) became increasingly available even in summer. Consequently, *H. convergens* skipped migration and reproductive diapause, and has become multivoltine (Hagen, 1962). Ovarian maturation of the high-altitude weevil *Acanthoscelides obtectus* depends on the availability of bean seed. The role of quality and quantity of food in terminating reproductive diapause and initiating reproduction has been described elsewhere (Section V D 4).

Intertidal midges have geared their reproductive cycles to lunar and tidal rhythms (Neumann, 1975). *Clunio marinus* (Chironomidae) is found in the Atlantic and Pacific shores from temperate to arctic zones. Larvae of *C. marinus* remain dormant during winter and begin pupation under the influence of rising temperature of late spring. Pupation follows a semilunar rhythm, taking place three to five days prior to full moon or new moon days. The semilunar rhythm in pupation facilitates emergence on full moon or new moon days when the spring tide reaches the lowest level and exposes a vast intertidal zone. The adult span of *C. marinus* lasts for two hours, within which mating and oviposition have to be completed. Synchronization of emergence with low tide guarantees a high concentration of short-lived adults and facilitates instantaneous mating and oviposition. Larvae reared in laboratory without exposure to moonlight pupate and emerge arhythmically (Neumann, 1969). Exposure of larvae to short pulses of moonlight for three days in a month or to a 12.4 hours tidal cycle of water turbulence restores emergence rhythm (Neumann, 1975). Briefly, the presence of sensitive mechanisms for receiving environmental signals and timing the periods of dormancy and reproduction has contributed largely to the success of insects.

ACKNOWLEDGEMENTS

I thank my students Dr. M. Senthamizhselvan, M/s S. Srinivasaperumal, R. Paramasivan, B. Ananthagowri and E. Pushpalatha for assistance in the preparation of the manuscript, Prof. T.J. Pandian, Madurai for encouragement and Prof. T.N. Ananthakrishnan, Director, Entomology Research Institute, Madras for his suggestions and for library facilities made available to me. I gratefully acknowledge all those who graciously granted permission to adopt figure(s) and table(s) from their publications in the original or modified form.

REFERENCES

Adams, T.S., and Gerst, J.W. (1991). 'The effect of pulse-feeding a protein diet on ovarian maturation, vitellogenin levels, and ecdysteroid titre in houseflies, *Musca domestica*, maintained on sucrose', *Inv. Reprod. Dev.*, **20**, 49–57.

Adiyodi, K.G., and Adiyodi, R.G. (1974). 'Comparative physiology of reproduction in arthropods', *Adv. comp. Physiol. Biochem.*, **5**, 37–107.

Adkisson, P.L., Bell, R.A., and Wellso, S.G. (1963). 'Environmental factors controlling the induction of diapause in the pink bollworm, *Pectinophora gossypiella* (Saunders)', *J. Insect Physiol.*, **9**, 299–310.

Adlakha, V., and Pillai, M.K.K. (1975). 'Involvement of male accessory gland substance in the fertility of mosquitoes', *J. Insect Physiol.*, **21**, 1453–1455.

Aiken, R.B., and Gibo, D.L. (1979). 'Changes in fecundity of *Drosophila melanogaster* and *D. simulans* in response to selection for competitive ability', *Oecologia*, **43**, 63–78.

Alcock, J. (1975a). 'Territorial behaviour by males of *Philanthus multimaculatus* (Hymenoptera, Sphecidae) with a review of territoriality in male sphecids', *Anim. Behav.*, **23**, 889–895.

Alcock, J. (1975b). 'Male mating strategies of some philanthine wasps (Hymenoptera: Sphecidae)', *J. Kansas Ent. Soc.*, **48**, 532–545.

Alcock, J., Eickwort, G.C., and Eickwort, K.R. (1977a). 'The reproductive behaviour of *Anthidium maculosum* (Hymenoptera: Megachilidae) and the evolutionary significance of multiple copulations by females', *Beh. Ecol. Sociobiol.*, **2**, 385–396.

Alcock, J., Jones, C.E., and Buchmann, S.L. (1977b). 'Male mating strategies in the bee *Centris pallida* Fox (Hymenoptera: Anthophoridae)', *Amer. Natur.*, **111**, 145–155.

Alcock, J., Barrows, E.M., Gordh, G., Huffard, L.J., Kiradendall, L., Pyle, D.G., Ponder, T.L., and Zalem, F.G. (1978). 'The ecology and evolution of male reproductive behaviour in the bees and wasps', *Zool. J. Linn. Soc.*, **64**, 293–326.

Alexander, G., and Hilliard, J.R. (1964). 'Life history of *Aeropedellus clavatus* (Orthoptera: Acrididae) in the Alpine tundra Colarado', *Ann. Ent. Soc. Amer.*, **57**, 310–317.

Alexander, R.D. (1975). 'Natural selection and specialized chorusing behaviour in acoustical insects', in *Insects, Science and Society* (Ed. D. Pimentel), Academic Press, New York, pp. 284–303.

Alexander, R.D., and Sherman, P.W. (1977). 'Local mate competition and parental investment in social insects', *Science*, **196**, 494–500.

Ambrose, D.P., and Livingstone, D. (1985). 'Impact of mating on adult longevity, oviposition pattern, hatchability and incubation period in *Rhinocoris marginatus*', *Env. Ecol.*, **3**, 99–102.

Ananthakrishnan, T.N., Raman, K., and Sanjayan, K.P. (1982). 'Comparative growth rate, fecundity and behavioural diversity of the dusky cotton bug, *Oxycarenus hyalinipennis* Coasta on certain malvaceous host plants', *Proc. Indian Nat. Sci. Acad.*, **48B**, 577–584.

Annadurai, R.S. (1987). 'Influence of leaf age on feeding and reproduction in *Raphidopalpa atripennis* F. (Coleoptera: Chrysomelidae). *Proc. Indian Acad. Sci.*, **96**, 207–215.

Arthur, A.P., and Wylie, H.G. (1959). 'Effects of host size on sex ratio, development time and size of *Pimpala turrionella*, L. (Hymenoptera: Ichneumonidae)', *Entomophaga*, **4**, 297–301.

Askew, R.R. (1971). *Parasitic Insects*, American Elsevier, New York.

Astaurov, B.L. (1940). *Artificial Parthenogenesis in* Bombyx mori *L.*, Akad. Nauk, U.S.S.R., Moscow.

Bacci, G. (1965). *Sex Determination*, Pergamon Press, Oxford.

Bahrmann, R. (1972). 'Untersuchungen zur Populationdynamik von *Aleyrodes assari* Schrank und *Aleurolobus asari* Wunn. (Homoptera, Aleyrodina)', *Zool. Jb. (Syst.)*, **99**, 82–106.

Bailey, G.G. (1976). 'A quantitative study of consumption and utilization of various diets in the bertha armyworm, *Mamestra configurata* (Lepidoptera: Noctuidae)', *Can. Ent.*, **108**, 1319–1326.

Bailey, C.G., and Singh, N.B. (1977). 'An energy budget for *Mamestra configurata* (Lepidoptera: Noctuiidae)', *Can. Ent.*, **109**, 687–693.

Baker, A.M., and Lambdin, P.L. (1985). 'Fecundity, fertility and longevity of mated and unmated spined soldier bug females', *J. Agr. Ent.*, **2**, 378–382.

Baker, R.R. (1966). 'A possible method of evolution of the migratory habit in butterflies', *Phil. Trans. roy. Soc. London*, **253(B)**, 309–341.

Baker, R.R. (1972). *The Evolution of Animal Migration*, English University Press, London.

Baltzi, G.O. (1974). 'Production, assimilation and accumulation of organic matter in ecosystems', *J. Theor. Biol.*, **45**, 205–217.

Barnard, D.R., and Mulla, M.S. (1977). 'Effect of photoperiod and temperature on blood feeding, oogenesis and fat body development in the mosquito, *Culiseta inornata*', *J. Insect Physiol.*, **23**, 1261–1266.

Barrett, B.A., and Brunner, J.F. (1990). 'Types of parasitoid-induced mortality, host stage preferences, and sex ratios exhibited by *Pnigalio flavipes* (Hymenoptera: Eulophidae) using *Phyllonorycter elmaella* (Lepidoptera: Gracillariidae) as a host', *Environ. Entomol.*, **19**, 803–807.

Barrows, E.W. (1976). 'Mating behaviour in halictine bees (Hymenoptera: Halictidae): I. Patrolling and age-specific behavior in males', *J. Kansas Ent. Soc.*, **49**, 105–119.

Bartlett, A.C., and Lewis, L.J. (1985). 'The initial male may contribute the majority of the sperm as in the pink bollworm, *Pectinophora gossypiella* (Lepidoptera: Gelechiidae): Reproductive and sperm use by wild-type and mutant moths', *Ann. Ent. Soc. Amer.*, **78**, 559–563.

Baust, J.G. (1972). 'Mechanism of insect freezing protection: *Pterostichus brevicornis*', *Nature, Lond.*, **236**, 219–220.

Baust, J.G., Grandee, R., Condon, G., and Morrissey, R.E. (1979). 'The diversity of overwintering strategies utilized by separate populations of gall insects', *Physiol. Zool.*, **52**, 572–580.

Baylis, M., and Pierce, N.E. (1991). 'The effect of host-plant quality on the survival of larvae and oviposition by adults of an ant-tended lycaenid butterfly, *Jalmenus evagoras*', *Ecol. Entomol.*, **16**, 1–9.

Beck, S.D. (1968). *Insect Photoperiodism*, Academic Press, London.

Beddington, J.R., Free, C.A., and Lawton, J.H. (1976a). 'Concepts of stability and resilience in predator-prey models', *J. Anim. Ecol.*, **45**, 791–816.

Beddington, J.R., Hassell, M.P., and Lawton, J.H. (1976b). 'The components of arthropod predation. II. Predator rate of increase', *J. Anim. Ecol.*, **45**, 165–185.

Bentur, J.S., and Mathad, S.B. (1973). 'Influence of pairing and grouping on the fecundity of the cricket, *Plebeiogryllus guttiventris* Walker', *Indian J. exp. Biol.*, **11**, 570–571.

Benz, G. (1969). 'Influence of mating, insemination, and other factors on oogenesis and oviposition in the moth *Zeiraphera diniana*', *J. Insect Physiol.*, **15**, 55–71.

Bessin, R.T., and Reagan, T.E. (1990). 'Fecundity of sugarcane borer (Lepidoptera: Pyralidae) as affected by larval development on graminous host plants'. *Environ. Entomol.*, **19**, 635–637.

Bhatnagar, M.S., Singh, S.P., Chandra, R., and Ganguly, S.K. (1982). 'Sterility effect of thiourea in cockroach, *Periplaneta americana* L.', *Indian J. Med. Res.*, **76**, 783–785.

Bick, G.H., and Bick, J.C. (1968). 'Demography of the damselfly, *Argia plana*', *Proc. Ent. Soc. Washington*, **70**, 197–203.

Bick, G.H., Bick, J.C., and Hornuff, L.E. (1976). 'Behaviour of *Chromagrion conditum* (Hagen) adults', *Odonatalogica*, **5**, 129–141.

Bick, G.H., and Hornuff, L.E. (1966). 'Reproductive behaviour in the damselflies *Enallagma aspersum* (Hagen) and *Enallagma exsulsans* (Hagen)', *Proc. Ent. Soc. Washington*, **68**, 78–85.

Biemont, J.C. (1975). 'Influence de modifications de courte durée de la temperature d'elevage sur la capacité reproductrice de la bruche du haircot *Acanthoscelides obtectus* (Col., Bruchidae)', *Int. Atom. Energy Ag., Wien.*, **186**, 279–288.

Blackman, R.L. (1971). 'Variations in the photoperiodic response within population of *Myzus persicae* (Sulzer) (Hem. Aphididae)', *Bull. Ent. Res.*, **60**, 533–546.

Blackman, R.L. (1974). 'Life cycle variations of *Myzus persicae* (Sulz.) (Hom., Aphididae) in different parts of the world, in relation to genotype and environment', *Bull. Ent. Res.*, **63**, 595–607.

Boggs, C.L., and Gilbert, L.E. (1979). 'Male contribution to egg production in butterflies; evidence for transfer of nutrients and at mating', *Science*, **206**, 83–84.

Boharat, G.E., Tochic, P.F., Macta, Y., and Rust, R.W. (1972). 'Notes on biology of *Emphoropsis pallida* Tiberlake', *J. Kansas Ent. Soc.*, **45**, 381–382.

Boorman, E., and Parker, G. (1976). 'Sperm (ejaculate) competition in *Drosophila melanogaster*, and the reproductive value of female to male in relation to female age and mating status', *Ecol. Entomol.*, **1**, 145–155.

Bouletreau, J. (1975). 'Influence de l'accouplement sur la physiologie reproductive des femelle de *Drosophila melanogaster* (Meig.)', Thèse Doctorat d'Etat, Lyon.

Bouletreau, M. (1976). 'Influence de la photoperiode subir par les adults sur la sex ratio de la descendance chez *Pteromalus puparum* (Hymenoptera: Chalcididae)', *Ent. Exp. App.*, **19**, 197–204.

Boyer, J.F. (1978). 'Reproductive compensation in *Tribolium castaneum*', *Evolution*, **32**, 519–528.

Bradshaw, W.E. (1969). 'Major environmental factors inducing the termination of larval diapause in *Chaoborus americanus* Johannsen (Diptera: Culicidae)', *Biol. Bull.*, **136**, 2–8.

Braman, S.K., and Yeargan, K.V. (1989). 'Reproductive strategy of *Trissolcus euschisti* (Hymenoptera: Scelionidae) under conditions of partially used host resources', *Ann. Ent. Soc. Amer.*, **82**, 172–176.

Brassilerio, V.L.F. (1982). 'Fecondidade e fertilidade da femea de *Triatoma brasiliensis* (Hemiptera: Reduvidae) 1. Influencia da copula e da longevidade', *Rev. brasil. biol.*, **42**, 1–13.

Brian, M.V. (1965). *Social Insect Population*, Academic Press, London.

Brinck, P. (1957). 'Reproductive system and mating in Ephemeroptera', *Opus. ent.*, **22**, 1–37.

Brower, J.H. (1975). 'Sperm precedence in Indian meal moth *Plodia interpunctella*', *Ann. Ent. Soc. Amer.*, **68**, 78–80.

Brower, L.P., Calvert, W.H., Hedrick, L.E., and Christian J. (1977). 'Biological observations on an overwintering colony of Monarch butterflies (*Danaus plexippus*, Danaidae) in Mexico', *J. Lepid. Soc.*, **31**, 232–242.

Brown, J.J., Jermy, T., and Butt, B.A. (1980). 'The influence of an alternate host plant on the fecundity of Colarado potato beetle, *Leptinotarsa decemlineata* (Coleoptera: Chrysomelidae)', *Ann. Ent. Soc. Amer.*, **73**, 197–199.

Brown, J.R., and Phillips, J.R. (1990). 'Diapause in *Microplitis croceipes* (Hymenoptera: Braconidae)', *Ann. Ent. Soc. Amer.*, **83**, 1125–1129.

Burnet, B., Connolly, K., Kearney, M., and Cook, R. (1973). 'Effects of male paragonial gland secretion on sexual receptivity and courtship behaviour of female *Drosophila melanogaster*', *J. Insect Physiol.*, **19**, 2421–2431.

Bursell, E. (1964a). 'Environmental aspects: temperature', in *Physiology of Insecta* (Ed. M. Rockstein), Academic Press, London, pp. 283–321.

Bursell, E. (1964b). 'Environmental aspects: humidity', in *Physiology of Insecta* (Ed. M. Rockstein), Academic Press, London, pp. 321–361.

Butler, C.G. (1964). 'Pheromones in sexual processes in insects', in *Insect Reproduction* (Ed. K.C. Highnam), Royal Entomological Society, London, pp. 66–77.

Butler, C.G. (1965). 'Sex attraction in *Andrena flavipes* Panzer (Hymenoptera: Apidae) with some observations on nest-site restriction', *Proc. roy. Ent. Soc. London*, **40A**, 77–80.

Butler, C.G., and Fairey, E.M. (1963). 'The role of the queen in preventing oogenesis in worker honey bees (*A. mellifera* L.)', *J. Apic. Res.*, **2**, 14–18.

Byers, G.W. (1969). 'Evolution of wing reduction in crane flies (Diptera:Tipulidae)', *Evolution*, **23**, 346–354.

Cairns, S.C. (1982). 'The life cycle energetics of *Rhopae verreauxi* (Coleoptera: Scarabaeidae)', *Oecologia*, **55**, 62–68.

Caldwell, R.L., and Hegmann, J.P. (1969). 'Heritability of flight duration in the milkweed bug *Lygaeus kalmii*', *Nature, Lond.*, **223**, 91–92.

Calow, P. (1977). 'Ecology, evolution and energetics; a study in metabolic adaptations', *Adv. Ecol. Res.*, **10**, 1–60.

Campanella, P.G., and Wolf, L.C. (1974). 'Temporal leks as a mating system in a temperate zone dragonfly — 1. *Plathemis lydia* (Drury)', *Behavior*, **51**, 49–87.

Campbell, A., Sing, N.B., and Sinha, R.N. (1976). 'Bioenergetics of the granary weevil *Sitophilus granarius* (L.) (Coleoptera: Curculionidae)', *Can. J. Zool.*, **54**, 786–798.

Campbell, A., and Sinha, R.N. (1978). 'Bioenergetics of granivorous beetles, *Cryptolestes ferrugineus* and *Rhyzopertha dominica* (Coleoptera: Cucujidae and Bostrichidae)', *Can. J. Zool.*, **56**, 624–633.

Carlsson, G. (1962). 'Studies on Scandinavian black flies', *Opusc. Entomol. Suppl.*, **21**, 168–186.

Carroll, S.P. (1988). 'Contrasts in reproductive ecology between temperate and tropical population of *Jadera haematoloma*, a mate-guarding hemipteran (Rhopalidae)', *Ann. Ent. Soc. Amer.*, **81**, 54–63.

Carson, H.L. (1967). 'Selection for parthenogenesis in *Drosophila mercatorum*', *Genetics*, **55**, 157–171.

Carson, H.L. (1973). 'The genetic system in parthenogenetic strains of *Drosophila mercatorum*', *Proc. natl Acad. Sci., U.S.A.*, **70**, 1772–1774.

Cassier, P. (1965). 'Interaction des effet du groupement et d'un facteur saisonnier chez *Locusta migratoria migratorioides* R & P', *Bull. Soc. Zool. Fr.*, **90**, 39–51.

Chandrakantha, J. (1985). 'Studies on seed-insect interactions: Bioenergetics and reproduction of *Callasobruchus maculatus* Fab.', Ph.D. thesis, Madurai Kamaraj University, Madurai.

Chapman, R.F. (1971). *The Insects*, 3rd Edn, Hodder and Stoughton, London.

Charnov, E.L. (1979). 'The genetical evolution of patterns of sexuality', *Amer. Natur.*, **113**, 465–480.

Charnov, E.L. (1982). *The Theory of Sex Allocation*, Princeton University Press, Princeton.

Chen, C.P., Denlinger, D.L., and Lee, R.E. (1991). 'Seasonal variation in generation time, diapause and cold hardiness in a central Ohio population of the fleshfly, *Sarcophaga bullata*', *Ecol. Entomol.*, **16**, 155–162.

Chippendale, G.M. (1978). 'The functions of carbohydrates in insect life processes', in *Biochemistry of Insects* (Ed. M. Rockstein), Academic Press, London, pp. 1–55.

Clark, A.M., and Rockstein, M. (1964). 'Aging in insects', in *The Physiology of Insects* (Ed. M. Rockstein), Academic Press, London, pp. 217–278.

Cognetti, G. (1961). 'Endomeiosis in parthenogenetic lines of aphids', *Experientia*, **17**, 168–169.

Cognetti, G., and Pagliai, A.M. (1963). 'Razze sessuali in *Brevicoryne brassicae* L. (Homoptera: Aphididae)', *Arch. Zool. Ital.*, **48**, 329–337.

Cohen, D. (1976). 'The optimal timing of reproduction', *Amer. Natur.*, **110**, 801–807.

Cook, W.C. (1967). 'Life history, host plants and migrations of the beet leaf hopper in the Western United States', *Tech. Bull. U.S. Dep. Agric.*, **1365**, 122 pp.

Coop, L.B., and Croft, B.A. (1990). 'Diapause and life history attributes of *Phytodietus vulgaris* (Hymenoptera: Ichneumonidae), a parasitoid of *Argyrotaenia citrana* (Lepidoptera: Totricidae)', *Ann. Ent. Soc. Amer.*, **83**, 1148–1151.

Corbeitt, T.S., and Hardie, J. (1985). 'Juvenile hormone effects on polymorphism in the pea aphid, *Acyrthosiphon pisum*', *Ent. Exp. Appl.*, **38**, 131–135.

Corbet, P.S. (1964). 'Autogony and oviposition in Arctic mosquitoes', *Nature, Lond.*, **203**, 669–670.

Corbet, P.S. (1980). 'Biology of Odonata', *A. Rev. Ent.*, **25**, 189–217.

Curtis, C.F., and Langley, P.A. (1972). 'Use of nitrogen chilling in the production of radiation induced sterility in the tsetse fly *Glossina morsitans*', *Ent. Exp. Appl.*, **15**, 360–376.

Cymborowski, B., and Giebultowicz, J.M. (1976). 'Effect of photoperiod on development and fecundity in the flour moth, *Ephestia kuehniella*', *J. Insect Physiol.*, **22**, 1213–1217.

Dadd, R.H. (1985). 'Nutrition: Organisms', in *Comprehensive Insect Physiology, Biochemistry, and Pharmacology*, Vol. 4 (Eds. G.A. Kerkut and L.I. Gilbert), Pergamon Press, Oxford, pp. 313–380.

Danilevskii, A.S. (1965). *Photoperiodism and Seasonal Development of Insects*, Oliver and Boyd, London.

Danilevskii, A.S., Goryshin, N.I., and Tyshchenke, V.P. (1970). 'Biological rhythms in terrestrial arthropods', *A. Rev. Ent.*, **15**, 201–244.

Danks, H.V., and Oliver, D.R. (1972). 'Diel periodicities of emergence of some high arctic Chironomidae (Diptera)', *Can. Ent.*, **105**, 903–916.

Danthanarayana, W., and Gu, H. (1991). 'Multiple mating and its effect on reproductive success of female *Epiphyas postvittana* (Lepidoptera: Tortricidae)', *Ecol. Entomol.*, **16**, 169–175.

David, H., Eswaramoorthy, S., and Subadhra, K. (1987). 'Influence of nutrition on the reproductive biology of sugarcane pests and their natural enemies', *Proc. Indian Acad. Sci. (Anim. Sci.)*, **96**, 245–251.

Davies, P.M., and Dadour, I.R. (1989). 'A cost of mating by male *Requena verticalis* (Orthoptera: Tettigoniidae)', *Ecol. Entomol.*, **14**, 467–469.

De Kort, C.A.D. (1990). 'Thirty-five years of diapause research with the Colorado potato beetle', *Ent. Exp. Appl.*, **56**, 1–13.

Delvi, M.R. (1972). 'Ecophysiological studies on the grasshopper *Poecilocerus pictus*', Ph.D. thesis, Bangalore University, Bangalore.

Delvi, M.R., and Pandian, T.J. (1971). 'Ecophysiological studies on the utilization of food in the paddy field grasshopper *Oxya velox*', *Oecologia*, **8**, 267–275.

Demetrius, L. (1975). 'Reproductive strategies and natural selection', *Amer. Natur.*, **109**, 243–249.

Dempster, J.P. (1975). *Animal Population Ecology*, Academic Press, London.

Denlinger, D.L. (1978). 'The developmental response of flesh flies (Diptera: Sarcophagidae) to tropical seasons', *Oecologia*, **35**, 105–107.

Denlinger, D.L. (1979). 'Pupal diapause in tropical flesh flies: Environmental and endocrine regulation, metabolic rate and genetic selection', *Biol. Bull.*, **156**, 31–46.

Denno, R.F., Douglass, L.W., and Jacobs, D. (1985). 'Crowding and host plant nutrition. Environmental determinants of wing-form in *Prokelisia marginata*', *Ecology*, **66**, 1588–1596.

Denno, R.F., Olmstead, K.L., and McCloud, E.S. (1989). 'Reproductive cost of flight capability: a comparison of life history traits in wing dimorphic planthoppers', *Ecol. Entomol.*, **14**, 31–44.

De Wilde, J. (1962). 'Photoperiodism in insects and mites', *A. Rev. Ent.*, **7**, 1–26.

De Wilde, J., and De Boer, J.A. (1961). 'Physiology of diapause in the adult Colorado beetle — II. Diapause as a case of pseudoallatectomy', *J. Insect Physiol.*, **6**, 152–161.

Dickinson, J.L., and Rutowski, R.L. (1989). 'The function of the mating plug in the chalcedon checkerspot butterfly', *Anim. Behav.*, **38**, 154–162.

Dimock, M.B., and Renwick, J.A.A. (1991). 'Oviposition by field populations of *Pieris rapae* (Lepidoptera: Pieridae) deterred by an extract of a wild Crucifer', *Environ. Entomol.*, **20**, 802–806.

Dingle, H. (1972). 'Migration strategies of insects', *Science*, **175**, 1327–1335.

Dingle, H. (1978). 'Migration and diapause in tropical, temperate and island milkweed bugs', in *Evolution of Insect Migration and Diapause* (Ed. H. Dingle), Springer-Verlag, New York, pp. 254–276.

Dingle, H. (1979). 'Adaptive variations in the evolution of insect migration', in *Movement of Highly Mobile Insects: Concepts and Methodology in Research* (Eds. R.L. Rabb and G.G. Kennedy), University Graphic, Raleigh, North Carolina, pp. 64–87.

Dingle, H., and Arora, G. (1973). 'Experimental studies of migration in bugs of the genus *Dysdercus*', *Oecologia*, **12**, 119–140.

Dingle, H., Brown, C.K., and Hegmann, P.G. (1977). 'The nature of genetic variance influencing photoperiodic diapause in a migrant insect, *Oncopeltus fasciatus*', *Amer. Natur.*, **111**, 1047–1059.

Dixon, A.F.G. (1959). 'An experimental study of the searching behaviour of the predatory coccinellid beetle *Adalia decempunctata* (L.)', *J. Anim. Ecol.*, **28**, 259–281.

Dixon, A.F.G. (1976). 'Reproductive strategies of the alate morphs of the bird cherry oat aphid *Rhopalosiphum padi* L.,' *J. Anim. Ecol.*, **45**, 817–830.
Dixon, A.F.G., and Glen, D.M. (1971). 'Morph determination in the bird cherry oat aphid *Rhopalosiphum padi* L.', *Ann. Appl., Biol.*, **68**, 11–21.
Donaldson, J.S., and Walter, G.H. (1991a). 'Brood sex ratios of a solitary parasitoid wasp, *Coccophagus atratus*', *Ecol. Entomol.*, **16**, 25–35.
Donaldson, J.S., and Walter, G.H. (1991b). 'Host population structure affects field sex ratios of the heteronomous hyperparasitoid, *Coccophagus atratus*', *Ecol. Entomol.*, **16**, 35–44.
Doods, K.S. (1939). 'Oogenesis in *Neuropterus baccarum* L.', *Genetica*, **21**, 177–190.
Downes, J.A. (1962). 'What is an arctic insect?', *Can. Ent.*, **94**, 143–162.
Downes, J.A. (1965). 'Adaptations of insects in the arctic', *A. Rev. Ent.*, **10**, 257–274.
Downes, J.A. (1969). 'The swarming and mating flight of Diptera', *A. Rev. Ent.*, **14**, 271–298.
Dunlap-Pianka, H.L. (1979). 'Ovarian dynamics in *Heliconius* butterflies: correlations among daily oviposition rates, egg weights, and quantitative aspects of oogenesis', *J. Insect Physiol.*, **25**, 741–749.
Dunlap-Pianka, H.L., Boggs, C., and Gilbert, L.E. (1977). 'Ovarian dynamics in heliconiine butterflies: programmed senescence versus eternal youth', *Science*, **197**, 487–490.
Eady, T.E. (1991). 'Sperm competition in *Callasobruchus maculatus* (Coleoptera: Bruchidae): a comparison of two methods used to estimate paternity', *Ecol. Entomol.*, **16**, 45–53.
Eertmoed, G.E. (1978). 'Embryonic diapause in the psocid, *Peripsocus quadrifasciatus*: photoperiod, temperature, ontogeny and geographic variation', *Physiol. Entomol.*, **3**, 197–206.
Ehrlich, A.H., and Ehrlich, P.R. (1978). 'Reproductive strategies in the butterflies: 1. Mating frequency, plugging and egg number', *J. Kansas Ent. Soc.*, **51**, 669–679.
Elliott, N.C., Gustin, R.D., and Hanson, S.L. (1990). Influence of adult diet on the reproductive biology and survival of the western corn rootworm, *Diabrotica virgifera virgifera*', *Ent. Exp. Appl.*, **56**, 15–21.
Emlen, S.T., and Oring, L.W. (1977). 'Ecology, sexual selection and evolution of mating system', *Science*, **197**, 215–223.
Engelmann, F. (1970). *The Physiology of Insect Reproduction*, Pergamon Press, New York.
Evans, H.F. (1973). A study of the predatory habits of *Anthocoris* species (Hemiptera: Heteroptera), D. Phil. thesis, University of Oxford.
Fielding, D.J. (1988). 'Photoperiodic induction of diapause in the squash bug, *Anasa tristis*', *Ent. Exp. Appl.*, **48**, 187–193.
Fisher, R.A. (1930). *The Genetical Theory of Natural Selection*, Clarendon Press, Oxford.
Fisher, R.C. (1971). 'Aspects of the physiology of endoparasitic Hymenoptera', *Biol. Rev.*, **46**, 243–278.
Fisher, R.C. (1977). 'Food conversion efficiency of a parasitic wasp, *Nemeritis canescens*', *Ecol. Entomol.*, **2**, 143–151.
Force, D.C. (1972). 'r and k-strategists in endemic host parasitoid communities', *Bull. Ent. Soc. Amer.*, **18**, 135–137.
Forrest, T.G. (1986). 'Oviposition and maternal investment in mole crickets (Orthoptera: Gryllotalpidae): Effects of season, size, and senescence', *Ann. Ent. Soc. Amer.*, **79**, 918–924.
Frank, S.A. (1986). 'Hierarchial selection theory and sex ratios. I. General solutions for structured populations', *Theor. Popul. Biol.*, **29**, 312–342.
Friend, W.G., Chey, C.T.H. and Cart Wright, E. (1965). 'The effect of nutrient intake on the development and the egg production of *Rhodnius prolixus* Stahl (Hemiptera: Reduviidae)', *Can. J. Zool.*, **43**, 891–904.
Fukuda, S. (1951). 'Production of the diapause eggs by transplanting the sub-oesophageal ganglion in the silkworm', *Proc. Imp. Acad. Japan*, **27**, 672–677.
Gaaboub, I.A., El-Sawab, S.K., and El-Latif, M.A. (1971). 'Effect of different relative humidity and temperatures on egg production and longevity of adults of *Anopheles* (*Myzomyia*) *pharoensis* Thoet', *Z. ang. Ent.*, **67**, 88–94.
Gadgil, M. (1972). 'Male dimorphism as a consequence of sexual selection', *Amer. Natur.*, **106**, 574–580.

Gadgil, M., and Bossert, W.H. (1970). 'Life historical consequences of natural selection', *Amer. Natur.*, **104**, 1–24.

Gadgil, M., and Solbrig, O.T. (1972). 'The concept of r- and K-selection: evidence from wild flowers and some theoretical consideration', *Amer. Natur.*, **106**, 14–31.

Gavin, J.A., and Williamson, J.H. (1976). 'Juvenile hormone-induced vitellogenesis in apterous, non-vitellogenic mutant in *Drosophila melanogster*', *J. Insect Physiol.*, **22**, 1737–1742.

Geden, C.J., Smith, L., Long, S.J., and Rutz, D.A (1992). 'Rapid deterioration of searching behaviour, host destruction and fecundity of the parasitoid *Muscidifurax raptor* (Hymenoptera: Pteromalidae) in culture', *Ann. Ent. Soc. Amer.*, **85**, 179–187.

Geldiay, S. (1966). 'Influence of photoperiod on imaginal diapause in *Anacridium aegyptium* L.', *Sci. Rep. Fac. Sci. Ege Univ.*, **40**, 1–19.

George, J.A., and Brown, A.W.A. (1967). 'Effect of the chemosterilant hempa on the yellow-fever mosquito and its liability to induce resistance', *J. Econ. Ent.*, **60**, 974–978.

Gerber, G.H. (1975). 'Reproductive behaviour and physiology of *Tenebrio molitor* (Coleoptera: Tenebrionidae) II. Egg development and oviposition in young females and the effects of mating', *Can. Ent.*, **107**, 551–559.

Gerber, G.H., and Church, N.S. (1973). 'Courtship and copulation in *Lytta nuttali* (Coleoptera: Meloidae)', *Can. Ent.*, **105**, 719–724.

Geyspitz, K.F. (1965). 'Photoperiodic and temperature reactions affecting the seasonal development of the pine moths *Dendrolimus pini* and *D. sibiricus* Tschetw', *Entomol. Rev.*, **44**, 316–325.

Ghiselin, M.T. (1974). *The Economy of Nature and the Evolution of Sex*, University of California Press, Berkeley.

Giesel, J.T. (1976). 'Reproductive strategies as adaptations to life in temporally heterogenous environments', *Ann. Rev. Ecol. Syst.*, **7**, 57–79.

Gilbert, L.E. (1972). 'Pollen feeding and reproductive biology of *Heliconius* butterflies', *Proc. natl Acad. Sci.*, **69**, 1407.

Gilbert, L.E. (1976). 'Postmating female odour in *Heliconius* butterflies: a male-contributed antiaphrodisiac', *Science*, **193**, 419–420.

Gillott, C., and Friedel, C. (1977). 'Fecundity-enhancing and receptivity-inhibiting substances produced by male insects: a review', in *Advances in Invertebrate Reproduction*, Vol. 1 (Eds. K.G. and R.G. Adiyodi), Peralam-Kenoth, Karivellur, pp. 199–218.

Glesener, R.R., and Tilman, D. (1978). 'Sexuality and the components of environmental uncertainty: Clues from geographic parthenogenesis in terrestrial animals', *Amer. Natur.*, **112**, 659–673.

Gordh, G. (1976). '*Geniozus alliola* Fouls, a parasite of moth larvae, with notes on other bethylids (Hymenoptera: Bethylidae, Lepidoptera: Gelechiidae)', *Agric. Res. Serv. Tech. Bull.*, **1524**, 1–27.

Gordh, G., and Evans, H.E. (1976). 'A new species of *Goniozus* Forester, 1856, imported into California from Ethiopia for biological control of pink bollworm (Hymenoptera: Bethylidae; Lepidoptera: Gelechiidae) and some notes on the taxonomic status of *Parasierola* Cameron, 1833 and *Goniozus*', *Proc. Ent. Soc. Washington*, **78**, 479–489.

Grafton-Cardwell, E.E., and Hoy, M.A. (1985). 'Short-term effects of permethrin and fenvalerate on oviposition by *Chrysopa carnea* (Neuroptera: Chrysopidae)', *J. Econ. Ent.*, **78**, 955–959.

Grison, P. (1952). 'Relations entre l'état physiologique de la fecondité du Doryphore, *Leptionotarsa decemlineata* Say', *Trans. IX Int. Congr. Entomol.*, **1**, 331–337.

Gromko, M.H., and Pyle, D.W. (1978). 'Sperm competition, male fitness and repeated mating by female *Drosophila melanogaster*', *Evolution*, **32**, 588–593.

Grover, K.K., and Pillai, M.K.K. (1970a). 'Mating ability and permanency of sterility in hempa-sterilized males of the yellow fever mosquito, *Aedes aegypti* (L.)', *J. Med. Ent.*, **7**, 198–204.

Grover, K.K., and Pillai, M.K.K. (1970b). 'The mating ability of males of *Culex pipiens fatigens* Wiedemann sterilized with apholate or tepa', *Bull. Org. Mond. Sante Bull. Wld Hlth Org.*, **42**, 807–815.

Grunin, K.Y. (1959). 'Aggregations of botfly males on the highest points in the locality and their cause', *Zool. Zh.*, **38**, 1683–1688.

Gwynne, D.T. (1984). 'Courtship feeding increases female reproductive success in bush crickets', *Nature, Lond.*, **307**, 361–363.

Hagen, K.S. (1962). 'Biology and ecology of predaceous Coccinellidae', *A. Rev. Ent.*, **7**, 289–326.

Hall, T.J., Sanders, S.M., and Cummings, M.R. (1976). 'A biochemical study of oogenesis in the housefly *Musca domestica*', *Insect Biochem.*, **6**, 13–18.

Hamilton, R.L., Cooper, R.A., and Schal, C. (1990). 'The influence of nymphal and adult dietary protein on food intake and reproduction in female brown-banded cockroaches', *Ent. Exp. Appl.*, **55**, 23–31.

Hamilton, R.L., and Schal, C. (1988). 'Effects of dietary protein levels on reproduction and food consumption in the German cockroach (Dictyoptera: Blatellidae)', *Ann. Ent. Soc. Amer.*, **81**, 976–986.

Hamilton, W.D. (1967). 'Extraordinary sex ratios: A sex ratio theory for sex linkage and inbreeding has new implication in cytogenetics and entomology', *Science*, **189**, 330–331.

Haniffa, M.A., Devaraj, M., Murugesan, A.G., and Thomas Punitham, M. (1986). 'Inhibition of pupal development and egg production in the silkworm *Bombyx mori* (L.) following applications of insecticide and secondary plant substance', *Mitt. Zool. Mus. Berl.*, **62**, 337–342.

Haniffa, M.A., and Stephen Jose, S. (1987a). 'Inhibition of egg production and fertility in the house cricket *Gryllodes sigillatus* Walker (Orthoptera, Gryllidae) following topical application of cythion', *Anz. Schädlingsk.*, **60**, 136–138.

Haniffa, M.A., and Stephen Jose, S. (1987b). 'Inhibition of egg production and fertility in the house cricket *Gryllodes sigillatus* Walker (Orthoptera: Gryllidae) following topical application of DDT', *Malaysian J. Sci.*, **9**, 41–45.

Haniffa, M.A., Thomas Punitham, M., and Arunachalam, S. (1988). 'Effect of larval nutrition on survival, growth and reproduction in the silkworm *Bombyx mori* L.', *Sericologia*, **28**, 563–575.

Hardie, J., and Lees, A.D. (1985). 'The induction of normal and tenatoid viviparae by a juvenile hormone analogue kinoprene in two species of aphids', *Physiol. Entomol.*, **10**, 65–74.

Hare, J.D. (1980). 'Impact of defoliation by the Colorado potato beetle on potato yields', *J. Econ. Ent.*, **73**, 369–373.

Harper, J.L., and Ogden, J. (1970). 'The reproductive strategy of higher plants, 1. The concept of strategy with special reference to *Senecio vulgaris* L.', *J. Ecol.*, **58**, 681–698.

Harwood, R.F., and Halfhill, E. (1964). 'The effect of photoperiod on fat body and ovarian development of *Culex tarsalis* (Diptera: Culicidae)', *Ann. Ent. Soc. Amer.*, **57**, 596–600.

Hassell, M.P., and May, R. (1973). 'Stability in insect host-parasite models', *J. Anim. Ecol.*, **42**, 693–726.

Higgins, C.H., and Myers, J.H. (1992). 'Sex ratio patterns and population dynamics of western flower thrips (Thysanoptera: Thripidae)', *Environ. Entomol.*, **21**, 322–330.

Highnam, K.C., and Hill, L. (1979). *The Comparative Endocrinology of Invertebrates*, ELBS and Edward Arnold, London.

Hinton, H.E. (1951). 'A new chironomid from Africa, the larva of which can be dehydrated without injury', *Proc. Zool. Soc. (Lond.)*, **121**, 371–380.

Hinton, H.E. (1952). 'Survival of a chironomid larvae after 20 months dehydration', *Int. Congr. Ent.*, **9**, 478–482.

Hinton, H.E. (1960). 'Cryptobiosis in the larvae of *Polypedixum vanderplanki* Hint. (Chironomidae)', *J. Insect Physiol.*, **5**, 286–300.

Hinton, H.E. (1981). *Biology of Insect Eggs*, Vol. 1, Pergamon Press, New York.

Hiratsuka, E. (1920). 'Researches on the nutrition of the silkworm', *Bull. Ser. Exp. Sta. Tokyo*, **1**, 257–315.

Hodek, I. (1962). 'Experimental influencing of the imaginal diapause in *Coccinella septempunctata* L. (Col., Coccinellidae)', *Acta Soc. Cechoslov.*, **59**, 297–313.

Hodek, I. (1967). 'Bionomics and ecology of predaceous Coccinellidae', *A. Rev. Ent.*, **12**, 79–104.

Hodek, I. (1973). *Biology of Coccinellidae*, Academic Press, Prague.

Hodek, I., and Honek, A. (1970). 'Incidence of diapause in *Aelia acuminata* (L.) populations from south west Slovakia', *Věst. čsl. zool. Spol.*, **33**, 170–183.

Hoelscher, C.E., and Vinson, S.B. (1971). 'The sex ratio of a hymenopterous parasitoid, *Campoletis perdistinctus* as affected by photoperiod, mating and temperature', *Ann. Ent. Soc. Amer.*, **64**, 1373–1376.

Hoffmann, K.H. (1974). 'Wirkung von konstanten und tagesperiodisch alternierend Temperaturen auf Lebensdauer, Nahrungsverwertung und Fertilitat adulter *Gryllus bimaculatus*', *Oecologia*, **17**, 39–54.

Holdsworth, R.P. (1941). 'The life history and growth of *Pteronarcys proteus* Newman', *Ann. Ent. Soc. Amer.*, **34**, 495–502.

Holt, G.G., and North, D.T. (1970). 'Effect of gamma irradiation on the mechanisms of sperm transfer in *Trichoplusia ni*', *J. Insect Physiol.*, **16**, 2211–2222.

Holter, P. (1979). 'Abundance and reproductive strategy of the dung beetle *Aphodius rufipes* (Scarabaeidae)', *Ecol. Entomol.*, **4**, 317–326.

Hopkins, A.D. (1909). 'The genus *Dendroctonus*', *U.S. Bur. Ent. Tech. Ser.*, **17**, 148–164.

Huignard, J., and Biemont, J.C. (1979). 'Vitellogensis in *Acanthoscelides obtectus* (Coleoptera: Bruchidae). II—The conditions of vitellogenesis in a strain from Colombia, comparative study and adaptive significance', *Int. J. Inv. Reprod.*, **1**, 233–244.

Hunter, M.S. (1989). 'Sex allocation and egg distribution in an autoparasitoid, *Encarsia pergandiella*, Hymenoptera:Aphelinidae', *Ecol. Entomol.*, **14**, 57–67.

Hurlbutt, B.L. (1987). 'Sexual size dimorphism in parasitoid wasps', *Biol. J. Linn. Soc.*, **30**, 63–89.

Hynes, H.B.N. (1970). *The Ecology of Running Waters*, Liverpool University Press, Liverpool.

Hynes, H.B.N. (1976). 'Biology of Plecoptera', *A. Rev. Ent.*, **21**, 135–153.

Ibrahim, Y.B., and Knowles, C.O. (1986). 'Influence of formamidines on reproduction in two spotted spider mite (Acari: Tetramychidae)', *J. Econ. Ent.*, **79**, 7–14.

Illies, J. (1959). 'Retardierte Schlupfzeit von *Baetis* Gelegen (Ins., Ephem.)', *Naturwissenschaften*, **46**, 119–120.

Istock, C.A. (1981). 'Natural selection and life history variation theory plus lessons from a mosquito', in *Insect Life History Patterns* (Eds. R.F. Denno and H. Dingle), Springer-Verlag, New York, pp. 113–127.

Jackson, C.H.N. (1944). 'The analysis of a tsetse population', *Ann. Eugen.*, **12**, 176–205.

Jackson, D.J. (1966). 'Observations on the biology of *Caraphractus cinctus* Walker (Hymenoptera Mymaridae), a parasitoid of the eggs of Bytiscidae (Coleoptera), III. The adult life and sex ratio', *Trans. Roy. Ent. Soc. London*, **118**, 23–49.

Jennings, J.B., and Calow, P. (1975). 'The relationship between high fecundity and the evolution of entoparasitism', *Oecologia*, **21**, 109–115.

Johannsson, O.E. (1978). 'Coexistence of larval Zygoptera common to the Norfolk Broads Cu.K. II. Temporal and spatial separation', *Oecologia*, **32**, 303–321.

Johnson, C. (1968). 'Seasonal ecology of the dragonfly *Oplonaeschna armata* Hagen', *Am. Midland Naturalist*, **80**, 449–457.

Johnson, M.D. (1988). 'The relationship of provision weight to adult weight and sex ratio in the solitary bee, *Ceratina calcarata*', *Ecol. Entomol.*, **13**, 165–170.

Jones, J.C. (1973a). 'A study of the fecundity of male *Aedes aegypti*', *J. Insect Physiol.*, **19**, 435–439.

Jones, J.C. (1973b). 'Are mosquitoes monogamous?', *Nature, Lond.*, **242**, 343–344.

Jones, J.C. (1974). 'Sexual activities during single and multiple co-habitations in *Aedes aegypti* mosquitoes', *J. Ent.*, **48(A)**, 185–191.

Jones, V.P., and Parrella, M.P. (1984). 'The sublethal effects of selected insecticides on life table parameters of *Panonychus citri* (Acari: Tetranychidae)', *Can. Ent.*, **116**, 1033–1040.

Jordan, A.M. (1972). 'The inseminating potential of male *Glossina austeni* Newst. and *G. morsitans morsitans* Westw. (Dipt., Glossinidae)', *Bull. Ent. Res.*, **62**, 319–325.

Jurzitza, G. (1974). 'A note on mating and oviposition behaviour of three Argentine Libellulidae', *Odonatologica*, **3**, 265–266.

Kaiser, E.W. (1961). 'Studier over do danske *Sialis*- arter II. Biologien hos *S. fuliginosa* Pict. og *S. nigripes* Ed. Pect', *Flora Fauna*, **67**, 74–96.

Kaiser, H. (1974a). 'Intraspezifische Aggression und reumlictic Verteilung bei der Libell *Onychogomphus forcipatus*', *Oecologia*, **15**, 223–234.

Kaiser, H. (1974b). 'Vertralteus Gefüge und temperialverhalten der Libell *Aeschna cyanea*', *Z. Tierpsychol.*, **34**, 398–429.

Kalyanam, N.P., Gurumani, N., Ayadurai, M., and Thangavelu, P. (1972). 'Study of sex-ratio in the natural population of *Pheropsophus hilaris* Fabr. (Carabidae: Coleoptera)', *Curr. Sci.*, **41**, 336–337.

Kamp, J.W. (1970). 'The cavernicolous Gryllobattoidea of the western United States', *Ann. Spelcol.*, **25**, 223–230.

Karanavar, G.K. (1972). 'Mating behaviour and fecundity in *Trogoderma granarium* (Coleoptera: Dermestidae)', *J. Stored Prod. Res.*, **8**, 65–69.

Kasule, F.K. (1986). 'Repetitive mating and female fitness in *Dysdercus cardinalis* (Hemiptera: Pyrrhocoridae)', *Zool. J. Linn. Soc.*, **88**, 191–199.

Kaufmann, T. (1969). 'Life history of *Upis ceramboides* at Fairbanks, Alaska', *Ann. Ent. Soc. Amer.*, **62**, 922–923.

Kaufmann, T. (1971). 'Hibernation in the arctic beetle, *Pterostichus brevicornis* in Alaska', *J. Kansas Ent. Soc.*, **44**, 81–92.

Kawada, K. (1990). 'Example of parthenogenesis among insect species of various orders', in *Advances in Invertebrate Reproduction*, Vol: 5 (Eds. M. Hoshi and O. Yamashita), Elsevier, Amsterdam, pp. 339–341.

Kehat, K., and Gordon, D. (1977). 'Mating ability, longevity and fecundity of the spiny bollworm, *Earias insulana* (Lepidoptera: Noctuidae)', *Ent. Exp. Appl.*, **22**, 267–273.

Kfir, R. (1991). 'Effect of diapause on development and reproduction of the stem borer *Busseola fusca* (Lepidoptera: Noctuidae) and *Chilo partellus* (Lepidoptera: Pyralidae)', *J. Econ. Ent.*, **84**, 1677–1680.

Khoo, S.G. (1964). 'Studies on the biology of stone flies', Ph.D. thesis, University of Liverpool, Liverpool.

Kibman, L.B., and Yeargan, K.V. (1985). 'Development and the reproduction of the predator *Orius insidiosus* (Hemiptera: Anthochoridae) reared on diets of selected plant material and arthropod prey', *Ann. Ent. Soc. Amer.*, **78**, 464–467.

Kind, T.V. (1965). 'Neurosecretion and voltinism in *Orgyia antigua* L. (Lepidoptera, Lymantridae)', *Ent. Obozr.*, **44**, 534–536.

King, B.H. (1987). 'Offspring sex ratios in parasitoid wasps', *Q. Rev. Biol.*, **62**, 367–396.

King, P.E. (1961). 'A possible method of sex-ratio determination in the parasitic hymenopteran, *Nasonia vitripennis*', *Nature, Lond.*, **189**, 330–331.

Kogure, M. (1933). 'The influence of light and temperature on certain characters of the silkworm *Bombyx mori*', *J. Dept. Agr. Kyushu Univ.*, **4**, 1–93.

Kolmp, H. (1966). 'The dynamics of the field population of the pine looper *Bupalus piniarius* L. (Lep., Geom.)', *Adv. Ecol. Res.*, **3**, 207–303.

Kopelman, A.H., and Chabora, P.C. (1992). 'Resource variability and life history parameters of *Leptopilina boulardi* (Hymenoptera: Eucoilidae)', *Ann. Ent. Soc. Amer.*, **85**, 195–199.

Krishna, S.S. (1987). 'Nutritional modulation of reproduction in two phytophagous insect pests', *Proc. Indian Acad. Sci. (Anim. Sci.)*, **96**(3), 153–169.

Krishna, S.S., and Misra, M.P. (1992). 'Ethophysiological studies on *Epiricania melanoleuca* (Fletcher) in relation to parasitization of *Pyrilla perpusilla* (Walker)', in *Emerging Trends in Biological Control of Phytophagous Insects* (Ed. T.N. Ananthakrishnan), Oxford & IBH, New Delhi, pp. 27–36.

Krishna, S.S., and Narain, A.S. (1976). 'Ovipositional programming in the rice moth, *Corcyra cephalonica* (Stainton) (Lepidoptera: Gelechidae) in relation to cetain extrinsic and intrinsic cues', *Proc. Indian Nat. Sci. Acad.*, **42B**, 325–332.

Krishna, S.S., and Thorsteinson, A.J. (1972). 'Ovarian development of *Melanoplus sanguinipes* (Fab.) (Acrididae: Orthoptera) in relation to utilization of water soluble food protein', *J. Insect Physiol.*, **50**, 1319–1324.

Krishna, S.S., Vishwaparemi, K.K.C., and Shahi, K.P. (1971). 'Studies on the reproduction in *Earias fabia* Stoll (Lepidoptera: Noctuidae): Oviposition in relation to adult nutrition, mating and some environmental factors', *Entomon*, **2**, 11–16.

Kunkel, J.G., and Nordin, J.H. (1985). 'Yolk proteins', in *Comprehensive Insect Physiology, Biochemistry, and Pharmacology*, Vol. I (Eds. G.A. Kerkut and L.I. Gilbert), Pergamon Press, Oxford, pp. 84–111.

Kuris, A.M. (1978). 'Trophic interactions; similarity of parasitic castrators to parasitoids', *Q. Rev. Biol.*, **49**, 129–148.

Labeyrie, V. (1964). 'Action selective de la frequence de l'hoto utilisable (*Acrolepiopsis assectella* Z.) sur *Diadromus pulchellus* Wsn. variabilité de la fecondité, en function de l'intensité de la stimulation', *C. R. Acad. Sci., Paris*, **259**, 3644–3647.

Labeyrie, V. (1978a). The significance of the environment in the control of insect fecundity', *A. Rev. Ent.*, **23**, 69–89.

Labeyrie, V. (1978b). 'Reproduction of insects and coevolution of insects in plants', *Ent. Exp. Appl.*, **24**, 496–504.

Labine, P.A. (1966). 'The population biology of the butterfly *Euphdryas editha* IV. Sperm precedence, a preliminary report', *Evolution*, **20**, 580–586.

Lamb, R.Y., and Willey, R.B. (1989). 'Parthenogenetic mechanism and its evolutionary potential in the cave cricket *Euhadenoecus insolitus* (Orthoptera: Rhaphidophoridae)', *Ann. Ent. Soc. Amer.*, **82**, 101–108.

Landhal, J.T., and Root, R.B. (1969). 'Differences in the life tables of tropical and temperate milkweed bugs, genus *Oncopeltus* (Hemiptera: Lygaeidae)', *Ecology*, **50**, 734–737.

Lang, C.A. (1956). 'The influence of mating on egg production by *Aedes aegypti*', *Am. J. Trop. Med. Hyg.*, **5**, 909–914.

Lang, J.T. (1978). 'Relationship of fecundity to the nutritional quality of larval and adult diets of *Wyeemyea smithii*', *Mosquito News*, **38**, 396–403.

Langley, P.A. (1977). 'Physiology of tsetse flies (*Glossina* spp.) (Diptera: Glossinidae): a review', *Bull. Ent. Res.*, **67**, 523–574.

Lanier, G.N., and Wood, D.L. (1968). 'Controlled mating karyology, morphology and sex ratio in the *Dendroctonus penderosae* complex', *Ann. Ent. Soc. Amer.*, **61**, 518–525.

Lauscher, M. (1961). 'Social control of polymorphism in termites', *Symp. Roy. Ent. Soc. London*, **1**, 57–67.

Lawrence, W.S. (1990). 'Effect of body size and repeated matings on female milkweed beetle's (Coleoptera: Cerambycidae) reproductive success', *Ann. Ent. Soc. Amer.*, **83**, 1096–1100.

Lawton, J.H., Hassell, M.P., and Beedington, J.R. (1975). 'Prey death rates and rate of increase of arthropod predator population', *Nature, Lond.*, **255**, 60–62.

Lea, A.O. (1964). 'Studies on the dietary and endocrine regulation of autogenous reproduction in *Aedes taeniorhynchus* (Wied.)', *J. Med. Ent.*, **1**, 40–44.

Lea, A.O., Breigel, H., and Lea, H.M. (1978). 'Arrest, resorption or maturation of oocytes in *Aedes aegypti*: dependence on the quantity of blood and the interval between blood meals', *Physiol. Entomol.*, **3**, 309–316.

Leepla, N.C. (1976). 'Circadian rhythms of locomotion and reproductive behaviour in adult velvetbean caterpillars', *Ann. Ent. Soc. Amer.*, **69**, 45–48.

Lees, A.D. (1965). 'Is there a circadian component in the *Megoura* photoperiodic clock?', in *Circadian Clocks* (Ed. J. Aschoff), North-Holland, Amsterdam, pp. 351–356.

Lees, A.D. (1966). 'The control of polymorphism in aphids', *Adv. Insect Physiol.*, **3**, 207–277.

Lees, A.D. (1989). 'The photoperiodic responses and phenology of an English strain of the pea aphid *Acyrthosiphon pisum*', *Ecol. Entomol.*, **14**, 69–78.

Legner, E.F., and Warkentin, R.W. (1988). 'Parasitization of *Goniozus legneri* (Hymenoptera: Bethylidae) at increasing parasite and host *Amyelois transitella* (Lepidoptera: Phycitidae) densities', *Ann. Ent. Soc. Amer.*, **81**, 744–776.

Leopold, R.A. (1976). 'The role of male accessory glands in insect reproduction', *A. Rev. Ent.*, **21**, 199–221.

Levine, E., Oloumi-Sadeghi, H., and Fisher, J.R. (1992). 'Discovery of multiyear diapause in Illinois and South Dakota Northern corn rootworm (Coleoptera: Chrysomelidae) eggs and incidence of the prolonged diapause trait in Illinois', *J. Econ. Ent.*, **85**, 262–267.

Lewontin, R.C. (1965). 'Selection for colonizing ability', in *The Genetics of Colonizing Species* (Eds. H.G. Baker and G.L. Stebbins), Academic Press, New York, pp. 77–91.

Lin, N., and Michener, C.D. (1972). 'Evolution of sociality in insects', *Q. Rev. Biol.*, **47**, 131–159.

Linsley, E.G., Macswain, J.W., and Raven, P.H. (1963a). 'Comparative behaviour of bees and Onagraceae, 1. Oenothera bees of the Colarado desert', *Univ. Calif. Publ. Ent.*, **33**, 59–98.

Linsley, E.G., Macswain, J.W., and Raven, P.H. (1963b). 'Comparative behaviour of bees and Onagraceae, 2. Oenothera bees of the Great Basin', *Univ. Calif. Publ. Ent.*, **33**, 25–58.

Linsley, J.R., and Hind, M.J. (1975). 'Quantity of the male ejaculate influenced by female unreceptivity in *Culicoides melleus* (Diptera)', *J. Insect Physiol.*, **21**, 281–285.

Liu Shu-Sheng, and Hughes, R.D. (1987). 'The infuence of temperature and photoperiod on the development, survival and reproduction of the sowthistle aphid, *Hyperomyzus lactucae*', *Ent. Exp. Appl.*, **43**, 31–38.

Llewllyn, M., and Qureshi, A.L. (1978). 'The energetics and growth efficiency of *Aphis fabae* Scop reared on different parts of the broad bean plant (*Vicia faba*)', *Ent. Exp. Appl.*, **23**, 26–39.

Lokki, J., Saura, A., Lankinen, P., and Suomalainen, E. (1976). 'Genetic polymorphism and evolution in parthenogenetic animals: VI. Diploid and triploid *Polydrosus mollis* (Coleoptera: Curculionidae)', *Herditas*, **82**, 209–216.

Lubzense, E., Moshitzky, P., and Applebaum, S.W. (1981). 'Active vitellogenosis in precocene treated *Locusta migratoria*', *Gen. Comp. Endocr.*, **43**, 178–183.

Lutz, P.E. (1974). 'Environmental factors controlling duration of larval instars in *Tetragoneuria cynosura*', *Ecology*, **55**, 630–637.

Lyons, L.A. (1976). 'Mating ability in *Neodiprion sertifer* (Hymenoptera: Diprionidae)', *Can. Ent.*, **108**, 321–326.

MacArthur, R.H., and Wilson, E.O. (1967). *The Theory of Island Biogeography*, Princeton University Press, Princeton, New Jersey.

Mackey, A.P. (1978). 'Growth and bioenergetics of the moth *Cyclophragma leucosticta* Grumberg', *Oecologia*, **32**, 367–376.

MacLean, S.F., Jr. (1973). 'Life cycle and growth energetics of the arctic crane fly *Pedicia hannai antenatta*', *Oikos*, **24**, 436–443.

MacLean, S.F., Jr. (1975). 'Ecological adaptations of tundra invertebrates', in *Physiological Adaptations to the Environment* (Ed. F.J. Vernberg), Intext Educational Publishers, New York, pp. 269–300.

Magnarelli, L.A., and Anderson, J.F. (1979). 'Oogenesis and oviposition in *Chrysops atlanticus* (Diptera: Tabanidae)', *Ann. Ent. Soc. Amer.*, **72**, 350–352.

Magnarelli, L.A., and Stoffolano, J.G. (1980). 'Blood feeding, oogenesis and oviposition by *Tabanus nigrovittatus* in the laboratory', *Ann. Ent. Soc. Amer.*, **73**, 14–17.

Maier, C.T. (1978). 'The immature stages of biology of *Mallota posticata* (Fabricius) (Diptera: Syrphidae)', *Proc. Ent. Soc. Washington*, **80**, 424–440.

Maier, C.T., and Waldbauer, G.P. (1979). 'Dual mate-seeking strategies in male syrphid flies (Diptera: Syrphidae)', *Ann. Ent. Soc. Amer.*, **72**, 54–61.

Marian, M.P., Pandian, T.J., and Muthukrishnan, J. (1982). 'Energy balance in *Sceliphron violaceum* (Hymenoptera) and use of meconium weight as an index of bioenergetics components', *Oecologia*, **55**, 264–267.

Markow, T.A. (1985). 'A comparative investigation of the mating system of *Drosophila hydei*', *Anim. Behav.*, **33**, 775–781.

Masaki, S., and Ohmachi, F. (1967). 'Divergence of photoperiodic response and hybrid development in *Teleogryllus* (Orthoptera: Gryllidae)', *Kontyu*, **35**, 83–105.

Mathavan, S. (1975). 'Ecophysiological studies in chosen insects', Ph.D. thesis, Madurai University, Madurai, India.

Matthey, R. (1941). 'Étude biologique et cytologique de *Saga peda* Pallas (Orthoptera: Tettigoniidae)', *Rev. Suisse Zool.*, **48**, 91–142.

Mau, R.F.L., and Mitchell, W.C. (1978). 'Development and reproduction of the oriental stink bug, *Plautia stali* (Hemiptera: Pentatomidae)', *Ann. Ent. Soc. Amer.*, **71**, 756–757.

Maynard Smith, J., and Parker, G.A. (1976). 'The logic of asymmetric contests', *Anim. Behav.*, **24**, 159–175.

Maynard Smith, J., and Price, C.R. (1973). 'The logic of animal conflict', *Nature, Lond.*, **246**, 15–88.

Mclain, D.D., Lanier, D.L., and Marsh, N.B. (1990). 'Effects of female size, male size, and number of copulations on fecundity, fertility, longevity of *Nezara viridula* (Hemiptera: Pentatomidae)', *Ann. Ent. Soc. Amer.*, **83**, 1130–1136.

Montgomery, M.E. (1982). 'Life-cycle nitrogen budget for the gypsy moth *Lymantria dispar* reared on artificial diet', *J. Insect Physiol.*, **28**, 437–442.

Moobola, S.M., and Cupp, E.W. (1978). 'Ovarian development in the stable fly *Stomoxys calcitrans* in relation to diet and juvenile hormone control', *Physiol. Entomol.*, **3**, 317–321.

Mordue, W., Highnam, K.C., Hill, L., and Limtz, A.J. (1970). 'Environmental effects upon endocrine-mediated processes in locusts', in *Hormones and Environment* (Eds. G.K. Benson and J.G. Philipps), Cambridge University Press, Cambridge, pp. 111–136.

Morrisey, R.E., and Baust, J.G. (1976). 'The ontogeny of cold tolerance in the gall fly, *Eurosta solidagensis*', *J. Insect Physiol.*, **22**, 431–438.

Morse, J.G., and Zareh, N. (1991). 'Pesticide induced homoligosis of citrus thrips (Thysanoptera: Thripidae) fecundity', *J. Econ. Ent.*, **84**, 1169–1174.

Mukerji, M.K., and Guppy, J.C. (1970). 'A quantitative study of food consumption and growth in *Pseudoletia unipuncta* (Lepidoptera: Noctuidae)', *Can. Ent.*, **102**, 1179–1188.

Müller, H.J. (1970). 'Forman der Dormanz bei Inseketen', *Nova Acta Leopold*, **35**, 7–27.

Murdoch, W.W. (1966). 'Population stability and life history phenomena', *Amer. Natur.*, **100**, 5–11.

Muthukrishnan, J. (1987). 'Effect of temperature on the energy balance of *Mantis religiosa* (Dictyoptera: Mantidae)', *J. Sing. natl Acad. Sci.*, **16**, 51–54.

Muthukrishnan, J., Mathavan, S., and Navarathina Jothi, V. (1978). 'Effects of the restriction of feeding duration on food utilisation, emergence and silk production in *Bombyx mori* (Lepidoptera: Bombycidae)', *Monit. Zool. Ital.*, **12**, 87–94.

Muthukrishnan, J., and Pandian, T.J. (1983). 'Effect of temperature on growth and bioenergetics of a tropical moth', *J. therm. Biol.*, **8**, 361–367.

Muthukrishnan, J., and Pandian, T.J. (1984). 'Effects of interaction of ration and temperature on growth and bioenergetics of *Achaea janata* Linnaeus (Lepidoptera: Noctuidae)', *Oecologia*, **62**, 272–278.

Muthukrishnan, J., and Pandian, T.J. (1987a). 'Insecta', in *Animal Energetics*, Vol. I (Eds. T.J. Pandian and F.J. Vernberg), Academic Press, New York, pp. 373–511.

Muthukrishnan, J., and Pandian, T.J. (1987b). 'Relation between feeding and egg production in insects', *Proc. Indian Acad. Sci. (Anim. Sci.)*, **96**, 171–179.

Muthukrishnan, J., and Sen Selvan, M. (1993). 'Fertilization affects leaf consumption and utilization by *Porthesia scintillans* Walker (Lepidoptera: Lymantridae)', *Ann. Entomol. Soc. Amer.*, **86**, 173–178.

Nadel, H., and Luck, R.F. (1985). 'Span of female emergence and male sperm depletion in the female-biased quasi-gregarious parasitoid, *Pachycrepoideus vindemiae* (Hymenoptera: Pteromalidae)', *Ann. Ent. Soc. Amer.*, **78**, 410–414.

Nadganda, D., and Pitre, H. (1983). 'Development, fecundity and longevity of the tobacco budworm (Lepidoptera: Noctuidae) fed soybean, cotton and artificial diet at three temperatures', *Environ. Entomol.*, **12**, 582–586.

Nakano, S. (1985). 'Effect of interspecific mating on female fitness in two closely related ladybirds (*Henosepilachna*)', *Kontyu*, **53**, 112–119.

Naranjo, S.E., Gibson, R.L., and Walgenback, D.D. (1990). 'Development, survival, and reproduction of *Scymnus frontalis* (Coleoptera: Coccinellidae), an imported predator of Russian wheat aphid at four fluctuating temperatures', *Ann. Ent. Soc. Amer.*, **83**, 527–531.

Narbel-Hofstetter, M. (1963). 'Cytologie de pseudogamie chez *Luffia lapidella* Gooze (Lepidoptera: Psychidae)', *Chromosoma*, **13**, 623–645.

Nielson, W.T.A. (1975). 'Fecundity of virgin and mated apple maggot (Diptera: Tephritidae) females confined with apple and black ceresin wax domes', *Can. Ent.*, **107**, 909–911.

Neumann, D. (1969). 'Die kombination verschiedener endogener Rhythmen bei der zeitlichen Programmierung van Entwicklung und Verhalten', *Oecologia*, **3**, 166–183.

Neumann, D. (1975). 'Lunar and tidal rhythms in the development of an intertidal organism', in *Physiological Adaptation to the Environment* (Ed. F.J. Vernberg), Intext Educational Publishers, New York, pp. 451–463.

Nielsen, A. (1942). 'Über die Entwicklung und Biologie der Trichopteran', *Arch. Hydrobiol., Suppl.*, **17**, 255–631.

Nilakhe, S.S. (1976). 'Over-wintering survival, fecundity and mating behaviour of the rice stink bug', *Ann. Ent. Soc. Amer.*, **70**, 717–720.

Nilakhe, S.S. (1977). 'Logevity and fecundity of female boll weevil placed with varying numbers of males', *Ann. Ent. Soc. Amer.*, **70**, 673–674.

Norris, M.J. (1950). 'Reproduction in the African migratory locust (*Locusta migratoria migratorioides* R. VF.) in relation to density and phase', *Anti-Locust Bull.*, **6**, 1–48.

Norris, M.J. (1952). 'Reproduction in the desert locust (*Schistocerca gregaria* Forskal) in relation to density and phase', *Anti-Locust Bull.*, **13**, 1–49.

Norris, M.J. (1959). 'Reproduction in the red locust (*Nomadacris septemfasciata* Serville) in the laboratory', *Anti-Locust Bull.*, **36**, 1–46.

Nozato, K. (1969). 'Biology of *Itoplectis cristatae* (Hymenoptera: Ichneumonidae), a pupal parasite of the pine-sheet moth in Japan', *Kontyu*, **37**, 134–146.

Numata, H., and Yamamoto, K. (1990). 'Feeding on seeds induces diapause in the cabbage bug *Eurydema rugosa*', *Ent. Exp. Appl.*, **57**, 281–284.

Okelo, O. (1979). 'Influence of male presence on clutch size in *Schistocerca vaga* Scudder (Orthoptera: Acrididae)', *Int. J. Inv. Reprod.*, **1**, 317–322.

Oldiges, H. (1959). 'Der Einfluss der Temperatur auf Stoffwechsel und Eiproduction von Lepidoptera', *Z. Angew. Entomol.*, **44**, 115–166.

Opp, S.B., Ziegner, J., Bui, N., and Prokopy, R.J. (1990). 'Factors influencing estimates of sperm competition in *Rhagoletis pomonella* (Walsh) (Diptera: Tephritidae)', *Ann. Ent. Soc. Amer.*, **83**, 521–526.

Orians, G.H., Gates, R.G., Rhoades, D.F., and Schultz, J.C. (1974). 'Producer-consumer interaction', *Proc. Int. Congr. Ecol.*, pp. 213–217.

Orshan, L., and Penner, M.P. (1979a). 'Termination and reinduction of reproductive diapause by photoperiod and temperature in males of the grasshopper, *Oedipoda miniata*', *Physiol. Entomol.*, **4**, 55–61.

Orshan, L., and Penner, M.P. (1979b). 'Repeated reversal of the reproductive diapause by photoperiod and temperature in males of the grasshopper, *Oedipoda miniata*', *Ent. Exp. App.*, **25**, 219–226.

Otte, D., and Joern, A. (1975). 'Insect territoriality and its evolution: Population studies of desert grasshoppers on creosote bushes', *J. Anim. Ecol.*, **44**, 29–54.

Ottens, R.J., and Todd, J.W. (1979). 'Effect of host plants on the fecundity, longevity, and oviposition rate of a whitefringed beetle', *Ann. Ent. Soc. Amer.*, **72**, 837–839.

Paarmann, W. (1973). 'Bedeutung der Larvenstadien fur die Fortpflanzungs rhythmik der Laufkafer *Broscus laevigatus* Dej. und *Orthomus atlanticus* Fairm (Col., Carab.) aus Nordafrika', *Oecologia*, **15**, 87–92.

Paarmann, W. (1977). 'Propagation rhythm of subtropical and tropical Carabidae (Coleoptera) and its control by exogenous factors', in *Advances in Invertebrate Reproduction* (Eds. K.G. and R.G. Adiyodi), Peralam-Kenoth, Karivellur, pp. 49–60.

Paarmann, W. (1979). 'Ideas about the evolution of the various annual reproduction rhythms in carabid beetles of the different climatic zones', in *On the Evolution of Behaviour in Carabid Beetles* (Eds. P.J. Den Boer, H.U. Thiele, and F. Weber), H. Veenman and Zonen, B.V. Wageningen, pp. 119–132.

Padghan, D.E. (1983). 'Flight fuels in the brown planthopper, *Nilaparvata lugens*', *J. Insect Physiol.*, **29**, 95–99.

Page, R.E., Jr. (1986). 'Sperm utilization in social insects', *A. Rev. Ent.*, **31**, 297–320.

Pair, S.D., Laster, M.L., and Martin, D.F. (1977). 'Hybrid sterility of the tobacco budworm; effects of alternate sterile and normal matings on fecundity and fertility', *Ann. Ent. Soc. Amer.*, **70**, 952–954.

Pajin, H.R., and Sukeska Sood (1974). 'Effect of pea pollen feeding on maturation and copulation in the beetle, *Bruchus pisorum*', *Indian J. exp. Biol.*, **13**, 202–203.

Pak, G.A., and Oatman, E.R. (1982). 'Biology of *Trichogramma brevicapillum*', *Ent. Exp. Appl.*, **32**, 61–67.

Palavesam, A., and Muthukrishnan, J. (1992). 'Influence of food quality and temperature on fecundity of *Kiefferlulus barbitarsis* (Kieffer) (Diptera: Chironomidae)', *J. Aquatic Insects*, **14**, 145–152.

Palm, N.B. (1948). 'Normal and pathological histology of the ovaries in *Bombus* Latr. (Hymenoptera)', *Opusc. Entomol. Suppl.*, **7**, 1–101.

Park, T. (1933). 'Studies in population physiology. 2. Factors regulating initial growth of *Tribolium confusum* populations', *J. Exp. Zool.*, **65**, 17–42.

Parker, E.D., and Orzack, S.H. (1985). 'Genetic variation for the sex ratio in *Nasonia vitripennis*', *Genetics*, **110**, 93–105.

Parker, G.A. (1970a). 'Sperm competition and its evolutionary consequences in the insects', *Biol. Rev.*, **45**, 525–567.

Parker, G.A. (1970b). 'Sperm competition and its evolutionary effect on copula duration in the fly *Scatophaga sterocoraria*', *J. Insect Physiol.*, **16**, 1301–1328.

Parker, G.A. (1972). 'Reproductive behaviour of *Sepsis cyniprea* (L.) (Diptera: Sepsidae). II. The significance of the precopulatory passive emigration', *Behavior*, **41**, 242–250.

Parker, G.A. (1974a). 'Assessment strategy and the evolution of animal conflicts', *J. Theor. Biol.*, **47**, 223–243.

Parker, G.A. (1974b). 'The reproductive behaviour and the nature of sexual selection in *Scatophaga sterocoraria* L. IX. Spatial distribution of fertilisation rates and evolution of male search strategies within reproductive area', *Evolution*, **28**, 39–108.

Parker, G.A., (1978a). 'Evolution of competitive mate searching', *A. Rev. Ent.*, **23**, 173–196.

Parker, G.A. (1978b). 'Searching for mates', in *Behavioural Ecology: an Evolutionary Approach* (Eds. J.R. Krebs and N.B. Davies), Blackwells, Oxford, pp. 214–244.

Parker, G.A., and Smith, J.L. (1975). 'Sperm competition and the evolution of the precopulatory passive phase behaviour in *Locusta migratoria migratorioides*', *J. Ent.*, **49(A)**, 155–171.

Patterson, J.W. (1979). 'The effect of larval nutrition on egg production in *Rhodnius prolixus*', *J. Insect Physiol.*, **25**, 311–314.

Peng, C., and Williams, R.N. (1991). 'Influence of food on development, survival, fecundity, longevity and sex ratio of *Glischrochilus quadrisigratus* (Coleoptera: Nitidulidae), *Environ. Entomol.*, **20**, 205–210.

Percival, E., and Whitehead, H. (1929). 'A quantitative study of some types of stream-bed', *J. Ecol.*, **17**, 282–314.

Peschken, D.P. (1972). '*Chrysomelina quadrigemina* introduced from California to British Columbia against the weed *Hypericum perforatum*: Comparison of behaviour, physiology and colour in association with post-colonization adaptation', *Can. Ent.*, **104**, 1689–1698.

Pianka, E.R. (1970). 'On r- and K-selection', *Amer. Natur.*, **104**, 592–597.

Pianka, E.R. (1974). *Evolutionary Ecology*, Harper and Row, New York.

Pittendrigh, C.S., and Minis, D.H. (1971). 'The photoperiodic time measurement in *Pectinophora gossypiella* and its relation to circadian system in that species', in *Biochemistry* (Ed. M. Menaker), National Academy of Science, Washington, pp. 212–250.

Plateaux, I. (1972). 'Sur les modifications produits chez un fourmi par le presence d'un parasite cestode', *Ann. Sci. nat. (Zool.), Paris*, **14**, 203–220.

Policansky, D. (1974). 'Sex ratio, meiotic drive, and group selection in *Drosophila pseudoobscura*', *Amer. Natur.*, **108**, 75–90.

Policansky, D., and Ellison, J. (1970). 'Sex ratio in *Drosophila pseudoobscura*: spermiogenic failure', *Science*, **196**, 888–889.

Price, P.W. (1970a). 'Biology and host exploitation by *Pleolophus indistinctus* (Hymenoptera: Ichneumonidae)', *Ann. Ent. Soc. Amer.*, **63**, 1502–1509.

Price, P.W. (1970b). 'Trail odors: Recognition by insects parasitic on cocoons', *Science*, **170**, 546–547.

Price, P.W. (1972). 'Parasitoids utilizing the same host: Adaptive nature of differences in size and form', *Ecology*, **53**, 190–195.

Price, P.W. (1973a). 'Reproductive strategies in parasitoid wasps', *Amer. Natur.*, **107**, 684–693.

Price, P.W. (1973b). 'Parasitoid strategies and community organisation', *Environ. Entomol.*, **2**, 623–626.

Price, P.W. (1974a). 'Energy allocation in ephemeral adult insects', *Ohio J. Sci.*, **74**, 380–387.

Price, P.W. (1974b). 'Strategies for egg production', *Evolution*, **28**, 76–84.

Price, P.W. (1975a). *Insect Ecology*, John Wiley, New York.

Price, P.W. (1975b). 'Reproductive strategies of parasitoids', in *Evolutionary Strategies of Parasitic Insects and Mites* (Ed. P.W. Price), Plenum Press, New York, pp. 87–111.

Price, R.E., and Brown, H.D. (1990). Ȑeproductive performance of the African migratory locust, *Locusta migratoria migratorioides* (Orthoptera: Acrididae), in a cereal crop environment in South Africa', *Bull. Ent. Res.*, **80**, 465–472.

Pritchard, G. (1967). 'Laboratory observations on the mating behaviour of the island fruitfly *Rioxa pornia* (Diptera: Tephritidae)', *J. Austr. Ent. Soc.*, **6**, 127–132.

Puttarudriah, M., and Basavanna, G.P.C. (1953). 'Beneficial coccinellids of Mysore', *Indian J. Ent.*, **15**, 87–96.

Pyle, D.W., and Gromko, M.H. (1978). 'Repeated mating by female *Drosophila melanogaster*: The adaptive importance', *Experientia*, **34**, 449.

Raabe, M. (1982). *Insect Neurohormones*, Plenum Press, New York.

Rabb, R.L., and Bradley, J.R. (1968). 'The influence of host plants on parasitism of eggs of the tobacco hornworm', *J. Econ. Ent.*, **61**, 1249–1252.

Rahn, R. (1969). 'Rôle de la plante-hôte sur l'activity sexuelle chez *Acrolepia assectella* (Lep. Plutel)', *C. R. Acad. Sci., Paris*, **266**, 2004–2006.

Ralph, C.P. (1976). 'Natural food requirements of the large milkweed bug, *Oncopeltus fasciatus* (Hemiptera: Lygaeidae), and their relation to gregariousness and host plant morphology', *Oecologia*, **24**, 157–175.

Raman, K. (1987). 'Nutritional value of malvaceous seeds and related life-table analysis in terms of feeding and reproductive indices in the dusky cotton bug *Oxycarenus laetus* Kirby', *Proc. Indian Acad. Sci. (Anim. Sci.)*, **96**, 195–206.

Raman, K., and Sanjayan, K.P. (1983). Quantitative food utilization and reproductive programming in the dusky cotton bug *Oxycarenus hyalinipennis* (Hemiptera: Lygaeidae)', *Proc. Ind. Nat. Sci. Acad.*, **49B**, 231–236.

Regis, L. (1979). 'The role of the blood meal in egg-laying periodicity and fecundity in *Triatoma infestans*', *Int. J. Inv. Reprod.*, **1**, 187–195.

Reid, R.W. (1962). 'Biology of the mountain pine beetle, *Dendroctonus monticolae* Hopkins, in the East Kootenay region of British Columbia. II. Behaviour in the host, fecundity, and internal changes in the female', *Can. Ent.*, **94**, 605–613.

Reinecke, J.P. (1985). 'Nutrition: artificial diets', in *Comprehensive Insect Physiology, Biochemistry and Pharmacology* (Eds. G.A. Kerkut and L.I. Gilbert), Vol. 4, Pergamon Press, New York, pp. 391–413.

Rempel, J.G. (1936). 'The life history and morphology of *Chironomus hyperboreus*', *J. Biol. Bd Can.*, **2**, 209–221.

Rempel, J.G. (1940). 'Intersexuality in Chironomidae induced by nematode parasitism', *J. Exp. Zool.*, **84**, 261–289.

Richards, O.W., and Davies, R.G. (1977). *Imm's General Text Book of Entomology*, John Wiley, New York.

Richmond, R.C. (1976). 'Frequency of multiple insemination in natural population of *Drosophila*', *Amer. Natur.*, **110**, 485–486.

Richmond, R.C., and Ehrman, L. (1974). 'The incidence of repeated mating in *Drosophila paulistorum*', *Experientia*, **30**, 489–490.

Ricklefs, R.E. (1973). *Ecology*, Nelson, Great Britain.

Riddiford, L.M., and Truman, J. (1978). 'Biochemistry of insect hormones and insect growth regulators', in *Biochemistry of Insects* (Ed. M. Rockstein), Academic Press, London, pp. 307–357.

Ridgway, N.M., and Mahr, D.L. (1990). 'Reproduction, development, longevity and host mortality of *Sympiesis marylandensis* (Hymenoptera: Eulophidae), a parasitoid of spotted tentiform leafminer (Lepidoptera: Gracillariidae), in the laboratory', *Ann. Ent. Soc. Amer.*, **83**, 795–797.

Ridley, M. (1988). 'Mating frequency and fecundity in insects', *Biol. Rev.*, **63**, 509–549.

Riemann, J.G., and Thorson, B.J. (1974). 'Viability and use of sperm after irradiation of the large milkweed bug', *Ann. Ent. Soc. Amer.*, **67**, 871–876.

Robertson, J.G. (1964). 'Effect of supernumerary chromosomes on sex ratio in *Calligrapha philadelphica* L. (Coleoptera: Chrysomelidae)', *Nature, Lond.*, **204**, 605.

Roff, D.A. (1986). The evolution of wing dimorphism in insects', *Evolution*, **40**, 1009–1020.

Roy, D.N. (1936). 'On the role of blood in ovulation in *Aedes aegypti*', *Bull. Ent. Res.*, **27**, 423.

Ruberson, J.R., Bush, J., and Kring, T.J. (1991). 'Photoperiodic effect on diapause induction and development in the predator *Orius insidiosus* (Heteroptera: Anthocoridae)', *Environ. Entomol.*, **20**, 786–789.

Rutowski, R.L., Dickinson, J.L., and Terkanian, B. (1989). 'The structure of the mating plug in the checkerspot butterfly', *Euphydryas chalacedona*', *Psyche*, **96**, 279–286.

Sakaluk, S.K., and Cade, W.H. (1983). 'The adaptive significance of female multiple matings in house and field crickets', in *Orthopteran Mating Systems* (Eds. T. Gwynne and G.K. Morris), Westview Press, Boulder, Colorado, pp. 319–336.

Salt, G. (1937). 'The egg-parasite of *Sialis lutaria*: a study of the influence of the host upon a dimorphic parasite', *Parasitology*, **29**, 539–553.

Sanburg, L.L., and Larson, J.R. (1973). 'Effect of photoperiod and temperature on ovarian development in *Culex pipiens pipiens*', *J. Insect Physiol.*, **19**, 1173–1190.

Sandlan, K. (1979). 'Sex ratio regulation in *Coccygomimus turionella* Linnaeus (Hymenoptera: Ichneumonidae) and its ecological implications', *Ecol. Entomol.*, **4**, 365–378.

Sanjayan, K.P., and Murugan, K. (1987). 'Nutritional influence on the growth and reproduction in two species of acridids (Orthoptera: Insecta)', *Proc. Indian Acad. Sci. (Anim. Sci.)*, **96**, 229–237.

Saul, S.H., Tam, S.Y.T., and McInnis, D.O. (1988). 'Relationship between sperm competition and copulation duration in the Mediterranean fruitfly (Diptera: Tephritidae)', *Ann. Ent. Soc. Amer.*, **81**, 498–502.

Saunders, D.S. (1976). *Insect Clocks*, Pergamon Press, Oxford.

Saunders, D.S., and Dodd, C.W.H. (1972). 'Insemination and ovulation in the tsetse fly *Glossina morsitans*', *J. Insect Physiol.*, **18**, 187–198.

Saura, A., Lokki, J., Lankinen, P., and Suomalainen, E. (1976a). 'Genetic polymorphism and evolution in parthenogenetic animals: III. Tetraploid *Otiorrhynchus scaber* (Coleoptera: Curculionidae)', *Hereditas*, **82**, 79–100.

Saura, A., Lokki, J., Lankinon, P., and Suomalainen, E. (1976b). 'Genetic polymorphism and evolution in parthenogenetic animals: II. Triploid *Otiorrhynchus salicis* Stroin (Coleoptera: Curculionida)', *Entomol. Scand.*, **7**, 1–6.

Schields, O. (1967). 'Hilltopping', *J. Lepid. Res.*, **6**, 69–178.

Schneider, F. (1969). 'Bionomics and physiology of aphidophagous Syrphidae', *A. Rev. Ent.*, **14**, 103–124.

Schoonhoven, L.M. (1962). 'Diapause and the physiology of host-parasite synchronization in *Bupalus piniarius* L. and *Eucarcelia rutilla* VIII', *Arch. néerl. Zool.*, **15**, 111–174.

Schwartz, H. (1932). 'Der chromosomenzyklue van *Tetraneura ulmi* De Geer', *Z. Zellforsch.*, **15**, 645–687.

Scott, A.C. (1941). 'Reversal of sex production in *Micromalthus*', *Biol. Bull.*, **81**, 420–431.

Senthamizhselvan, M. (1987). 'Ecophysiological studies on pests, predators and parasites', Ph.D. thesis, Madurai Kamaraj University, Madurai.

Senthamizhselvan, M., and Murugan, K. (1988). 'Bioenergetics and reproductive efficiency of *Atractomorpha crenulata* F. (Orthoptera:Insecta) in relation to food quality', *Proc. Indian Acad. Sci. (Anim. Sci.)*, **97**, 505–517.

Senthamizhselvan, M., and Muthukrishnan, J. (1988). 'Interspecific differences in the host plant utilization of three coexisting lepidopteran larvae', *Proc. Indian Nat. Sci. Acad.*, **54**, 307–314.

Senthamizhselvan, M., and Muthukrishnan, J. (1989). 'Effect of feeding tender and senescent leaf by *Eupterote mollifera* and tender leaf and flower by *Spodoptera litura* on food utilization', *Proc. Indian Acad. Sci. (Anim. Sci.)*, **98**, 77–84.

Shapiro, A.M. (1970). 'The role of sexual behaviour in density related dispersal of pierid butterflies', *Amer. Natur.*, **104**, 367–372.

Shivashankar, T., and Veeresh, G.K. (1987). 'Impact of differential feeding on the reproduction of tiger beetle *Cicindela cancellata* De Jean (Cicinidelidae: Coleoptera)', *Proc. Indian Acad. Sci. (Anim. Sci.)*, **96**, 317–321.

Shorey, H.H. (1963). 'The biology of *Trichoplusia ni* (Lepidoptera: Noctuidae) II. Factors affecting adult fecundity and longevity', *Ann. Ent. Soc. Amer.*, **56**, 476–480.

Simmons, L.W. (1988). 'The contribution of multiple mating and spermatophore consumption to the lifetime reproductive success of female field cricket effects *Gryllus bimaculatus*', *Ecol. Entomol.*, **13**, 57–69.

Singh, N.B., Campbell, A., and Sinha, R.H. (1976). 'An energy budget of *Sitophilis oryzae* (Coleoptera: Curculionidae)', *Ann. Ent. Soc. Amer.*, **69**, 503–512.

Sivaramakrishnan, K.G., and Venkataraman, K. (1987). 'Observations on feeding propensities, growth rate and fecundity in mayflies (Insecta: Ephemeroptera)', *Proc. Indian Acad. Sci. (Anim. Sci.)*, **96**, 305–309.

Skinner, S.W. (1985). 'Son-killer — a third chromosomal factor affecting the sex ratio in the parasitoid wasp *Nasonia (Normoniella) vitripennis*', *Genetics*, **109**, 745–759.

Slansky, F. (1980). 'Effect of food limitation on food consumption and reproductive allocation by adult milkweed bugs, *Oncopeltus fasciatus*', *J. Insect Physiol.*, **26**, 79–84.

Slansky, F., and Rodriguez, J.G. (1986). *Nutritional Ecology of Insects, Mites, Spiders and Related Invertebrates: An Overview*, Wiley, New York.

Smith, C.C., and Fretwell, S.D. (1974). 'The optimal balance between size and number of offspring', *Amer. Natur.*, **108**, 499–506.

Smith, G.J.C., and Pimental, D. (1969). 'The effects of two host species on the longevity and fertility of *Nasonia vitripennis*', *Ann. Ent. Soc. Amer.*, **62**, 305–308.

Solomon, J.D. (1976). 'Sex ratio of the carpenterworm moth (*Prionoxystus robiniae*) (Lepidoptera: Cossidae)', *Can. Ent.*, **108**, 317–318.

Solomon, J.D., and Neel, W.W. (1974). 'Fecundity and oviposition behaviour in the carpenterworm, *Prionoxystus robiniae*', *Ann. Ent. Soc. Amer.*, **67**, 238–240.

Southwood, T.R.E. (1977). 'Habitat, the template for ecological strategies', *J. Anim. Ecol.*, **46**, 337–365.

Southwood, T.R.E., May, R.M., Hassell, M.P., and Conway, C.R. (1974). 'Ecological strategies and population parameters', *Amer. Natur.*, **108**, 791–804.

Spates, G. (1979). 'Fecundity of the stable fly: Effect of soyabean trypsin inhibitor and phospholipase A inhibitor on the fecundity', *Ann. Ent. Soc. Amer.*, **72**, 845–849.

Srinivasaperumal, S., Samuthiravelu, P., and Muthukrishnan, J. (1992). 'Host plant preference and life table of *Megacopta cribraria* (Fab.) (Hemiptera: Plataspidae)', *Proc. Indian Nat. Sci. Acad.*, **58B**, 333–340.

Stalker, H.D. (1954). 'Parthenogenesis in *Drosophila*', *Genetics*, **39**, 4–34.

Stall, G.B. (1986). 'Anti-juvenile hormone agents', *A. Rev. Ent.*, **31**, 391–429.

Stark, J.D., Vargas, R.I., and Thalman, R.K. (1990). 'Azadirachtin: Effects on metamorphosis, longevity and reproduction of three tephritid fly species (Diptera: Tephritidae)', *J. Econ. Ent.*, **83**, 2168–2174.

Strong, D.R., Jr., and Wang, M.D. (1977). 'Evolution of insect life histories and host plant chemistry: Hispine beetles on *Heliconia*', *Evolution*, **31**, 854–862.

Subramaniam, M., Haniffa, M.A., and Pandian, T.J. (1988). 'Effect of multiple matings on egg production in *Gryllodes sigillatus* (Walker) (Orthoptera: Gryllidae)', *Entomon*, **13**, 317–320.

Sundararaj, R., and David, B.V. (1990). 'Influence of biochemical parameters of host plants on the biology of *Dialeurodes vulgaris* Singh', *Proc. Indian Acad. Sci.*, **99**, 137–140.

Suomalainen, E. (1955). 'A further instance of geographical parthenogenesis and polypoidy in the weevils, Curculionidae', *Arch. Soc. Vanamo. Suppl.*, **9**, 350–354.

Suomalainen, E. (1962). 'Significance of parthenogenesis in the evolution of insects', *A. Rev. Ent.*, **7**, 349–366.

Suomalainen, E. (1966). 'The first known case of polyploidy in a parthenogenetic curculionid native of America', *Hereditas*, **56**, 213–216.

Suomalainen, E., Saura, A., and Lokki, J. (1976). 'Evolution of parthenogenetic insects', *Evol. Biol.*, **9**, 209–257.

Suzuki, Y., and Iwasa, Y. (1980). 'A sex ratio theory of gregarious parasitoids', *Res. Popul. Ecol.*, **22**, 366–382.

Suzuki, Y., Tsuji, H., and Sasakawa, M. (1984). 'Sex allocation and effects of superparasitism on secondary sex ratios in the gregarious parasitoid, *Trichogramma chilonis* (Hymenoptera: Trichogrammatidae)', *Anim. Behav.*, **32**, 478–484.

Svensson, B.G., and Peterson, E. (1987). 'Sex-role reversed courtship behaviour, sexual dimorphism and nuptial gifts in the dance fly, *Empis borealis* (L.)', *Ann. Zool. Fenn.*, **24**, 323–334.

Taketo, A. (1971). 'Studies on the life history of *Tanypteryx pryeri* Selys II. Habitat and habit of the nymph', *Kontyu*, **39**, 299–310.

Tauber, M.J., and Tauber, C.A. (1969). 'Diapause in *Chrysopa carnea* (Neuroptera: Chrysopidae) II. Effect of photoperiod on reproductively active adults', *Can. Ent.*, **101**, 364–370.

Tauber, M.J., and Tauber, C.A. (1972). 'Geographic variation in critical photoperiod and in diapause intensity of *Chrysopa carnea* (Neuroptera)', *J. Insect Physiol.*, **18**, 25–29.

Tauber, M.J., and Tauber, C.A. (1973). 'Nutritional and photoperiodic control of the seasonal reproductive cycle in *Chrysopa mohave*', *J. Insect Physiol.*, **19**, 729–736.

Tauber, M.J., and Tauber, C.A. (1974). 'Thermal accumulations, diapause, and oviposition in a conifer-inhabiting predator, *Chrysopa harrisii* (Neuroptera)', *Can. Ent.*, **106**, 969–978.

Tauber, M.J., and Tauber, C.A. (1976). 'Insect seasonality: diapause maintenance, termination and postdiapause development', *A. Rev. Ent.*, **21**, 81–107.

Tauber, M.J., Tauber, C.A., and Masaki, S. (1986). *Seasonal Adaptations of Insects*, Oxford University Press, New York.

Tauber, M.J., Tauber, C.A., Obrycki, J.J., Gollands, B., and Wright, R.J. (1988). 'Voltinism and the induction of aestival diapause in the Colorado potato beetle, *Leptinotarsa decemlineata* (Coleoptera: Chrysomelidae)', *Ann. Ent. Soc. Amer.*, **81**, 748–754.

Taylor, L.R. (1975). 'Longevity, fecundity and size; control of reproductive potential in a polymorphic migrant, *Aphis fabae* Scope', *J. Anim. Ecol.*, **44**, 135–159.

Testerink, G.J. (1982). 'Strategies in energy consumption and partitioning in Collembola', *Ecol. Entomol.*, **7**, 341–351.

Teth, R.S., and Chew, R.M. (1972). 'Development and energetics of *Notonecta undulata* during predation on *Culex taraslis*', *Ann. Ent. Soc. Amer.*, **65**, 1270–1279.

Thanh-Xuan, N. (1967). 'Influence de facteurs extérnes sur l'élimination anticipés de la diapause de *P. pyri* dans les conditions naturelles', *C. R. Acad. Sci., Paris*, **264**, 1445–1448.

Thibout, E. (1979). 'Stimulation of reproductive activity of females of *Acrolepiopsis assectella* (Lepidoptera: Hyponomentoidea) by the presence of eupyrene spermatozoa in the spermathecae', *Ent. Exp. Appl.*, **26**, 279–290.

Thiele, H.U. (1969). 'The control of larval hibernation and adult aestivation in the carabid beetles *Nebria brevicollis* F. and *Patrobus atrorufus*', *Oecologia*, **2**, 347–361.

Thiele, H.U. (1971). 'Die Steuerung der Jahresrhythmik von Carabidon durch exogene und endogene Faktoren', *Zool. Jb. (Syst.)*, **98**, 341–371.

Thomas Punitham, M., Haniffa, M.A., and Arunachalam, S. (1987). 'Effect of mating duration on fecundity and fertility of eggs in *Bombyx mori* L. (Lepidoptera: Bombycidae)', *Entomon*, **12**, 55–58.

Thompson, D.J. (1978). 'Towards a realistic predator-prey model: the effect of temperature on the functional response and life history of larvae of the damselfly, *Ischnura elegans*', *J. Anim. Ecol.*, **47**, 757–767.

Thompson, S.N. (1986). 'Nutrition and *in vitro* culture of insect parasitoids', *A. Rev. Ent.*, **31**, 197–219.

Thornhill, A.R. (1974). 'Evolutionary ecology of Mecoptera (Insecta)', Ph.D. thesis, University of Michigan, Michigan.

Thornhill, R. (1976). 'Sexual selection and paternal investment in insects', *Amer. Natur.*, **110**, 153–163.

Tomlinson, J. (1966). 'The advantage of hermaphroditism and parthenogenesis', *J. Theor. Biol.*, **11**, 54–58.

Toscano, N.C., and Stern, V.M. (1976). 'Development and reproduction of *Euschistus conspersus* at different temperatures', *Ann. Ent. Soc. Amer.*, **69**, 839–840.

Trimble, R.M. (1979). 'Laboratory observations on oviposition by the predaceous tree hole mosquito *Toxorhynchites rutilus septentrionalis* (Diptera: Culicidae)', *Can. J. Zool.*, **57**, 1104–1108.

Tripathi, S.R., and Singh, R. (1989). 'Effect of different pulses on development, growth and reproduction of *Heliothis armigera* (Hubner) (Lepidoptera: Noctuidae)', *Insect Sci. Applic.*, **10**, 145–148.

Trivers, R.L. (1972). 'Parental investment and sexual selection, in *Sexual Selection and the Descent of Man* (Ed. B. Campbell), Aldine-Atherton, Chicago, pp. 136–179.

Trivers, R.L., and Hare, H. (1976). 'Haplodiploidy and the evolution of the social insects', *Science*, **191**, 249–263.

Tschinkel, W.R. (1987). 'Relationship between ovariole number and spermathecal sperm count in ant queens: A new allometry', *Ann. Ent. Soc. Amer.*, **80**, 208–211.

Tsiropoulos, G.J. (1978). 'Holidic diets and nutritional requirements for survival and reproduction of the adult walnut husk fly', *J. Insect Physiol.*, **24**, 239–342.

Tsiropoulos, G.J. (1980). 'Major nutritional requirements of adult *Dacus oleae*', *Ann. Ent. Soc. Amer.*, **73**, 251–253.

Turner, N. (1960). 'The effect of inbreeding and crossbreeding on numbers of insects', *Ann. Ent. Soc. Amer.*, **53**, 686–688.

Tuskes, P.M., and Brower, L.P. (1978). 'Overwintering ecology of the monarch butterfly *Danaus plexippus* L., in California', *Ecol. Entomol.*, **3**, 141–153.

Ueda, T. (1979). 'Plasticity of the reproductive behaviour in a dragonfly *Sympetrum parvulum* Bartenef, with reference to the social relationship of males and the density of territories', *Res. Popul. Ecol.*, **21**, 58–72.

Unnithan, G.C., and Pye, S.O. (1991). 'Mating, longevity, fecundity and egg fertility of *Chilo partellus* (Lepidoptera: Pyralidae): Effects of delayed or successive matings and their relevance to pheromonal control methods', *Environ. Entomol.*, **20**, 150–155.

Van den Assem, J. (1974). 'Male courtship patterns and female receptivity signal of Pteromalinae (Hym. Pteromalidae), with a consideration of some evolutionary trends and a comment on the taxonomic position of *Pachycrepoideus vindemiae*', *Neth. J. Zool.*, **24**, 253–278.

Van den Assem, J. (1976). 'Male courtship behaviour, female receptivity signal, and size differences between the sexes in Pteromalinae, and comparative notes on other chalcidoides', *Neth. J. Zool.*, **26**, 535–548.

Van den Assem, J., and De Bruzin, E.F. (1977). 'Second matings and their effect on the sex ratio of the offspring in *Nasonia vitripennis* (Hymenoptera: Pteromalidae)', *Ent. Exp. Appl.*, **21**, 23–28.

Van den Assem, J., Gijswijt, M.J., and Nubel, B.K. (1980). 'Observations on courtship and mating strategies in a few species of parasitic wasps (Chalcidoidea)', *Neth. J. Zool.*, **30**, 208–227.

Van der Kraan, C., and Van der Straten, M. (1988). 'Effects of mating rate and delayed mating on the fecundity of *Adoxophyes orana* (Lepidoptera: Tortricidae)', *Ent. Exp. Appl.*, **48**, 15–23.

Van Dijk, Th.S. (1979). 'On the relationship between reproduction, age and survival in two carabid beetles: *Calathus melanocephalus* L. and *Pterostichus coerulescens* L. (Coleoptera, Carabidae)', *Oecologia*, **40**, 63–80.

Van Dijken, J., Nauenschwander, P., Van Alpher, J.M., and Hammomd, N.O. (1991). 'Sex-ratios in field population of *Epidinocarsis lopezi*, an exotic parasitoid of the cassava mealybug, in Africa', *Ecol. Entomol.*, **16**, 233–240.

Van Emden, H.F. (1966). 'Studies on the relations of insect and host plants, III. A comparison of the reproduction of *Brevicoryne brassicae* and *Myzus persicae* (Hemiptera: Amphididae) on brussels sprout plants supplied with different rates of nitrogen and potassium', *Ent. Exp. Appl.*, **9**, 444–460.

Van Hook, R.I., Nielsen, M.G., and Shugart, H.H. (1980). 'Energy and nitrogen relations for a *Macrosiphum liriodendri* (Homoptera: Aphididae) population in an east Tennessee *Liriodendron tulipifera* stand', *Ecology*, **61**, 960–975.

Vargas, R.I., and Chang, H.B. (1991). 'Evaluation of oviposition stimulants for mass production of melon fly, oriental fruit fly, and mediterranean fruit fly (Diptera: Tephritidae)', *J. Econ. Ent.*, **84**, 1695–1698.

Varley, G.C. (1961). 'Conversion rates in hyperparasitic insects', *Proc. roy. Soc. London*, **26**, 1–11.

Veazey, J.N., Kay, C.A.R., Walker, T.J., and Whitcoms, W.H. (1976). 'Seasonal abundance, sex-ratio, and macroptery of field crickets in Northern Florida', *Ann. Ent. Soc. Am.*, **69**, 374–380.

Velayudhan, R. (1987). 'Host preferences in some pentatomids and related impact on the fecundity of their parasitoids', *Proc. Indian Acad. Sci. (Anim. Sci.)*, **96**, 281–291.

Victorov, G.A. (1968). 'The influence of the population density upon the sex ratio in *Trissolcus grandis* Thoms (Hymenoptera: Scelionidae)', *Zool. Zh.*, **47**, 1035–1039.

Villavaso, E.J. (1975). 'Functions of the spermathecal muscle of the boll weevil, *Anthonomus grandis*', *J. Insect Physiol.*, **21**, 1275–1278.

Virag, J.C., and Darvas, B. (1983). 'Effectiveness of kinoprene and hydroprene on pests of greenhouse ornamentals; Mealybugs (Pseudococcidae), soft scales (Coccidae) and armored scales (Diaspididae)', *Proc. Int. Conf. Integr. Plant Prot.*, **4**, 198–202, Budapest.

Voegtlin, D.J., and Halbert, S.E. (1990). 'Lifecycle and hosts of *Rhopalosiphum cerasifoliae* (Homoptera: Aphididae)', *Ann. Ent. Soc. Amer.*, **83**, 43–45.

Voss, R.H., Ferro, D.N., and Logan, J.A. (1988). 'Role of reproductive diapause in the population dynamics of the Colorado potato beetle (Coleoptera: Chrysomelidae) in western Massachusetts', *Environ. Entomol.*, **17**, 863–871.

Waage, J.K. (1973). 'Reproductive behaviour and its relation to territoriality in *Calopteryx maculata* (Beauvois)', *Behaviour*, **47**, 240–256.

Waage, J.K. (1979a). 'Dual function of the damselfly penis: sperm removal and transfer', *Science*, **203**, 916–918.

Waage, J.K. (1979b). 'Adaptive significance of postcopulatory guarding of mates and nonmates by male *Calopteryx maculata*', *Beh. Ecol. Sociobiol.*, **6**, 147–154.

Waage, J.K. (1986). 'Family planning in parasitoids: adaptive patterns of progeny and sex allocation', in *Insect Parasitoids* (Eds. J. Waage and D. Greathead), Academic Press, London, pp. 63–95.

Walker, W.F. (1976). 'Juvenoid stimulation of egg production in *Oncopeltus fasciatus* on non-host diets', *Environ. Entomol.*, **5**, 599–603.

Weaver, N. (1978). 'Chemical control of behaviour-intraspecific', in *Biochemistry of Insects* (Ed. M. Rockstein), Academic Press, New York, pp. 360–389.

Webb, R.E., and Smith, F.F. (1968). 'Fertility of eggs of Mexican bean beetles from females mated alternatively with normal and apholate-treated males', *J. Econ. Ent.*, **61**, 521–523.

Welch, H.E. (1965). 'Entomophilic nematodes', *A. Rev. Ent.*, **10**, 275–302.

Wellings, P., Morton, W.R., and Hart, P.T. (1986). 'Primary sex ratio and differential progeny survivorship in solitary haplo-diploid parasitoids', *Ecol. Entomol.*, **11**, 341–348.

Wellso, F.W., and Adkisson, P.L. (1966). 'A long-day short-day effect in the photoperiodic control of the pupal diapause of the bollworm, *Heliothis zea* (Boddie) (Lepidoptera: Noctuidae)', *J. Insect Physiol.*, **12**, 1445–1465.

Werren, J.H. (1980). 'Sex ratio adaptations to local mate competition in a parasitic wasp', *Science*, **208**, 1157–1159.

Werren, J.H., and Charnov, E.L. (1978). 'Facultative sex ratios and population dynamics', *Nature, Lond.*, **272**, 349–350.

White, M.J.D. (1977). *Animal Cytology and Evolution*, Cambridge University Press, New York, 931 pp.

White, N.D.G., and Sinha, R.N. (1981). 'Energy budget of *Oryzaephilus surinamensis* (Coleoptera: Cucujidae) feeding on rolled oats', *Environ. Entomol.*, **10**, 320–326.

White, N.D.G. and Sinha, R.N. (1987). 'Bioenergetics of *Cynaeus angustus* (Coleoptera: Tenebrionidae) feeding on stored grain', *Ann. Ent. Soc. Amer.*, **80**, 184–190.

Whittier, T.S., and Kaneshiro, K.Y. (1991). 'Male mating success and female fitness in the mediterranean fruit fly (Diptera: Tephritidae)', *Ann. Ent. Soc. Amer.*, **84**, 608–611.

Wiklund, C. (1974). 'Oviposition preference in *Papilio machon* in relation to the host plants of the larvae', *Ent. Exp. Appl.*, **17**, 189–198.

Wildish, D.J. (1977). 'Biased sex ratio in invertebrates', in *Advances in Invertebrate Reproduction*, Vol. 1 (Eds. K.G. and R.G. Adiyodi), Peralam-Kenoth, Karivellur, pp. 8–24.

Williams, G.C. (1966). *Adaptation and Natural Selection*, Princeton University Press, Princeton, New Jersey.

Williams, K.L., Browne, L.B., and Van Gerwen, A.C.M. (1979). 'Quantitative relationships between the ingestion of protein-rich material and ovarian development in the Australian sheep blowfly, *Lucilia cuprina* (Wied)', *Int. J. Invert. Reprod.*, **1**, 75–88.

Witteveen, J., and Joosse, E.N.G. (1987). 'Growth, reproduction and mortality in marine littoral Collembola at different salinities', *Ecol. Entomol.*, **12**, 459–469.

Witteveen, J., Verhoet, A., and Huipen, T.E.A.M. (1988). 'Lifehistory strategy and egg diapause in the intertidal collembolan *Anurida maritima*', *Ecol. Entomol.*, **13**, 443–451.

Worke, P.A. (1937). 'Comparative effects of man and cannery on egg production of *Culex pipiens* Linn.', *J. Parasitol.*, **23**, 311–313.

Wright, J.W., and Lawe, C.H. (1968). 'Weeds, polyploids, parthenogenesis, and the geographical and ecological distribution of all female species of *Cnemidophorus*', *Copeia*, **1**, 128–138.

Wyatt, G.R. (1991). 'Gene regulation in insect reproduction', *Inv. Reprod. Dev.*, **20**, 1–35.

Wyatt, I.J. (1961). 'Pupal paedogenesis in the Cecidomyiidae (Diptera)', *Proc. roy. Ent. Soc. London*, **36(A)**, 133–143.

Wylie, H.G. (1965). 'Discrimination between parasitized and unparasitized housefly pupae by females of *Nasonia vitripennis* (Walk.) (Hymenoptera: Pteromalidae)', *Can. Ent.*, **97**, 279–286.

Wylie, H.G. (1966). 'Some mechanisms that affect the sex ratio of *Nasonia vitripennis* (Walk.) (Hymenoptera: Pteromalidae) reared from superparasitized housefly pupae', *Can. Ent.*, **98**, 645–653.

Wylie, H.G. (1967). 'Some effects of host size on *Nasonia vitripennis* and *Muscidifurax raptor* (Hymenoptera: Pteromalidae)', *Can. Ent.*, **99**, 742–748.

Wylie, H.G. (1971a). 'Observations on intraspecific larval competition in three hymenopterous parasites of fly puparia', *Can. Ent.*, **103**, 137–142.

Wylie, H.G. (1971b). 'Oviposition restraint of *Muscidifurax raptor* (Hymenoptera: Pteromalidae) on parasitized housefly pupae', *Can. Ent.*, **103**, 1537–1544.

Wylie, H.G. (1972). 'Oviposition restraint of *Spalangia cameroni* (Hymenoptera: Pteromalidae) on parasitized housefly pupae', *Can. Ent.*, **104**, 209–214.

Wylie, H.G. (1973). 'Control of egg fertilization of *Nasonia vitripennis* (Hymenoptera: Pteromalidae) when laying on parasitized housefly pupae', *Can. Ent.*, **105**, 709–718.

Wylie, H.G. (1976a). 'Interference among females of *Nasonia vitripennis* (Hymenoptera: Pteromalidae) and its effect on sex ratio of their progeny', *Can. Ent.*, **108**, 655–661.

Wylie, H.G. (1976b). 'Observation on life history and sex ratio variability of *Eupteromalus dubius* (Hymenoptera: Pteromalidae), a parasite of cyclorrhaphous Diptera', *Can. Ent.*, **108**, 1267–1274.

Wylie, H.G. (1977). 'Preventing and terminating pupal diapause in *Anthrycia cinerea* (Diptera: Tachinidae)', *Can. Ent.*, **109**, 1083–1090.

Wylie, H.G. (1979). 'Sex ratio variability of *Muscidifurax raptor* (Hymenoptera: Pteromalidae)', *Can. Ent.*, **111**, 105–109.

Yeargan, K.V. (1982). 'Reproductive capability and longevity of the parasitic wasps *Telenomus podisi* and *Trissolcus euschisti*', *Ann. Ent. Soc. Amer.*, **75**, 181–183.

Yin, C.M., and Chippendale, G.M. (1970). 'Juvenile hormone regulation of the larval diapause of the southwestern corn borer, *Diatrae grandiosella*', *J. Insect Physiol.*, **19**, 2403–2420.

Yokoyama, V.Y., and Pritchard, J. (1984). 'Effect of pesticides on mortality, fecundity and egg viability of *Geocoris pallens* (Hemiptera: Lygaeidae)', *J. Econ. Ent.*, **77**, 876–879.

5. PENTASTOMIDA

JOHN RILEY
*Department of Biological Sciences, University of Dundee,
Dundee DD1 4HN, U.K.*

I. INTRODUCTION

The phylum Pentastomida is divided into two orders, the Cephalobaenida (approx. 40 species) and the Porocephalida (approx. 65 species) (Table 1): all are dioecious and reproduce sexually.

Despite the considerable volume of literature that has accumulated on the group since the pioneering studies of Leuckart (1860) (see reviews of Self, 1969; Riley, 1986) the life cycles of a large number of species (and indeed entire genera) remain to be elucidated. Adult parasites inhabit the respiratory tract of tetrapod definitive hosts and their eggs, released into the lungs and nasopharynx, gain access to the oesophagus. Eggs, normally deposited in faeces, are immediately infective and contain distinctive primary larvae which are specialized for tissue migration (Self, 1969). Each is equipped with two pairs of stumpy legs furnished with double hooks, and a variety of penetration spines and stylets on the dorsal cephalothorax which enable it to penetrate and traverse tissues, a process which may be entirely mechanical. Under natural conditions eggs remain viable for months and can withstand low temperatures, but not, apparently, desiccation (Keegan, 1943). In the majority of life cycles one or more intermediate hosts, which may be invertebrate (cephalobaenids only) or vertebrate (both orders), are implicated, although in at least three species a direct life cycle is known to exist (Banaja *et al.*, 1975; Böckeler, 1984; Haugerud, 1988).

Development to an adult is through a series of larval instars, each separated by a moult, and there is no metamorphosis. In most indirect life cycles, eggs must gain access to the alimentary tract of a suitable intermediate host where appropriate stimuli can initiate hatching. Liberated primary larvae penetrate the stomach or intestinal wall, encyst in or on the viscera, and grow to an infective stage; a total of six and 10 moults respectively occur during this ontogenesis in the porocephalids *Porocephalus crotali* (see Esslinger, 1962) and *Linguatula serrata*

Table 1

An outline classification of the phylum Pentastomida (slightly modified from Riley, 1986)

Order	Family	Genus	No. of species	Definitive host	Intermediate host	References*
Cephalobaenida	Cephalobaenidae	*Cephalobaena*	1	Snakes	?	von Haffner and Rack (1971)
		Raillietiella	> 35	Snakes, lizards, amphisbaenians, amphibians	Direct (?), insects, amphibians, lizards	Ali *et al*. (1985)
	Reighardiidae	*Reighardia*	2	Marine birds	Direct	Banaja *et al*. (1975); Dyck (1975)
Porocephalida	Sebekidae	*Sebekia*	10	Crocodilians (Chelonians)	Fish (snakes, lizards?)	Riley *et al*. (1990)
		Alofia	5	Crocodilians	Fish	Giglioli (1922)
		Leiperia	2	Crocodilians	Fish	Fain(1961)
		Diesingia	1	Chelonians	?	Fonseca and Ruiz (1956)
	Subtriquetridae	*Subtriquetra*	3(?)	Crocodilians	Fish	Vargas (1975)
	Sambonidae	*Sambonia*	4	Monitor lizards	Direct (?)	Fain and Mortelmans (1960)
		Elenia	2	Monitor lizards	Amphibians, mammals	Bosch and Frank (1986)
		Waddycephalus	10	Snakes	Amphibians, reptiles, mammals	Riley and Self (1981b)
		Parasambonia	2	Snakes	Amphibians, reptiles, mammals	Riley and Self (1982)

Porocephalidae	*Porocephalus*	8	Snakes	Snakes and/or mammals	Riley and Self (1979)
	Kiricephalus	5	Snakes	Amphibians or lizards or mammals, and snakes	Riley and Self (1980)
Armilliferidae	*Armillifer*	7	Snakes	Mammals	Fain (19661); Riley and Self (1981a)
	Cubirea	2	Snakes	?	Heymons (1935)
	Gigliolella	1	Snakes	Mammals	Chabaud and Choquet (1954)
Linguatulidae	*Linguatula*	6	Mammals	Mammals	von Haffner *et al.* (1969)
				Direct	Haugerud (1988)

* The references offer the most complete description of a genus (or taxon) and/or contain information of life cycles.

(see Leuckart, 1860), whereas three moults are necessary in *Raillietiella* (see Bosch, 1986; Mehlhorn, 1988). The life cycle is completed when an infected intermediate host is consumed by the definitive host. Clearly, in the case of direct life cycles eggs are infective to definitive hosts and it is now known that some of these are sustained by autoreinfection (Deakins, 1973; Banaja *et al.*, 1976; Böckeler, 1984).

In terms of gross morphology, the Pentastomida appears to comprise a moderately homogeneous taxon but, as I shall show, this is not necessarily reflected in life-cycle strategies.

II. FECUNDITY

Differences in total egg output, by as much as three orders of magnitude, exist within the taxon, but in general the majority of cephalobaenids produce relatively few eggs, whereas egg production in most porocephalids is massive. It is important that estimates of egg production should be considered in terms of not only generation time (this may not be significantly different between the two orders: see below) but also the length of the patent period which can be very variable.

Cephalobaenid females possess a saccate uterus. During the patent period, eggs, shed continuously from the ovary, accumulate within the uterus, which clearly must contain eggs in varying stages of maturity. In at least two species, the pre-patent period terminates when *all* of the eggs in the uterus become fully infective and females die following a patent period of just a few days' duration: thus, the total egg counts of dissected gravid females of *Reighardia sternae* and *R. lomviae* are 2,900 and 2,600 to 3,800 respectively (Banaja *et al.*, 1975, 1976; Dyck, 1975), which respresents the total produced in a lifetime (see also Böckeler, 1984). In most cephalobaenids, however, the patent period probably extends from months to years and a surprising discovery was that patency commences when the uterus contains a *mixture* of eggs of various sizes and states of maturity when as few as 25 per cent of eggs contain primary larvae (Ali and Riley, 1983; Winch and Riley, 1985). This is possible because the vagina is equipped with a selective filter which allows through large, mature embryonated eggs and retains smaller, more recently fertilized eggs (Riley, 1983). Small (< 1 cm), lizard-infecting raillietiellids contain less than 9,000 eggs, which are deposited at a rate of about 10/day (Ali and Riley, 1983). In intermediate-sized (< 2 cm) raillietiellids from amphisbaenians this rate is around 100/day (Winch and Riley, 1985), whereas the largest species from snakes, which comprise about one-third of raillietiellid spp. (see Ali *et al.*, 1985), contain up to 200,000 eggs (Ali *et al.*, 1982), suggesting a correspondingly higher fecundity.

The porocephalid uterus is tubular and elongate (Riley, 1983; also see Nørrevang, Volume I of this series) and eggs mature en route to the vagina. As in cephalobaenids, the number of eggs carried is roughly correlated with female size. Gravid, patent specimens of the larger genera such as *Linguatula*, *Porocephalus*, and *Kiricephalus* contain at least 5×10^5 eggs (Pillars, 1925; Salazar, 1964; Riley, 1983) and in *P. crotali* are deposited at the rate of 520 to 2,300/female/day (Riley, 1981) over a patent period extending to six years or more (Riley, 1981, 1986). Egg production in *L. serrata*, which extends for up to 21 months (Hobmaier and Hobmaier, 1940 as *L. rhinaria*), is estimated by Baer (1952) in millions, whereas *L. arctica*, which may be unique among porocephalids in having a direct life cycle, produces only 1.2×10^6 eggs/patent female over a short patent period of four months (Haugerud, 1988). Thus, in most pentastomids the theoretical upper limit of fecundity, that is the total number of eggs produced in a lifetime, is set by the number of sperm stored in the spermathecae. The minimum generation time for cephalobaenids, such as *Reighardia sternae* and small raillietiellids (< 1 cm) is 160 to 180 days (Banaja *et al.*, 1976; Ali and Riley, 1983; Böckeler, 1984), whereas in large porocephalids (e.g. *P. crotali*, *L. arctica*) it is around one year (Riley, 1981 and Haugerud, 1988 respectively).

In all of the above life cycles, eggs are acquired by intermediate or definitive hosts as contaminants of food or water, and prodigious egg production simply reflects massive larval mortality. In one porocephalid, the crocodilian-infecting *Subtriquetra subtriquetra*, larvae are uniquely free-living and larval behaviour is crucial in transmission to fish intermediate hosts. Because transmission depends to a large extend on the activity of the parasite itself, and this behavioural component greatly increases the probability of host contact, larval production in this species may be very low, possibly less than 50 to 100/female/week (see Winch and Riley, 1986). It should be emphasized that this life cycle is atypical.

A. Factors Influencing Fecundity

Little is known of the factors which influence fecundity although there are likely to be several. For example, host immune responses, acting in a density-dependent manner, can adversely affect parasite growth rate and egg production. Pentastomids, like helminths, are likely to be distributed within the host population in an overdispersed manner (see Anderson, 1982), which means that relatively few hosts support the bulk of the parasite population. Records of pentastomids in natural infections indicate occasional high intensities; thus, Self and Kuntz (1967) recorded over 100 adult *Porocephalus crotali* from the lung of a rattlesnake (Riley, 1981 has since found that egg output and worm burden in this species may be inversely correlated); Fantham and Porter (1950) reported the blockage of the lung of a viper (*Bitis gabonica*) by several large specimens of *Armillifer armillatus* and Riley (unpublished observation) has examined infections comprising 40 to 50 *Elenia australis* from the lungs of a monitor lizard (*Varanus salvator*) and 24 *A. mazzai* (up to 80 mm long) from a python (*Liasis amethystinus*). In such heavy infections intraspecific competition for space, impairment of feeding sites (see Riley, 1986), and other factors may also significantly reduce growth and fecundity.

In pentastomids the mechanical and physiological processes involved in the continuous fertilization of oocytes from sperm stored for long periods in spermathecae are complex (Riley, 1983, 1986; also see Riley, Volumes III and V of this series) and there is evidence that these become less efficient with increasing age (unpublished observations). As sperm stores become progressively depleted, the proportion of unfertilized, undeveloped eggs increases dramatically: in a six-year-old infection of *Porocephalus crotali*, as many as 40 per cent of the eggs produced fall into this category.

One remaining factor, seasonal fluctuations in temperature, must profoundly affect fecundity in those species infecting reptiles in temperate zones (e.g. *Porocephalus* spp. in North American rattlesnakes). These poikilothermic hosts overwinter in shallow dens where body temperature (and therefore metabolism) falls dramatically (Klauber, 1972), and this must in turn both reduce egg output and increase the generation time of lung-dwelling pentastomids.

III. REPRODUCTIVE STRATEGIES (LIFE-HISTORY TACTICS)

A. Sex Ratios

Leuckart (1860) and Hett (1924) noted that in definitive hosts, females tend to predominate in mature infections despite the fact that the sex ratio of infective nymphs in intermediate hosts is unity. In the case of *Linguatula serrata* in the nasal sinuses of dogs, Leuckart (1860) accounted for this disparity by assuming that the smaller males were more easily dislodged and expelled by sneezing, but as Hett (1924) has pointed out, this does not explain the odd sex ratio in lung-inhabiting pentastomids.

Riley (1972), Banaja (1975), Dyck (1975), and Böckeler (1984) found that gulls and guillemots harbouring maturing infections of *Reighardia sternae* and *R. lomviae* contained mostly females, although even in these single-sex infections males had once been present as evidenced by sperm in the spermathecae. Similarly, Haugerud (1988) found that the sex ratio in *Linguatula arctica* infections shifted from around 5.3 : 1.0 (males : females) at copulation to $< 0.7 : 1.0$ at patency. In all three species the patent period is short; the life cycle is direct and is synchronized with certain host behavioural traits which favour transmission (see Banaja *et al.*, 1976; Böckeler, 1984). Opportunities for males to copulate with females maturing from successive infections do not arise; hence, males rarely survive much beyond copulation.

A most unusual variation on the porocephalid life cycle is reported by John and Nadakal (1986, 1988) in the case of *Kiricephalus pattoni*, which has a snake definitive host and two vertebrate intermediate hosts, the second being a snake. Males mature precociously, copulation occurs within the latter host, and most males do not reach the definitive host (John and Nadakal, 1988).

Nonetheless it is apparent that in long-lived species, males can copulate with generations of maturing females (see Riley, 1981) and here natural selection has promoted male survival. Self (see Volume II of this series) has found that male and female *Raillietiella* spp. occur simultaneously over long periods and has surmised that autoreinfection may be important in this process: this theme is explored in the next section.

B. Life Cycles

Salazar (1964) and Deakins (1969) first described massive infections of *Raillietiella orientalis* in Philippine cobras (*Naja naja philippinensis*) resulting from autoreinfection. Furthermore, Deakins' (1973) findings of encysted first-, second-, and third-stage larvae, actually located within the uterus and haemocoel of an adult female *R. orientalis*, is good evidence that autoreinfection is an essential part of the natural life cycle. Banaja *et al.* (1976) have reported similar findings in experimental infections of *Reighardia sternae* in herring gulls—in one case 11 per cent of the total number of eggs deposited hatched within the host to produce a massive infection of 2,600 larvae. Böckeler (1984) has since shown that a possible crowding effect may drastically limit the number of females eventually attaining maturity. Examples of endoparasites commonly indulging in autoreinfection to supplement a direct life cycle are extremely rare (see Tinsley, 1983) and in the case of *R. sternae* it ensures success in a seemingly tenuous life cycle, where parasite fecundity and the incidence and intensity of infection in natural populations are low (Riley, 1972; Böckeler, 1984). In *Reighardia lomviae* the process of autoreinfection could perpetuate the infection until chicks mature into breeding adults, in which case the mode of transmission suggested by Dyck (1975) would be possible.

Other claims of direct life cycles in cephalobaenids (Hett, 1915, 1924; Fain, 1961, 1964) are rather less convincing and await experimental verification. All are based on the frequent findings of immature nymphs, sometimes in ecdysis, either free or encysted within a host known to be a definitive host for that particular species. At first sight this evidence seems sound but it could be that the host species has recently consumed an intermediate host harbouring immature pentastomids of the appropriate stage. It is known, for example, that transplanted pentastomids will survive in unnatural hosts for long periods: Stiles (1891) fed larvae of *Porocephalus clavatus* to two totally unnatural hosts, a grass snake (*Natrix natrix*) and a viper (*Vipera berus*) and discovered that the larvae survived but failed to grow; Penn (1942) experimentally infected two snake species and a frog with nymphs of *P. crotali*; and Nadakal and Nayar (1968) transplanted infective nymphs of *Kiricephalus pattoni* from snakes into frogs and recovered living nymphs up to 60 days later. Clearly, only long-term studies are of value in transplant experiments.

Pentastomid life cycles have recently been reviewed (Riley, 1986) and here it is only necessary to present an overview and to take stock of more recent findings.

The outline classification of pentastomids (Table 1) broadly categorizes definitive and intermediate hosts of the 18 recognized genera.

1. Cephalobaenida

Nothing is known of the life cycle of the sole representative of the genus *Cephalobaena* but one of the two *Reighardia* spp. which infect marine birds is known to have a direct life cycle (i.e. eggs are infective to definitive hosts) (Banaja *et al.*, 1975, 1976; Böckeler, 1984). It has been postulated that parasite behaviour facilitates transmission in *R. sternae* where migrating patent females possibly induce gull hosts to vomit, and egg-contaminated vomit may be subsequently ingested by other gulls (Banaja *et al.*, 1976). Many aspects of the reproductive biology of *R. lomviae*, which infects puffins and guillemots, mirror those of *R. sternae* (see Dyck, 1975; Böckeler, 1984) and a similar mode of transmission may operate in this case.

The largest cephalobaenid genus *Raillietiella* (Table 1) has been divided into six taxonomic groupings based on a combination of size, hook morphology, and host preferences (Ali *et al.*, 1985). Two groups (I and II), comprising 14 well-characterized, small (6–32 mm) species with sharp or blunt-tipped posterior hooks, infect insectivorous lizards. Investigations of the life cycles of five species have revealed that coprophagous blattids can serve as experimental intermediate hosts (Ali and Riley, 1983; Bosch, 1986), confirming observations of natural infections (Rajamanickam and Lavoipierre, 1965; Jeffery *et al.*, 1985). Bosch (1986) demonstrated that it is the fourth larval stage, encysted in the midgut or fat body of cockroaches, which is infective to lizards, not the third stage as reported by Ali and Riley (1983). Strong ecological (predominantly trophic) links between egg-contaminated lizard faeces, insects, and lizards (see Lim and Yong, 1977) ensure successful transmission, often culminating in very high prevalences and intensities (Pence and Selcer, 1988; Riley *et al.*, 1988; Yiqiang *et al.*, 1988; Powell, personal communication) even though parasite fecundity is comparatively low (see above). Insects are also probable intermediate hosts in raillietiellids belonging to groups VI (from amphisbaenians) and V (from toads) but in only one case has this been investigated experimentally. Winch and Riley (1985) showed that *R. gigliolii*, infecting the lungs of the South American worm-lizard *Amphisbaena alba* [a facultative inquiline of nests of leaf-cutting ants (Riley *et al.*, 1987b)] utilizes larvae of the three-horned rhinoceros beetle, itself an inquiline of ant nests. Ants play a crucial role in transmission by delivering parasite eggs, deposited in amphisbaenian faeces, directly to the larvae, which reside in subterranean refuse dumps serving as repositories for spent compost.

No complete life cycle of the remaining species (group III from varanid lizards and group VI from snakes) is known. But 'infective' larvae from intermediate hosts such as toads (Larrousse, 1925), skinks, and snakes (together with hooks and cuticles of the two preceding larval stages) (Fain, 1961, 1964) indicate that there may also be as many as three vertebrate hosts in group VI raillietiellids (Ali *et al.*, 1982). It remains to be clarified, however, whether the 'infective' larva of Fain is an equivalent instar to that of Bosch (1986). Autoreinfection is known to be important in captive hosts (see review in Riley, 1986; Bosch, 1986) but its significance in natural infections requires elucidation (Ali *et al.*, 1982).

A most unusual new species of cephalobaenid has recently come to light which inhabits the lungs and nasal sinuses of the arboreal, nectar- and pollen-feeding sugar glider, *Petaurus breviceps* (Petauridae: Marsupialia) (Spratt, personal communication). Clearly, a direct life cycle must operate in this system and it is a reasonable speculation that patent females might deposit eggs on the tongue which would contaminate flowers and thus infect animals subsequently visiting flowers. Apparently, the lungs contain nymphs moulting to adults, whereas the nasal sinuses harbour minute (1.5 mm) gravid females with relatively massive eggs (150 μm): this singular distribution lends support to the above hypothesis. This diminutive species, the only cephalobaenid so far known to infect mammals, clearly may not represent an isolated success.

2. Porocephalida

(a) Sebekidae

Three of the genera within the Sebekidae family are known to use fish intermediate hosts (Table 1). Recent morphological observations of *Diesingia megastoma*, a lung parasite of piscivorous turtles, and the sole representative of the genus *Diesingia*, suggest that this too should be included in the family Sebekidae (Riley, unpublished observations). The likelihood is that fishes are also intermediate hosts in this species (Riley, 1986). Winch and Riley (1986) followed egg hatching and subsequent larval development of *Sebekia oxycephala* from South American caiman. Unusually the first five larval stages remain encapsulated on the viscera and all these stages are devoid of hooks but these are reconstituted in the infective seventh stage which becomes lightly encapsulated. In *Leiperia* at least, early development in the crocodile definitive host is unusual in that early instars are located within the circulatory system before females become established in the bronchi (see Riley, 1986 for references). Copulation may occur at the former site.

(b) Subtriquetridae

One life cycle, that of *Subtriquetra subtriquetra* from the nasopharynx of South American caiman, has been looked at in detail. It is of unusual interest because it is the only species known to possess a free-living stage (Vargas, 1975; Winch and Riley, 1986). Laboratory studies of naturally infected caiman reveal that larvae hatch in the nasal cavity (the site occupied by patent females) although the significance of this is unclear. Following release into water they remain infective to fish intermediate hosts for four to five days. During this period larvae exhibit very stereotyped behaviour, alternating between a characteristic stationary fishing posture, in which hooks are maximally disposed to optimize latching onto passing fish, and very brief periods of movement aiding dispersal (Winch and Riley, 1986). Host location is passive, and subsequent larval development occurs in the swim bladder. In this system, a larval behavioural component facilitates transmission and offsets an exceedingly low fecundity.

(c) Sambonidae

Aspects of the life cycle of a single species belonging to the genus *Sambonia, S. lohrmanni* from African, Asian, and Indonesian monitors, based on incidental findings at autopsies of both wild and captive definitive hosts, have been reported by Fain and Mortelmans (1960) and Fain (1961). Clear evidence of autoreinfection was found, indicative of a direct life cycle, and some support for direct development (from eggs) in reptiles was later obtained by Deakins (1972); the relevance of these observations to natural transmission remains unclear.

Riley and Self (1981b) speculated, from the viewpoint of host dietary regimen, that members of the large and distinctive genus *Waddycephalus* use terrestrial vertebrates as intermediate hosts, and the characteristic double-hooked nymphs have since been described from marsupials, snakes, lizards, and frogs: life cycles in the closely related genus *Parasambonia* are similar (Riley *et al.*, 1985; Riley and Spratt, 1987).

Experimental infection with eggs, shed by naturally infected *Varanus gouldii* very heavily infested in the palate and throat region (not in the lungs!) by hundreds of red *Elenia* specimens, resulted in larval development in frogs (*Hyla* spp.) and small laboratory mammals (*Mus*, *Rattus*, *Meriones*) (Bosch and Frank, 1986). Again, autoreinfection was suspected and this may prove to be a recurrent theme throughout the family, possibly supplementing indirect life cycles which otherwise use a variety of vertebrate intermediate hosts.

(d) Porocephalidae

Both of the genera making up the family Porocephalidae contain species which mature in either mammal-eating or snake-eating definitive hosts, and this has clear implications for the life cycle (see review in Riley, 1986). Those in the latter category are the more interesting in that recent experimental confirmation of an obligate three-vertebrate life cycle (John and Nadakal, 1986, 1988) originally proposed for *Kiricephalus* by Riley and Self (1980), which may now also extend to some *Porocephalus* spp. (see Riley and Self, 1979), reveals such life cycles to have no counterpart among other endoparasites. Interestingly, infective nymphs in snake second intermediate hosts migrate freely in the body cavity (John and Nadakal, 1988), and when such hosts are roughly handled or killed, nymphs exit through rents in the epidermis or through the mouth and nares (Self and Kuntz, 1967) in the expectation that they will emerge into the stomach of the ophiophagous definitive host: similar behaviour is seen when *Porocephalus*-infected rodents are killed (Riley and Self, 1979).

The confusion surrounding the specific identity of *Porocephalus* nymphs in mammals (reviewed in Riley and Self, 1979) has been compounded by Rogers *et al.* (1985), who reported that *P. crotali*, normally a parasite of small rodent intermediate hosts, apparently can become infective in dogs.

(e) Armilliferidae

Armilliferidae are of particular interest because three species are known to infect humans, but no new information concerning life cycles within this family has appeared since the review of Riley (1986).

(f) Linguatulidae

Most adult *Linguatula* infest the nasal sinuses and nasopharynx of carnivorous mammals belonging to the families Canidae, Hyaenidae, and Felidae and eggs are infective to a number of mammal intermediate hosts, most notably large grazing herbivores (Sachs *et al.*, 1973) but including humans (reviewed by Haugerud, 1988). However, the most recently described member of the taxon, *L. arctica*, parasitic in the nasopharynx of semidomesticated yearling reindeer in northern Norway (Riley *et al.*, 1987a), is the only species to infect a herbivore intermediate host, and early speculation concerning a direct life cycle (Riley, 1986; Riley *et al.*, 1987a) has recently been experimentally confirmed (Haugerud, 1989). Reindeer are essentially migratory animals, returning annually to traditional calving grounds and summer pastures; parasite transmission is perfectly synchronized with this activity. Complete development to patency takes a little less than one year, and eggs are deposited on pasture when susceptible calves are available for infection; increasing host resistance precludes the possibility of adult animals becoming infected (Haugerud, 1988).

IV. CONCLUSION

It is apparent from the above account that although few pentastomid life cycles have been experimentally investigated, it is nonetheless possible to draw some general conclusions. The vast majority of species, of both orders, which use vertebrate intermediate hosts liberate comparatively large numbers of eggs into the environment which are subsequently acquired as contaminants of food and water. Infection under these circumstances is clearly purely fortuitous. In most cases larvae appear to encyst in or on the viscera of intermediate hosts but there are some exceptions (e.g. *Kiricephalus* larvae remain free in the body cavity of snake second intermediate hosts, and *Subtriquetra subtriquetra* nymphs remain unencapsulated in fish). The possible influence of such free larvae on host vulnerability (i.e. its liability to predation) is at present unknown. There are, however, many examples from helminths where parasites within intermediate hosts are able to alter host behaviour, thereby facilitating transmission.

Those cephalobenids which depend upon insect intermediate hosts for transmission present a rather uniform picture; low fecundity generally seems to reflect strong trophic links between the component hosts of the life cycle. By contrast, the few variations on the theme of direct life cycles in pentastomids provide fascinating insights into the ways in which a simple strategy can be used to sustain transmission in apparently unlikely circumstances. Thus, *Reighardia sternae* infects ecologically wide-ranging hosts with extremely catholic diets (gulls and terns) despite an extremely low fecundity, and *Linguatula arctica* sustains a high prevalence and intensity in yearling reindeer, which are highly migratory. In the latter example, precise synchronization of host and parasite life cycles is paramount. The role of autoreinfection in natural direct and indirect modes of transmission awaits clarification and clearly provides an interesting avenue for future research. This is equally true of the newly discovered species of both orders which attain sexual maturity in the nasopharynx of mammals: as has already been mentioned, these are unlikely to represent isolated successes.

REFERENCES

Ali, J.H., and Riley, J. (1983). 'Experimental life-cycle studies of *Raillietiella gehyrae* Bovien, 1927 and *Raillietiella frenatus*, Ali, Riley and Self, 1981; pentastomid parasites of geckos utilizing insects as intermediate hosts', *Parasitology*, **86**, 147–160.

Ali, J.H., Riley, J., and Self, J.T. (1982). 'A revision of the taxonomy of *Raillietiella boulengeri* (Vaney and Sambon, 1910) Sambon, 1910, *R. orientalis* (Hett, 1915) Sambon, 1922 and *R. agcoi* Tubangui and Masilungan, 1956 (Cephalobaenida: Pentastomida)', *Syst. Parasit.*, **4**, 285–301.

Ali, J.H., Riley, J., and Self, J.T. (1985). 'A review of the taxonomy and systematics of the pentastomid genus *Raillietiella* Sambon, 1910 with a description of a new species', *Syst. Parasit.*, **7**, 111–123.

Anderson, R.M. (1982). 'Epidemiology', in *Modern Parasitology* (Ed. F.E.G. Cox), Blackwell, Oxford, pp. 204–251.

Baer, J.G. (1952). *Ecology of Animal Parasites*, University of Illinois Press, Urbana.

Banaja, A.A. (1975). 'Aspects of the biology of the pentastomid *Reighardia sternae* Diesing 1864', Ph.D. thesis, University of Dundee.

Banaja, A.A., James, J.L., and Riley, J. (1975). 'An experimental investigation of a direct life-cycle in *Reighardia sternae* (Diesing, 1864), a pentastomid parasite of the herring gull (*Larus argentatus*)', *Parasitology*, **71**, 493–503.

Banaja, A.A., James, J.L., and Riley, J. (1976). 'Some observations on egg production and autoreinfection of *Reighardia sternae* (Diesing, 1864), a pentastomid parasite of the herring gull', *Parasitology*, **72**, 81–91.

Böckeler, W. (1984). 'Der Entwicklungszyklus von *Reighardia sternae* (Pentastomida) nach Untersuchungen an natürlich und experimentell infestierten Möwen', *Zool. Anz.*, **213**, 374–394.

Bosch, H. (1986). 'Experimental life-cycle studies of *Raillietiella* Sambon, 1910 (Pentastomida: Cephalobaenida): The fourth-stage larvae is infective for the definitive host', *Z. Parasitenkde.*, **72**, 673–680.

Bosch, H., and Frank, W. (1986). 'The Australian pentastomid genus *Elenia* Heymons, 1932—development in the intermediate and definitive host', *Proc. VI Int. Congr. Parasitol. Brisbane*, p. 97.

Chabaud, A.G., and Choquet, M.T. (1954). 'Nymphes du pentastome *Gigliolella* (n.gen.) *brumpti* chez un lemurien', *Riv. Parasit.*, **15**, 331–336.

Deakins, D.E. (1969). '*Raillietiella orientalis* (Pentastomida) hyperinfection in *Naja naja philippinensis* (Reptilea: Serpentes), M.Sc. thesis, University of Oklahoma, 42 pp.

Deakins, D.E. (1972). 'Pentastome pathology in captive reptiles', Ph.D. thesis, University of Oklahoma.

Deakins, D.E. (1973). 'Occurrence of encysted larvae of *Raillietiella* (Pentastomida) in an adult female', *Trans. Am. Microsc. Soc.*, **92**, 287–288.

Dyck, J. (1975). '*Reighardia lomviae* sp. nov., a new pentastomid from guillemot', *Norwegian J. Zool.*, **23**, 97–109.

Esslinger, J.H. (1962). 'Development of *Porocephalus crotali* (Humboldt, 1808) (Pentastomida) in experimental intermediate hosts', *J. Parasitol.*, **48**, 631–638.

Fain, A. (1961). 'Les pentastomides de l'Afrique Centrale', *Musée Royal de l'Afrique Central-Tervuren, Belg. Annales Series 8. Sciences Zoologiques*, **92**, 1–115.

Fain, A. (1964). 'Observation sur le cycle l'évolutif du genre *Raillietiella* (Pentastomida)', *Bull. Acad. Roy. Sci. Belgique, Series* 5, **50**, 1036–1060.

Fain, A., and Mortelmans, J. (1960). 'Observations sur le cycle évolutif de *Sambonia lohrmanni* chez le varan. Preuve d'un development direct chez les Pentastomida', *Bull. Acad. Roy. Sci. Belgique*, Series 5, **46**, 518–531.

Fantham, H.B., and Porter, A. (1950). 'The endoparasites of certain South African snakes, together with some remarks on their structure and effects on the Ophidia', *Proc. Zool. Soc. (Lond.)*, **120**, 599–647.

Fonseca, F., and Ruiz, J.M. (1956). 'Was ist eigentlich *Pentastoma megastomum* Diesing, 1836? (Porocephalida, Porocephalidae)', *Senck. Biol.*, **37**, 469–485.

Giglioli, G.S. (1922). 'The new genus *Alofia* of the family Linguatulida. An anatomical account of *A. ginae*', *J. Trop. Med. Hyg.*, **25**, 371–377.

Haugerud, R.E. (1989). 'Evolution in the Pentastomids', *Parasitol. Today*, **5**, 126–132.

Haugerud, R.E. (1988). 'A life history approach to the parasite-host interaction—*Linguatula arctica* Riley, Haugerud and Nilssen, 1987—*Rangifer tarandus* (Linneaus, 1758)', Ph.D. thesis, University of Tromso.

Hett, M.L. (1915). 'On some new pentastomids from the Zoological Society's gardens', *Proc. Zool. Soc. (Lond.)*, **1915**, 115–121.

Hett, M.L. (1924). 'On the family Linguatulidae', *Proc. Zool. Soc. (Lond.)*, **1**, 107–159.

Heymons, R. (1935). 'Pentastomida', in *Klassen und Ordnungen des Tierreichs* (Ed. H.G. Bronn), Vol. 5, Part 4, Book 1, Akademische Verlagsgesellschaft M.B.H., Leipzig, 160 pp.

Hobmaier, H., and Hobmaier, N. (1940). 'On the life-cycle of *Linguatula rhinaria*', *Am. J. Trop. Med.*, **20**, 199–210.

Jeffery, J., Krishnasamy, M., Oothuman, P., Ali, J., Bakar, E.A., and Singh, I. (1985). 'Preliminary observations on the cockroach intermediate host of a house gecko raillietiellid in Peninsular Malaysia', *Mai. J. med. Lab. Scs.*, **2**, 82–84.

John, M.V., and Nadakal, A.M. (1986). 'Juvenile precocity in a pentastome, *Kiricephalus pattoni* (Stephens, 1908) Sambon, 1922, in the reptilian host *Tropidonotus piscator*', *J. Parasitol.*, **72**, 194–195.

John, M.V., and Nadakal, A.M. (1988). 'Juvenile precocity and maintenance of juvenile features in the males of the pentastome *Kiricephalus pattoni* (Stephens, 1908) Sambon, 1922', *Int. J. Inv. Reprod. Develop.*, **14**, 295–298.

Keegan, H.L. (1943). 'Observations on the pentastomid *Kiricephalus coarctatus* (Diesing) Sambon 1910', *Trans. Am. Microsc. Soc.*, **62**, 194–199.

Klauber, L.M. (1972). *Rattlesnakes*, Vols. I and II, University of California Press, Berkeley.

Larrousse, F. (1925). 'Larve de Linguatulidae parasite de *Bufo mauritanicus*', *Arch. Inst. Pasteur, Tunis*, **14**, 101–104.

Leuckart, R. (1860). *Bau und Entwicklungsgeschichte der Pentastomen nach Untersuchungen besonders von* Pent. taenioides *und* P. dendiculatum, C.F. Winter'sche Verlagshandlung, Leipzig, **6**, 160 pp.

Lim, B.L., and Yong, H.S. (1977). 'Pentastomid infections in house geckoes from Sarawak, Malaysia', *Med. J. Sci. Malaysia*, **32**, 59–62.

Mehlhorn, H. (1988). *Parasitology in Focus*, Springer-Verlag, Heidelberg, 925 pp.

Nadakal, A.M., and Nayar, K.K. (1968). 'Transplantation of pentastomids from reptilian to amphibian hosts', *J. Parasitol.*, **54**, 189–190.

Pence, D.B., and Selcer, K.W. (1988). 'Effects of pentastome infection on reproduction in a Southern Texas population of the Mediterranean gecko, *Hemidactylus turcicus*', *Copeia*, **3**, 565–572.

Penn, G.H. (1942). 'The life history of *Porocephalus crotali*, a parasite of the Louisiana muskrat', *J. Parasitol.*, **28**, 277–283.

Pillars, A.W.N. (1925). '*Linguatula serrata* Frohlich, 1789, in the nasal cavity of a bull bitch', *Vet. J.*, **81**, 126–130.

Rajamanickam, C., and Lavoipierre, M.M.J. (1965). '*Periplaneta australasiae* as an intermediate host of the pentastomid *Raillietiella hemidactyli*', *Med. J. Malaya*, **20**, 171.

Riley, J. (1972). 'Some observations on the life-cycle of *Reighardia sternae* Diesing, 1864 (Pentastomida)', *Z. Parasitenkde.*, **40**, 49–59.

Riley, J. (1981). 'Some observations on the development of *Porocephalus crotali* (Pentastomida: Porocephalida) in the western diamondback rattlesnake (*Crotalus atrox*)', *Int. J. Parasit.*, **11**, 127–131.

Riley, J. (1983). 'Recent advances in our understanding of pentastomid reproductive biology', *Parasitology*, **86**, 59–83.

Riley, J. (1986). 'The biology of the Pentastomida', *Adv. Parasitol.*, **25**, 46–128.

Riley, J., and Self, J.T. (1979). 'On the systematics of the pentastomid genus *Porocephalus* (Humboldt, 1811) with descriptions of two new species', *Syst. Parasit.*, **1**, 25–42.

Riley, J., and Self, J.T. (1980). 'On the systematics and life cycle of the pentastomid genus *Kiricephalus* Sambon, 1922 with descriptions of three new species', *Syst. Parasit.*, **1**, 127–140.

Riley, J., and Self, J.T. (1981a). 'Some observations on the taxonomy and systematics of the pentastomid genus *Armillifer* (Sambon, 1922) in South East Asian and Australian snakes', *Syst. Parasit.*, **2**, 171–179.

Riley, J., and Self, J.T. (1981b). 'A redescription of *Waddycephalus teretiusculus* (Baird, 1862) Sambon, 1922, and a revision of the taxonomy of the genus *Waddycephalus* (Sambon, 1922) with descriptions of eight new species', *Syst. Parasit.*, **3**, 243–257.

Riley, J. and Self, J.T. (1982). 'A revision of the pentastomid genus *Parasambonia* Stunkard and Gandal, 1968; a new generic character, a description of the male, and a new species', *Syst. Parasit.*, **4**, 125–133.

Riley, J., and Spratt, D. (1987). 'Further observations on pentastomids (Arthropoda) parasitic in Australian reptiles and mammals', *Rec. S. Aust. Mus.*, **21**, 139–147.

Riley, J., Haugerud, R.E., and Nilssen, A.C. (1987a). 'A new pentastomid from the nasal passages of the reindeer (*Rangifer tarandus*) in northern Norway, with speculation about its life cycle', *J. Nat. Hist.*, **21**, 707–716.

Riley, J., McAllister, C.T., and Reed, P.S. (1988). '*Raillietiella teagueselfi* n.sp. (Pentastomida: Cephalobaenida) from the Mediterranean gecko, *Hemidactylus turcicus* (Sauria: Gekkonidae), in Texas', *J. Parasitol.*, **74**, 481–486.

Riley, J., Spratt, D.M., and Presidente, P.J.A. (1985). 'Pentastomids (Arthropoda) parasitic in Australian reptiles and mammals', *Aust. J. Zool.*, **33**, 39–53.

Riley, J., Spratt, D.M., and Winch, J.M. (1990). 'A revision of the genus *Sebekia* Sambon, 1922 (Pentastomida) from crocodilians with descriptions of five new species', *Syst. Parasit.*, **16**, 1–25.

Riley, J., Stimpson, A.F., and Winch, J.M. (1987b). 'On the association of *Amphisbaena alba* (Squamata: Amphisbaenidae) with the leaf-cutter ant *Atta cephalotes* in Trinidad', *J. Nat. Hist.*, **20**, 459–469.

Rogers, K.S., Miller, G., Prestwood, A.K., Bjorling, D.E., and Latimer, K.S. (1985). 'Aberrant nymphal pentastomiasis in a dog', *J. Am. Anim. Hosp. Assoc.*, **21**, 417–420.

Sachs, R., Rack, G., and Woodford, M.H. (1973). 'Observations on pentastomid infestation of east African game animals', *Bull. epizoot. Dis. Afr.*, **21**, 401–409.

Salazar, N.P. (1964). 'A newly reported arthropod parasite in the lungs and body cavity of some Philippine snakes', *Philipp. J. Sci.*, **93**, 171–178.

Self, J.T. (1969). 'Biological relationships of the Pentastomida; a bibliography on the Pentastomida', *Exp. Parasitol.*, **24**, 63–119.

Self, J.T., and Kuntz, R.E. (1967). 'Host-parasite relations in some Pentastomida', *J. Parasitol.*, **53**, 202–206.

Stiles, C.W. (1891). 'Bau und Entwicklungsgeschichte von *Pentastomum proboscideum* Rud. und *Pentastomum subcylindricum* Dies', *Z. wiss. Zool.*, **52**, 85–157.

Tinsley, R.C. (1983). 'Ovoviviparity in platyhelminth life-cycles', *Parasitology*, **86**, 161–196.

Vargas, M.V. (1975). 'Description del huevecillo, larva y ninfa de *Subtriquetra subtriquetra* Sambon, 1922 (Pentastomida), y. algunas observaciones sobre su ciclo de vida', *Rev. Biol. Trop.*, **23**, 67–75.

von Haffner, K., and Rack, G. (1971). 'Beiträg zur Anatomie, Entwicklung und systematischen Stellung der *Cephalobaena tetrapoda* Heymons, 1922', *Zool. Jb. (Abt.) Anat.*, **88**, 505–526.

von Haffner, K., Rack, G., and Sachs, R. (1969). 'Verschiedene Vertreter der Familie *Linguatulidae* (Pentastomida) als parasiten von Säugetieren der Serengeti (Anatomie, Systematik, Biologie)', *Mitt. Hamburg. Zool. Mus.*, **66**, 93–144.

Winch, J.M., and Riley, J. (1985). 'Experimental studies on the life cycle of *Raillietiella gigliolii* (Pentastomida: Cephalobaenida) in the South American worm-lizard *Amphisbaena alba*: a unique interaction involving two insects', *Parasitology*, **91**, 471–481.

Winch, J.M., and Riley, J. (1986). 'Studies on the behaviour and development in fish of *Subtriquetra subtriquetra*: a uniquely free-living pentastomid larva from a crocodilian', *Parasitology*, **93**, 81–98.

Yiqiang, Z., Shunxiang, Z., Kaipang, Li., Chen, H., and Hongliao, X. (1988). 'Reports of pentastomiasis in *Gekko gekko* in China', *J. Guangxi Agric. Coll.*, **7**, 61–64.

6. BRYOZOA ENTOPROCTA

CLAUS NIELSEN
Zoological Museum, Universitetsparken 15,
DK-2100 Copenhagen, Denmark

I. INTRODUCTION

All entoprocts are capable of asexual reproduction through a budding process, which in some families leads to the formation of colonies. In the solitary species the liberated buds usually attach to the host animal close to the maternal zooid so that a 'pseudocolony' is formed. Some of the colonial species may form thick-walled cysts presumably able to survive periods of adverse conditions. The literature on budding and asexual reproductive bodies was summarized by Nielsen (1971), which may be consulted for references.

II. BUDDING

New zooids are formed through budding from a stolon or basal disc in colonial species (Fig. 1) and from the frontal side of the calyx in solitary species (Fam. Loxosomatidae, Fig. 2A) (for classification, see Systematic Résumé). In some of the loxosomatids budding starts already in the larvae, which release one or more buds and then degenerate (Fig. 2B and C). In adult *Loxosomella bocki* and larvae of *L. vivipara* and *Loxosoma jaegersteni* (Fig. 2D and E) the buds develop from the bottom of an ectodermal invagination and thus appear to be internal.

The buds form in an almost identical manner in all species studied (see Fig. 1). The first stage is the development of a small internal vesicle from the ectoderm of the budding area. This vesicle becomes partly divided into an 'upper' part (closest to the ectoderm of the budding area), which gives rise to the atrium, and a 'lower' part, which gives rise to the gut. The opening between the two compartments represents the mouth, whereas the anus is formed at a later stage. The gut becomes U-shaped and a ganglion is formed from an ectodermal invagination of the atrial epithelium covering the concave side of the gut. The protonephridia are formed through a proliferation from one of the ectodermal cells of the atrium just in front

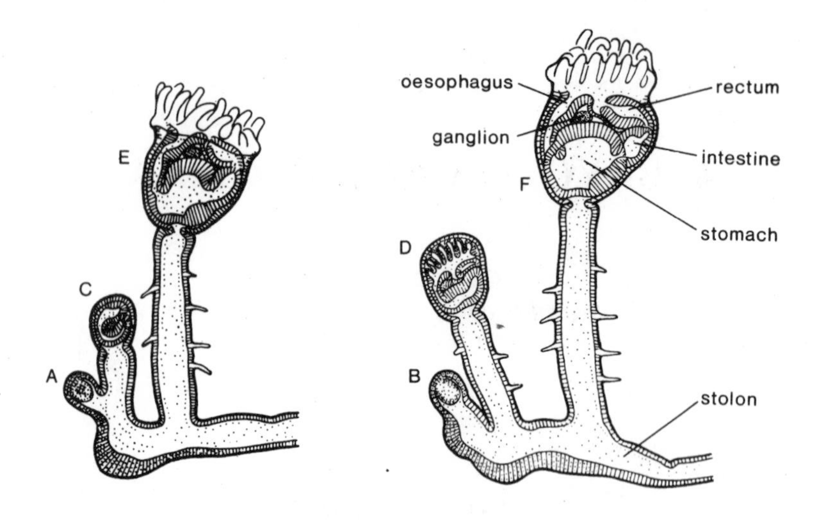

Fig. 1. Stolon tips of *Pedicellina cernua* showing the development of the buds; the relative ages of the buds are indicated by the lettering. (After Brien, 1957; from Nielsen, 1971.)

of the ganglionic invagination. A horseshoe of thickenings of the upper wall of the atrium forms the rudiments of the tentacles and finally, the upper atrial wall opens and the tentacles may expand.

In the loxosomatids an attachment organ in the shape of a sucking disc or a foot-like expansion with gland cells is formed at the end of a shorter or longer stalk and the buds are then ready for liberation (Fig. 2A).

III. ASEXUAL REPRODUCTIVE BODIES

Thick-walled cysts have been described in several species of barentsiids. They are usually formed from short branches of the stolons, but in *Urnatella gracilis* single muscular joints of the stalk may become cysts. The stolonal cysts may either be one-chambered as in *Loxosomatoides japonicum* and *Barentsia* (formerly *Arthropodaria*) *kovalevskii* (Fig. 3A–B) or multichambered as in *B. matsushimana* (Fig. 3C–D). Each chamber has a small, thin-walled area 'pore' on the upper side, and at germination this 'pore' opens and the new zooid grows out of the chamber and develops like a normal bud.

In *Urnatella gracilis* newly formed stolon tips with two to three zooids may detach from the parent colony and act as propagation units (Emschermann, 1987).

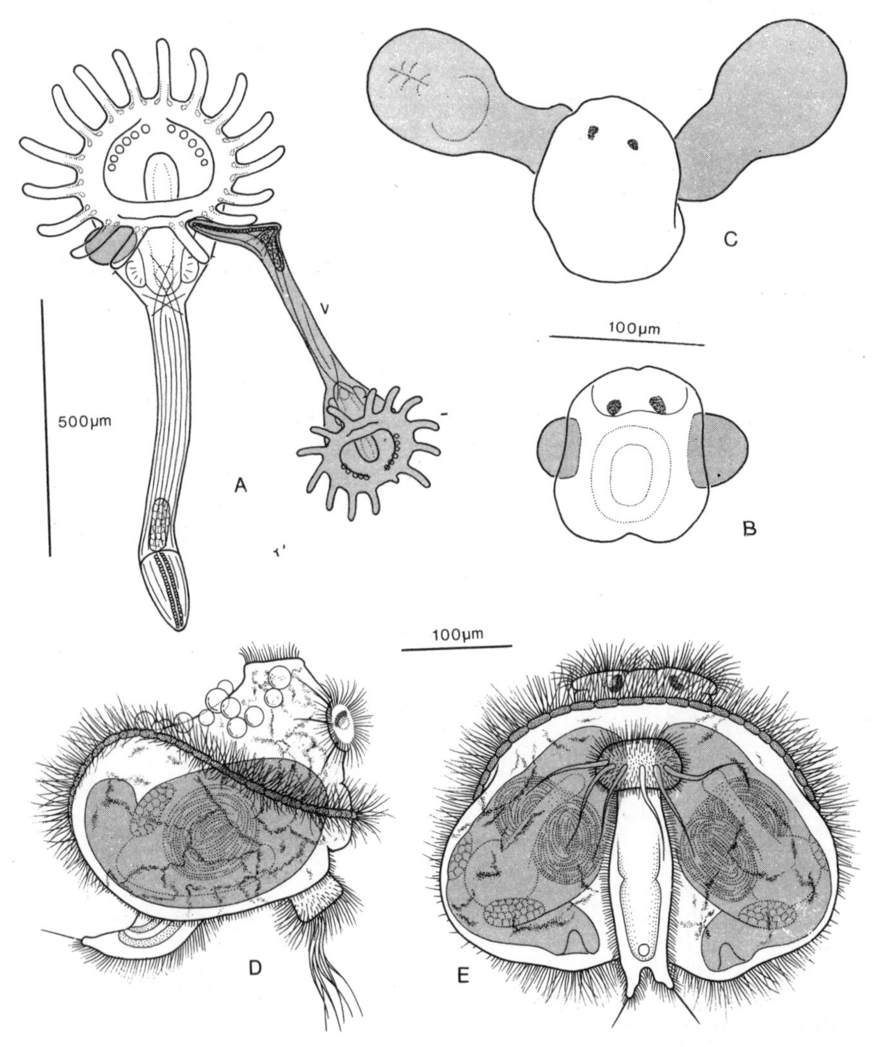

Fig. 2. Budding in entoprocts. **A**: *Loxosomella vivipara*, adult specimen with two buds of different ages; **B–C**: *Loxosomella leptoclini*, larvae with a pair of lateral buds, just after settling (the larva is still able to creep) and at a stage where the buds are almost ready for liberation; **D–E**: *Loxosoma jaegersteni*, larva with two 'internal' buds. The buds are dotted in all the drawings. (From Nielsen, 1966, 1971.)

IV. REGENERATION

In accordance with their ability to reproduce asexually the entoprocts show considerable powers of regeneration, but only the barentsiids have been studied in some detail.

Single tentacles are regenerated within a few days (Nasonov, 1926). Stalks which have lost the calyx may start regenerating after two to three days and after

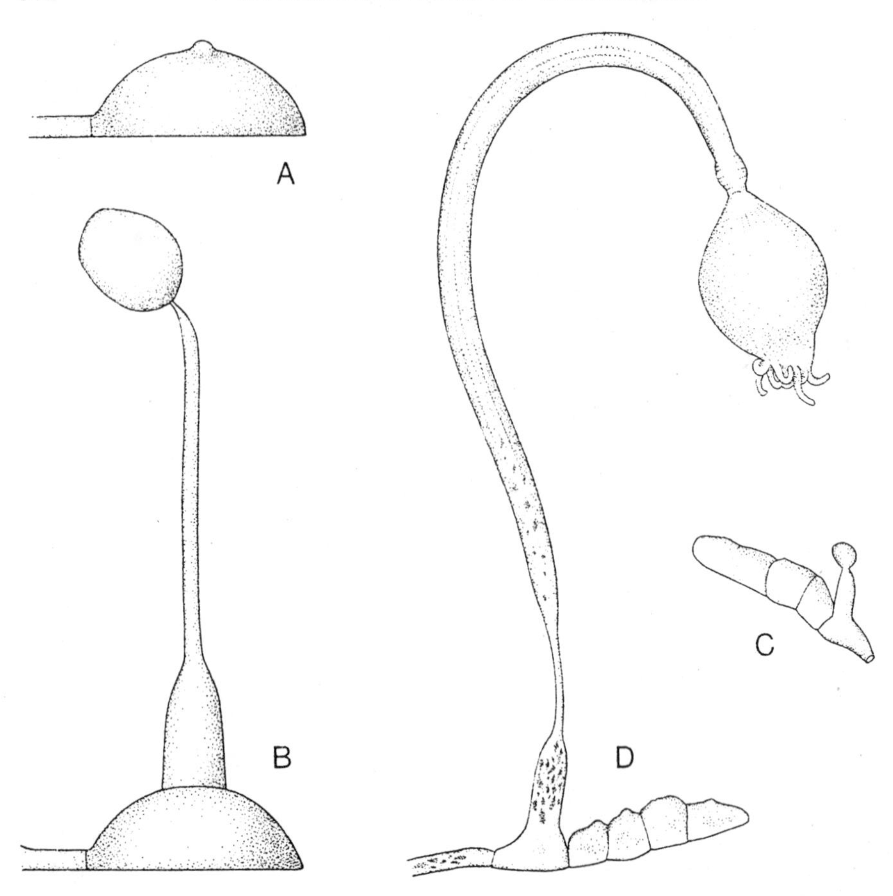

Fig. 3. Entoproct cysts. **A–B**: *Barentsia* (= *Arthropodaria*) *kovalevskii*, one-chambered cysts before and after germination; **C–D**: *Barentsia matsushimana*, multichambered cysts in various stages of germination. (**A–B** redrawn from Valkanov, 1951; **C–D** based on Toriumi, 1951.)

another two or three days a small, complete calyx may expand the lophophore (Mukai and Makioka, 1978). It is not rare to find colonies of *Barentsia* in which several zooids lack the calyx, and in damaged colonies several of the zooids may shed their calyx. It appears that this type of regeneration is fairly common in nature and that it may, for example, take place when a nudibranch has eaten a calyx. A periodic regeneration, like that occurring in ectoproct bryozoans, has not been reported in entoprocts.

Mukai and Makioka (1978) made a series of experiments to demonstrate the regenerative abilities of *Barentsia discreta*. They showed that almost any piece of stalk and stolon is able to regenerate a new zooid, but formations of "monsters" were observed several times, e.g. formation of double calyxes and the formation of a new calyx at both ends of the elastic part (rod) of the stalk.

V. REPRODUCTIVE STRATEGIES

The single entoproct zooid produces only few larvae (Nielsen, 1964, 1971). Some of the small loxosomatid species, such as *Loxosomella polita* and *Loxosoma agile*, have only one or two embryos or larvae in the atrium at a time and large species, such as *Loxosomella phascolosomata* and *Loxosoma pectinaricola*, may have up to 12 (a very large, undescribed species of *Loxosoma* from Phuket, Thailand, had up to 33 embryos and larvae in the atrium). The pedicellinids and barentsiids generally have larger zooids than the loxosomatids but the variation in the numbers of embryos and larvae in the atrium is of the same order of magnitude as that of the loxosomatids. The developmental time of the embryos and the length of the reproductive period of the single zooid are not known, so the total number of larvae produced per zooid cannot be ascertained.

Two larval types are represented within the family Loxosomatidae: mainly lecithotrophic larvae with a very short semipelagic phase (probably lasting about one day), such as those of *Loxosomella harmeri, L. leptoclini*, and *L. vivipara*; and planktotrophic larvae with a long pelagic phase, such as those of *Loxosoma jaegersteni* and probably also *L. pectinaricola* and *Loxosomella elegans*. Only short-lived larvae have been described in the families Barentsiidae and Pedicellinidae, but it cannot be excluded that long-lived, planktonic larvae exist in other species.

In many animal groups there is a marked difference between the low number of eggs in species with direct development or lecithotrophic larvae with a short free phase and the high number of eggs in species with planktotrophic larvae. Interestingly, this difference is not seen among the entoprocts. One reason for this is probably the brood protection which is found in all species, but one would still have expected to find differences in number of eggs between the species with a quite short, semipelagic larval phase and the species with a long, planktotrophic larval phase.

The low production of larvae is compensated for by the budding process which gives rise to a number of sexually reproducing zooids from one successfully metamorphosed larva. However, fertilization has not been studied in detail in entoprocts (see Nielsen, Volume IV B of this series) and it is not known if there is a sterility barrier between zooids in a colony or the likewise genetically identical zooids in an asexually produced 'pseudocolony' of loxosomatids, such as that which apparently exists in ectoproct bryozoans (Gooch and Schopf, 1971).

Both sexual reproduction (for details, see Franzén, Volumes I and II of this series) and budding may take place all year round, but at least in temperate waters the tendency is that both types of reproduction slow down or stop completely in winter (Nielsen, 1964).

The colonial species grow on various substrata and it is probable that the colonies may live several years. In these species the colonies may produce larvae over long periods until their habitat is eventually destroyed (by stones or shells becoming covered by sand or silt, by algae disintegrating, etc.), and the larvae are therefore primarily of importance for the colonization of new areas. Almost all the solitary species (and *Loxokalypus socialis*) are, however, associated with a host organism which produces water currents carrying food and oxygen to the zooids. As the host animals live only a few years and are usually not colonized by the entoprocts before they have reached a certain size, these species must be able to produce so many larvae that new hosts can be colonized rapidly. In some species it has been observed that the newly released buds can swim by means of the tentacular cilia, but this is probably of minor importance for the colonization of new hosts.

It is thus clear that more knowledge of both sexual and asexual reproduction must be gathered before the relative importance of the two processes for the survival of the species can be assessed. The characterization of the species as r- or K-strategists—or perhaps better the evaluation of the localization of the species in the r–K continuum—cannot be undertaken at present, and it must be considered whether this concept without modification can be used with advantage for colonial species.

REFERENCES

Brien, P. (1957). 'Le bourgeonnement des endoproctes et leur phylogenese. A propos du bourgeonnement chez *Pedicellina cernua*', *Ann. Soc. Roy. zool. Belg.*, **87**, 27–43.

Emschermann, P. (1987). 'Creeping propagation stolons—an effective propagation system of the freshwater entoproct *Urnatella gracilis* Leidy (Barentsiidae)', *Arch. Hydrobiol.*, **108**, 439–448.

Gooch, J.L., and Schopf, T.J.M. (1971). 'Genetic variation in the marine ectoproct *Schizoporella errata*', *Biol. Bull.*, **141**, 235–246.

Mukai, H., and Makioka, T. (1978). 'Studies on the regeneration of an entoproct, *Barentsia discreta*', *J. Exp. Zool.*, **205**, 261–275.

Nasonov, N. (1926). '*Arthropodaria kovalevskii* n. sp. (Entoprocta) und die Regeneration ihrer Organe', *Trudy osob. zool. Lab. Sevastop. biol. Sta.*, 2. Ser., **5**, 1–38, pl. 1 (in Russian with German summary).

Nielsen, C. (1964). 'Studies on Danish Entoprocta', *Ophelia*, **1**, 1–76.

Nielsen, C. (1966). 'On the life cycle of some Loxosomatidae (Entoprocta)', *Ophelia*, **3**, 221–247.

Nielsen, C. (1971). 'Entoproct life cycles and the entoproct/ectoproct relationship', *Ophelia*, **9**, 209–341.

Toriumi, M. (1951). 'Some entoprocts found in Matsushima Bay', *Sci. Rep. Tôhoku Univ., Biol.*, **19**, 17–22.

Valkanov, A. (1951). 'Eigentümlichkeiten in dem Bau und der Organisation von *Arthropodaria kovalevskii* Nasonov in Zusammenhang mit ihrer Überwinterung', *Trud. morsk. biol. Sta. Stalin*, **16**, 47–64, pls 8–12 (in Russian with German summary).

7. BRACHIOPODA

SHOU HWA CHUANG
144 Pasir Ris Road, Singapore 1851, Singapore

I. INTRODUCTION

An extensive radiation of the brachiopods during geologic time, especially the Palaeozoic Era, gave rise to many bizarre forms without pedicles, such as the richthofeniids, gigantoproductids, and lyttoniids, that left no living representatives or relatives (Williams *et al.*, 1965). The richthofeniids had a pedicle valve in the form of an inverted cone that grew upwards. The living tissues inside also moved up by sealing off one compartment after another at the older proximal part. The brachial valve was a flat operculum covering the rounded aperture at the distal end of the cone. The gigantoproductids with huge valves, almost half a metre in width, lay free on the sea bed. How did they attain that size with any of the usual modes of life that Recent brachiopods have adopted? The lyttoniids had flattened brachial valves with closely encasing pedicle valves. What strategies did these bizarre forms adopt to reproduce in the primaeval seas? There were also bizarre inarticulate fossils that left no direct descendants in modern seas (Chuang, 1977b). Recent brachiopods are not representative relics of their fossil ancestors. Evidently the reproductive strategies adopted by Recent brachiopods could not encompass all the variations that fossil brachiopods might conceivably have adopted.

II. REGENERATION

Not all tissues of a brachiopod can regenerate lost parts. Observations of regeneration are restricted to the more accessible organs such as the pedicle, lophophore, mantle, and shell.

A. Regeneration of Lophophore

Altogether two specimens of *Lingula anatina* among about two thousand specimens dissected during experiments had lost their lophophore on one side of the body where it was in the form of a blind stump. This indicates that only the growing point can form or differentiate new whorls of a lophophore and that a fully formed whorl of a spirolophe together with the growing point, when severed or otherwise lost by injury, seals itself up at the cut end during the process of healing but does not regenerate a new whorl. The spirolophe of a mature postlarva in *Lingula* also does not regenerate when its distal portion together with the growing tip is severed.

However, the spirolophe of the rhynchonellid brachiopod *Notosaria nigricans* regenerates when severed (Hoverd, 1985; Chuang, unpublished). The regenerated portion is shorter than the other brachium of the spirolophe.

B. Regeneration of the Mantle and Shell

In nature, many large specimens of *Lingula anatina* in a large sample have been found with the anterior edge of the shell bitten off or notched by an unknown predator. By virtue of the natural process of healing of the mantle edge and continuous life-long shell secretion in the lingulids (Chuang, 1961), the gap at the anterior edge of both mantle and shell is soon sealed by regeneration of mantle and the subsequent secretion of new shell material by the regenerated mantle. Similar regenerated injury notches have been observed in another lingulid, *Glottidia pyramidata* (see Paine, 1963), and on the calcareous shells of *Terebratalia transversa*, *Hemithiris psittacea*, and other articulates (Chuang, unpublished), indicating a similar regeneration. Presumably this regenerative ability to repair a slight injury of both shell and mantle is generally present in brachiopods. Injury even to the large mantle sinus in a lingulid may not lead to a great loss of coelomic fluid because of the valve-like fold that guards the entrance of the main mantle sinus into the visceral cavity. Moreover, the mantle sinus, which is distended with coelomic fluid under pressure imparted by contraction of the body-wall muscles (Chuang, 1964), collapses when it is punctured and mechanically impedes and reduces the flow of coelomic fluid.

When a small part of the shell together with its mantle at the anterior or lateral edge of the postlarval shell in *Lingula anatina* is removed in growth experiments, the resulting gap in the shell border is soon patched up with a thinner shell deposited by the regenerated mantle that grows to span the gap in a few days to a few weeks depending on the size of the notch cut and the age of the animal (Chuang, 1961). Younger postlarvae regenerate mantle and shell more rapidly than older ones.

C. Regeneration of the Pedicle

The pedicle in *Lingula anatina* may accidentally snap in nature. It breaks off at its narrowest weakest region at the proximal end where its cuticular coat is thinnest (Fig. 1a). It may be considered to have undergone autotomy. If this happens when the animal between the shell valves is not far from the upper end of the burrow, it has access to the dissolved oxygen in the water aerating the burrow and survives to regenerate a new pedicle in a few days. This gradually grows in length until it is big enough to push the shelled animal to the upper end of the burrow to resume its normal feeding position. If the shelled animal lies deep inside the burrow when its pedicle snaps, the animal fails to maintain the upper part of the burrow, which duly collapses to deny the animal its supply of food and oxygen. The trapped animal later dies and the snapped pedicle invariably rots away inside the burrow. Some rotting specimens have been dug out frequently. Occasional pedicle-less specimens have been washed up along the shore: how they lost their pedicles is not known. In nature, specimens with regenerated pedicles are often found. The smaller size and lighter colour of the regenerated pedicle distinguish it from the one which the lingulid larva formed before settlement to the postlarval infaunal mode of life.

When the pedicle of *Lingula anatina* is accidentally severed between its proximal end and its distal bulbous tip during collection, the stump is usually discarded by autotomy at the thinnest proximal constriction in a few hours to a few days. Presumably this autotomy minimizes the loss of coelomic fluid and also disposes of that region of the pedicle that cannot heal its wound. For some time, after autotomy, a minute amount of coelomic fluid with coelomic corpuscles may escape from the narrow wound, which contracts. The new stump appears as a flat tube, a few millimetres long. A conical pedicle bud with a thin chitinous coat grows out from the posterior end of the ventral mantle in a few days (Fig. 1b). The distal end of the regenerating pedicle forms a bulb with a sticky outer surface. This bulb seems to move away as the intervening part of the pedicle between it and the shell elongates.

A *Lingula anatina* specimen which had entirely lost its pedicle on collection regenerated a short new pedicle within four days. This new pedicle continued to grow posteriorly for several weeks (Trueman and Wong, 1987). However, no accounts or experiments involving other inarticulates or articulates exist. From the existence of a narrow, thinly chitinized region at the proximal end of the pedicle in the other lingulid genus *Glottidia* and from the presence of regenerated pedicles in our collection of *G. pyramidata*, it can be concluded that regeneration of the pedicle also occurs in this lingulid.

D. Reorganization of the Pedicle

A large postlarva of *Lingula anatina* with a well-developed pedicle was dug out of its burrow by us and reared in a dish in the laboratory for a few weeks. The pedicle formed a new bulb about halfway along its length. Distal to this newly formed bulb all the pedicle tissues, except the chitinous covering, shrank and degenerated (Fig. 1c). Gradually the chitinous lining around the new bulb swelled up and separated from the more distal degenerated part of the pedicle to complete the reorganization. Presumably the reduced circulation of the coelomic fluid along its

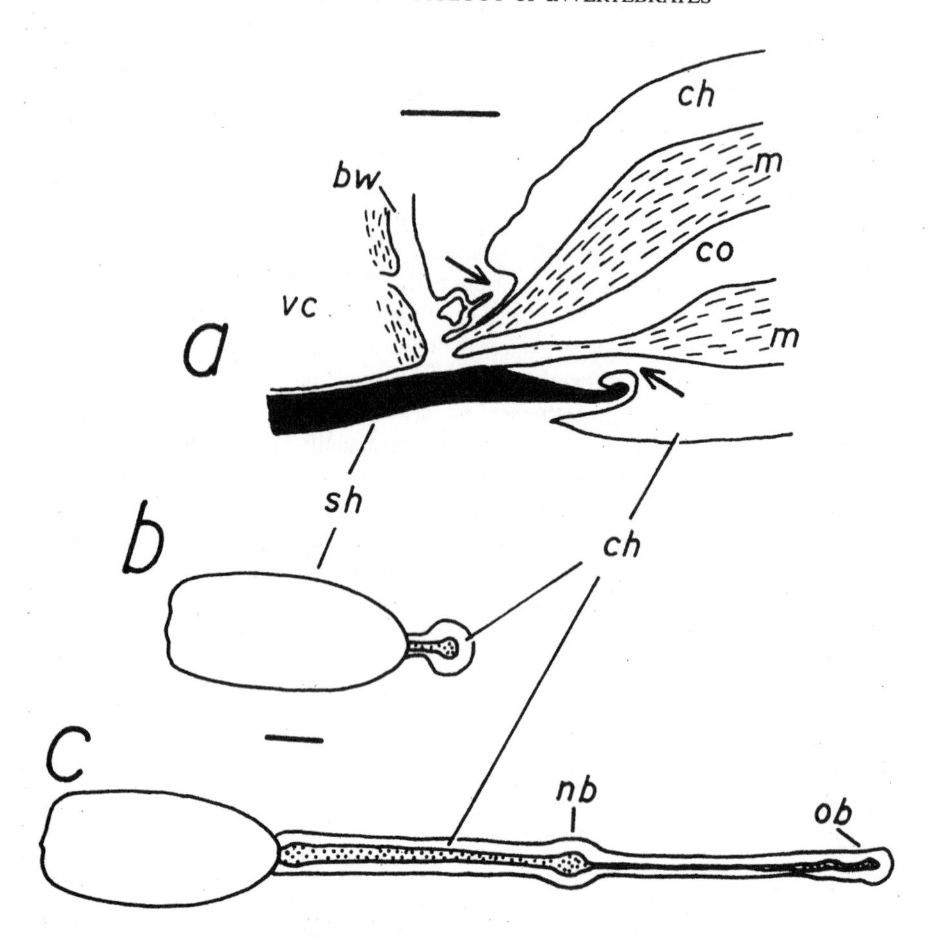

Fig. 1. *Lingula anatina*. **a:** Part of a sagittal section showing the junction of the pedicle with the ventral shell valve. Arrows indicate the thinnest region of the chitinous coat along which pedicle autotomy occurs. **b:** Diagram of mature individual with a regenerated pedicle about seven days after pedicle autotomy. **c:** Diagram of mature individual with a reorganized pedicle and a new bulb. bw, Posterior body wall; ch, chitinous coat of the pedicle; co, coelomic diverticulum inside the pedicle; m, muscles inside the pedicle; nb, new bulb; ob, original bulb; sh, shell; vc, visceral cavity. All scale bars = 1 cm.

length, as a result of inactivity in the laboratory dish, caused this reorganization of the pedicle. Although there was circulation of coelomic fluid within the pedicle canal, as demonstrated by the ciliary currents seen through the translucent pedicle cuticle, apparently it was no substitute for the inrush and outflow of large quantities of coelomic fluid that normally occurred with each relaxation and contraction of the pedicle during the antagonistic contraction and relaxation of the body-wall muscles of the shelled animal in nature (Chuang, 1964). Presumably the ability to reorganize the pedicle enables a lingulid to discard the distal region of its pedicle without total pedicle autotomy and may be useful in adjusting to alterations of its burrow in any disturbance or movement of the shore level.

III. REPRODUCTIVE STRATEGIES

A. Disexuality and Evolution of Hermaphroditism

Disexuality (dioecism) is presumably a primitive trait of the brachiopods, since it occurs in the living fossil *Lingula* and in the majority of Recent brachiopods. Hermaphroditism, rare among articulates, is unknown among inarticulates. It occurs among small articulates that brood their large ova into lecithotrophic larvae. Presumably the smaller articulates not only need to maintain an adequate ovum size for successful development into lecithotrophic larvae but also have to economize on sperm, large quantities of which may be necessary for successful fertilization of the ova retained in the mantle cavity of other individuals, which may be some distance away. The evolution of hermaphroditism is a strategy to overcome these constraints especially when self-fertilization occurs. Presumably it evolved independently in many lines of articulates, since it occurs in *Argyrotheca* of family Megathyridae and in *Pumilus* of Kraussinidae among Recent brachiopods. *Argyrotheca* first appeared in Upper Cretaceous of the Mesozoic Era: hermaphroditism in this genus presumably appeared then.

Simultaneous hermaphrodites such as *Argyrotheca cordata* (see Shipley, 1883; Kowalevsky, 1883; Senn, 1934), *A. cuneata* (see Senn, 1934), *Lacazella mediterranea* (see Lacaze-Duthiers, 1861), and *Pumilus antiquatus* (see Rickwood, 1968) have less testicular tissues than ovarian tissues in their small gonads. Either from exigency of space or extreme economy of successful hermaphroditism, both *A. cordata* and *A. cuneata* have no gonads in their ventral mantle (Senn, 1934). Hermaphrodites produce few ova and are apparently successful in their strategy of reducing energy outlay in reproduction apparently with no adverse effects on postlarval recruitment into the population.

B. Age, Size, Sexual Maturity, and Longevity

The faster growth rate in a more suitable habitat enabled *Lingula anatina* to attain a larger size leading to greater fecundity, i.e. more ova per female, a longer life span, and also more years of breeding activity than another in a less favourable habitat (Chuang, 1961). Articulate *Waltonia inconspicua*, growing at various rates in different localities in New Zealand, reached sexual maturity at the same age but at different sizes (Rickwood, 1977). Inarticulates reach sexual maturity at an earlier age than articulates (see Chuang, Volume V).

Among lingulids the females of the palaeontologically more recent genus *Glottidia* begin to spawn at the age of six to seven months (Paine, 1963), i.e. at half the age of the females of the more ancient genus *Lingula*, but *L. anatina* lives four to six times as long as the two years for *G. pyramidata* (see Chuang, Volume V). Among articulates the smaller females of hermaphrodite *Pumilus* reach sexual maturity at an earlier age than the disexual *Waltonia* but the latter lives three to five times as long as the former (Rickwood, 1968, 1977). Presumably the reproductive strategy is to attain sexual maturity at an earlier age even at the expense of longevity.

C. Strategy of Year-round Breeding

Both inarticulate and articulate brachiopods breed all the year round in the tropical western Pacific (Chuang, 1959, 1968, 1977a; also see Chuang, Volume V of this series). Presumably this is a primitive trait realizable under optimum conditions of food, physiology, and oceanography. At higher latitudes the breeding season is generally short. Thus even the brooding hermaphrodite *Pumilus* has a short breeding season (Rickwood, 1968). However, presumably under the influence of the Gulf Stream and the staggering of spawning in a large population the breeding season of *Glottidia pyramidata* lasts seven to nine months off north Florida and throughout the year off south Florida (Paine, 1963). Other brachiopods at higher latitudes extend their breeding season in various ways. The females of articulate *Waltonia* (as *Terebratella) inconspicua* off North Island of New Zealand save more than 60 per cent of their apparently mature ova for the August–September breeding season after spawning in April–May (Doherty, 1979). The gonads of *Terebratulina retusa* off Scotland pass through two development cycles per year to spawn semiannually (Curry, 1982).

D. Strategy of Staggering the Release of Larvae

Some articulates with short breeding season at higher latitudes space out the release of larvae so as to stagger the recruitment of the larvae into the ranks of the epifaunal postlarval population. In spite of the short breeding season, the recruitment occurs throughout the year for *Terebratalia transversa*, from R.V. Best's seasonal sampling of intertidal population (Thayer, 1977). *Terebratalia transversa* is not a brooder: how this staggering of the recruitment is accomplished is unexplained.

Some brooding articulates stagger the release of their larvae to increase their chances of successful recruitment into the sedentary postlarval population. While hermaphrodite *Argyrotheca cuneata* broods embryos or larvae all at the same stage of development, *A. cordata* broods them at different stages of development presumably by staggering the maturation of its oocytes and the release of its ova from the visceral cavity (Senn, 1934). The female of Japanese *Frenulina sanguinolenta* reaches the highest perfection in this stratagem of staggering: with an annual outlay of about 250 ova, she spawns them one by one throughout the year and broods them into larvae for continuous year-round recruitment into the population (Mano, 1960). Year-round breeding is thus accomplished in spite of the high latitudes.

E. From Planktotrophy to Lecithotrophy and Brooding

The ovum size, the number of ova spawned (estimated or recovered per animal), the number of embryos per brood, the type of fertilization, and the type of larvae in brachiopods are summarized in Table 1. The tabulated data indicate the following trends:

(1) Lingulids spawn small ova in great numbers into the ambient sea water for external fertilization and development into planktotrophic larvae.
(2) Articulates spawn larger ova that develop into lecithotrophic larvae.
(3) A few articulates spawn large numbers of ova for external fertilization and embryonic development in the sea water outside.
(4) Most articulates spawn small numbers of large ova for internal fertilization and brooding inside the mantle cavity.
(5) Articulate species with tiny shell dimensions at sexual maturity produce the largest ova in the smallest numbers for internal fertilization and brooding inside the spawners' mantle cavity.

Table 1 shows that the reproductive strategy is towards economy in nutrients and energy by reducing the number of ova spawned but at the same time increasing their size. It is presumed that spawning large numbers of small ova and spermatozoa into sea water for external fertilization with great waste of nutrients and energy, as in lingulids *Lingula* and *Glottidia* (see Yatsu, 1902; Chuang, 1959; Paine, 1963) and discinids *Discina, Discinisca*, and *Pelagodiscus* (see Chuang, 1977a), is a primitive feature. Moderately large, externally fertilized ova are spawned in moderately large numbers into sea water by articulates *Coptothyris grayi, Terebratalia coreanica* (see Hirai and Fukushi, 1960), *T. transversa* (see Long, 1964), and *Terebratulina retusa* (see Franzén, 1969). These articulate ova develop in the plankton into sluggish larvae with only a gut rudiment devoid of mouth or anus, exposed to the predations and hazards of planktonic mode of life. These lecithotrophic larvae do not feed (Chuang, 1977a). Lecithotrophy frees the articulate larvae from unnecessarily long stay in the plankton and allows them to search promptly for a suitable substratum for settlement, which they generally find in a few hours. The strategy now shifts to the retention of a smaller number of spawned ova for internal fertilization in the safety of the mantle cavity away from planktonic predators. In the more confined space of the mantle cavity of the spawner ova are less spread out to give better chances of successful fertilization.

Presumably brooding naturally follows when developing embryos adhere to the mantle or become entangled among the cirri by mucus, which is normally secreted, and are not forced out when the valves clap shut. Brooding in articulates has evolved in different ways into:

Table 1

Size and numbers of ova, mode of fertilization, and type of larvae in brachiopods

Fertilization	Larval mode of life	Species of brachiopod	Ovum diameter, μm	Number of ova per spawn/brood	Author, year
		INARTICULATA			
External	Planktotrophic	*Lingula anatina*	95.6	2,300–28,600	Chuang, 1959
		Glottidia pyramidata	90	2,200–47,000	Paine, 1963
		Discina and *Discinisca*			Chuang, 1977a
		Pelagodiscus			Chuang, 1977a
Unknown	Unknown	*Crania anomala*	96 × 120		Senn, 1934
Unknown	Unknown	*Crania* sp. (Singapore)	125		Chuang, Volume II
		ARTICULATA			
External	Lecithotrophic	*Coptothyris grayi*	130 × 90		Hirai and Fukushi, 1960
		Terebratalia coreanica	130 × 90		Hirai and Fukushi, 1960
		Terebratalia transversa	150		Long, 1964
		Terebratulina retusa	120		Senn, 1934; Franzén, 1969
		Terebratulina septentrionalis	144 × 160		Conklin, 1902
Internal, in mantle cavity and brooded	Lecithotrophic	*Argyrotheca cordata* and *A. cuneata*			Senn, 1934
		Frenulina sanguinolenta	130		Chuang, Volume II
		Frenulina sanguinolenta		250 per year	Mano, 1960
		Gwynia capsula		2 per brood	Swedmark, 1967
		Hemithiris psittacea	190		Long, 1964
		Liothyrina antarctica			Blochmann, 1906
		Notosaria nigricans	160	8,680	Percival, 1960
		Notosaria nigricans	200		Tortell, 1981
		Pumilus antiquatus	225	100 per year	Rickwood, 1968

		Terebratulina septentrionalis			Webb *et al.*, 1976
		Terebratulina unguicula	170	X × 1,000/brood	Long, 1964
		Waltonia inconspicua		21,500	Doherty, 1979
		Waltonia inconspicua		18,000	Rickwood, 1977
		Waltonia inconspicua	180		Percival, 1944; Tortell, 1981
Visceral cavity and brooded	Lecithotrophic	*Lacazella mediterranea*			Lacaze-Duthiers, 1861

(1) Adherence of embryos to unmodified general surface of the mantle as in *Waltonia* (see Percival, 1944).
(2) Adherence of embryos to two modified longitudinal strips, the nurse-ridges, on the dorsal mantle of *Frenulina sanguinolenta* (see Mano, 1960).
(3) Confinement of developing embryos in the meshes formed by the cirri as in *Pumilus* (see Rickwood, 1968), *Notosaria* (see Percival, 1960), *Hemithiris* and *Terebratulina unguicula* (see Long, 1964).
(4) Attachment of embryos to a pair of enlarged cirri housed in a brood pouch as in *Lacazella mediterranea* (see Lacaze-Duthiers, 1861).
(5) Confinement of embryos in the space formed by the spirals of the lophophore as in *Liothyrina antarctica* (see Blochmann, 1906).
(6) Brooding of embryos in brood pouches in *Argyrotheca* (see Shipley, 1883; Senn, 1934) and *Gwynia capsula* (see Swedmark, 1967).

The evolution of internal fertilization in mantle cavity of the spawner avoids the wastage of ova. The evolution of lecithotrophic larvae and brooding eliminate predation and hazards of the planktonic mode of life. These evolutions enable some diminutive articulates to maintain successfully their populations by brooding a small number of large ova.

F. Abbreviation of the Life History

The life history of Recent brachiopods is abbreviated from the more primitive inarticulates, the lingulids, via the discinids to the articulates. A lingulid hatches out from the embryo as a completely formed larva with an external cover of a pair of chitinous protegula, functional gut, an extensile lophophore as swimming and food-collecting organ, and a pair of statocysts. After leading a long planktotrophic life, it settles down into a sedentary postlarva (Brooks, 1879; Yatsu, 1902; Chuang, 1959; Paine, 1963). When a lingulid larva fails to find a suitable substratum for settlement, it remains in the plankton as a drift larva and continues to add cirri to its lophophore (Paine, 1963; Chuang, 1977a). Other parts of the body such as the gut, pedicle, mantle, shell, and setae also grow and differentiate.

A discinid hatches out as a naked but otherwise complete larva with a functional gut, an extensile lophophore as swimming and food-collecting organ, and a pair of statocysts. It also leads a long planktotrophic life, and secretes a pair of oval chitinous larval valves, which gradually come to cover completely the soft parts. If it fails to become attached at the end of the usual larval stage, it remains in the plankton as drift larva and continues to differentiate until settlement (Chuang, 1977a).

A Recent articulate larva hatches out as a naked lecithotrophic larva. Having only a gut rudiment with no mouth and no anus, it does not feed (Chuang, 1977a). With only cilia as locomotor organ, the larva of an articulate spends only a few hours in the plankton, long enough to find a suitable substratum for settlement. On attachment to the substratum and after metamorphosis into a sedentary postlarva, it immediately differentiates gut, lophophore, skeleton, and other organs (Lacaze-Duthiers, 1861, Kowalevsky, 1883; Shipley, 1883; Percival, 1944, 1960; Long, 1964). An articulate larva that fails to find a suitable substratum is presumed to starve and die.

Lecithotrophy results in a non-feeding, poorly differentiated larva that dispenses with such organs used in fast swimming as lophophore and statocysts. It postpones the formation of lophophore, mouth, shell, and other organs to the postlarval stage (Chuang, 1977a).

IV. CONCLUSIONS

From observations on *Lingula* with the cut but not regenerated stump of a brachium, it is presumed that only the growing tip of a spirolophe can differentiate brachial whorl and cirri in this genus. However, *Lingula, Glottidia*, articulates, and presumably also discinids and craniids rapidly regenerate mantle and shell when a small portion of each of these is cut or otherwise injured.

Lingula and *Glottidia* can autotomize their pedicle and regenerate it rapidly within a few days. Autotomy is of adaptive value since the break at the narrowest part minimizes loss of coelomic fluid and presumably facilitates regeneration of a new pedicle. Rapid pedicle regeneration has adaptive value because it enables the animal to re-establish itself in its burrow rapidly. Lingulid specimens with a regenerated pedicle occur in nature. The power to reorganize the pedicle by the formation of a new bulb between the terminal bulb and the proximal end may have adaptive value in allowing *Lingula* to remove the distal portion of its pedicle without autotomy of the entire pedicle in adjusting to any changes of the burrow or shore level.

The reproductive strategies among Recent brachiopods are directed towards more efficient reproduction. From the presumably primitive trait of disexuality evolved hermaphroditism, which ensures success in fertilization especially when self-fertilization occurs. Inarticulates generally reach sexual maturity at an earlier age than articulates, presumably due partly to the larger ova articulates spawn and partly to the reduced digestive system, resulting in a longer time required to accumulate the nutrients and energy to be packed into the ova. A female of the palaeontologically younger lingulid genus *Glottidia* reaches sexual maturity at an earlier age than the more ancient lingulid genus *Lingula* but does not live as long as the latter. Similarly hermaphroditic articulate *Pumilus* becomes sexually mature earlier but has a shorter life span than the larger disexual *Waltonia inconspicua* (see Rickwood, 1968, 1977). Apparently the strategy is towards early maturity.

Year-round breeding is presumably a primitive trait realizable only under optimum conditions of food, physiology, and oceanography. From it appears to have evolved the short breeding season which prevails at higher latitudes. However, several brachiopods manage to extend their short breeding season in various ways, such as:

(1) by repeating gametogenesis during the year;
(2) by shedding a portion of their ripe ova in one spawning and saving the rest for later spawning some time later in the year;
(3) by staggering the release of their ova so that brooded embryos are at different stages of development; or
(4) by spawning their ova one by one continuously throughout the year and brooding them into larvae for continuous year-round recruitment into the population of postlarvae.

From small ova of inarticulates appear to have evolved the larger ones of articulates. The small ova of lingulids are spawned in great numbers for external fertilization and embryonic development in sea water outside the body into planktotrophic larvae exposed to the predation and hazards of the long planktonic mode of life. Articulates spawn large ova that develop into non-feeding lecithotrophic larvae; they promptly seek a suitable substratum for settlement and generally find one within a few hours. A few articulates still retain external fertilization but most articulates spawn smaller numbers of much larger ova which are fertilized inside the female mantle cavity, where the enclosed space prevents spreading of the ova and ensures better chances of fertilization. The reproductive strategy is for smaller number of larger ova with internal fertilization inside the mantle cavity of the spawner.

Presumably brooding naturally evolved in articulates when ova fertilized in the mantle cavity of the female adhered to the surface of the mantle or became entangled among the cirri of the lophophore and were not forced out when the shell valves snapped shut. The various modes of mantle brooding and cirral brooding are enumerated.

Lecithotrophy and brooding free the embryos and larvae from the predation and hazards of the planktonic mode of life. They make it unnecessary for the majority of articulates to spawn large quantities of ova. They have important adaptive value in enabling diminutive articulates with short life span to maintain their populations by spawning and brooding a small number of large ova.

One other consequence of lecithotrophy is the abbreviation of the life history of the articulates by the omission of the long planktotrophic larval stage, which characterizes the life history of the lingulids and discinids. One disadvantage of this life-history tactic, however, is that while the planktotrophic larvae of lingulids and discinids can remain in the plankton as drift larvae for a long period of time when a suitable substratum for settlement is not available, the lecithotrophic larvae of articulates must promptly find one before their store of energy endowed by the large ovum is exhausted. Lecithotrophic larvae of articulates are presumed to starve and die if they fail to find a suitable substratum for settlement but manage to avoid predation.

REFERENCES

Blochmann, F. (1906). 'Neue Brachiopoden der Valdivia- und Gauss-Expedition', *Zool. Anz.*, **30**, 690–702.

Brooks, W.K. (1879). 'Development of *Lingula* and the systematic position of the Brachiopoda', *Chesapeake Zool. Lab. Sci., Results of Session of 1878*, 35–112.

Chuang, S.H. (1959). 'The breeding season of the brachiopod, *Lingula unguis*', *Biol. Bull.*, **117**, 202–207.

Chuang, S.H. (1961). 'Growth of the postlarval shell in *Lingula unguis* (L.) (Brachiopoda)', *Proc. zool. Soc., Lond.*, **137**, 299–310.

Chuang, S.H. (1964). 'The circulation of coelomic fluid in *Lingula unguis*', *Proc. zool. Soc., Lond.*, **143**, 221–237.

Chuang, S.H. (1968). 'The larvae of a discinid (Inarticulata, Brachiopoda)', *Biol. Bull.*, **135**, 263–272.

Chuang, S.H. (1977a). 'Larval development in *Discinisca* (Inarticulate brachiopod)', *Amer. Zool.*, **17**, 39–53.

Chuang, S.H. (1977b). 'The morphology and palaeobiology of *Linnarssonella girtyi* Walcott (Acrotretid inarticulate brachiopod)', *J. Palaeontol. Soc., India*, **20**, 261–267.

Conklin, E.G. (1902). 'The embryology of a brachiopod, *Terebratulina septentrionalis* Couthouy', *Proc. Am. Phil. Soc.*, **41**, 41–76.

Curry, G.B. (1982). 'Ecology and population structure of the Recent brachiopod *Terebratulina* from Scotland', *Palaeontology*, **25**, 227–246.

Doherty, P.J. (1979). 'A demographic study of a subtidal population of the New Zealand articulate brachiopod *Terebratella inconspicua*', *Mar. Biol.*, **52**, 331–342.

Franzén, Å. (1969). 'On larval development and metamorphosis in *Terebratulina*, Brachiopoda', *Zool. Bidr. Upps.*, **38**, 155–174.

Hirai, E., and Fukushi, T. (1960). 'The development of two species of lamp-shells, *Terebratalia coreanica* and *Coptothyris grayi*', *Bull. Mar. Biol. Stat., Asamushi*, **10**, 77–80.

Hoverd (1985). 'Histological and ultrastructural observations of the lophophore and larvae of the brachiopod, *Notosaria nigricans* (Sowerby 1846)', *J. Nat. Hist.*, **19**, 831–850.

Kowalevsky, A.O. (1883). 'Observations sur le développement des brachiopodes, Analyse par Oehlert et Deniker', *Arch. Zool. exp. gén.*, ser. 2, **1**, 57–76.

Lacaze-Duthiers, H. (1861). 'Histoire naturelle des brachiopodes vivants de la Méditerranée. I. Histoire naturelle de la Thécidie (*Thecidium mediterraneum*)', *Ann. Sci. nat.* (*Zool.*), *Paris*, **15**, 259–330.

Long, J.A. (1964). 'The embryology of three species representing three superfamilies of articulate Brachiopoda', Ph.D. dissertation, University of Washington, 185 pp.

Mano, R. (1960). 'On the metamorphosis of a brachiopod, *Frenulina sanguinolenta* (Gmelin)', *Bull. Mar. Biol. Stat., Asamushi*, **10**, 171–175.

Paine, R.T. (1963). 'Ecology of the brachiopod *Glottidia pyramidata*', *Ecol. Monogr.*, **33**, 187–213.

Percival, E. (1944). 'A contribution to the life history of the brachiopod, *Terebratella inconspicua* Sowerby', *Trans. Roy. Soc. New Zealand*, **74**, 1–23.

Percival, E. (1960). 'A contribution to the life history of the brachiopod *Tegulorhynchia nigricans*', *Q. Jl microsc. Sci.*, **101**, 439–457.

Rickwood, A.E. (1968). 'A contribution to the life history and biology of the brachiopod *Pumilus antiquatus* Atkins', *Trans. Roy. Soc. New Zealand* (*Zool.*), **10**, 163–182.

Rickwood, A.E. (1977). 'Age, growth and shape of the intertidal brachiopod *Waltonia inconspicua* Sowerby, from New Zealand', *Am. Zool.*, **17**, 63–73.

Senn, E. (1934). 'Die Geschlechtsverhaeltnisse der Brachiopoden, im besonderen die Spermato- und Oogenese der Gattung *Lingula*', *Acta Zool., Stockholm*, **15**, 1–154.

Shipley, A.E. (1883). 'On the structure and development of *Argiope*', *Zool. Stat. Neapel, Mitt.*, **4**, 494–520.

Swedmark, B. (1967). '*Gwynia capsula* (Jeffreys), an articulate brachiopod with brood protection', *Nature, Lond.*, **213**, 1151–1152.

Thayer, C.W. (1977). 'Recruitment, growth, and mortality of a living articulate brachiopod, with implications for the interpretation of survivorship curves', *Paleobiology*, **3**, 98–109.

Tortell, P. (1981). 'Notes on the reproductive biology of brachiopods from southern New Zealand', *N.Z. J. Zool.*, **8**, 175–182.

Trueman, E.R., and Wong, T.M. (1987). 'The role of the coelom as a hydrostatic skeleton in lingulid brachiopods', *J. Zool. London.*, **213**, 221–232.

Webb, G.R., Logan, A., and Noble, J.P.A. (1976). 'Occurrence and significance of brooded larvae in a Recent brachiopod, Bay of Fundy, Canada', *J. Paleont.*, **50**, 869–871.

Williams, A., Rowell, A.J., Muir-Wood, H.M., Pitrat, C.W., Schmidt, H., Stehli, F.G., Ager, D.V., Wright, A.D., Elliott, G.F., Arnsden, T.W., Rudwick, M.J.S., Hatai, K., Biernat, G., McLaren, D.J., Boucot, A.J., Johnson, J.G., Staton, R.D., Grant, R.E., and Jope, H.M. (1965). 'Part H: Brachiopoda', in *Treatise on Invertebrate Paleontology* (Ed. R.C. Moore), Geological Society of America, Inc. and University of Kansas Press, Lawrence.

Yatsu, N. (1902). 'On the development of *Lingula anatina*', *J. Coll. Sci., Tokyo*, **17**, 1–112.

8. CHAETOGNATHA

A. ALVARIÑO
National Marine Fisheries Service,
Southwest Fisheries Science Center,
8604 La Jolla Shores Drive,
P.O.Box 271, La Jolla, California 92038-0271, U.S.A.

I. INTRODUCTION

Asexual propagation does not occur in chaetognaths. When a chaetognath is broken, regeneration occurs only of the part that includes the head and the ventral ganglion. If the head is missing, the ventral ganglion assumes the direction of the regenerative process. However, there is no production of other individuals, only a regeneration of the missing part of the animal.

II. REGENERATION

Few experiments have been conducted on regeneration in chaetognaths. Kulmatycki (1918) and Ghirardelli (1968) found that *Spadella cephaloptera* could regenerate the whole caudal section, and *Sagitta helenae* and *S. enflata* the head (Pierce, 1951).

In animals there appears to be a close correlation between regenerative power and the stage at which segregation of the germinal line occurs (Ghirardelli, 1968). Animals with precocious determination of the primordial germ cell have usually limited regenerative ability, insufficient for the reconstruction of complete new individuals from a portion of the body, and they cannot regenerate important and large portions of the body. They also lack the power of asexual reproduction by fission or budding (Ghirardelli, 1956, 1959a). In Chaetognatha, segregation of the germinal-line determinant occurs early in development of the egg (see Alvariño, Volume IVB of this series). Deficiency of regenerative power seen in chaetognaths is shared by animals with RNA-rich ova, a characteristic associated with absence of asexual reproduction (Ghirardelli, 1959b, 1965). Different animals of analogous evolution (e.g. Rotifera, Cladocera, Copepoda, Chaetognatha, Anura Amphibia) do not have the totipotent, or at least pluripotent cells which constitute the so-called embryonic reserve (Ghirardelli, 1968). Based on these observations, Ghirardelli (1958) stated that the regenerative ability of chaetognaths could not be as marked and effective as claimed by Kulmatycki (1918) and Pierce (1951).

Ghirardelli (1968) performed experiments on *Sagitta enflata* and *Spadella cephaloptera* from Villefranche-sur-Mer and the north Adriatic Sea, by cutting the animals at various levels of the body. Regeneration of the caudal fin is always rapid (from two to four days) if the posterior part of the body is not damaged. Epithelial cells in mitosis migrate towards the wound, and reconstruct the fin rays (Ghirardelli, 1959a). The lateral fins may be regenerated in seven to 10 days. The caudal region, if sectioned at a level *anterior* to the seminal vesicles, does not regenerate. The sectioned muscles fold in a scar made of alveolar stratified epidermis, reminiscent of an epithelial collarette. In some cases, the lateral fins fold towards the wound and fuse with the midline of the body, and there is no formation of seminal vesicles in the wounded caudal coelom, although normal development of the sperm proceeded. In such animals the sperm, however, cannot be extruded due to want of seminal vesicles. Therefore, no new cycle of male gametes would be possible as implied by Kulmatycki (1918) and Pierce (1951). On the contrary, if the cut passes through immediately *behind* the seminal vesicles, only a rudimentary regeneration of the caudal fin occurs.

Ghirardelli (1958) obtained regeneration of the caudal fin of *Spadella cephaloptera* beheaded immediately after amputation of the posterior end of the body, 56 per cent fin regeneration in decapitated animals against 36 per cent in those not decapitated. This may suggest that the presence of the head inhibits the regenerative process of the caudal fin, and that regeneration is only possible via the ventral ganglion.

Regeneration of the head does not occur, according to Ghirardelli (1956, 1959a), in *Spadella cephaloptera* and *Sagitta* (probably *S. bipunctata*). However, Pierce (1951) and other authors have observed specimens of *Sagitta setosa, S. minima, S. enflata*, and *S. euneritica* with heads that are abnormally small, and bearing no teeth or hooks. It is not clear from their accounts if that condition was produced by regeneration or what they saw was only a wounded head. Species of *Sagitta* may be beheaded by predators and, if the amputation occurred at the level of the ventral ganglion or in front of it, a sharp stump could be formed with histological features of a scar. This would indicate a lack of differentiation and of the regeneration process. Duration of the regeneration process and the chances of survival of the wounded animal are not known. The stumpy head is more rounded and the eyes can be seen, and there is doubt whether it is a case of true regeneration or only one of repair of a lesion on the cephalic region.

Pierce (1951) indicated that *Sagitta helenae* and *S. enflata* found with their heads cut off, together with a short section of the trunk, were at various stages of head regeneration. This condition was first attributed to accidental damage encountered in the net, but analyses of plankton samples have shown many specimens at various degrees of head regeneration. No regrowth of the posterior part of the body and tail was found. After the loss of the head, the trunk tissues contracted tightly together at the wounded end, giving the appearance of a sausage. The head began to form inside the contracted end, the eyes appearing first, followed by the mouth and the hooks (Alvariño, 1965).

Pierce (1941) stated that although he had examined hundreds of *Sagitta setosa* and *S. elegans* from the northern latitudes, no evidence of head regeneration was observed in these species.

Ghirardelli (1959c) reported young chaetognaths, probably *Sagitta bipunctata* and *S. enflata*, showing the beginnings of regeneration of the head. I have observed regeneration of the head in *S. elegans, S. enflata, S. euneritica, S. pacifica, S. neglecta*, and *S. pulchra* (Alvariño, unpublished).

III. PARTHENOGENESIS

This form of reproduction has not been investigated in chaetognaths. However, laboratory experiments with isolated individuals (Reeve, 1966), in which no copulation would occur, have shown that bursting seminal vesicles ejected sperm to the water, some of which would enter the receptaculum seminis. Only a few extruded ova were fertilized and those produced larvae, while the unfertilized ova invariably degenerated. It could therefore be assumed that parthenogenesis proper does not occur in Chaetognatha.

IV. FECUNDITY

The fecundity of *Sagitta elegans* has been calculated (McLaren, 1963, 1966) by setting ovary length against body length, and in turn relating the number of ova to ovary length. It appears that egg production by *S. elegans* is the function of adult size. The brood size is the result of the number of eggs produced by a functional female population. Egg number is equivalent to the number of ova carried by the parent, and it is a function of the size of the animal and the capacity of the ovary filled with ova. Comparisons must be made between animals of the same size, if a calibration were to be set up between parent size and egg number. The standardized number of eggs may then be used to compare the reproduction potential of a given species under different environmental conditions, within a restricted locality or over various areas in the wide distributional region of the species.

Breeding may be restricted to one or several seasons, or a succession of broods may be produced, as in *Sagitta enflata*. Therefore, simply put, the rate at which eggs are produced will be termed 'fecundity'. The number of eggs produced per standard animal per species per unit of time (season) could be determined. Different species reach various adult sizes, in each of which the number of eggs is a function of the size of the animal. Fecundity can thus be defined as 'the number of eggs produced per individual per species and geographic location'.

Determination of the number of ova is easy in chaetognaths, the animals being transparent. The number of ova can be plotted against standard body weight or body length. Variability within species, time of year, and locality and over the years can so be observed. Differences are likely an expression of long-term cyclical fluctuations in reproduction level from year to year, related basically to food supply. Such differences and annual variations in egg production may be of importance in competition and dominance of chaetognath populations. The number of ova produced by chaetognaths is usually higher in 'poor' areas and in species living under stress (e.g. bathypelagic and cold-water species). Small numbers of ova — flat, elongated, and scattered along the ovaries — are observed in tropical populations carried out of their normal environmental living areas because of either scarcity of food or infestation with parasites (e.g. *Sagitta bierii* off Central American Pacific, *S. bedfordii* in some Indonesian locations). This confirms that food supply is a basic factor in the success of reproduction in chaetognaths as in many other animals. The number of ova in epipelagic populations is related to the size of the population: in large populations, the number of ova produced could be lower than in small populations.

To compare relative reproductive efficiency, there is a need to take into account the metabolic process, the number and size of eggs produced, and the time required for the production and hatching of eggs. Certain species can withstand severe climatic conditions advantageously by a marked seasonal breeding period, and by production of a relatively small number of large eggs. However, the same annual cycle could be maintained with a relatively large number of small eggs, leading to greater fecundity, and consequently greater competitive power. Large eggs produce larvae which are almost as large and as advanced in development as the juveniles — strong, able to feed, and better suited to survive.

V. STERILITY

Sterility, or poor reproductive ability, of *Sagitta scrippsae* off California (Alvariño, 1983) appears to be related to lack of adequate food to provide for the development of the gonads. *Sagitta bierii* inhabits the eastern Pacific from the United States to Chile, but populations off Central America do not seem to reproduce there due to high temperature and/or lack of the right kind of food in the area; the animals are all small with few flat ova (Alvariño 1964b, 1965, 1967). This observation leads us to suggest that *S. bierii* of Central American Pacific are likely descendants of specimens inhabiting the north and south of the region, mainly from the California- and Peru-Currents domain of the species (Alvariño, 1965, 1967). How higher temperatures can inhibit breeding is illustrated by *S. hispida*, a species unable to breed in the Biscayne Bay during the high water temperature of the summer.

Parasitism is another cause of arrest in gonadal development of chaetognaths, at least in some areas of the world: e.g. Gulf of Siam (*Sagitta bruuni, S. bedoti, S. pulchra*), the Caribbean (*S. enflata*), the Mediterranean (*S. enflata*), and off California mainly in the neighbourhood of Los Angeles (Alvariño, unpublished; and *S. scrippsae*:Alvariño, 1983). A high percentage of Chaetognatha here are infested with parasites (trematodes, cercarias), a strong indication of man-made pollution.

VI. REPRODUCTIVE CYCLES AND STRATEGIES

In *Sagitta friderici*, there is only one reproductive cycle, at the end of which the animals die (Cavalieri, 1963). Studies of abundance, distribution, and seasonal reproductive cycles of *S. elegans* indicate a life span of three months in Georges Bank (Clarke *et al*., 1943); the animals die soon after breeding. In Georges Bank there is only one main breeding period per year. However, two populations of *S. elegans* may coexist there at least during part of the year, one indigenous to the area, and the other carried south from northern waters with the Labrador Current. Therefore, a wide size-range of maturity stages is encountered for *S. elegans* in the American North Atlantic. In the Georges Bank area, reproduction of *S. elegans* generally starts in April. Stage I animals present in September may represent the end of the spring and summer reproduction; those overwintering will produce a new generation next spring.

Two generations of *Sagitta elegans* were found in Bedford Basin by Zo (1973) with different life-history tactics: a spring generation which reached maturity in mid-September and reproduced during the rest of the year and an autumn generation (born from the spring generation) which apparently overwintered and reproduced during the next spring. Most of the young developed to maturity between June and September. This suggests that the successive hatching of eggs in spring provides a continuous recruitment of adults. The autumn generation, i.e. animals hatched during the autumn-winter period, grow slowly and reach maturity in the spring with increase in temperature and food availability. The larvae measure 1.28–1.30 mm. Autumn adults developed at 3.8°C have a mean length of 19.7 mm, while spring adults developed at a mean water temperature of 2.3°C are more than 25 mm long.

Sagitta enflata, which is abundant in temperate and tropical oceanic waters and has also invaded neritic regions, is probably the only epipelagic chaetognath to resort to the tactics of multiple cycles of maturity during its life span. As a result of this, the fecundity reaches a high level and the population becomes large and agglomerated in space leading to severe competition for food among the individuals. *Sagitta enflata* is also the chaetognath presenting the highest degree of cannibalistic behaviour, a possible regulator of population size.

Uninterrupted reproduction takes place in tropical and subtropical chaetognath populations; the rate of survival of the fall generation, however, is higher than that of the winter generation. Breeding index in the tropics is not constantly maintained through the year, and variations in density of the populations result from biotic and abiotic changes in the environment. Neritic species inhabit a region of drastic environmental fluctuations along the year, and consequently they show greater variation in density of population than oceanic species distributed over wide ranges. *Sagitta euneritica*, off California, exhibits great seasonal variation through the year (Alvariño, 1967). In species studied over a wide latitudinal range of their distribution, such as *S. bipunctata* and *Pterosagitta draco*, reproduction occurs throughout the year in the tropics, but only in spring and summer at temperate locations. This means that the reproductive span shortens coincidentally with the termic variations, which also affect changes in the food supply.

Annual breeding seasons are common to both Arctic and Antarctic regions. A large proportion of polar species brood their embryos and larvae (e.g. *Eukrohnia*) and release them as juveniles, when food is available. In species inhabiting deep waters, maturity takes place once a year as in polar species, but their life span may include, at least occasionally, several maturity cycles.

A. Reproductive and Developmental Delays

One mode of ensuring optimum reproductive and developmental success for a species is by temporally scheduling these processes to take place when conditions for survival of the progeny are favourable. Coming to strategic delays in reproduction and development, in *Eukrohnia*, 'diapause' in development of eggs and larvae may be regulated to favour hatching process at the right time for survival of the species as we have just seen above. Overwintering population of *Sagitta crassa* delay reproduction until food is plentiful for the larvae. Russel (1936) suggested that normal delay in development of gonads of late summer broods of many planktonic animals, including chaetognaths, is frequently accompanied by a descent of that population to deep strata, provoked by the removal of some essential factor "from the water during the summer, which becomes available again after a period of time or is perhaps lacking in the diminishing food supply". The significance of carotenoids, vitamins, and related substances in the plankton has only been tentatively investigated, but it is known that the carotenoid, vitamin and sterol constant varies greatly, not only in relation to the plankton but also to the season. Carotenoids and vitamin A within the plankton (Gillam *et al.*, 1939) begin to rise just about or at the spring diatom peak, the maximum being during the autumn. The peak coincides with the main breeding period of the plankton (Lucas, 1947).

It is well known that water-borne metabolites may influence reproduction. The conditioning of the environment through external products may bring about community integration through adaptation of animals to the products of others. The production of certain compounds may be related with reproduction of planktonic animals and induce particular stages of their life history. Local concentration of such compounds might tend to stimulate the development of maturing animals and the fertilization of eggs (Alvariño, 1989). Carnivores do not eat plants, but some of the substances available in plants are obtained by them through their prey. It would be of interest to develop research on the biological factors responsible for the variation and constitution of plankton communities and their influence in reproduction, spawning and breeding, hatching, and survival.

Water masses may have similar parameters of salinity and temperature but, in keeping with different plankton populations they harbour, differences in rate of reproduction between chaetognath species are not uncommon. These variations may depend on vital metabolites and other substances released to, or removed from, the waters. This is best illustrated by *Sagitta elegans* and *S. setosa*, dominant in plankton off England. Abundance of *S. elegans* is accompanied by plankton assemblages different from those of *S. setosa*; also, successful fish-larvae populations are related to dominance of *S. elegans* in plankton of the region. Appropriate combination of experiments suggests the presence of a 'beneficial' substance in both Celtic and Biscay Bay waters, and its lack in the local waters characterized by an abundance of *S. setosa*. The English Channel waters are characterized usually by *S. setosa* community, and the Celtic Sea and Bay of Biscay by *S. elegans* and other different planktonic communities. It appears that the Celtic and Biscay waters supply some necessary ingredients for successful reproduction, resulting also in an abundant fish population (Wilson and Armstrong, 1952). Alvariño (1989) showed that certain plankton assemblages ensure high reproductive success and high survival of anchovy larvae.

B. Breeding and Spawning Strategies

Reproductive success can be guaranteed only if suitable strategies are evolved by the species to synchronize breeding and spawning activities of its members and thereby to increase the chances of fertilization and production of the young. The simplest method of estimating reproductive activity of any species is the observation of breeding of animals in the field or in the laboratory. Such observations are imprecise but they produce important data on the presence of mature gametes, sperm, and ova. Spawning is not easy to induce when gonads are not gravid, but when the animal is gravid almost any stimulus may cause spawning, and also egg laying.

There is need to determine the factors inducing spawning in nature. It may be argued (see Giese and Pearse, 1973 for discussion) that in nature no spawning stimuli *per se* are required, and that spawning and breeding are spontaneous after the gonads have become mature. This hypothesis is, however, disputed as changes in biotic and abiotic factors and hormonal milieu are definitely known to play a role in inducing spawning in animals (see Volumes I and II of this series). *Sagitta tasmanica* specimens reaching the Gulf of Maine area grow larger than normal with large ovaries and well-developed full seminal vesicles, but there is no evidence of breeding perhaps because there the spawning stimulus is wanting. The animals may die and disappear either with no breeding or with a young population that does not survive. In any case, no *S. tasmanica* eggs have been reported from that area.

It is well known that light induces spawning and breeding (Knight-Jones, 1951; Kume and Dan, 1968; Segal, 1971) in animals. Exudates from green algae (Myazaki, 1938), diatom blooms (Barnes, 1957), and chemical exudates from males and females of the same species have been shown to synchronize breeding. Probably some of the most mature individuals, the best receptors of stimuli, are first induced to breed in nature and produce exogenous stimulants to other individuals in the population, until the synchronization phenomenon reaches epidemic proportions ensuring maximum fertilization.

Studies on bathymetric distribution of chaetognath species indicate that young and small specimens inhabit the upper oceanic layers, while full-mature and large individuals extend into deep waters. In other words, there is an ontogenic vertical distribution of the population down the water mass (Alvariño, 1964a, 1965). Our analysis of the bathymetric distribution of *Sagitta scrippsae* (open-closing net collections), off California, included specimens 10 to 50 mm long. Small-sized individuals appeared in the upper 225 m, and the largest in strata below. Young, Stage I individuals of *S. gazellae* inhabit the upper 250 m (Alvariño *et al.*, 1983) whereas Stage II appears mainly at 500 m to 1,000 m depth (David, 1955). Distribution of *S. elegans* in Bedford Basin (Zo, 1973) also showed an ontogenic bathymetric pattern of the populations. Observations on the ontogenic stratification of the chaetognath populations (Alvariño, 1964a) indicate that fecundation and laying of eggs take place in deep layers, and eggs and young are brought up to upper layers. Studies by Reeve and Cosper (1975) on *S. hispida* populations in the shallow waters of Biscayne Bay, Florida, indicate that surface waters rarely contain mature animals. In the eastern Canadian region, young *S. elegans* and *Eukrohnia* appear in the upper layers, while large individuals are at deep strata (Huntsman, 1919).

REFERENCES

Alvariño, A. (1964a). 'Bathymetric distribution of Chaetognaths', *Pac. Sci.*, **18**(1), 64–82; *Contrib. Scripps Inst. Oceanogr.*, **34**(1616), 39–57.

Alvariño, A. (1964b). 'Zoogeography of Chaetognatha, specially in the California Region. (Zoogeografía de los Quetognatos, especialmente de la region de California)', *Ciencia*, **23**(2), 51–74, *Contrib. Scripps Inst. Oceanogr.*, **34**(1705), 1677–1702.

Alvariño, A (1965). 'Chaetognaths', *Oceanogr. mar. Biol. Ann. Rev.*, **3**, 115–194.

Alvariño, A. (1967). 'Zoogeography of California: Chaetognatha (Zoogeografía de California: Quetognatos)', *Rev. Soc. Mex. Hist. Nat.*, **27**, 199–243, *Contrib. Scripps Inst. Oceanogr.*, **37**(2139), 487–531.

Alvariño, A. (1983). 'The depth distribution, relative abundance and structure of the population of the Chaetognatha *Sagitta scrippsae* Alvariño 1962, in the California Current off California and Baja California', *Anales del Instituto de Ciencias del Mar y Limnologia, Universidad Nacional Autonoma de Mexico (UNAM)*, **10**(1), 47–84.

Alvariño, A., Hosmer, S.C., and Ford, R.F. (1983). 'Antarctic Chaetognatha: United States Antarctic Research Program ELTANIN Cruises 8–28, Part I. American Geophysical Union. Biology of the Antarctic Seas XI', *Antarctic Research Series*, **34**, 129–338.

Alvariño, A. (1989). 'Abundance of zooplankton species, females and males, eggs and larvae of holoplanktonic species. Zooplankton assemblages and changes in the zooplankton communities related to *Engraulis mordax* spawning and survival of the larvae', *Mem. III Encontro Brasileiro de Plancton*, pp. 63–149.

Barnes, H. (1957). 'Processes of restoration of synchronization in marine ecology: The spring diatom increases and the "spawning" of the common barnacle, *Balanus balanoides* (L)', *Ann. Biol.*, **33**, 67–85.

Cavalieri, F. (1963). 'Nota preliminar sobre *Sagitta* (Chaetognatha) del litoral Atlantico Argentino', *Physis*, **24**(67), 223–236.

Clarke, G.L., Pierce, E.L., and Bumpus, D.F. (1943). 'The distribution and reproduction of *Sagitta elegans* on Georges Bank in relation to the hydrographical conditions', *Biol. Bull.*, **85**, 201–226.

David, P.M. (1955). 'The distribution of *Sagitta gasellae* Ritter-Zahony', *Discovery Rept.*, **27**, 235–278.

Ghirardelli, E. (1956). 'La rigenerazione in *Spadella cephaloptera* Busch', *Boll. Zool.*, **18**(2), 597–608.

Ghirardelli, E. (1958). 'La rigenerazione in *Spadella cephaloptera* Busch: Influenza del capo sulla rigenerazione della regione caudale', *Riv. Biol.*, **50**(2), 169–177.

Ghirardelli, E. (1959a). 'La struttura della pinne e la istogenesi rigenerativa in *Spadella cephaloptera* Busch.', *Pubbl. Staz. zool. Napoli*, **31**(1), 1–14.

Ghirardelli, E. (1959b). 'Osservazioni sulla deficienza dei poteri rigenerativi nei Chaetognati. Considerazioni sui rapporti fra riproduzione agamica e determinazione del ceppogerminale', *Atti Accad. Sci. Ist. Bologna*, Rc. 11, **6**, 1–15.

Ghirardelli, E. (1959c). 'Habitat e biologia della riproduzione nei Chetognati', *Arch. Oceanogr. Limnol.*, **11**(3), 1–18.

Ghirardelli, E. (1965). 'Differentiation of the germ cells and regeneration of the gonads in Planarias', in *Regeneration in Animals and Related Problems* (Eds. V. Kiortsis and H.A.L. Trampusch), pp. 177–184.

Ghirardelli, E. (1968). 'Some aspects of the biology of the chaetognaths', *Adv. Mar. Biol.*, **6**, 271–375.

Giese, A.C., and Pearse, J.S. (1973) (Eds). *Reproduction in Marine Invertebrates*, Vol. 1, Academic Press, New York, pp. 1–48.

Gillam, A.E., El Ridi, M.S., and Wimpenny, R.S. (1939). 'The seasonal variation in biological composition of certain plankton samples from the North Sea in relation to their content of vitamin A, carotenoids, chlorophyll and total fatty matter', *J. Exp. Biol.*, **16**, 71–88.

Huntsman, A.G. (1919). 'Some quantitative and qualitative plankton studies of the Eastern Canadian plankton. 3. A special study of the Canadian Chaetognatha, their distribution, etc. in the waters of the eastern coast', *Can. Fish. Exped. 1914–1915, Ottawa*, pp. 421–485.

Knight-Jones, E.W. (1951). 'Gregariousness and some other aspects of settling behaviour in *Spirorbis*, *J. mar. biol. Ass. U.K.*, **30**, 201–222.

Kulmatycki, W.J. (1918). 'Bericht uber die Regenerations-fahigkeit der *Spadella cephaloptera*', *Zool. Anz.*, **49**, 281–284.

Kume, M., and Dan, K. (1968). 'Introduction', in *Invertebrate Embryology*, Nobit Publ. House, Belgrade. pp. 1–7.

Lucas, C.E. (1947). 'The ecological effects of external metabolites', *Biol. Rev.*, **22**(3), 270–295.

McLaren, Ian A. (1963). 'Effects of temperature on growth of zooplankton and the adaptive value of vertical migration', *J. Fish. Res. Bd. Canada*, **20**(3), 685–727.

McLaren, Ian A. (1966). 'Adaptive significance of large size and long life of the chaetognath *Sagitta elegans* in the Arctic', *Ecology*, **47**, 852–855.

Myazaki, J. (1938). 'On the substance which is contained in green algae and induces spawning action of the male oyster (preliminary note)', *Bull. Japan. Soc. Sci. Fish.*, **7**, 137–138.

Pierce, E.L. (1941). 'The occurrence and breeding of *Sagitta elegans* Verrill and *Sagitta setosa* J. Muller in parts of the Irish Sea', *J. mar. biol. Ass. U.K.*, **25**, 113–124.

Pierce, E.L. (1951). 'The Chaetognatha of the West Coast of Florida', *Biol. Bull.*, **100**(3), 206–228.

Reeve, M.R. (1966). 'Observations on the biology of a chaetognath', in *Some Contemporary Studies in Marine Science* (Ed. H. Barnes), George Allen and Unwin, London, pp. 613–630.

Reeve, M.R., and Cosper, T.C. (1975). 'Chaetognatha', in *Reproduction of Marine Invertebrates*, Vol. II, *Entoprocts and lesser Coelomates* (Eds. A.C. Giese and J.S. Pearse), Academic Press, New York, pp. 157–181.

Russell, F.S. (1936). 'A review of some aspects of plankton research', *Rapp. Proc. Verb. Cons. Perm. Intern. Explor. Mer.*, **95**, 5–30.

Segal, E. (1971). 'Light, animals, invertebrates', *Mar. Ecol.*, **1**(1), 159–211.

Wilson, D.P., and Armstrong, F.A.J. (1952). 'Further experiments on biological differences between natural sea waters', *J. mar. biol. Ass. U.K.*, **311**, 335–349.

Zo, Z. (1973). 'Breeding and growth of the chaetognath *Sagitta elegans* in Bedford Basin', *Limnol. Oceanogr.*, **18**, 750–756.

9. ECHINODERMATA: ASEXUAL PROPAGATION

PHILIP V. MLADENOV
Department of Marine Science, University of Otago, PO Box 56, Dunedin, New Zealand

ROBERT D. BURKE
Department of Biology, University of Victoria, Victoria, British Columbia V8W 2Y2, Canada

I. INTRODUCTION

The topic of asexuality in echinoderms is not an overly familiar one to most biologists. The asexual abilities of other groups of metazoans like corals, hydras (see Shostak, this volume), and turbellarians (see Benazzi and Lentati, this volume) seem to overshadow those of echinoderms. Nonetheless, asexuality is relatively widespread among certain groups of echinoderms, like brittle stars and sea stars (for classification of Echinodermata, see Systematic Résumé). Furthermore, some of the asexual species are among the most successful of echinoderms as measured by both their widespread geographic distribution and their often local abundance.

In this review we explore the current state of knowledge of echinoderm asexuality on a class-by-class basis. We are by no means the first to review this topic. A.M. Clark (1967) discussed the phenomenon of fission in asteroids and ophiuroids and Emson and Wilkie (1980) produced a thorough overview of fission and autotomy (the voluntary loss of body parts) in echinoderms. Our review incorporates the abundant literature that has emerged in the ten years following Emson and Wilkie's (1980) important publication. It also addresses the topic in a wider context to include such aspects as naturally occurring parthenogenesis, asexuality in larval echinoderms, and the population genetics of asexual echinoderms.

Asexual propagation, autotomy, and regeneration are closely linked phenomena in the echinoderms. Autotomy, in combination with the remarkable ability of echinoderms to manipulate the mechanical properties of their mutable collagenous tissues (MCTs), often initiates the process of asexual propagation by facilitating the fragmentation of body parts. Regenerative growth, on the other hand, completes the process by·restoring the lost structures. In this review, we touch on the subjects of autotomy and MCTs only as background to such topics as the mechanism or regulation of asexual processes. Autotomy and MCTs have been well reviewed by Emson and Wilkie (1980), Motokawa (1984, 1985), and Wilkie (1984), and the reader is referred to these papers for details. We have also considered the process of regeneration to be largely outside the scope of this review. This is because echinoderm regeneration is a large and important topic that would in itself make the subject of a comprehensive review.

II. CRINOIDEA

There is no convincing evidence that crinoids propagate asexually in nature, either by apomictic parthenogenesis or by fragmentation. Freshly spawned, unfertilized eggs of the feather star *Comanthus japonica* are susceptible to parthenogenetic activation by mechanical agitation in the laboratory; about 50 per cent of such eggs develop into swimming larvae, but they then die (Dan and Dan, 1941). Danielssen (1892) noted several individuals of the sea lily *Bathycrinus carpenteri* in the process of regenerating a new crown from an old stalk at the level of the basal plates (i.e. the skeletal plates that form the lower part of the calyx). It thus seems that some sea lilies can lose the crown under certain circumstances and then replace it. The fate of the shed crown, however, is unknown. Danielssen (1892) suggested that repeated shedding and regeneration of crowns might result in asexual proliferation by budding, but this has not been substantiated.

III. ASTEROIDEA

A. Asexual Propagation by Fission

1. Occurrence

Some sea stars propagate asexually in nature by fission (Steenstrup, 1857; Kowalevsky, 1872; Crozier, 1920; Fisher, 1925; Bennett, 1927; Edmondson, 1935; A.M. Clark, 1967; Marsh, 1977; Ottesen and Lucas, 1982; Rowe and Marsh, 1982; Jangoux, 1984; Achituv and Sher, 1991). In these fissiparous species, the sea star divides into two, occasionally three, parts each of which is capable of regenerating into a whole animal. Individuals of such species are usually distinctly asymmetrical, with a set of large and a set of smaller regenerating arms; they typically have more than five arms, with six-, seven-, and eight-armed forms being particularly common; and they often possess more than one madreporite and anus.

Emson and Wilkie (1980) listed as fissiparous 19 species of sea star from two families, the Asteriidae and Asterinidae. Since then, one new species of fissiparous sea star has been discovered, *Seriaster regularis*, from the family Solasteridae

(Jangoux, 1984; Guille *et al.*, 1986). In addition, a new species of asterinid sea star, *Nepanthia fisheri*, is recorded as potentially fissiparous by Rowe and Marsh (1982). These authors also suggest that the fissiparous *Nepanthia brevis* and *N. variabilis* are synonyms of the variable species *N. belcheri*. The known number of fissiparous sea stars thus varies between 18 and 21 (Table 1). Although a few more fissiparous sea stars will undoubtedly come to light in the future, it is clear that only a small proportion of the approximately 1,600 extant species of sea star have evolved the capacity to exploit fission as a regular means of asexual proliferation. Nonetheless, it is evident from Table 1 that some fissiparous sea stars, notably *N. belcheri*, *Coscinasterias calamaria*, and *Stephanasterias albula*, have a broad geographic distribution. Moreover, some species can be locally very abundant. For example, densities of *S. albula* of 40/m^2 and 28/m^2 were recorded at North Lubec, Maine (Mladenov *et al.*, 1986), and Pigeon Hill, Massachusetts (Hulbert *et al.*, 1983), respectively.

Table 1

Fissiparous sea stars and their geographic distribution

Family	Species	Distribution	References
Asterinidae	*Asterina anomala* H.L. Clark	Australia to Hawaiian Islands	Bennett (1927), Yamaguchi (1975), Marsh (1977)
	A. burtoni Gray[1]	Mediterranean, Red Sea to Maldive Island	A.M. Clark (1967), Achituv (1969)
	A. corallicola Marsh	Palau	Marsh (1977)
	A. heteractis H.L. Clark	Lord Howe Island	A.M. Clark (1967), Marsh (1977)
	Nepanthia belcheri (Perrier)	NE Australia to SE Asia	Rowe and Marsh (1982), Kenny (1969), Ottesen and Lucas (1982)
	N. brevis (Perrier)[2]	N Australia	A.M. Clark (1967)
	N. briareus (Bell)	China Sea	A.M. Clark (1967)
	N. fisheri (Rowe and Marsh)[3]	Philippines to Timor Sea	Rowe and Marsh (1982)
	N. variabilis H.L. Clark[2]	NW Australia	A.M. Clark (1967)
Asteriidae	*Coscinasterias acutispina* (Stimpson)	N Pacific	Edmondson (1935)
	C. calamaria (Gray)	Mauritius to Australia	Bennett (1927), A.M. Clark (1967)
	C. tenuispina (Lamarck)	Mediterranean to Bermuda	Kowalevsky (1872)
	Sclerasterias alexandri (Ludwig)	Panama	Fisher (1925)
	S. euplecta (Fisher)	Hawaiian Islands	Fisher (1925)
	S. heteropes (Fisher)	California	Fisher (1925)
	S. richardi (Perrier)	Biscay to Mediterranean	A.M. Clark (1967)

Contd.

Table 1 Contd.

Family	Species	Distribution	References
	Stephanasterias albula (Stimpson)	Bering Sea to Southern Alaska, Greenland to Iceland; NW Atlantic to South Carolina	Fisher (1930), Mladenov *et al.* (1986)
	Allostichaster inaequalis (Perrier)	Patagonia, South Africa	A.M. Clark (1967)
	A. insignis (Farquhar)	New Zealand	Bennett (1927)
	A. polyplax (Müller and Troschel)	New Zealand to S Australia	Bennett (1927)
Solasteridae	*Seriaster regularis* Jangoux	China Sea (Macclesfield Bank) and New Caledonia	Jangoux (1984), Guille *et al.* (1986)

[1]The fissiparous form of this species has also been referred to as *A. wega* (Achituv, 1969, 1973b). See text for further discussion.

[2]Rowe and Marsh (1982) consider *N. brevis* and *N. variabilis* to be junior synonyms of the variable *N. belcheri*.

[3]Potentially fissiparous (Rowe and Marsh, 1982).

2. Process

There appear to be three distinct patterns of fission in sea stars. Commonly, splitting of the disc begins with a furrow forming in each of two more or less opposing interradii (Fig. 1); the two furrows progress across the disc as the two halves of the sea star pull apart (Kowalevsky, 1872; Emson, 1978). A second pattern involves the formation of a furrow on just one side of the disc, which then progresses across the disc and transects the sea star. In *Stephanasterias albula*, five-armed individuals generally employ the single furrow mode of fission, and six-armed individuals the double furrow mode (Mladenov *et al.*, 1986). A third pattern of fission has been reported in *Nepanthia belcheri* (see Ottesen and Lucas, 1982). In this predominately seven-armed species, a shallow furrow forms across the centre of the disc, defining the plane of fission, and the sea star pulls itself into four- and three-armed halves; the four-armed individuals then immediately split into a pair of two-armed individuals. Fission thus results in three new individuals. Fission in sea stars is evidently a relatively lengthy process, taking about one hour to as long as 24 hours to accomplish, depending on the species and its structure (Emson, 1978; Emson and Wilkie, 1980; Ottesen and Lucas, 1982).

Fission may occur repeatedly during the life of probably all fissiparous sea stars as demonstrated by the fact that repeated fission has been observed in captive sea stars. As well, many fissiparous sea stars possess sets of arms of three different sizes, the result of successive fissions along different planes in the disc, prior to complete regeneration of the arms (Fisher, 1928; Achituv, 1969; Emson, 1978; Ottesen and Lucas, 1982; Johnson and Threlfall, 1987). Although successive

Fig. 1. The fissiparous sea star *Stephanasterias albula* undergoing fission (scale bar = 1 mm) (sketch by K. Brady).

divisions can occur along different planes in some species, there is a tendency for fission to occur mainly along the line of previous fission in others (Emson, 1978; Johnson and Threlfall, 1987; Mladenov *et al.*, 1986). Fission may begin before the new arms are fully regenerated. This has been observed in *Allostichaster polyplax* (see Emson, 1978), *Stephanasterias albula* (see Mladenov *et al.*, 1986) and *Coscinasterias calamaria* (see Johnson and Threlfall, 1987) and may maximize the rate of clonal growth by decreasing the interval between successive fissions.

3. Regulation of fission

The factors that regulate the onset of fission in sea stars are still poorly understood. Discussion has focused on both exogenous and endogenous stimuli. Fission can be triggered in some sea stars by shock associated with laboratory holding conditions, including temperature elevation, interruption of seawater circulation, and lack of aeration (Kowalevsky, 1872; Crump, 1969; Emson, 1978). On the other hand, fission could not be initiated in fully regenerated individuals of *Stephanasterias albula* subjected to various combinations of temperature and salinity stress, as well as to mechanical stress in the form of vigorous shaking (Carson, 1984). External stimuli of more natural kinds, such as storms, and even damage resulting from feeding on sharp-edged items of food, have also been cited as initiating fission (Yamazi, 1950; Tartarin, 1953). Such random stimuli alone could not, however, be responsible for the periodic nature of fission observed in many fissiparous populations (Table 2). Endogenous factors must therefore be involved.

Mladenov *et al.* (1986) demonstrated a temporal correlation between seasonal

Table 2

Periodicity of fission in sea stars

Species	Locality	Seasonal temperature range (°C)	Period of highest incidence of fission J F M A M J J A S O N D	Source
Allostichaster polyplax	Raglan, New Zealand	?	? ?[1]	Emson, 1978
A.insignis	Otago Harbour, New Zealand	12–20		Barker *et al*., 1992
Coscinasterias acutispina	Wakayama Prefecture, Japan		[2]	Yamazi (1950)
C.calamaria	New Zealand	?		Crump (1969), Crump and Barker (1985)
C. tenuispina	Bermuda	17–28		Crozier (1920)
C. tenuispina	Mediterranean	?		Tartarin (1953)
Nepanthia belcheri	Queensland, Australia (27° 31′S, 154°24′ E)	14.5–27.5		Kenny (1969)
N. belcheri	North Queensland, Australia (19°15′S, 146°7′E)	21.8–31.2		Ottesen and Lucas (1982)
Stephanasterias albula	N Lubec, Maine, USA (44°53′N, 67° 06′W)	1–13		Mladenov *et al.* (1986)

[1]Data available for summer only.
[2]No quantitative data presented.

changes in the incidence of fission and changes in day length in the sea star, *Stephanasterias albula*. They showed that more individuals undergo fission in the spring and summer than at other times of the year. Increasing day length may stimulate the synthesis of an endogenous chemical substance that, upon reaching some threshold level, either initiates fission directly, or makes a sea star more responsive to external, fission-inducing stimuli. Recent work has revealed that humoral and nervous factors play an important role in the control of arm autotomy in sea stars (Mladenov *et al.*, 1989), a phenomenon that is closely related to fission. Similar chemical factors are therefore likely involved in the regulation of fission. One of the ultimate effects of these chemicals may be to promote a reduction in the tensile strength of the collagenous component of the body wall in the plane of fission, thereby facilitating splitting. However, the presence of mutable connective tissues in the disc of fissiparous sea stars has yet to be demonstrated. Putative fission-inducing chemicals may also promote behaviours that are specifically associated with fission, such as stretching. In this regard, it has been suggested by several authors that splitting results from the simultaneous establishment of two rival centres of dominance, such that the sea star tugs in two opposite directions and pulls itself in half (Preyer, 1886–87; Hopkins, 1926; A.M. Clark, 1967; Emson, 1978). Cutting of the nerve ring of fissiparous species in two places will invoke such activity (Hopkins, 1926; Emson, 1978; Carson, 1984). Unfortunately, centres of dominance have never been demonstrated in more than a few asteroids (Reese, 1966) and there is only circumstantial evidence that arm dominance is a characteristic of asteroid nervous systems. Nonetheless, the fact that a tug-of-war is established among two sets of arms during fission in many sea stars demonstrates that putative fission-promoting chemicals may result, ultimately, in coordinated arm activity appropriate to the process.

4. Seasonality

Some populations of fissiparous sea star, such as *Coscinasterias calamaria* at New Zealand and *C. tenuispina* in the Mediterranean, exhibit no apparent seasonal variation in incidence of fission (Table 2). Some species, however, are distinctly seasonal (Table 2). This has been clearly demonstrated in *Stephanasterias albula* at North Lubec, Maine, which splits most frequently in the spring and summer (Mladenov *et al.*, 1986). It may be advantageous for this population to split in the summer since there is a large seasonal range in sea temperature and the higher summer sea temperatures may maximize rates of regeneration thereby decreasing recovery time and thus the interval between fissions.

5. Frequency of fission

The rate of clonal growth due to fission depends, ultimately, upon the period of the year over which fission can occur and the rate of regeneration of the lost half of the sea star. The latter may be affected by physical and biotic factors such as sea temperature and food supply. Rates of arm regeneration in *Stephanasterias albula* at North Lubec, Maine, are about 0.5 mm per month; this is sufficient to allow individuals to split every one or two years (Mladenov *et al.*, 1986). *Asterina burtoni* from the Mediterranean coast of Israel may split about once every year (Achituv and Sher, 1991). Published information on rates of regeneration is available for two other fissiparous sea stars, *Coscinasterias calamaria* (see Crump and Barker, 1985; Johnson and Threlfall, 1987) and *Allostichaster polyplax* (see Emson, 1978). These species live in warmer water than *S. albula* and regenerate arms more rapidly (perhaps up to 1 mm per week). They may consequently undergo fission more frequently than once a year. In this regard, Johnson and Threlfall (1987) estimated that the time between successive fissions in *C. calamaria* was 12 to 25 weeks based on observation of fed animals in the laboratory, which implies a potential for two to four divisions a year in the field.

6. Intraspecific variation in fission

In some populations of fissiparous sea star (e.g. *Nepanthia belcheri* from Townsville, Queensland, Australia), the propensity for fission is unrelated to body size (Ottesen and Lucas, 1982). It has often been reported, however, that fission is most common in the smaller 'juvenile' members of a population (Fisher, 1925; Yamazi, 1950; A.M. Clark, 1952; Tortonese, 1960). Yamazi (1950) found that large specimens of *Coscinasterias acutispina* from Japan showed no evidence of recent regeneration and thus fission. Johnson and Threlfall (1987) reported that there was an overall negative correlation between incidence of fission and body size in *C. calamaria* from Western Australia, although at specific sites there was generally no tendency for smaller sea stars to divide more frequently. Cases involving other species such as *Sclerasterias euplecta*, *S. heteropes* (see Fisher, 1925), *S. richardi* (see A.M. Clark, 1967), *Asterina burtoni* (see Achituv, 1969), *N. belcheri* (see Rowe and Marsh, 1982), and *Stephanasterias albula* (see Mladenov *et al.*, 1986) are more confusing. In these instances, large five-armed non-fissiparous forms and smaller multi-armed fissiparous forms have been described from the same population. In some of these species, geographic variation in the relative proportions of the two forms have been documented. In *A. burtoni*, for instance, there is an apparent gradation from five-armed, non-fissiparous forms in the tropical Red Sea to multi-armed, fissiparous forms in the more temperate Gulf of Suez and Mediterranean (James and Pearse, 1969; Achituv, 1969). In *N. belcheri*, there is a preponderance of small, multi-radiate, fissiparous forms along the east coast of Australia but a greater proportion of larger, five-armed, non-fissiparous forms along the north coast (Kenny, 1969; Rowe and Marsh, 1982). In *C. acutispina* small fissiparous forms were reported from the island of Maui, Hawaii, whereas larger non-fissiparous forms were present in Kaneohe Bay, Oahu, Hawaii (Edmondson, 1935). As an explanation for the sympatric distribution of morphologically similar fissiparous and non-fissiparous forms of sea star, it has been proposed that the fissiparous "juveniles" transform into non-fissiparous "adults" through adjustment of the number of regenerating arms following the final fission event (A.M. Clark, 1952; Tortonese, 1960). There is very little evidence, however, that such a conversion ever normally takes place in sea stars (A.M. Clark, 1967). Another possibility is that a single species may exhibit two forms, a multiarmed one that remains small and fissiparous throughout life, and a five-armed form that grows and never splits (Emson and Wilkie, 1980). An alternative explanation is that the two forms are sympatric, sibling species with contrasting sexual and asexual life histories. In this regard, Achituv (1969, 1973b) suggested that the fissiparous form of *A. burtoni* was a separate species, *A. wega*, but this remains controversial (Clark and Rowe, 1971). An electrophoretic analysis of allozymes may help to clarify the confusing taxonomic status of such closely related multiradiate and five-armed forms of sea stars.

7. Interrelation of fission and sex

Gonads have been recorded in many fissiparous sea stars and it appears that most or all species are capable of sexual reproduction as well as fission (Crozier, 1920; Cognetti and Delavault, 1962; Achituv, 1969; Falconetti *et al.*, 1976, 1977; Febvre *et al.*, 1981; Ottesen and Lucas, 1982; Crump and Barker, 1985). The nature of the interaction between sexual reproduction and fission is variable and complex. In *Coscinasterias tenuispina* from the Gulf of Naples and the Livorno Coast, the proportion of animals with gonads was reported to increase with body size (Cognetti and Delavault, 1962). The implication was that smaller animals were generally fissiparous and larger ones generally sexual. In populations of other species, such as *Nepanthia belcheri*, the propensity for fission is independent of body size, so that splitting individuals also commonly contain gonads.

In *Coscinasterias tenuispina* at Bermuda, sex and fission appear to alternate seasonally, sexual reproduction occurring in January and February and fission in May, June, and July (Crozier, 1920). In *Nepanthia belcheri* at Townsville, Australia, sexual reproduction occurs in early summer (October–November) after a peak of fission in the fall (April–June) (Ottesen and Lucas, 1982). Details of the interaction between sex and fission in this species have been thoroughly documented by Ottesen and Lucas (1982). This sea star is a protandric hermaphrodite with functional females being significantly larger than functional males. Regeneration following fission causes regression or retardation of gonadal growth; hence, regeneration occurs at the expense of gonads. Regeneration also has a masculinizing effect such that ovaries may change to testes. This reduces the number of mature females in the population during the breeding period with resultant reduction in potential fecundity.

Some field observations suggest a correlation between environmental factors and levels of sexual reproduction and fission in sea stars (Emson, 1978; Crump and Barker, 1985). In general, well-fed subtidal populations of *Allostichaster polyplax* and *Coscinasterias calamaria* have a larger body size, a lower incidence of fission, and larger gonads than intertidal populations subjected to more stress and poorer food conditions. These interesting observations invite the design of field and laboratory experiments that manipulate physical factors and food supply. Such experiments may provide greater insight into the ecological correlates of sexual reproduction and fission in these and other sea stars.

8. The role of sex in fissiparous sea stars

The level of sexual activity in some populations of fissiparous sea star is apparently very low. Achituv (1969) noted that only a small number of specimens of the fissiparous form of *Asterina burtoni* from the Gulf of Elat, Red Sea possessed well-developed ovaries and testes. In the Mediterranean at Shikmona, this form undergoes fission in the summer and spawning occurs at about the same time of the year (Achituv and Sher, 1991). Only males are present in this particular population. Also, Achituv and Malik (1985) discovered that the spermatozoa of the fissiparous form of this sea star are asymmetric, with a flagellum that coils around the head. These authors suggested that the motility of the spermatozoa may be impaired, thereby further reducing successful sexual reproduction. The structure of the spermatozoa of the five-armed, sexual forms was not reported, however. Emson (1978) believed that the attainment of sexual maturity was uncommon in *Allostichaster polyplax* at Raglan, New Zealand, and that the small size of the sea stars resulted in very low fecundities on a per individual basis. Fecundity in the fissiparous *Sclerasterias richardi* from the Mediterranean Sea was also low, with 400–500 ova per individual female (Febvre *et al.*, 1981). The potential for successful sexual recruitment might be further compromised by the fact that for those fissiparous sea stars whose mode of development is known or inferred from egg size, the eggs develop into pelagic larvae which would be at risk in the plankton (Falconetti *et al.*, 1977; Emson, 1978; Ottesen and Lucas, 1982; Crump and Barker, 1985). (However, see Section III A 9 for a discussion on the possible role of clonal growth in *increasing* genotypic fecundity.)

Obligate asexuality is known for some populations of fissiparous sea star. Examination of 1 μm plastic sections from even the largest specimens of *Stephanasterias albula* at North Lubec, Maine, revealed no trace of a differentiated reproductive system (Mladenov *et al.*, 1986). No gonads were recorded in the same species from deeper water at Jeffrey's Ledge, Gulf of Maine (Hulbert *et al.*, 1983). It is not known whether these obligately fissiparous *S. albula* still retain primordial germ cells somewhere in their bodies. In any case, it appears that in at least some populations of the species the genetic programme for sexual reproduction is not turned on. It would be interesting to learn whether this depends on environmental or genetic factors and whether the reproductive system of these sea stars can be stimulated to develop under certain conditions. In this regard, it is noteworthy that

Lütken (1873, pp. 331–332) records well-developed gonads in *Asterias problema* from Greenland, a fissiparous sea star that may be a synonym of *S. albula* (Verrill, 1866, p. 351; Fisher, 1930, p. 157). A population of the fissiparous form of *Asterina burtoni* at Wadi el Dom in the Gulf of Suez, like *S. albula* populations, may also fail to develop gonads (Pearse, 1968, 1982; James and Pearse, 1969).

The problem of long-range dispersal for populations of fissiparous sea stars in which gonads are absent or poorly developed and which consequently lack a dispersive larval phase may be somewhat alleviated by occasional episodes of adult rafting on seaweed or other floating objects. Direct evidence for this is provided by Grainger (1966, p. 43), who observed three specimens of *S. albula* attached to floating seaweed (*Laminaria*) in Frobisher Bay in the Canadian arctic, and by Achituv (1973a), who reported the occurrence of several specimens of the fissiparous form of *Asterina burtoni* on floating algae in the Mediterranean Sea.

9. Population genetics of fissiparous sea stars

The high incidence of fission in many populations of fissiparous sea star (for example, see Emson, 1978; Ottesen and Lucas, 1982; and Mladenov *et al.*, 1986) is suggestive of a substantial asexual contribution to the maintenance of local populations. So far, however, few studies have been performed that attempt to assess the relative contributions of larval and clonal recruitment to fissiparous sea star populations. The only detailed study has been that of Johnson and Threlfall (1987) on *Coscinasterias calamaria* sampled at 14 different sites on Rottnest Island and the adjacent mainland of Western Australia. An electrophoretic analysis of six polymorphic enzymes revealed only small numbers of distinct six-locus genotypes at each site, and thus low levels of genotypic diversity, thereby confirming that local populations had a highly clonal structure. Thus, in the short term at least, clonal proliferation predominates over recruitment by sexually produced larvae. Large differences in clonal composition over distances as small as 50 m emphasized the very localized scale of clonal mixing in this species.

A priori, one might expect to find lower genotypic diversity at sites with higher measures of fission. One of the interesting outcomes of Johnson and Threlfall's (1987) study, however, was that local genotypic diversity was not correlated with fission at their intertidal sites. Also, subtidal sites with the lowest apparent levels of fission also showed the lowest values of genotypic diversity. Apparently, these subtidal sites had low levels of both sexual and asexual recruitment. As the authors point out, these results demonstrate the potential pitfalls of assessing the relative contributions of larval and clonal recruitment to a population in the absence of direct estimates of recruitment and mortality for both modes of reproduction.

Some information on the genetic structure of the obligately fissiparous population of *Stephanasterias albula* at North Lubec, Maine (Mladenov *et al.*, 1986), is also available. Using isoelectric focusing techniques, Sundaram (1986) detected no differences in allozyme patterns among any of the individuals (number analysed ranged from 23 to 30) of *S. albula* analysed at all five loci studied. Fixed heterozygosity was noted at the *Pgi* and *Idh* loci and fixed homozygosity was observed at the *Pgm, Mdh-1*, and *Mdh-2* loci. This suggests that the *S. albula* population at North Lubec, Maine, which in 1984 consisted of about 5,500 sea stars (Mladenov *et al.*, 1986), may have constituted a single clonal aggregation.

B. Asexual Propagation by Autotomy

1. Occurrence

A few sea stars are capable of propagation by arm autotomy (Steenstrup, 1857; Sars, 1859; Martens, 1866; Hirota, 1895; Monks, 1904; Yamaguchi, 1975; Marsh, 1977). In these species, an autotomized arm that contains no portion of the disc can regenerate a new disc and arms to form a complete new animal. This represents a specialized capacity for regeneration since in most sea stars a piece of the disc is necessary for complete regeneration. Species capable of autotomous propagation are frequently found in the 'comet' form, which consists of one large arm that is regenerating a small disc and four or five small arms from its base (Fig. 2).

Propagation through arm autotomy in sea stars is restricted to a very small number of tropical species comprising mostly closely related and often taxonomically confusing linckids and ophidiasterids from the family Ophidiasteridae, although *Echinaster luzonicus* from the family Echinasteridae also exploits this special ability (Table 3). Emson and Wilkie (1980) listed an asterinid, *Nepanthia belcheri*, as capable of propagation by arm autotomy, but it has since been shown that the small number of single-armed individuals in this species arise by fission of two- or three-armed individuals rather than by the shedding of single arms (Ottesen and Lucas, 1982).

Sea stars that propagate through arm autotomy are a common sight on coral reefs and are often extremely abundant in localized areas (H.L. Clark, 1933; Yamaguchi, 1975; Marsh, 1977). Furthermore, most of them are widely distributed, particularly *Linckia guildingi*, which is circumtropical.

2. Process

Arm autotomy in these species follows the same pattern as non-propagative arm autotomy in other sea stars and is, in some ways, reminiscent of the more drastic process of fission. Generally, an arm pulls away from the remainder of the animal, which remains more or less stationary (Monks, 1904). This process is accompanied by a narrowing of the arm at the point of breakage (Rideout, 1978), the body wall in this region becoming soft and elastic. The tissues are stretched,

Fig. 2. Comet forms of the sea star *Linckia guildingi* from Barbados (photo by P. Mladenov).

often to great lengths (25 mm), and eventually rupture. Sometimes, however, autotomy is accomplished in the absence of any obvious softening and stretching of tissues in the plane of breakage. In this instance, a crack appears on the dorsal arm surface which slowly spreads ventrally to separate the arm from the rest of the sea star. In either case, the pyloric duct seems to be the toughest structure, and the last to rupture. Autotomy has been reported to require anywhere from 5 to 10 minutes to several hours to accomplish (Monks, 1904; Rideout, 1978), with the arm usually separating 1 to 2 cm from the disc, although rupture may take place anywhere along the length of the arm (Monks, 1904; MacGinitie and MacGinitie, 1949).

Rideout (1978) has found in *Linckia multifora* that once the comet is formed, the principal or founding arm is autotomized. The arms of the comet itself are then sequentially autotomized according to a particular pattern for at least the first few autotomies. Rideout (1978) also found that autotomies will only occur after the

Table 3

Sea stars capable of propagation through arm autotomy and their geographic distribution

Family	Species	Distribution	Reference
Ophidiasteridae	*Linckia columbiae* Gray[1]	Tropical E Pacific	Monks (1904), H.L. Clark (1940)
	L. guildingi Gray[2]	Circumtropical	H.L. Clark (1913, 1933), Yamaguchi (1975)
	L. multifora (Lamarck)	Indo-West Pacific	Hirota (1895), Edmondson (1935)
	Ophidiaster cribrareus de Loriol[3]	Tropical Pacific	Yamaguchi (1975), Marsh (1977)
	O. robillardi de Loriol	Indo-West Pacific	Yamaguchi (1975), Marsh (1977)
Echinasteridae	*Echinaster luzonicus* (Gray)	Tropical Pacific	Yamaguchi (1975), Marsh (1977)

[1]Called *Phataria unifascialis* by Monks (1904), although H.L. Clark (1940) lists *P. unifascialis* as distinct from *L. columbiae*.
[2]Probably synonymous with both *L. diplax* and *L. pacifica* (Fisher, 1919; Yamaguchi, 1975, p. 21).
[3]A morphologically similar congener, *O. lorioli*, does not propagate asexually by arm autotomy (Marsh, 1977).

regenerated arms of the comet have reached a certain length. He concluded that continuous reproduction by autotomy alone can account for the high population level of *L. multifora* in Guam.

3. Mechanism

One can assume that the process is facilitated by a softening of mutable connective tissues in the autotomy plane, but this remains to be demonstrated.

4. Regulation

Several authors report that the process of propagative arm autotomy is sometimes inducible by exogenous stimuli such as mechanical disturbance, temperature elevation, or ligation of an arm, although not all individuals respond to such treatment (Monks, 1904; Edmondson, 1935; Davis, 1967).

Davis (1967) showed that arms of *Linckia multifora* that were parasitized by the gastropod, *Stylifer linckiae* were significantly less likely to autotomize, either spontaneously or in response to ligation, than unparasitized arms. Davis (1967) suggested that the parasite, which encysts in the body wall, somehow suppresses autotomy of the host arm, perhaps by interfering with an unknown endocrinological controlling mechanism. No information on the propensity for autotomy of unparasitized arms on parasitized sea stars was provided in this study, so it is difficult to judge whether the effect of the parasite is a generalized one, or localized to the parasitized arm itself. The discovery of chemical factors that promote arm autotomy in sea stars (Chaet, 1962; Mladenov *et al.*, 1989) also suggests that the process may be under some form of internal chemical control.

5. Interrelation with sexual reproduction

Sexual reproduction is known to occur in *Linckia columbiae* and *L. multifora* (Monks, 1904; Mortensen, 1937, 1938). In these species, gonads can be present in animals that also practise propagative autotomy, the gonads associated with the larger arms. In *L. multifora*, sexual reproduction results in planktotrophic larvae. Clutch size in the *L. multifora* examined by Rideout (1978) was small, about 200 eggs per individual. Thus, on a per individual basis, sexual fecundity appears limited.

6. Cloning potential

The potential for asexual propagation by arm autotomy depends on both the rate of mortality and the rate of regeneration of the shed arms, as well as the frequency with which arms are shed. Edmondson (1935) and Davis (1967) suggested that mortality of shed arms in *Linckia multifora* was very high, although Rideout (1978) reported that mortality was very low in the same species. Rate of regeneration of the shed arms seems to be relatively slow. Rideout (1978) found that single arms of *L. multifora* regenerated a new mouth in 27 days and small new arms (mean length 1.3 mm) by 40 days. Valentine (1925–26) reported that arm buds were present after 25–30 days and Edmondson (1935) reported that regeneration to the comet stage took about three months. There appear to be no estimates of the frequency with which a mature comet sheds arms in the field.

There is thus very little information on which to base an estimate of the potential for clonal growth in sea stars capable of propagative arm autotomy. Furthermore, no population genetic studies of such sea stars seem to have been reported, so nothing can be said about the relative contribution of sexual and asexual recruitment in structuring local populations.

C. Asexual Propagation by Parthenogenesis

Greef (1876) reported that the eggs of *Asterias glacialis* would divide and produce larvae without fertilization and MacBride (1896) found that isolated female *Asterina gibbosa* would spawn eggs which develop "with perfect regularity up to the conclusion of metamorphosis". Loeb (1905) and Newman (1921) found that eggs of *Patiria miniata* could undergo spontaneous parthenogenetic development. Generally, only a very small percentage of mature eggs begin development this way, although in one case up to 75 per cent of the eggs of a spawning began to develop parthenogenetically (Newman, 1921). The development of parthenogenetic embryos was slow and the bipinnariae which resulted were always abnormal. It should be noted that both males and females exist in these species and they naturally reproduce sexually by fusion of sperm and eggs. The extent to which these reports on parthenogenesis represent laboratory phenomena alone is not known, but it is very unlikely that parthenogenesis is of frequent occurrence in natural populations of these species. Furthermore, the exact mechanism of parthenogenesis is not known. Only apomictic (ameiotic) parthenogenesis, in which the egg possesses an unrecombined maternal genome, would be an example of asexual propagation. However, automictic mechanisms of parthenogenesis are possible, in which meiotic reduction occurs during oocyte maturation, which is compensated for in some manner (Bell, 1982, table 1). Automictic parthenogenesis represents a sexual process with genetic consequences akin to self-fertilization.

An interesting study by Yamaguchi and Lucas (1984) provides the first evidence for natural parthenogenetic development in an echinoderm. Populations of the Indo-West Pacific coral reef sea star, *Ophidiaster granifer*, from Micronesia, New Guinea, and the Great Barrier Reef consisted only of females; no testes were ever found, even in individuals examined histologically. Proliferating interstitial cells were, however, noted in some gonads, sometimes assuming a form that resembled spermatogenic columns, but there was no evidence that these cells differentiated into spermatozoa. Eggs (0.6 mm in diameter) released by females, either spontaneously or in response to injection of 1-methyladenine, developed into lecithotrophic brachiolaria larvae. Some larvae were planktonic, while others remained attached to the bottom. Yamaguchi and Lucas (1984) also noted developing embryos within the ovaries of some individuals.

Yamaguchi and Lucas (1984) were unable to karyotype material from *Ophidiaster granifer* so the mechanism of parthenogenesis is not known. They did not, however, observe polar bodies so apomictic parthenogenesis remains a possibility. An electrophoretic study of the genetics of *O. granifer* populations would be helpful in this regard. Apomictic parthenogenesis should result in clonally structured populations with high levels of heterozygosity. On the other hand, automictic parthenogenesis should result in populations with reduced levels of heterozygosity. Electrophoretic comparisons of adults and their offspring would also be useful in defining the mechanism of parthenogenesis.

D. Asexual Propagation by Larvae

Proliferation by budding has been implied for one sea star, *Luidia sarsi*. During the metamorphosis of the unusually large bipinnaria larva of this species, the rudiment may detach itself from the larval body, which includes the larval arms and the preoral lobe (Koren and Danielssen, 1847; Delap and Delap, 1907; Tattersall and Sheppard, 1934). As Wilson (1978) has more recently demonstrated, another form of metamorphosis, in which the larval body is partly or almost completely resorbed, may also occur in this species. In instances where metamorphosis occurs by detachment, the discarded larval body contains the larval mouth and anus and short, unconnected pieces of the oesophagus and intestine. In the laboratory, these discarded larval bodies can remain alive and swimming for up to three months after metamorphosis (Delap and Delap, 1907; Wilson, 1978). Thorson (1961) mooted the possibility that the discarded larva may be able to regenerate and undergo a second metamorphosis. However, Wilson (1978) found that the cast-off larvae soon died under laboratory conditions, even in the presence of suitable food, and indicated that regeneration and a second metamorphosis "most probably never occurs".

Asexual propagation by paratomic fission has been convincingly described for a second, unknown species of *Luidia* (Bosch *et al.*, 1989). About 30 per cent of bipinnariae collected from the Gulf Stream and the western Sargasso Sea during summer had highly modified posterolateral arms that were in the process of transformation into secondary larvae (Fig. 3). The archenteron of these secondary larvae is established by invagination of the underside of the arm. Both posterolateral arms can undergo such transformation, sometimes simultaneously. Bosch *et al.* (1989) observed separation of secondary larvae from primary larvae with subsequent development into feeding bipinnariae in ship-board cultures. About 10 per cent of freshly collected bipinnariae were regenerating one or both posterolateral arms, indicating that primary larvae can regrow new arms following paratomy. It is not known if these regenerated posterolateral arms can produce more secondary larvae. Bosch *et al.* (1989) speculate that paratomic fission in *Luidia* sp. may represent a strategy by which cloning can prolong the planktonic existence of a larval genotype and thus increase the possibility of some individuals of the clone reaching regions suitable for settlement and metamorphosis.

IV. OPHIUROIDEA

A. Asexual Propagation by Fission

1. Occurrence

Many brittle stars are fissiparous, capable of asexual propagation in nature by binary fission (Steenstrup, 1857; Sars, 1859; Lütken, 1872; H.L. Clark, 1914b; A.M. Clark, 1967; Emson and Wilkie, 1980). Individuals of such species are readily recognizable because they are usually distinctly asymmetrical, often possessing a set of three large and a set of three smaller regenerating arms (Fig. 4).

At this time, about 45 of the roughly 2,000 extant species of brittle star are reported to show evidence of being fissiparous (Table 4), and further examples will undoubtedly come to light. Fissiparous species are known from 11 of 17 extant families of brittle star (Table 4). They thus represent a systematically di-

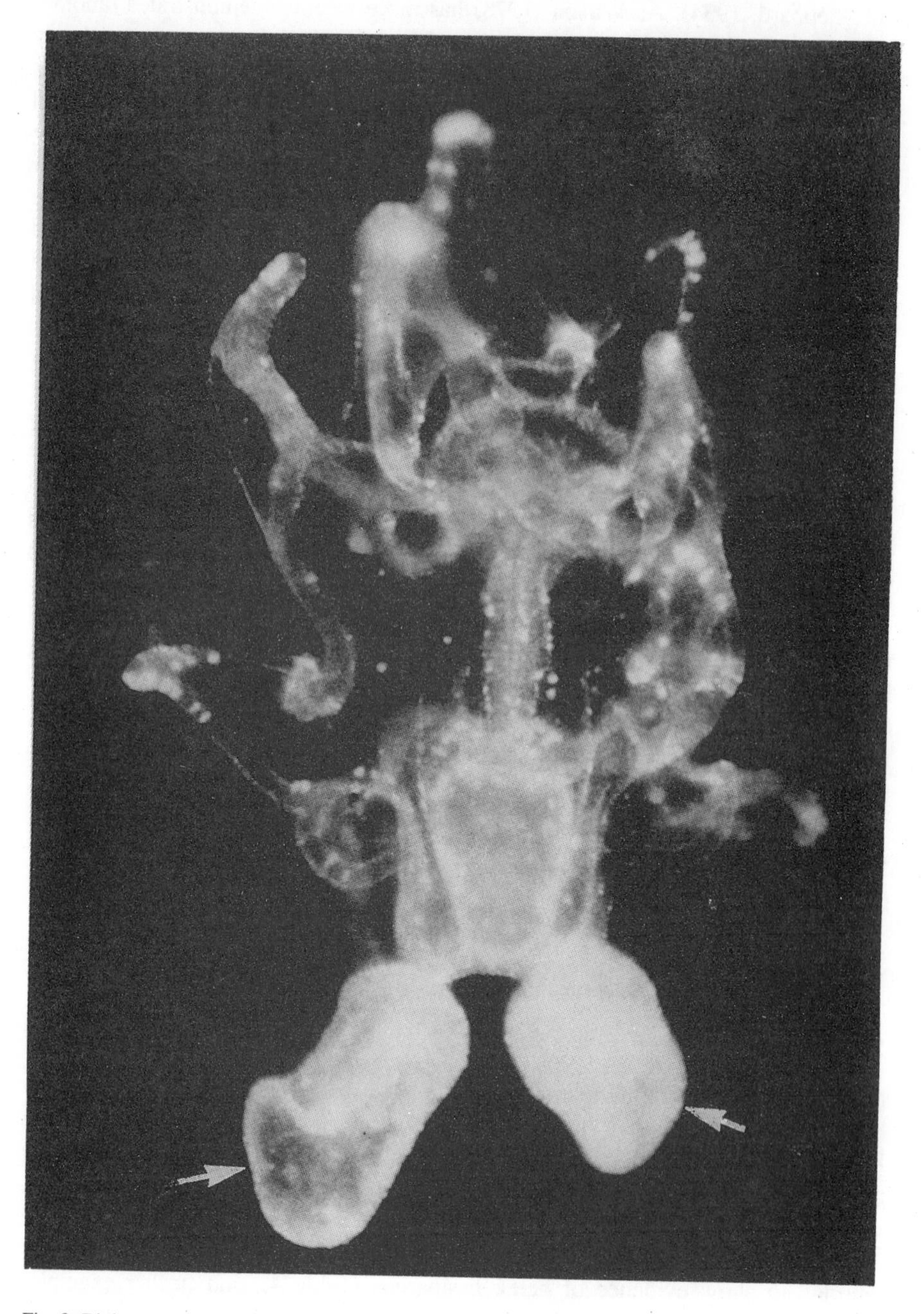

Fig. 3. Bipinnaria larva of the sea star *Luidia* sp. showing posterolateral arms (arrows) in the process of transformation into secondary larvae (photo courtesy of I. Bosch).

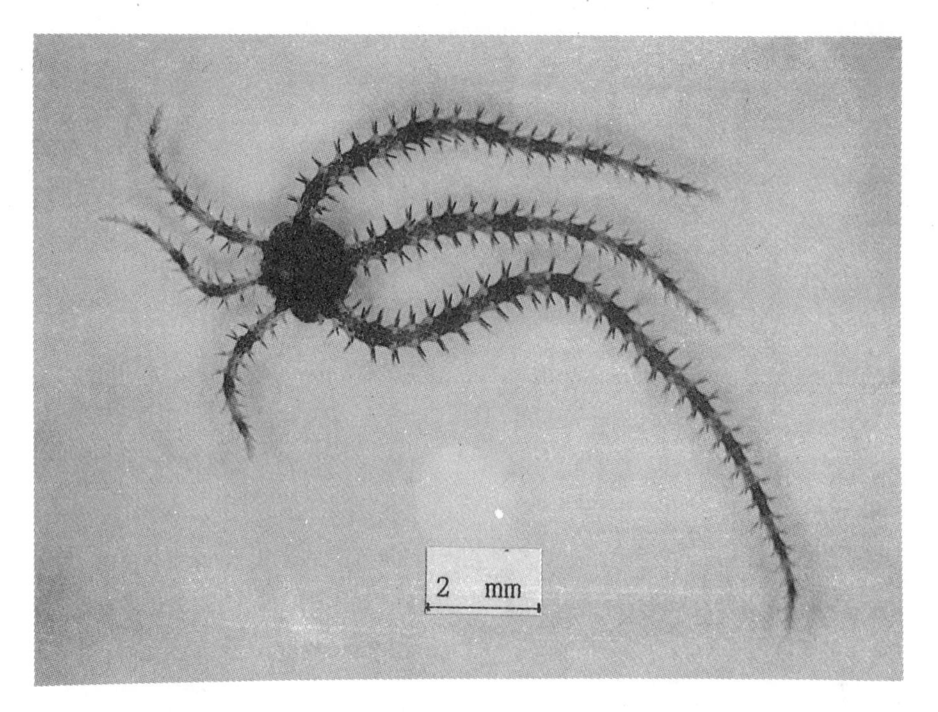

Fig. 4. Typical example of the fissiparous brittle star *Ophiocomella ophiactoides* (photo by P.V. Mladenov).

verse assemblage resulting from the concurrent evolution of fissiparity within the Ophiuroidea. Some fissiparous species are both geographically widespread and locally very abundant. For example, *Ophiactis savignyi* is a circumtropical species (Müller and Troschel, 1842; H.L. Clark, 1914b, 1933; Hyman, 1955) that is often associated with seaweeds and sponges. It attains extraordinary densities in sponges: Boffi (1972) and Mladenov and Emson (1988) recorded maximum densities of 1,892 individuals/100 g of dry sponge and 281 individuals/100 ml of living sponge, respectively. Also, the fissiparous *Ophiocomella ophiactoides* is widespread in the West Indies (Parslow and A.M. Clark, 1963) and can be locally very abundant in seaweed, achieving maximum densities of about 50 individuals/100 ml of algae (Mladenov *et al*., 1983; Mladenov and Emson, 1988).

There is a tendency for fissiparous brittle stars to fall into clusters of morphologically closely similar 'species' whose taxonomy can be notoriously difficult to unravel. Fissiparous ophiactids, ophiocomids, and ophiotrichids illustrate this point well. The fissiparous ophiactids, which form the largest group of fissiparous brittle stars, comprise at least 20 different closely related species of *Ophiactis*. The fissiparous ophiocomids comprise at least three closely related species of *Ophiocomella* whose taxonomy is further complicated by the belief of many early taxonomists

Table 4

Fissiparous brittle stars and their geographic distribution

Family	Species	Distribution	References
Astroschematidae	*Astrogymnotes catasticta* H.L. Clark	W Australia	H.L. Clark (1914a)
	Astrocharis ijimai Matsumoto	Japan	Matsumoto (1917)
	A. virgo Koehler	East Indies	Matsumoto (1917)
Euryalidae	*Astroceras annulatum*[1] Mortensen	Japan	Mortensen (1933a)
	A. nodosum Koehler	East Indies	Koehler (1930), Mortensen (1933a)
	Asteromorpha perplexum Koehler	East Indies	Koehler (1930), Mortensen (1933a)
Gorgonocephalidae	*Schizostella*[2] *bifurcata* A.H. Clark	West Indies	A.H. Clark (1952)
	S. bayeri A.H. Clark	?	A.H. Clark (1952)
Hemieuryalidae	*Ophioholcus sexradiata* Koehler	West Indies	Koehler (1914)
Ophiomyxidae	*Ophiostiba hidekii* Matsumoto	Japan	Matsumoto (1917)
	Ophiovesta granulata Koehler	Philippines	Koehler (1930)
Amphiuridae	*Amphiacantha dividua* Matsumoto	Japan	Matsumoto (1917)
	Amphiodia dividua Mortensen	Mauritius	Mortensen (1933a)
	Amphipholis torelli Ljungmann	North Atlantic and Arctic Ocean	Mortensen (1933b)
	Amphiura sexradiata Koehler	Gulf of Thailand	Koehler (1930)
	Ophiostigma isocanthum[3] (Say)	West Indies	Koehler (1913), H.L. Clark (1933, 1942), Thomas (1962), Hotchkiss (1982), Hendler and Littman (1986)
Ophiactidae	*Ophiactis acosmeta* (H.L. Clark)	West Indies	H.L. Clark (1933), Mladenov and Emson (1988)
	O. arenosa Lütken	W North America	MacGinitie and MacGinitie (1949)
	O. cyanosticta H.L. Clark	Tobago, West Indies	H.L. Clark (1933)
	O. hirta Lyman	New South Wales, Australia and Cook Strait, New Zealand	Lyman (1879)

Contd.

Table 4 Contd.

Family	Species	Distribution	References
	O. lymani Ljungmann	West Indies, South Atlantic, Tropical W Africa	Boffi (1972), Madsen (1970), Mladenov and Emson (1988)
	O. maculosa (von Martens)	Taiwan	A.M. Clark and Rowe (1971), Applegate (1984)
	O. modesta Brock	Indo-West Pacific	Matsumoto (1917), Yamazi (1950)
	O. mülleri Lütken	South Carolina to Brazil	Lütken (1857), H.L. Clark (1933)
	O. nidarosiensis Mortensen	N Europe to S Africa	Mortensen (1920)
	O. parva Mortensen	Red Sea	Mortensen (1926)
	O. plana Lyman	West Indies and Mauritius (Indian Ocean)	Lyman (1869)
	O. profundi Lütken and Mortensen	W Central America, New Zealand	Lütken and Mortensen (1899), Mortensen (1936)
	O. rubropoda Singletary	Florida	Singletary (1974)
	O. savignyi (Müller and Troschel)	Circumtropical	Müller and Troschel (1842), H.L. Clark (1933), Mladenov and Emson (1988)
	O. seminuda Mortensen	Tristan da Cunha	Mortensen (1936)
	O. simplex (Le Conte)	W Central America	Le Conte (1851)
	O. versicolor H.L. Clark	Red Sea to Mauritius	A.H. Clark (1939)
	O. virens (M. Sars)	Mediterranean	Sars (1859)
	Ophiactis sp.	Hong Kong	A.M. Clark (1980a)
Ophiocanthidae	*Ophiologimus hexactis* Clark	Japan	H.L. Clark (1911), Matsumoto (1917)
Ophiocomidae	*Ophiocomella ophiactoides* (H.L. Clark)	West Indies	H.L. Clark (1901), Parslow and Clark (1963), Mladenov *et al.* (1983), Mladenov and Emson (1988)
	O. schmitti A.H. Clark	E Pacific	A.H. Clark (1939), Parslow and Clark (1963)
	O. sexradia[4] (Duncan)	Indo-West Pacific	Duncan (1879), Parslow and Clark (1963, p. 42)

Contd.

Table 4 Contd.

Family	Species	Distribution	References
Ophionereididae	*Ophionereis dictydisca*[5] (H.L. Clark)	Japan	H.L. Clark (1921)
	O. dubia Müller and Troschel	Japan	Müller and Troschel (1842)
	O. sexradia Mortensen	W Africa to Canary Islands	Mortensen (1936)
Ophiotrichidae	*Ophiothela danae*[6] Verrill	Indonesia, Philippines, S Japan	Verrill (1869), Yamazi (1950)
	O. hadra H.L. Clark	Torres Strait and New South Wales, Australia	H.L. Clark (1921)
	O. mirabilis Verrill	Panama	Verrill (1869)

[1]Matsumoto (1917) referred to this as the young, fissiparous form of *A. pergamenum*. However, Mortensen (1933a) indicates that *A. annulatum* and *A. pergamenum* are distinct species, the former fissiparous, the latter non-fissiparous.
[2]Diagnostic features of this genus include seven arms and fissiparity (A.H. Clark, 1952, p. 451).
[3]Both six-armed and five-armed forms have been reported. Hotchkiss (1982) suggests that the fissiparous forms may be specifically distinct and refers to such specimens as *Ophiostigma* sp.
[4]Synonyms include *O. parva, O. schultzi, Ophiocnida sexradia, Amphiacantha sexradia.*
[5]*Ophionereis dictydisca* may be synonymous with *O. dubia* (A.M. Clark, 1967).
[6]Probable synonyms include *O. isidicola* and *O. verrilli* (Matsumoto, 1917; H.L. Clark, 1921; A.M. Clark, 1980b).

that fissiparous brittle stars represented the juvenile phase of larger, non-fissiparous adults (see Section IV A 8 below). The fissiparous ophiotrichids comprise at least three similar species of *Ophiothela*. The literature on these three genera of fissiparous forms is replete with synonyms reflecting *ad hoc* creation and subsequent pooling of species. A review of Matsumoto (1917, pp. 230–232), H.L. Clark (1921, p. 117), Parslow and A.M. Clark (1963, pp. 37–43), and A.M. Clark (1967, pp. 151–153) serves to illustrate the difficulty that echinoderm taxonomists have faced in dealing with these fissiparous forms using traditional techniques. The impression that emerges is that these genera contain a large number of locally differentiated fissiparous 'species' (long-lived clones?) whose taxonomy may only be traceable using comparative biochemical genetic approaches.

Adding further complexity to the subject of fissiparous brittle star systematics is the existence of closely related and often sympatric species pairs comprising both fissiparous and non-fissiparous representatives, the fissiparous representative often, though not always, larger than the non-fissiparous member. Examples include: *Ophiactis quinqueradia* (non-fissiparous) and *O. savignyi* (fissiparous) (West Indies); *O. picteti* (non-fissiparous) and *O. savignyi* (fissiparous) (Indonesia); *Ophiocoma alexandri* (non-fissiparous) and *Ophiocomella schmitti* (fissiparous) (East Pacific); *Ophiocoma pumila* (non-fissiparous) and *Ophiocomella ophiactoides* (fissiparous) (West Indies); *Amphipholis squamata* (non-fissiparous) and *A. torelli* (fissiparous) (Arctic); *Ophiostigma* sp. (non-fissiparous) and *O. isocanthum* (fissiparous) (Hotchkiss, 1982) (West Indies).

Despite their evolutionarily diverse background, fissiparous brittle stars in general share a suite of traits that binds them into an ecologically homogeneous group:

(1) All fissiparous brittle stars have a small adult body size, most individuals having a disc diameter that is less than 5–6 mm. It is unclear why larger species cannot or do not exploit fission (Mladenov and Emson, 1984), but there may be functional disadvantages to being half a brittle star, including problems associated with movement, food capture, and predator escape, that are not shared by small species. Alternatively, there may be energetic constraints, namely that it may be too costly for a large brittle star to periodically regenerate a large portion of its body to a functional size. It is also possible that large-bodied brittle stars may be mechanically or physiologically unsuited for fission.

(2) Most fissiparous brittle stars are six-armed. Although five-armed and seven-armed individuals may be present in a population at low frequency (Mladenov *et al.*, 1983), the result of natural variability in the fission and regeneration processes, most individuals of most fissiparous species are hexamerous. One exception is the fissiparous gorgonocephalid genus, *Schizostella*, which is seven-armed (A.H. Clark, 1952). The adaptive significance of mainly hexamerous symmetry may be related to the ability of such forms to divide into two symmetrical halves, rather than into a three-armed and a perhaps less viable two-armed portion. Surprisingly, as pointed out by A.M. Clark (1967), no one has carried out the appropriate breeding experiments to determine whether fissiparous brittle stars metamorphose as six-armed juveniles or five-armed juveniles that then convert in some manner to hexamerous forms.

(3) Fissiparous brittle stars have a largely tropical and subtropical distribution (Table 4). Noteworthy exceptions to the trend are: *Amphipholis torelli* from the Arctic Ocean, *Ophiactis nidarosiensis* from northern European and South African waters, and *O. profundi* from New Zealand waters. It remains unclear why fission is mainly a phenomenon of warm-water species, but it seems that there may be some significant adaptive advantage to cloning in brittle stars living in tropical habitats that is not shared by temperate and polar species. It may be relevant that the majority of brooding brittle stars (which are also usually small in size) are found in polar and temperate waters (Hendler, 1979).

(4) Fissiparous brittle stars are very often epiphytic or epizoic, occupying habitats such as clumps of algae (Boffi, 1972; Mladenov *et al.*, 1983; Mladenov and Emson, 1988), the inside of sponges (Caspers, 1985; Mladenov and Emson, 1988), the surface of gorgonians (Verrill, 1869; A.H. Clark, 1952), and turtle grass (Emson *et al.*, 1985). Hendler and Littman (1986) refer to such microhabitats as 'refuge-substrata' (microhabitats affording enhanced survival) and suggest that they select for species of brittle stars with small body size, such as fissiparous and brooding forms.

2. Process of fission

Fission occurs by division of the disc through two opposite interradii. The process has only been directly observed in one species, *Ophiocomella ophiactoides* (see Mladenov *et al.*, 1983; Wilkie *et al.*, 1984). In this species, fission commences with a softening and furrowing of the disc integument in one interradius. The furrow progresses across the disc to the opposite interradius and creates two new individuals (Fig. 5). During this process, the arms of the brittle star often coil about filaments of algae, as if exerting a pull in opposite directions to aid in fission. Healing of the torn parts of the disc occurs by infolding and rapid fusion of the dorsal and ventral edges of the disc and is accomplished as fission progresses. Based on laboratory observation, fission takes roughly 24 hours to accomplish although fission of undisturbed animals in the field may occur more rapidly.

Mladenov *et al.* (1983) found that recently split individuals of *Ophiocomella ophiactoides* with arms of three different magnitudes, the result of fission prior to complete arm regeneration, were about as common as recently split individuals with arms of two different magnitudes. This indicates that successive planes of fission

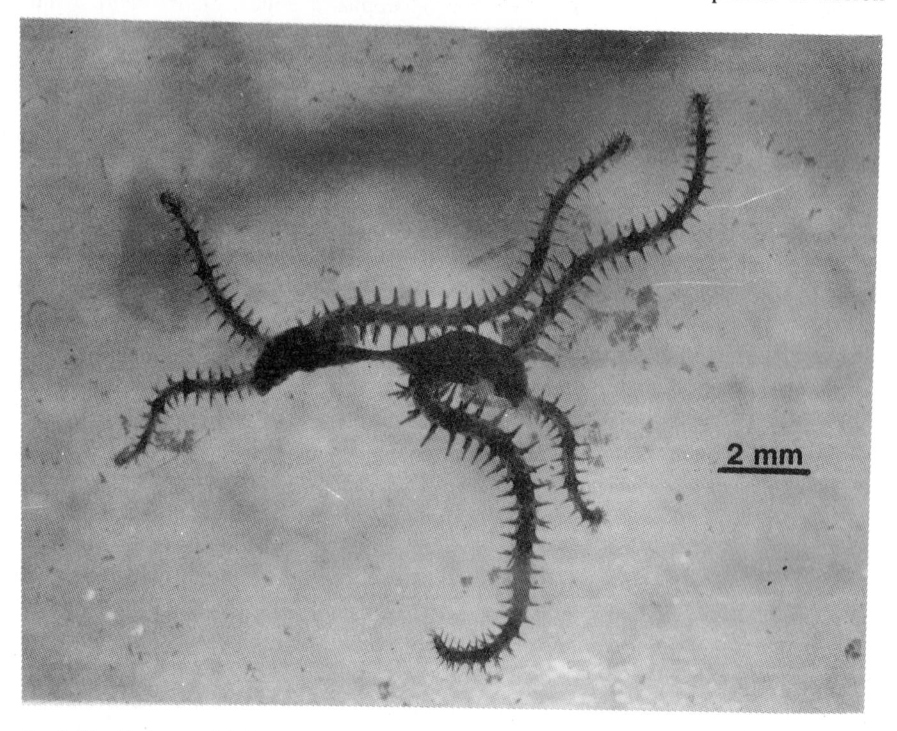

Fig. 5. The fissiparous brittle star *Ophiocomella ophiactoides* undergoing fission (photo courtesy of R. Emson).

are not always coincident, and that a split along a new plane in the disc occurs about as often as a split along the previous plane of fission. These authors also observed a few individuals that split unequally (i.e. to form two- and four-armed halves), and also some animals that were regenerating an inappropriate number of arms (i.e. three-armed specimens regenerating two arms, etc.). The presence of small numbers of non-hexamerous individuals in populations can generally be accounted for by such 'errors' in fission and regeneration.

3. Mechanism of fission

The principal structures transected by the fission plane are the disc integument, jaw frame, circumoral nerve and water vascular rings, and the gut. The fission plane is always interradial, i.e. it passes through and not between the jaws (Wilkie *et al.*, 1984), substantiating the claim of Simroth (1877). In contrast, the plane of breakage in narcotized animals ruptured by force is radial or interradial (usually radial) (Wilkie *et al.*, 1984). Thus, fission is not facilitated by a predetermined plane of weakness. This suggests that fission is restricted to the interradial plane by a physiological mechanism.

The plane of fission is subtended mainly by connective tissue elements that include the dermis of the integument, the collagenous ligaments of the jaw frame, and the tendons of the jaw muscles (Wilkie *et al.*, 1984). aSeveral lines of morphological and experimental evidence indicate that these connective tissues are capable of nervously mediated changes in their mechanical properties. First, cell processes resembling juxtaligamental cell processes are present in the integument dermis and ligaments of the jaw frame. Second, based on the responses of isolated preparations of the disc integument and jaw frame to ionic manipulation, it is clear that their behaviour is dominated by mutable connective tissues. Wilkie *et al.* (1984) thus conclude that fission is facilitated by changes in the mechanical behaviour of certain mutable connective tissues in the fission plane.

4. Regulation of fission

The factor or factors that initiate fission in brittle stars have not been identified. Emson and Wilkie (1980) noted that many fissiparous brittle stars occupy intertidal or shallow water habitats where they would be subject to environmental fluctuations. It is thus possible that stressful exogenous stimuli might trigger fission in such populations. However, for *Ophiocomella ophiactoides* occupying a relatively stressful shallow water habitat, temperature and salinity stress and hypoxia failed to initiate fission in the laboratory (Mladenov *et al.*, 1983). It is thus unlikely that exogenous factors are important initiators of fission in the field. Mladenov *et al.* (1983) and Wilkie *et al.* (1984) hypothesize that fission in *O. ophiactoides* may be controlled by an endogenous chemical factor that accumulates in the brittle star during regeneration. They further speculate that when this factor reaches a threshold concentration, collagenous tissues in one interradius soften and fission is invoked. In this context, Motokawa (1982) has shown that the coelomic fluid of the brittle star *Ophiarchna incrassata* contains factors that affect the mechanical properties of echinoderm mutable connective tissues. Furthermore, Emson *et al.* (1985) observed that individuals of *Ophiocomella ophiactoides*, parasitized by the siphonostome copepod *Ophiopsyllus reductus*, were significantly less likely to undergo fission, perhaps because the parasite somehow interferes with the chemical basis of fission. The observation by Mladenov *et al.* (1983) that surgical incision in one interradius can trigger fission, especially in those animals that have fully regenerated after their previous fission, may indicate that softening of the connective tissues leading to fission is a nervously mediated event.

5. Fission and body size

Beginning with Lyman (1865), early workers assumed that small six-armed fissiparous brittle stars represented the juvenile phase of larger, sympatric, non-fissiparous five-armed adults (see Parslow and A.M. Clark, 1963 and Devaney, 1970 for a detailed discussion of this point). It was suggested that a pentamerous adult could arise from a hexamerous juvenile either through a three-armed half regenerating only two arms or a two-armed half (result of an unequal fission) regenerating three arms. As an example of this, it was claimed that the young of *Ophiocoma pumila* and *O. alexandri* were asexually propagating hexamerous forms, which later gave rise to the pentamerous adult form (Lütken, 1872; H.L. Clark, 1921, 1946). In this context, the small fissiparous *Ophiocomella ophiactoides* has sometimes been referred to *Ophiocoma pumila*. However, Parslow and A.M. Clark (1963) and Devaney (1970) suggested, on the basis of both morphological criteria and geographic distribution, that the two forms, although closely related, are separate species. Parslow and A.M. Clark (1963) and A.M. Clark (1967) also suggested, however, that a generic distinction seemed too drastic. Further evidence has since accumulated in favour of separating the two forms, including (1) distributional differences at the microhabitat level (P.V. Mladenov and R.H. Emson, unpublished), (2) lack of evidence for a permanent transformation from the six-armed to the five-armed condition (Mladenov *et al.*, 1983), (3) the presence of sexual *Ophiocomella ophiactoides*, which belies their status as a juvenile form of *Ophiocoma pumila* (Mladenov and Emson, 1984), (4) the extensive occurrence of a parasitic copepod on *Ophiocomella ophiactoides* but not on *Ophiocoma pumila* (Emson and Mladenov, 1987a), and (5) allozymic data indicative of a close sibling species relationship (Mladenov and Emson, 1990). Mladenov and Emson (1990) speculate that *Ophiocomella ophiactoides* represents a relatively recently derived, miniaturized, six-armed version of *Ophiocoma pumila*, the result perhaps of heterochrony caused by an acceleration of attainment of sexual maturity.

Although there is little evidence that fissiparous species represent the juvenile phase of large, pentamerous non-fissiparous adults, there is evidence that the incidence of fission is correlated with body size within some well-defined populations. In *Ophiocomella ophiactoides* living in algae in protected Jamaican coves, Mladenov and Emson (1988) found that the incidence of fission varied inversely with body size. However, even some of the largest individuals in this population still showed evidence of fission. In a population of the same species living in algae at Belize, incidence of fission also varied inversely with body size, although the incidence of fission was comparably higher at all body sizes at this location. Similarly, incidence of fission varied inversely with body size for several West Indian alga- and sponge-dwelling populations of *Ophiactis savignyi* (see Mladenov and Emson, 1988). This trend was particularly evident among certain sponge-dwelling aggregations in which most of the largest individuals showed no evidence of recent fission. As will be discussed below, sexual reproduction, when present in a population, occurs in the largest individuals. Thus, though not entirely exclusive events, fission tends to predominate in small members and sex in the larger members of a population.

6. Potential for clonal growth

The potential rate of clonal growth in populations of fissiparous brittle stars depends ultimately on the rate of regeneration following fission and the period of the year over which fission can occur. Boffi (1972) reports that in *Ophiactis savignyi* on the Brazilian coast the proportion of individuals displaying evidence of fission (three arms and half a disc, or three shorter and three longer arms) was between 80 per cent and 95 per cent between October and July, indicating that asexual propagation probably occurs throughout the year at this location. Boffi (1972) also found that in *O. lymani* a high percentage of specimens showed signs of fission throughout the year, although the incidence of fission was slightly lower in February, March, and April. Boffi (1972) observed that *O. lymani* was most abundant from July to December, which corresponded to the period of highest incidence of fission. Recently split individuals of *Ophiocomella ophiactoides* at Discovery Bay, Jamaica, were present in both winter and summer suggesting that fission in this population also occurs on a more or less continuous basis (Mladenov *et al.*, 1983). On average, about 11 per cent of this population showed evidence of recent fission. There is thus evidence that fission in populations of tropical brittle stars occurs throughout much or all of the year and that the potential for clonal growth may not be much constrained by this factor.

In *Ophiocomella ophiactoides*, the fission plane is variable and fission takes place before the new arms are fully regenerated (Mladenov *et al.*, 1983). This results in recently split individuals with long arms of slightly different length. By comparing the length of the shorter and longer of these arms, Mladenov *et al.* (1983)

were able to determine that, on average, fission occurs when the regenerating arms are about 62 per cent of the length of the old arms. They suggested that initiation of fission prior to complete regeneration serves to minimize the interval between successive fissions and so maximize the rate of cloning. Based on measures of rates of regeneration in the laboratory, Mladenov *et al.* (1983) also determined that the interval between successive fissions was of the order of 89 days (i.e. about four fissions per year assuming little or no seasonal variation). A single specimen could thus potentially produce about 15 ($2^4 - 1$) clonemates in a year, or about 4,095 ($2^{12} - 1$) clonemates over a three-year period. This represents a large potential rate of clonal growth. Unfortunately, estimates of natural mortality are lacking for this or any species of fissiparous brittle star.

7. Interrelation of fission and sex

All or most fissiparous species are heterogonic, having a mixed life history that includes both sexual reproduction and clonal proliferation. Gonads have been noted in *Ophiocomella ophiactoides* (see Mladenov and Emson, 1984), *Ophiactis savignyi* (see Boffi, 1972; Emson *et al.*, 1985; Mladenov and Emson, 1988), *Ophiostigma isocanthum* (see Emson *et al.*, 1985), and *Amphipholis torelli* (see Mortensen, 1933b).

Sexual reproduction and its interrelationship with fission have been studied in detail in *Ophiocomella ophiactoides* (see Mladenov and Emson, 1984, 1988). Gonads are present throughout much or all of the year in a population at Discovery Bay, Jamaica. Gonads were found in many recently split individuals indicating that sexual reproduction and fission can occur simultaneously. In general, however, fission predominated among the small individuals and gonads were present only in the largest individuals. Individuals without gonads were significantly smaller than males, which in turn were significantly smaller than females. This relationship between sexual status and size may indicate that *O. ophiactoides* is a protandric hermaphrodite. However, no transitional individuals bearing both male and female elements were ever encountered and hermaphroditism has never been recorded among the ophiocomids. The largest female *O. ophiactoides* contained a clutch of no more than about 7,400 eggs of 80 μm diameter. These eggs develop into typical planktotrophic ophiopluteus larvae of a form virtually indistinguishable from the larvae of *Ophiocoma pumila* and have a pelagic existence of at least one month, and probably much longer (Mladenov, 1985). On a per individual basis, clutch size is very low for a brittle star with planktotrophic development (Hendler, 1975; Mladenov, 1979). However, clutch size per group of clonemates may be considerably higher. The long-lived planktotrophic larvae of this species clearly provide a means of dispersal from deteriorating habitats and the potential to colonize new habitats (see Emson and Wilkie, 1984).

The scattered information available on sexual reproduction in *Ophiostigma isocanthum* and *Ophiactis savignyi* reveals similar sexual patterns. Per individual clutch size is low and egg diameter is well within the size range for planktotrophic development in both species (Emson *et al.*, 1985). Adult specimens of *O. savignyi* with evidence of fission can also contain mature gonads in the complete half of the disc (Boffi, 1972; Mladenov and Emson, 1988). However, fission predominates among the smaller individuals and sex among the larger individuals of a population (Mladenov and Emson, 1988).

The sexual pattern in the fissiparous *Amphipholis torelli* is noteworthy. Mortensen (1933b, pp 65–67, fig. 39) reported the presence of many small blastulae (approximately 82 μm in diameter) within the bursae of this species which probably escape to become pelagic larvae. This is in contrast to the closely related *A. squamata*, which is a non-fissiparous, hermaphroditic brooder (Fell, 1946). It is not known if *A. torelli* is hermaphroditic.

8. Intraspecific variation in levels of sexual reproduction and fission

Mladenov and Emson (1988) obtained evidence for within- and between-habitat intraspecific variation in levels of sexual reproduction and fission from a study of two West Indian brittle stars, *Ophiocomella ophiactoides* and *Ophiactis savignyi*. A population of *Ophiocomella ophiactoides* occupying algal turf in protected coves at Jamaica was present at moderate to high densities and contained both sexual and fissiparous individuals. Here, clonal propagation predominated among the many small individuals in the population while gonads were present mainly in the small number of large individuals present. However, a population of the same species living in algal turf in more exposed backreef and reef crest areas at Belize was present at much lower densities, showed a higher incidence of fission at all body sizes, and did not possess individuals with gonads. The case for intraspecific variation was also clear in *Ophiactis savignyi*. This species was collected from both algal turf and sponge habitats at Jamaica, Belize, and Bermuda. At each location, the mean body size of *O. savignyi* in sponges was significantly larger than that of *O. savignyi* in algae. Furthermore, at each location, *O. savignyi* in algae occurred at lower densities and were exclusively or almost exclusively fissiparous, whereas *O. savignyi* in sponges occurred at much higher densities and sexual as well as fissiparous individuals were present.

The absence of experimental manipulative studies of these populations makes it difficult to explain the observed trends in body size, density, and relative levels of sex and fission. Correlations between high population density, a decline in rate of asexual reproduction, and appearance of sexual individuals have been experimentally demonstrated for the polychaete *Pygiospio elegans* (see Wilson, 1983) and, in freshwater habitats, for *Daphnia* spp. and *Hydra pseudoligactis* (see Bell and Wolfe, 1985). These results are compatible with those reported above and also with the predictions of the 'tangled bank' hypothesis (Ghiselin, 1974; Bell, 1982) that proposes that heterogonic animals should reproduce asexually when population density is low, thereby saturating a localized habitat with genetically identical copies, but switch to sexual reproduction when population density increases, to produce genetically diverse progeny capable of dispersal and colonization. In this regard, stable, predator-free habitats should favour the creation of high density populations and the switch to sexual reproduction. However, the observed trends may be a function of factors other than habitat stability and density. For example, nutrient levels may be higher in sponges than in algal turf, as well as higher in protected than in exposed algal turf. Higher nutrient levels could in turn result in higher growth rates, larger body size, and the development of gonads (Emson and Wilkie, 1984). Also, recent work hints at the possibility of a genetic basis to the observed differences in population characteristics. Mladenov and Emson (1990) have observed striking differences in the allelic composition of sponge- and algal-dwelling *O. savignyi* at Jamaica. Only one of the 17 alleles at the five loci studied was common to both populations. They speculate that sponge- and algal-dwelling *O. savignyi* may represent hidden species.

9. Genetics and clonal structure of fissiparous populations

Indirect evidence suggests that clonal proliferation may make a significant contribution to the genetic structure of populations of fissiparous brittle stars. First, information on size structure and frequency of fission in various populations of *Ophiocomella ophiactoides* indicates that rates of recruitment of planktonic larvae are low over the short term and that incidence of fission is high (Mladenov *et al.*, 1983; Mladenov and Emson, 1988). Second, information on the interval between splitting in this species derived from measures of rate of regeneration point to a high potential rate of cloning (Mladenov *et al.*, 1983). A third line of evidence is derived from information on sex ratios collected by Mladenov and Emson (1988). These authors noted extremely biased sex ratios and the frequent occurrence of all-male or all-female clusters within aggregations of *O. ophiactoides* at Jamaica, which would be consistent with the notion of clonal aggregations. They also discovered that sponges at Belize and Jamaica were frequently occupied by all-male, and sometimes by all-female, aggregations of *O. savignyi*. Furthermore, the *O. savignyi* population in Harrington Sound, Bermuda, seemed to be all-male.

An electrophoretic analysis of five polymorphic enzymes by Mladenov and Emson (1990) provided more direct information on the clonal structure of fissiparous populations. These authors determined whether the observed genotypic diversity values differed significantly from expectations under the conditions of Hardy-Weinberg equilibrium and random mating by computer simulation (Stoddart and Taylor, 1988). They found that multilocus genotypic diversity in a population of *Ophiocomella ophiactoides* at Discovery Bay, Jamaica, was significantly lower than expected for a sexually reproducing population, due largely to the predominance of clonal proliferation over larval recruitment. In contrast, genotypic diversity of the closely related but obligately sexual species, *Ophiocoma pumila*, at the same site conformed to expectations for a sexually reproducing population. They also noted large variation in clonal composition in *Ophiocomella ophiactoides* over a short distance (50 m), indicative of a very localized scale of cloning mixing. Mladenov and Emson (1990) also found that genotypic diversity of *Ophiactis savignyi* aggregations in two separate sponges was very low and that each sponge was dominated by a single genotype. Although the sponges were separated by a distance of only 2 m, a large difference in clonal composition was observed. Genotypic diversity of *O. savignyi* in algae was higher, though still significantly lower than expectations for a sexually reproducing population.

These data thus confirm that in nature fission in brittle stars can serve to create a large number of genetically identical modules (ramets) forming a dispersed clone (genet). Hendler and Littman (1986), Pearse *et al.* (1989), and Mladenov and Emson (1990) suggest that 'replicate copy growth' of this kind might increase the fitness of genets by increasing genotype-specific biomass and, hence, fecundity. Viewed in this light, fission may represent an alternate form of growth (as opposed to growth of genetically unique individuals) that may be adaptive under certain circumstances. Mladenov and Emson (1990) suggest that sibling pairs of brittle stars with contrasting sexual and fissiparous patterns of propagation (like *Ophiocoma pumila* and *Ophiocomella ophiactoides*) may provide useful models for critically examining the life-history consequences and ecological correlates of replicate copy growth by allowing comparison of total biomass and clutch size of electrophoretically distinguished genets with that of non-fissiparous individuals as well as comparison of the attributes of their respective microhabitats.

B. Possible Propagation by Parthenogenesis

Mortensen (1936) reported on some interesting observations of three species of Antarctic ophiuroids that apparently exist largely or exclusively as females. *Ophiacantha vivipara* is an intrabursally brooding, proterogynic hermaphrodite in which males are extremely rare; *Amphiura microplax* and *A. eugeniae* are intraovarial brooders that exist only as females. Mortensen (1936) suggested that all three species may be parthenogenetic and, hence, may be propagating asexually if apomictic parthenogenesis is involved.

C. Possible Asexual Propagation by Larvae

Asexual propagation by paratomy may occur in some ophiuroid larvae. Ophioplutei of the genus *Ophiothrix* display a Type I pattern of metamorphosis (Mladenov, 1985) in which the larvae typically retain the outer pair of longer postero-lateral arms during metamorphosis. At the completion of metamorphosis the rudiment hangs suspended from these arms. Although the arms are cast off during settlement of the juvenile, they remain afloat and active for some time afterwards (Mortensen, 1921, 1937; Mladenov, 1979). Mortensen (1921) claimed that the abandoned postero-lateral arms of an unassigned *Ophiothrix*-type ophiopluteus, *Ophiopluteus opulentus*, showed signs of regenerating a new larval mouth and oesophagus in preparation, apparently, for the development of another larval body and, ultimately, a second metamorphosis; small granules present in the vibratile band of the postero-lateral arms were thought to furnish the necessary nourishment. Such a phenomenon does not occur in the reduced lecithotrophic ophiopluteus of *Ophiothrix oerstedi*. In this species, the postero-lateral arms remain afloat for at least four days following their separation from the juvenile but they eventually sink to the bottom and die (Mladenov, 1979). However, the possibility of asexual propagation in some *Ophiothrix*-type ophioplutei under certain circumstances cannot be entirely ruled out, especially in light of recent observations by Bosch *et al.* (1989) of asexual propagation by paratomy in larvae of the sea star *Luidia* sp. (see Section IIID).

V. ECHINOIDEA

There are no reports of fission as a means of asexual propagation in echinoids. Lack of fission in this group is probably related to the presence of a hard test and a spacious coelom which would make the sealing of wounds following fission impossible to accomplish.

Parthenogenetic activation of echinoid eggs by artificial means has given rise to a considerable amount of literature (e.g. Hertwig and Hertwig, 1887; Harvey, 1956; Ishikawa, 1975). Although development of parthenogenetically activated eggs is usually slower than normally developing eggs, with considerable mortality, there have been several successful attempts to rear larvae through metamorphosis in the laboratory (e.g. Delage 1909; Shearer and Lloyd, 1913; Brandiff *et al.*, 1975). The young sea urchins generally displayed a large number of abnormalities, however, and died soon after metamorphosis.

Naturally occurring parthenogenesis appears to be very rare in echinoids. Gladfelter (1978) reported that embryos of the small cassiduloid sea urchin, *Cassidulus caribbearum*, developed from eggs spawned by females isolated in aquaria from males. He suggested that fertilization is internal in this species, but pointed out that parthenogenesis could not be excluded as a possibility. Bak *et al.* (1984) found numerous blastulae in the gonads of two individuals of the sea urchin, *Diadema antillarum*, following mass mortalities of this species on the coral reefs at Curacao. They suggested that under conditions of extremely low population density following a mass mortality, parthenogenetic production of embryos could facilitate population recovery. It is not known if the mechanism of parthenogenesis is apomictic (= ameiotic) and, hence, asexual.

VI. HOLOTHUROIDEA

A. Asexual Propagation by Fission

1. Occurrence

Some holothuroids are capable of asexual propagation by transverse fission. Individuals of such species are recognizable because they are often distinctly asymmetrical, regenerating a new oral or anal end (Fig. 6). Asexual propagation by fission is of rather limited occurrence in the holothuroids. At present, only six species from two genera are known to be capable of transverse fission (Table 5) and it is only in the species of the genus *Holothuria* that fission is sufficiently common to be potentially an important means of asexual propagation. Fissiparous species of the genus *Holothuria* have a tropical distribution frequenting shallow rocky shores and rock pools; the two fissiparous species of *Ocnus* are present in shallow waters of the temperate Atlantic Ocean and the Mediterranean Sea.

2. Process of fission

Dalyell (1851) was the first to note a holothurian undergoing self-division and subsequent regeneration of a complete animal by each of the resulting parts. Details of the process were later provided by Chadwick (1890), Monticelli (1896), Crozier (1917), and Bonham and Held (1963). Fission can apparently take place by a number of different processes. One involves the stretching and rupture of the middle portion of the body due to the pull of anal and oral ends in opposite directions (Fig. 7). A second method involves the vigorous twisting of the body with resulting constriction and rupture at two places to produce three new body parts. The time required for the completion of the process seems to vary from 24 hours to 5 days.

Table 5

Fissiparous sea cucumbers and their geographic distribution

Family	Species	Distribution	Reference
Holothuridae	*Holothuria difficilis* Semper	Hawaiian Islands	Deichmann (1921)
	H. parvula (Selenka)	Bermuda, West Indies	Benham (1911), Crozier (1917), Deichmann (1921)
	H. surinamensis (Ludwig)	Bermuda	Crozier (1917)
	H. atra (Jaeger)	Pacific	Bonham and Held (1963), Harriot (1982)
Cucumaridae	*Ocnus planci* Brandt	Atlantic	Dalyell (1851), Chadwick (1890)
	O. lactea (Forbes and Goodsir)	Atlantic, Mediterranean	Dalyell (1851)

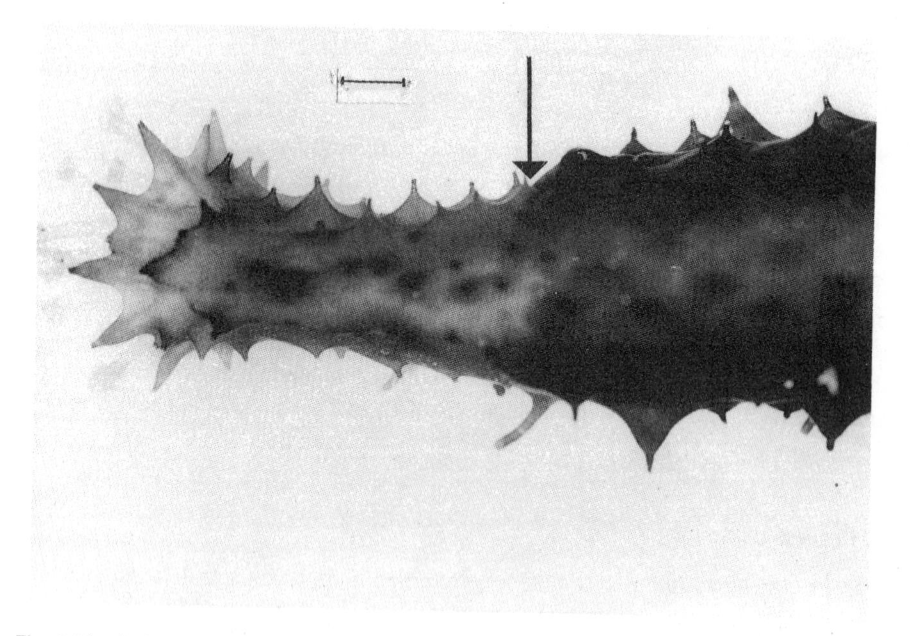

Fig. 6. The fissiparous sea cucumber *Holothuria parvula* regenerating a new oral half following transverse fission. The position of the plane of fission is indicated by an arrow (scale bar = 1 mm) (photo by P. Mladenov).

Emson and Mladenov (1987b) compared the mean lengths of newly split oral and anal ends of *Holothuria parvula* from Bermuda and found them to be very similar. Furthermore, the number of individuals regenerating oral and anal ends was virtually identical in this population. They concluded that individuals of *H. parvula* split into more or less equal parts and that there is little difference in the rate of survival of the two halves. This is in close correspondence to the conclusions of Crozier (1917) and Deichmann (1922), who worked with the same species. In *H. atra* from New Caledonia and Papua New Guinea, Conand and De Ridder (1990) concluded that the plane of fission is located in the anterior 45 per cent of individuals as determined from direct observations of splitting individuals and the sizes of recently split individuals.

3. Mechanism of fission

Based on the observations of Chadwick (1890), Monticelli (1896), and Crozier (1917), it appears that fission begins with a deep insinking of the body wall along with a strong contraction of circular muscles to form a constriction at the point where rupture of the tissues ultimately occurs. While this contraction is taking place,

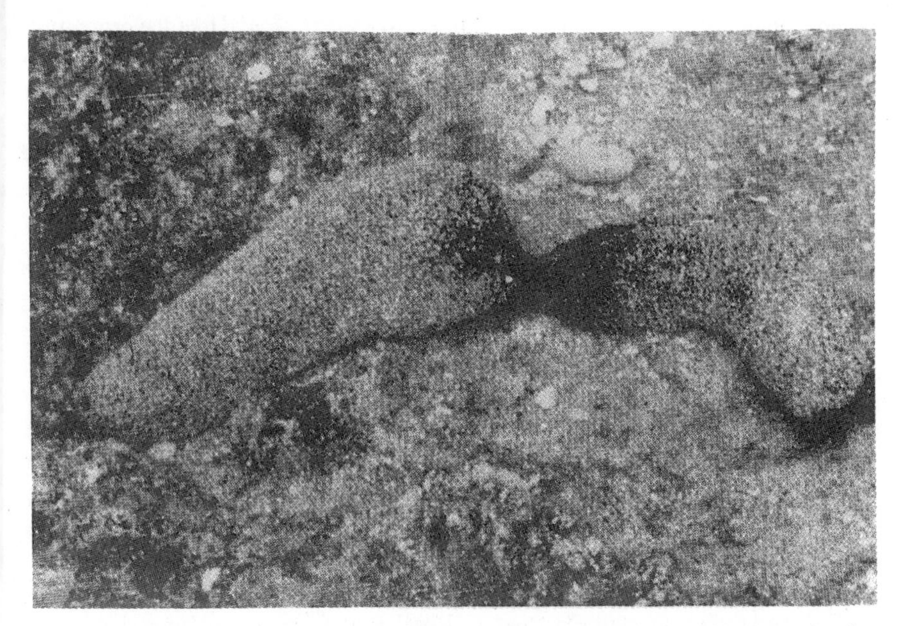

Fig. 7. The sea cucumber *Holothuria atra* undergoing transverse fission in the field (photo taken from Bonham and Held, courtesy *Pacific Science*, **17**, 305–314 [1963]).

the tube feet at each end of the body are pulling in opposite directions. A localized disintegration of tissues of the body wall at the point of constriction effects the separation of the two body parts. Often the two parts remain joined for some time only by the gut, which also eventually ruptures. It is quite possible that a softening of mutable collagenous tissues of the body wall facilitates the fission process.

4. Fission and body size

Crozier (1917) stated that fission is a phenomenon associated with young (small) individuals of *Holothuria parvula* from Bermuda, as larger animals observed by him showed no signs of fission. Deichmann (1921) also believed fission to be characteristic of young animals in this species. However, Emson and Mladenov (1987b) found little evidence to support this idea, noting evidence for recent fission in all sizes of individuals sampled from a Bermudan population. Conand and De Ridder (1990) also observed fission in all sizes of individuals of *H. atra* sampled in New Caledonia.

5. Incidence and seasonality of fission

Deichmann (1921) demonstrated that fission can be a common phenomenon in *Holothuria parvula* as 65 per cent of the preserved specimens studied by her showed evidence of fission. Emson and Mladenov (1987b) found that the proportion of individuals of *H. parvula* showing evidence of regeneration varied from 43 to 83 per cent over the year, with fission taking place principally in the summer, and subsequent regeneration occurring during the late summer, autumn, and winter. Conand and De Ridder (1990) found that the mean incidence of dividing individuals was about 10 per cent in a population of *H. atra* at New Caledonia, although the incidence of fission varied seasonally, peaking during the cool season. On the basis of size-frequency data, Bonham and Held (1963) believed that fission in *H. atra* at Rongelap Atoll occurred throughout the year.

6. Interrelationship of fission and sex

Emson and Mladenov (1987b) found individuals of *Holothuria parvula* with large gonads only during the summer months, which was also the period when fission was most common. Such sexual individuals formed only a small (< 10 per cent) proportion of the population, however, suggesting that the level of potential gamete production was quite small. It is possible that regeneration constantly inhibits the development of gonads in most individuals of this population. In contrast to findings for *H. parvula*, Conand and De Ridder (1990) found that sexual reproduction in *H. atra* took place during the warm season (the gonad index peaked in December), which was the period when fission was least common.

7. Regulation of fission

The periodic nature of fission in *Holothuria parvula* (see Emson and Mladenov, 1987b) and *H. atra* (see Conand and De Ridder, 1990) suggests that some form of endogenous control mechanism, perhaps a chemical factor, regulates fission (see also Section III A 3). Such a factor might initiate fission directly or make the sea cucumber more responsive to external fission-inducing stimuli. In this context, Bonham and Held (1963) noted *H. atra* splitting in the field when the water temperature was 37°C and Conand and De Ridder (1990) hypothesize that fission in this species is triggered by desiccation and thermal stress during spring low tides. Similarly, Emson and Mladenov (1987b) hypothesize that fission in *H. parvula* is triggered by high summer sea temperature and physical disturbance at low tide.

8. Cloning potential

There is little available information on which to base an assessment of the potential for asexual propagation in fissiparous holothuroids. The incidence of regenerating individuals is known to be quite high in *Holothuria atra* (see Ebert, 1978; Conand and De Ridder, 1990) and *H. parvula* (see Deichmann, 1921; Emson and Mladenov,

1987b), suggesting that fission can be of common occurrence in populations of these species. Field studies of the relative proportion of regenerating to parent tissue in *H. parvula* demonstrated that individuals may regenerate fully in one year and that fission is an annual event for many individuals. Conand and De Ridder (1990) found that complete regeneration could take place in as little as two months in *H. atra*. The high incidence of fission in combination with relatively rapid regeneration rates suggest that the potential for clonal growth is quite high in these species. Studies of the population genetics of fissiparous holothuroids would likely provide more definitive information on the relative contribution of sexual and asexual recruitment in structuring local populations.

VII. CONCLUSIONS

About 19 species of asteroid, 45 species of ophiuroid, and six species of holothuroid propagate asexually by fission; fission does not occur in crinoids and echinoids. In addition, about six species of sea star propagate asexually by arm autotomy. Thus, about 76 (1.3 per cent) of the roughly 5,900 extant species of echinoderm are capable of asexual propagation by fission or autotomy. In addition, asexual propagation by paratomic fission is known for the larva of at least one species of asteroid, and asexual propagation by budding is thought to be possible for the larvae of a few species of asteroid and ophiuroid. There is also evidence that a few species of asteroid, ophiuroid, and echinoid are naturally parthenogenetic, but it is not known if the mechanism is amictic or automictic. The former would be an example of asexual propagation; the latter involves a process functionally equivalent to self-fertilization.

Asexual propagation in echinoderms is associated with the much more widespread ability to autotomize and regenerate body parts. As pointed out by Ghiselin (1987), this may represent an evolutionary sequence in which regeneration came first to provide a means of dealing with predation, followed by autotomy, which refined the defensive adaptation. Asexual propagation by arm autotomy and fission then arose in a few species.

Asexual species of echinoderm can be widely spread geographically and often occur in very high densities in certain habitats (e.g. Mladenov and Emson, 1988). On the whole, fissiparous echinoderms seem well adapted for saturating local habitats, particularly those that favour small body size (Hendler and Littman, 1986).

Most, perhaps all, fissiparous echinoderms have a mixed life history that includes both sexual reproduction and asexual proliferation. The relative levels of sex and fission can vary on a geographical basis and also between habitats within a given location (Crump and Barker, 1985; Mladenov and Emson, 1988). There is some evidence that fission predominates in environments undersaturated with organisms, whereas sex occurs in high-density situations and results in the production of genetically diverse, dispersive larvae. However, the onset of sex may also be linked to higher food availability in certain kinds of habitats, which would favour individual growth and the production of gonads.

There is good evidence that populations of fissiparous ophiuroids and asteroids are often highly clonal in structure, each clone, or genet, consisting of many discrete, genetically identical bodies, or ramets (Johnson and Threlfall, 1987; Mladenov and Emson, 1990). The effective body size of a genet may potentially be very large, in terms of both the total number and the total biomass of the contributing ramets. Furthermore, although the fecundity of a single ramet is very low, the aggregate fecundity of all the ramets comprising the genet may be very high. Pearse *et al.* (1989) have argued persuasively that clonal replication in animals may best be viewed as an alternate form of growth, as opposed to reproduction. In the case of fissiparous echinoderms, the genotype is packaged into many small ramets, not a single individual. Collectively, the ramets may enjoy high fitness in terms of size, survival, and fecundity compared to genotypes confined to a single body.

There is still much to learn about the biology and ecology of asexual echinoderms. Future topics of study should include the population genetics of asteroids that propagate by arm autotomy and by parthenogenesis; work aimed at identifying the characteristics of habitats that promote fission and sex in echinoderms; studies of asexual propagation in larval echinoderms; research on the physicochemical basis and environmental regulation of fission and autotomy; and comparison of life-history traits of closely related fissiparous and non-fissiparous species to test ideas concerning the adaptive significance of sex and clonal growth in echinoderms.

REFERENCES

Achituv, Y. (1969). 'Studies on the reproduction and distribution of *Asterina burtoni* (Gray) and *Asterina wega* (Perrier) (Asteroidea) in the Red Sea and the eastern Mediterranean', *Israel J. Zool.*, **18**, 329–342.

Achituv, Y. (1973a). 'The genital cycle of *Asterina burtoni* (Gray) (Asteroidea) from the Gulf of Elat, Red Sea', *Cah. Biol. mar.*, **14**, 547–553.

Achituv, Y. (1973b). 'On the distribution and variability of the Indo-Pacific sea star *Asterina wega* (Echinodermata: Asteroidea) in the Mediterranean Sea', *Mar. Biol.*, **18**, 333–336.

Achituv, Y., and Malik, Z. (1985). 'The spermatozoa of the fissiparous starfish, *Asterina burtoni*', *Int. J. Inv. Reprod. Develop.*, **8**, 67–72.

Achituv, Y., and Sher, E. (1991). 'Sexual reproduction and fission in the sea star *Asterina burtoni* from the Mediterranean coast of Israel', *Bull. mar. Sci.*, **48**(3), 670–678.

Applegate, A.L. (1984). 'Echinoderms of south Taiwan', *Bull. Inst. Zool.*, **23**, 93–118.

Bak, R.P.M., Carpay, M.J.E., and De Ruyter van Steveninck, E.D. (1984). 'Densities of the sea urchin *Diadema antillarum* before and after mass mortalities on the coral reefs of Curaçao', *Mar. Ecol. Prog. Ser.*, **17**, 105–108.

Barker, M.F., Scheibling, R., and Mladenov, P.V. (1992). 'Seasonal changes in population structure of the fissiparous asteroids *Allostichaster insignis* (Farquahr) and *Coscinasterias calamaria* (Gray)', in *Echinoderm Research* 1991 (Eds. L. Scalera-Liaci and C. Canicatti), A.A. Balkema, Rotterdam, pp. 191–196.

Bell, G. (1982). *The Masterpiece of Nature*, Croom Helm, London and Canberra, 365 pp.

Bell, G., and Wolfe, L.M. (1985). 'Sexual and asexual reproduction in a natural population of *Hydra pseudoligactis*', *Can. J. Zool.*, **63**, 851–856.

Benham, W.B. (1911). 'Report on sundry invertebrates from the Kermadec Islands', *Trans. New Zealand Inst.*, **44**, 135–138.

Bennett, E.W. (1927). 'Notes on some New Zealand seastars and on autotomous reproduction', *Rec. Canterbury (N.Z.) Mus.*, **3**, 125–149.

Boffi, E. (1972). 'Ecological aspects of ophiuroids from the phytal of S.W. Atlantic Ocean warm waters', *Mar. Biol.*, **15**, 316–328.

Bonham, K., and Held, E.E. (1963). 'Ecological observations on the sea cucumbers, *Holothuria atra* and *H. leucospilota* at Rongelap Atoll, Marshall Islands', *Pac. Sci.*, **17**, 305–314.

Bosch, I., Rivkin, R.B., and Alexanders, S.P. (1989). 'Asexual reproduction by oceanic planktotrophic echinoderm larvae', *Nature, Lond.*, **337**, 169–170.

Brandiff, B., Hinegardner, R.T., and Steinhardt, R. (1975). 'Development and life cycle of parthenogenetically activated sea urchin embryos', *J. Exp. Zool.*, **192**, 13–24.

Carson, S.F. (1984). 'Asexual propagation in the sea star *Stephanasterias albula*', Honours thesis, Mount Allison University, Canada, 58 pp.

Caspers, H. (1985). 'The brittle star, *Ophiactis savignyi* (Müller & Troschel), an inhabitant of a Pacific sponge, *Damiriana hawaiiana* de Laubenfels', in *Echinodermata* (Eds. B.F. Keegan and B.D.S. O'Connor), A.A. Balkema, Rotterdam, pp. 603–608.

Chadwick, H.C. (1890). 'Notes on *Cucumaria planci*', *Proc. and Trans. Liverpool Biol. Soc.*, **5**, 81–82.

Chaet, A.B. (1962). 'A toxin in the coelomic fluid of scalded starfish (*Asterias forbesi*)', *Proc. Soc. exp. Biol. Med.*, **109**, 791–794.

Clark, A.H. (1939). 'Echinoderms (other than holothurians) collected on the presidential cruise of 1938', *Smith. misc. Coll.*, **99**(11), 1–18, 5 plates.

Clark, A.H. (1952). '*Schizostella*, a new genus of brittle-star (Gorgonocephalidae)', *Smith. Inst. U.S. natl Mus.*, **102**, 451–454.

Clark, A.M. (1952). 'Echinodermata. Report of the Manihine Expedition of the Gulf of Aqaba', *Bull. Brit. Mus. Nat. Hist. Zool.*, **1**, 203–214.

Clark, A.M. (1967). 'Variable symmetry in fissiparous Asterozoa', *Symp. zool. Soc. London.*, **20**, 143–157.

Clark, A.M. (1980a). 'Echinoderms of Hong Kong', in *Proceedings of the First International Marine Biological Workshop: The Marine Flora and Fauna of Hong Kong and Southern China, Hong Kong* (Eds. B. Morton and C.K. Tseng), Hong Kong University Press, Hong Kong, pp. 485–501.

Clark, A.M. (1980b). 'Some Ophiuroidea from the Seychelles Islands and Inhaca, Mozambique', *Rev. Zool.*, **94** (3), 534–558.

Clark, A.M., and Rowe, F.W.E. (1971). *Monograph of Shallow-water Indo-West Pacific Echinoderms*, Trustees of the British Museum (Natural History), London, 238 pp., 31 plates.

Clark, H.L. (1901). 'The echinoderms of Porto Rico', *Bull. U.S. Fish. Comm.*, **19** (1899), 233–263, plates 14–17.

Clark, H.L. (1911). 'North Pacific ophiurans in the collection of the United States National Museum', *Bull. U.S. natnl Mus.*, **75**, 1–302.

Clark, H.L. (1913). 'Autotomy in *Linckia*', *Zool. Anz.*, **42**, 156–159.

Clark, H.L. (1914a). 'Growth changes in brittle stars', *Publ. Carnegie Inst., Washington*, **182**, 91–126.

Clark, H.L. (1914b). 'The echinoderms of the Western Australian Museum', *Rec. West. Aust. Mus.*, **1**, 132–173.

Clark, H.L. (1921). 'The echinoderm fauna of Torres Strait: Its composition and its origin', *Publ. Carnegie Inst., Washington*, **294**, 1–223.

Clark, H.L. (1933). 'A handbook of the littoral echinoderms of Porto Rico and other West Indian islands', *Surv. P. Rico*, **16**, 1–60.

Clark, H.L. (1940). 'Eastern Pacific expeditions of the New York Zoological Society XXI. Notes on echinoderms from the west coast of Central America', *Zoologica*, **25**, 331–352, and 2 plates.

Clark, H.L. (1942). 'The echinoderm fauna of Bermuda', *Bull. Mus. Comp. Zool. Harvard*, **89**, 367–391, and 1 plate.

Clark, H.L. (1946). 'The echinoderm fauna of Australia, its composition and its origin', *Carnegie Inst. Washington Publ.*, **566**, 1–567.

Cognetti, G., and Delavault, R. (1962). 'La sexualité des asterides', *Cah. Biol. mar.*, **3**, 157–182.

Conand, C., and De Ridder, C. (1990). 'Reproduction asexuée par scisson chez *Holothuria atra* (Holothuroidea) dans des populations de platiers récifaux', in *Echinoderm Research* (Eds. C.D. Ridder, P. Dubois, and M.-C. Lahayeet), A.A. Balkema, Rotterdam, pp. 71–76.

Crozier, W.J. (1917). 'Multiplication by fission in holothurians', *Amer. Natur.*, **51**, 560–566.

Crozier, W.J. (1920). 'Notes on some problems of adaptation. 2. On the temporal relations of asexual propagation and gametic reproduction in *Coscinasterias tenuispina*: With a note on the direction of progression and on the significance of the madrepores', *Biol. Bull.*, **34**, 116–129.

Crump, R.G. (1969). 'Aspects of the biology of two New Zealand echinoderms', Ph.D. thesis, University of Otago, New Zealand.

Crump, R.G., and Barker, M.F. (1985). 'Sexual and asexual reproduction in geographically separated populations of the fissiparous asteroid *Coscinasterias calamaria* (Gray)', *J. exp. mar. Biol. Ecol.*, **88**, 109–127.

Dalyell, J.G. (1851). *The Powers of the Creator Displayed in the Creation, etc.*, Vol. I, Van Voorst, London, 286 pp.

Dan, J.C., and Dan, K. (1941). 'Early development of *Comanthus japonicus*', *Jap. J. Zool.*, **9**, 565–574, with 1 table and 3 plates.

Danielssen, D.C. (1892). 'Crinoida. Norske Nordhavs-Expedition 1876–1878', *Zoology*, **5** (21), 1–28, and plates 1–V.

Davis, L.V. (1967). 'The suppression of autotomy in *Linckia multiflora* (Lamark) by a parasitic gastropod *Stylifer linckiae*', *Veliger*, **9**, 343–346.

Deichmann, E. (1921). 'On some cases of multiplication by fission and coalescence by holothurians: With notes on the synonymy of *Actinopyga parvula* (Sel)', *Vid. Medd. Dansk Naturh. Foren.*, **73**, 183–191.

Deichmann, E. (1922). 'On some cases of multiplication by fission and coalescence in holothurians', *Vid. Medd. Dansk Naturh. Foren.*, **73**, 199–213.

Delage, Y. (1909). 'Le sexe chez les oursins issus du parthénogénèse expérimental', *C. R. Acad. Sci., Paris*, **148**, 453–455.

Delap, M., and Delap, C. (1907). 'Notes on the plankton of Valencia Harbour 1902–1905', *Fisheries, Ireland, Sci. Invest.*, **7**, 1–22.

Devaney, D.M. (1970). 'Studies on ophiocomid brittlestars. I. A new genus *Clarkcoma* of Ophiocominae with a re-evaluation of the genus *Ophiocoma*', *Smithsonian Contrib. Zool.*, **51**, 1–41.

Duncan, P.M. (1879). 'On some Ophiuridae from the Korean Seas', *J. Linn. Soc. Lond.*, **114**, 445–483.

Ebert, T.A. (1978). 'Growth and size of the tropical sea cucumber, *Holothuria* (Holodeima) *atra* Jaeger at Enewetak Atoll, Marshall Islands', **32** (2), 183–191.

Edmondson, C.H. (1935). 'Autotomy and regeneration in Hawaiian starfishes', *Occas. Pap. Bernice Pauachi Bishop Mus. Honolulu*, **11**, 1–29.

Emson, R.H. (1978). 'Some aspects of fission in *Allostichaster polyplax*', in *Physiology and Behavior of Marine Organisms* (Eds. D.S. McLusky and A.J. Berry), Pergamon Press, Oxford, pp. 321–329.

Emson, R.H., and Mladenov, P.V. (1987a). 'Brittlestar host specificity and apparent host discrimination by the parasitic ccpepod *Ophiopsyllus reductus*', *Parasitology*, **94**, 7–15.

Emson, R.H., and Mladenov, P.V. (1987b). 'Studies of the fissiparous holothurian *Holothuria parvula* (Selenka) (Echinodermata: Holothuroidea)', *J. exp. mar. Biol. Ecol.*, **111**, 195–211.

Emson, R.H., Mladenov, P.V., and Wilkie, I.C. (1985). 'Patterns of reproduction in small Jamaican brittle stars: Fission and brooding predominate', in *The Ecology of Coral Reefs, NOAA Symposium Series for Undersea Research* 3 (Ed. M.L. Reaka), NOAA Undersea Research Program, Rockville, Maryland, pp. 87–100.

Emson, R.H. and Wilkie, I.C. (1980). 'Fission and autotomy in echinoderms', *Oceanogr. mar. Biol. Ann. Rev.*, **18**, 155–250.

Emson, R.H., and Wilkie, I.C. (1984). 'An apparent instance of recruitment following sexual reproduction in the fissiparous brittlestar *Ophiactis savignyi* (Müller & Troschel)', *J. exp. mar. Biol. Ecol.*, **77**, 23–28.

Falconetti, C., Fredj-Reygrobellet, D., and Fredj, G. (1976). 'Sexualité et fissiparite concomitantes chez l'asterie *Sclerasterias richardi*: Premieres données', *Mar. Biol.*, **34**, 247–257.

Falconetti, C., Fredj-Reygrobellet, D., and Fredj, G. (1977). 'Introduction de l'émission des gametes et premiers stades du développement larvaire chez l'asterie fissipare *Sclerasterias richardi*', *Mar. Biol.*, **39**, 171–178.

Febvre, M., Fredj-Reygrobellet, D., and Fredj, G. (1981). 'Reproduction sexuee d'une asterie fissipare, *Sclerasterias richardi* (Perrier, 1882)', *Int. J. Inv. Reprod.*, **3**, 193–208.

Fell, H.B. (1946). 'The embryology of the viviparous ophiuroid *Amphipholis squamata* Delle Chiaje', *Trans. Roy. Soc. New Zealand*, **75** (4), 419–464.

Fisher, W.K. (1919). 'Starfishes of the Philippine Seas and adjacent waters', *Bull. U.S. natnl Mus.*, **100(3)**, xii + 711, 156 plates.

Fisher, W.K. (1925). 'Asexual reproduction in the starfish *Sclerasterias*', *Biol. Bull.*, **48**, 171–175.

Fisher, W.K. (1928). 'Asteroidea of the North Pacific and adjacent waters', *Bull. U.S. natnl Mus.*, **75** (2), 1–245.

Fisher, W.K. (1930). 'Asteroidea of the North Pacific and adjacent waters', *Bull. U.S. natnl. Mus.*, **76**, 3, 349.

Ghiselin, M.T. (1974). *The Economy of Nature and the Evolution of Sex*, Univ. of California Press, Berkeley.

Ghiselin, M.T. (1987). 'Evolutionary aspects of marine invertebrate reproduction', in *Reproduction of Marine Invertebrates* (Eds. A.C. Giese, J.S. Pearse, and V.B. Pearse), Blackwell Scientific and The Boxwood Press, California, IX, pp. 609–665.

Gladfelter, W.B. (1978). 'General ecology of the cassiduloid urchin *Cassidulus caribbearum*', *Mar. Biol.*, **47**, 149–160.

Grainger, E.H. (1966). 'Sea stars (Echinodermata: Asteroidea) of arctic North America', *Bull. Fish. Res. Board, Can.*, **152**, 70.

Greeff, R. (1876). 'Uber den Bau und die Entwicklung der Echinodermens. Mitteilung, Parthenogenesis bei den Seesternen', *Sitz. Gesells. Beford. Gesammt. Naturwiss. Marburg*, 83–85.

Guille, A., Laboute, P., and Menou, J.-L. (1986). *Guide des étoiles de Mer, oursins et autres Échinodermes du Lagon de Nouvelle-Calédonie*, Orstom, Paris, 238 pp.

Harriot, V.J. (1982). 'Sexual and asexual reproduction of *Holothuria atra* Jaeger at Heron Island Reef, Great Barrier Reef', *Aust. Mus. Mem.*, **16**, 53–66.

Harvey, E.B. (1956). *The American* Arbacia *and Other Sea Urchins*, Princeton University Press, Princeton, New Jersey, 298 pp.

Hendler, G. (1975). 'Adaptational significance of the patterns of ophiuroid development', *Am. Zool.*, **15**, 691–715.

Hendler, G. (1979). 'Sex-reversal and viviparity in *Ophiolepis kieri*, n.sp., with notes on viviparous brittlestars from the Caribbean (Echinodermata: Ophiuroidea)', *Proc. biol. Soc. Wash.*, **92**, 783–795.

Hendler, G., and Littman, B.S. (1986). 'The ploys of sex: Relationships among the mode of reproduction, body size, and habitats of coral-reef brittlestars', *Coral Reefs*, **5**, 31–42.

Hertwig, O., and Hertwig, R. (1887). 'Uber den Befruchtrungsund Teilungsvorgang des tierischen Eis unter dem Einfluss ausserer Argentiea', *Jena Z. Naturf.*, **13**, 120–241.

Hirota, S. (1895). 'Anatomical notes on the comet of *Linckia multifora*, Lamarck', *Zool. Mag. Tokyo*, **7**, 67–76.

Hopkins, A.E. (1926). 'On the physiology of the central nervous system in the starfish *Asterias tenuispina*', *J. Exp. Zool.*, **46**, 263–275.

Hotchkiss, F.H.C. (1982). 'Ophiuroidea (Echinodermata) from Carrie Bow Cay, Belize', in *The Atlantic Barrier Reef Ecosystem at Carrie Bow Cay, Belize* (Eds. K. Rutzler and I.G. Macintyre), Smith. Contr. Mar. Sci. Washington, Smithsonian Institution Press, 12, pp. 387–412.

Hulbert, A.W., Pecci, K.J., and Witman, J.D. (1983). 'Ecosystem definition and community structure of the macrobenthos of the NEMP Monitoring Station at Pigeon Hill in the Gulf of Maine', *NOAA Technical Memorandum NMFS-F/NEC-14*.

Hyman, L.H. (1955). The Invertebrates, Vol. 4, *Echinodermata*, McGraw-Hill, New York, 763 pp.

Ishikawa, M. (1975). 'Parthenogenetic activation and development', in *The Sea Urchin Embryo. Biochemistry and Morphogenesis* (Ed. G. Czihak), Springer-Verlag, Berlin, pp. 148–169.

James, D.B., and Pearse, J.S. (1969). 'Echinoderms from the Gulf of Suez and the northern Red Sea', *J. mar. biol. Ass. India*, **11**, 78–125.

Jangoux, M. (1984). 'Les asterides littoraux de Nouvelle-Calédonie', *Bull. Mus. nat. Hist. natur. Paris*, **6**, 279–293.

Johnson, M.S., and Threlfall, T.J. (1987). 'Fissiparity and population genetics of *Coscinasterias calamaria*', *Mar. Biol.*, **93**, 517–525.

Kenny, R. (1969). 'Growth and asexual reproduction of the starfish *Nepanthia belcheri* (Perrier)', *Pac. Sci.*, **23**, 51–56.

Koehler, R. (1913). 'Ophiures', *Zool. Jb.*, **11** (Suppl.), 351–380, and plates 20, 21.

Koehler, R. (1914). 'Ophiurans of the U.S. National Museum', *Bull. U.S. natnl Mus.*, **84**, 1–173.

Koehler, R. (1930). 'Papers from Dr. Th. Mortensen's Pacific Expedition 1914–1916 LIX. Ophiures recueillis par le Docteur Th. Mortensen dans l'Archipel Malais', *Vid. Medd. Dansk Naturh. Foren.*, **89**, 1–295.

Koren, and Danielssen, D.C. (1847). *Am. Sci. Nat.*, **3**(VII), 347.

Kowalevsky, A. (1872). 'Uber die Vermehrung der Seesterne durch Theilung und Knospung', *Z. wiss. Zool.*, **22**, 283–384.

Le Conte, J.L. (1851). *Proc. Acad. Nat. Sci., Philadelphia*, **5**, 318–319.

Loeb, J. (1905). 'Artificial membrane-formation and chemical fertilization in a starfish (*Asterina*) 1', *Physiology*, **2**(16), 147–158.

Lütken, C.F. (1857). 'Oversit over de Vestindiske Ophiurer', *Vid. Medd. Dansk Naturh. Foren.*, **4** (1856), 1–26.

Lütken, C.F. (1872). 'Nogle nye eller mindre bekjendte Slangestjerner beskrevnemed. Nogle Bemaerkkninger om Selvdelingen hos Straaledyrene', *Overs. K. Danske Vid. Selsk. Forh.*, 75–168.

Lütken, C.F. (1873). 'On spontaneous division in the Echinodermata and other Radiata', *Ann. Mag. Nat. Hist.*, **12**, 323–337, 391–399.

Lütken, C.F., and Mortensen, T. (1899). 'The Ophiuridae. Report of the Explorations of the Panamic Albatross, 1891', *Mem. Mus. comp. Zool. Harvard*, **23**, 97–208.

Lyman, T. (1865). 'Ophiuridae and Astrophytidae', *Illus. Cat. Mus. comp. Zool.*, **1**, i–viii, 1–200, plates I, II.

Lyman, T. (1869). 'Preliminary report on the Ophiuridae and Astrophytidae dredged in deep water between Cuba and the Florida Reef etc.', *Bull. Mus. comp. Zool. Harvard*, **1**, 309–354.

Lyman, T. (1879). 'Ophiuridae and Astrophytidae of the 'Challenger' Expedition II', *Bull. Mus. comp. Zool. Harvard*, **6**, 17–83.

MacBride, E.W. (1896). 'The development of *Asterina gibbosa*', *Q. Jl microsc. Sci.*, **38**, 339–412.

MacGinitie, G.E., and MacGinitie, N. (1949). *Natural History of Marine Animals*, McGraw-Hill, New York, 473 pp.

Madsen, F.J. (1970). 'West African Ophiuroids. Scientific results of the Danish expedition to the coast of tropical West Africa 1945–1946', *Zool. Mus., Copenhagen*, Danish Science Press Ltd., Copenhagen, **11**, 152–243.

Marsh, L.M. (1977). 'Coral reef asteroids of Palau, Caroline Islands', *Micronesia*, **13**, 251–281.

Martens, E.V. (1866). 'Ueber ostasiatische Echinodermen', *Arch. Naturgesch.*, **32**, 57–88.

Matsumoto, H. (1917). 'A monograph of Japanese Ophiuroidea', *J. Cell. Sci. imp. Univ. Tokyo*, **38**(2), 1–408.

Mladenov, P.V. (1979). 'Unusual lecithotrophic development of the Caribbean brittle star *Ophiothrix oerstedi*', *Mar. Biol.*, **55**, 55–62.

Mladenov, P.V. (1985). 'Development and metamorphosis of the brittle star *Ophiocoma pumila*: Evolutionary and ecological implications', *Biol. Bull.*, **168**, 285–295.

Mladenov, P.V., Carson, S.F., and Walker, C.W. (1986). 'Reproductive ecology of an obligately fissiparous population of the sea star *Stephanasterias albula* (Stimpson)', *J. exp. mar. Biol. Ecol.*, **96**, 155–175.

Mladenov, P.V., and Emson, R.H. (1984). 'Divide and broadcast: Sexual reproduction in the West Indian brittle star *Ophiocomella ophiactoides* and its relationship to fissiparity', *Mar. Biol.*, **81**, 273–282.

Mladenov, P.V., and Emson, R.H. (1988). 'Density, size structure and reproductive characteristics of fissiparous brittle stars in algae and sponges: Evidence for interpopulational variation in levels of sexual and asexual reproduction', *Mar. Ecol. Prog. Ser.*, **42**, 181–194.

Mladenov, P.V., and Emson, R.H. (1990). 'Genetic structure of populations of two closely related brittle stars with contrasting sexual and asexual life histories, with observations on the genetic structure of a second asexual species', *Mar. Biol.*, **104**, 265–274.

Mladenov, P.V., Emson, R.H., and Colpitts, L.V. (1983). 'Asexual reproduction in the West Indian brittle star *Ophiocomella ophiactoides* (H.L. Clark) (Echinodermata: Ophiuroidea)', *J. exp. mar. Biol. Ecol.*, **72**, 1–23.

Mladenov, P.V., Igdoura, S., and Asotra, S. (1989). 'Purification and partial characterization of an autotomy-promoting factor from the sea star *Pycnopodia helianthoides*', *Biol. Bull.*, **176**, 169–175.

Monks, S.D. (1904). 'Variability and autotomy of *Phantaria*', *Proc. Acad. nat. Sci., Philadelphia*, **56**, 596–600.

Monticelli, F.S. (1896). 'Sull'autotomia delle *Cucumaria planci* (Br.)', *Atti Accad. Naz. Lincei Cl. Sci. Fis. Mat. Nat. Rendi. Ser.*, **55**, 231–239.

Mortensen, T. (1920). 'Notes on the development and the larval forms of some Scandinavian echinoderms', *Vid. Medd. Dansk Naturh. Foren.*, **71**, 133–160.

Mortensen, T. (1921). 'Studies on the development and larval forms of echinoderms', *G.E.C. Gad., Copenhagen*, 1–261.

Mortensen, T. (1926). 'Zoological results of the Cambridge Expedition to the Suez Canal 1924, No. 6. Report on the echinoderms', *Trans. Zool. Soc. Lond.*, **1**, 117–131.

Mortensen, T. (1933a). 'Papers from Dr. Th. Mortensen's Pacific Expedition (1914–1916). LXVI. The echinoderms of St. Helena (other than crinoids)', *Vid. Medd. Dansk Naturh. Foren.*, **93**, 401–472.

Mortensen, T. (1933b). 'Ophiuroidea', *Dan. Ingolf-Exped.*, **4**(8), 1–121.

Mortensen, T. (1936). 'Echinoidea and Ophiuroidea', *Discovery Rept.*, **12**, 199–348.

Mortensen, T. (1937). 'Contributions to the study of the development and larval forms of echinoderms III', *D. Kgl. Danske Vid. Selsk. Skrifter, Naturv. og Math. Afd.*, **9**, 7(1).

Mortensen, T. (1938). 'Contributions to the study of the development and larval forms of echinoderms IV, *D. Kgl. Danske Vid. Selsk. Skrifter, Naturv. og Math. Afd.*, **9**, 7(3).

Motokawa, T. (1982). 'Rapid change in mechanical properties of echinoderm connective tissues caused by coelomic fluid', *Comp. Biochem. Physiol.*, **73C**, 223–229.

Motokawa, T. (1984). 'Connective tissue catch in echinoderms', *Biol. Rev.*, **59**, 255–270.

Motokawa, T. (1985). 'Catch connective tissue: The connective tissue with adjustable mechanical properties', in *Echinodermata* (Eds. B.F. Keegan and B.D.S. O'Connor), A.A. Balkema, Rotterdam, pp. 69–73.

Müller, J., and Troschel, F.H. (1842). *System der Asteriden*, Vieweg, Braunschweig, 134 pp.

Newman, H.H. (1921). 'On the development of spontaneously parthenogenetic eggs of *Asterina* (*Patiria*) *miniata*', *Biol. Bull.*, **40**, 186–204.

Ottesen, P.O., and Lucas, J.S. (1982). 'Divide or broadcast: Interrelation of asexual and sexual reproduction in a population of the fissiparous hermaphroditic seastar *Nepanthia belcheri* (Asteroidea: Asterinidae)', *Mar. Biol.*, **69**, 223–233.

Parslow, E.R., and Clark, A.M. (1963). 'Ophiuroidea of the Lesser Antilles', *Stud. Fauna Curaçao*, **15**, 24–50.

Pearse, J.S. (1968). 'Reproductive periodicities of marine animals of tropical middle east waters', *Final Report for Contract N62558-5022 (NR 104-889) between the Office of Naval Research and the American University of Cairo, Egypt*, 71 pp.

Pearse, J.S. (1982). 'The Gulf of Suez: Signs of stress on a tropical biota', Proc. International Conf. Mar. Sci. Red Sea, *Bull. Inst. Oceanogr. Fish. (Egypt)*, **9**, 148–159.

Pearse, J.S., Pearse, V.B., and Newberry, A.T. (1989). 'Telling sex from growth: Dissolving Maynard Smith's paradox', *Bull. mar. Sci.*, **45**, 433–446.

Preyer, W. (1886–87). 'Uber die Bewegungen der Seesterne', *Mitt. Zool. Sta. Neapel*, **7**, 27–127, 191–233.

Reese, E.S. (1966). 'The complex behavior of echinoderms', in *Physiology of Echinoderms* (Ed. R.A. Boolootian), Interscience Publishers, New York, pp. 157–218.

Rideout, R.S. (1978). 'Asexual reproduction as a means of population maintenance in the coral reef asteroid *Linckia multifora* on Guam', *Mar. Biol.*, **47**, 287–295.

Rowe, F.W.E., and Marsh, L.M. (1982). 'A revision of the asterinid genus *Nepanthia* (Gray, 1840) (Echinodermata: Asteroidea), with the description of three new species', *Aus. Mus. Mem.*, **16**, 89–120.

Sars, M. (1859). *Ngt. Mag. Naturvid*, **10**, 1–99.

Shearer, C., and Lloyd, D. (1913). 'On methods of producing artificial parthenogenesis in *Echinus esculentus* and the rearing of parthenogenetic plutei through metamorphosis', *Q. Jl microsc. Sci.*, **58**, 523–551.

Simroth, H. (1877). 'Anatomie und Schizogonie der *Ophiactis virens* II', *Z. wiss. Zool.*, **28**, 419–526.

Singletary, R.L. (1974). 'A new species of brittle star from Florida', *Florida Scientist*, **36**, 175–178.

Steenstrup, F. (1857). *Fork. Skand. Naturf. Mote*, 229–230.

Stoddart, J.A., and Taylor, J.F. (1988). 'Genotypic diversity: Estimation and prediction in samples', *Genetics, Austin, Tex.*, **118**, 705–711.

Sundaram, S. (1986). 'Genetic structure of a population of the obligately asexual sea star *Stephanasterias albula*', Honours thesis, Mount Allison University, Sackville N.B., Canada, 42 pp.

Tartarin, A. (1953). 'Observations sur les mutilations, le régénération, les néoformations et l'anatomie de *Coscinasterias tenuispina*', *Recl. Trav. Sta. Mar. Endoume Bull.*, **5** (10), 1–107.

Tattersall, W.M., and Sheppard, E.M. (1934). 'Observations on the bipinnaria of the asteroid genus *Luidia*', in *James Johnstone Memorial Volume* (Ed. R.J. Daniel), University of Liverpool Press, Liverpool, pp. 35–61.

Thomas, L.P. (1962). 'The shallow water amphiurid brittle stars (Echinodermata: Ophiuroidea) of Florida', *Bull. mar. Sci. Gulf Caribb.*, **12**, 623–694.

Thorson, G. (1961). 'Length of pelagic larval life in marine bottom invertebrates as related to larval transport by ocean currents', in *Oceanography* (Ed. M. Sears), Am. Assn. Adv. Sci. Publication No. 67, Washington, D.C., pp. 455–474.

Tortonese, E. (1960). 'Echinoderms from the Red Sea. I. Asteroidea', *Bull. Sea Fish Res. Stn, Israel*, **29**, 17–23.

Valentine, J.N. (1925–26). 'Regeneration in the starfish *Linckia*', *Yb. Carnegie Instn., Wash.*, **25**, 257–258.

Verrill, A.E. (1866). 'On the polyps and echinoderms of New England with descriptions of new species', *Proc. Boston Soc. Nat. Hist.*, **10**, 333–357.

Verrill, A.E. (1869). 'On new and imperfectly known echinoderms and corals', *Boston Soc. nat. Hist.*, **12**, 381–396.

Wilkie, I.C. (1984). 'Variable tensility in echinoderm collagenous tissues: A review', *Mar. Behav. Physiol.*, **11**, 1–34.

Wilkie, I.C., Emson, R.H., and Mladenov, P.V. (1984). 'Morphological and mechanical aspects of fission in *Ophiocomella ophiactoides* (Echinodermata: Ophiuroida)', *Zoomorphology*, **104**, 310–322.

Wilson, D.P. (1978). 'Some observations on bipinnariae and juveniles of the starfish genus *Luidia*', *J. mar. biol. Ass. U.K.*, **58**, 467–478.

Wilson, W.H. (1983). 'The role of density dependence in a marine infaunal community', *Ecology*, **64**, 295–306.

Yamaguchi, M. (1975). 'Coral reef asteroids of Guam', *Biotropica*, **7**, 12–23.

Yamaguchi, M., and Lucas, J.S. (1984). 'Natural parthenogenesis, larval and juvenile development, and geographical distribution of the coral reef asteroid *Ophidiaster granifer*', *Mar. Biol.*, **83**, 33–42.

Yamazi, I. (1950). 'Autotomy and regeneration in Japanese sea stars and ophiurans: 1. Observations on a sea star *Coscinasterias acutispina* and four species of ophiurans', *Ann. Zool. Japon.*, **23**, 175–186.

10. HEMICHORDATA

J.A. PETERSEN*
Universidade de São Paulo, Instituto de Biociências e Centro de Biologia Marinha
Caixa Postal 11.461 CEP 05422-970, São Paulo SP, Brasil

I. INTRODUCTION

Hemichordates have been known for more than 150 years and hundreds of reports on them have been added to the scientific literature since the original discovery of an enteropneust in 1825. The first comprehensive review on the group was presented by Spengel as early as 1893; it was followed by the extensive work published by Van der Horst from 1932 to 1939, which continues to be a major reference on the Hemichordata. Two additional reviews deserve our attention: the one by Dawydoff (1948) and the more critical one by Hyman (1959).

Different aspects of the reproduction of hemichordates have been summarized more recently by Hadfield (1975). Hadfield's comprehensive review also contains a considerable amount of new data on reproduction and early development of individuals of a Hawaiian population of *Ptychodera flava*. Additional information on growth and metamorphosis of planktonic larvae of *P. flava* can be found in Hadfield (1978).

It seems advisable, in the present review, to separate the information available for the two classes of the Hemichordata (for classification, see Systematic Résumé). Such a view is justified by the several morphological, developmental, and behavioral differences to be found between the Enteropneusta and the Pterobranchia. Also, in agreement with the work of Damas and Stiasny (1961) and Hadfield (1975), the third 'class' of the Hemichordata (Planctosphaeroidea) will be considered and treated here just as an unusually large and aberrant group of enteropneust larvae.

As adequate illustrations can be found in many of the more recent reviews on the group, we have refrained from reproducing them again in the present publication. Throughout the text, however, reference will be made to the origi-

*Deceased.

nal illustrations which, we feel, can best illustrate the information under discussion.

II. ASEXUAL PROPAGATION IN THE ENTEROPNEUSTA

Asexual propagation is known only in one enteropneust family, the Ptychoderidae. Detailed information is available only in three species: *Balanoglossus capensis* (see Gilchrist, 1923), *B. australiensis* (see Packard, 1968), and *Glossobalanus crozieri* (see Petersen and Ditadi, 1971). If we take into consideration the observations of Packard (1968) in the Bay of Naples, a fourth species, *G. minutus*, could be added to this list. However, as previously pointed out (Petersen and Ditadi, 1971), it still remains debatable whether *G. minutus* and *G. crozieri* are co-specific or not. Whatever be the answer to this question, the fact remains that we now have records of the phenomenon of asexual reproduction from four parts of the world which could hardly be farther apart: the East coast of Africa, the Pacific coast of New Zealand, the Mediterranean, and the South Atlantic.

An analysis of the process of asexual propagation in these different species indicates that the same basic pattern of reproduction is common to them all. Such a process is initiated by a rupture of the body of an adult animal, just anterior to the hepatic region. A good description for *Balanoglossus australiensis*, valid also for the other species mentioned above, is to be found in Packard (1968). Figure 4 in Packard (1968) depicts the general scheme of asexual reproduction: following breakage in front of the first liver sacs, the hind 'hepatic' half regenerates a new proboscis and collar, as well as the other missing structures. The anterior 'branchiogenital' half attenuates posteriorly, dividing eventually into a number of portions (up to 15 fragments). It is important to notice that such 'regenerands' (fragments) represent a new generation, though originally they have been a structural part of the preceding generation.

The fate of these 'regenerands' is shown in fig. 7 A–G of Packard's paper (1968); such a reconstruction was based on specimens encountered in the field. In *Glossobalanus crozieri*, the process of development of 'regenerands' has been followed not only in fragments collected in the field but also on those obtained in the laboratory by cutting small pieces of the elongated genital portion of 'branchiogenital' individuals. In both situations the stages in redifferentiation of these forms closely follow the observations of Dawydoff (1909) in his studies of regeneration in *G. minutus*. The importance of the gonadal yolk reserves for the nourishment of these reconstituting fragments has been stressed by different authors. For detailed descriptions and chronology of the process of 'regenerand' development, the reader is referred to the papers by Packard (1968) and Petersen and Ditadi (1971).

The 'branchiogenital' individuals, after giving rise to 'regenerands' by the process of architomy, elongate again. After about 10 days they show the first signs of the formation of internal hepatic sacculations, which continue to develop throughout the following weeks, together with the formation of an intestinal region similar to that of normal adults.

III. REGENERATION IN THE ENTEROPNEUSTA

The regenerative capacity of the Enteropneusta has been known since the preliminary observations of Hill (1895), Willey (1899), Spengel (1893), Cori (1902), and Assheton (1908). However, details of the regenerative processes were well documented only in a series of papers by Dawydoff (1902, 1907a, 1907b) which culminated in his very comprehensive paper of 1909, describing the results on *Glossobalanus minutus*. A summary of such investigations can be found in Dawydoff (1948) and in Hyman (1959). More recent studies on the subject include observations on *Ptychodera flava* (see Rao, 1954; Hadfield, 1975; Nishikawa, 1977), *Saccoglossus kowalevskii* (see Tweedel, 1961), and *G. crozieri* (see Petersen and Ditadi, 1971).

According to Dawydoff (1909; text fig. 1), if an adult specimen of *Glossobalanus minutus* is cut into several pieces ('troncons'), each one of these fragments may regenerate the missing parts. Dawydoff (1909) also noted that the anterior portions of these fragments were more apt to regenerate than the posterior ones, and that it generally took a fairly long time for the hind portions of these fragments to acquire the missing parts, as already observed by Cori (1902) and Kuwano (1902).

The processes which take place during regeneration are similar to those which occur during embryonic development and very likely also during asexual reproduction by transverse fission: transformations can be due both to dedifferentiation ('morphallaxis') and to redifferentiation of the tissues of the original piece to give rise to the missing parts.

An appraisal of the information available indicates that regenerative powers vary not only in different species but also on the level of section (or rupture) along the longitudinal axis of the animal. Even in closely related species (e.g. *Glossobalanus minutus* and *G. crozieri*) differences are to be found, particularly in relation to regeneration of pieces of the intestinal region.

As rightly pointed out by Hadfield (1975), Tweedel's studies (1961) on *Saccoglossus kowalevskii*, a species which shows only limited powers of regeneration, "throw doubt on the suggestion that the capacity for regeneration automatically implies a capacity for asexual multiplication of a species".

IV. ASEXUAL REPRODUCTION IN THE PTEROBRANCHIA

Two pterobranch genera deserve our attention as far as asexual reproduction (vegetative growth) is concerned: *Cephalodiscus* and *Rhabdopleura*. No information is available on the possible modes of reproduction of the third genus, *Atubaria*.

In *Cephalodiscus* an outstanding feature is the formation of buds from the sucker-like proximal end of the stalk, on either side of the median line. According to the available evidence, all the zooids of a coenecium originate by budding from an original sexually produced animal, with a tendency for constant increase in size and number of zooids; also, such buds do not leave the parent coenecium to found new coenecia.

The main accounts of the development of buds are those of Masterman (1898), Harmer (1905), Ridewood (1907), Schepotieff (1908), and John (1932). As observed, the number of buds which may originate at any one time varies, in different species, from two to 15. Such buds begin as an outgrowth of the stalk epidermis which contains an extension of the stalk coelom, subdivided by a dorsoventral mesentery. The distal end of the bud flattens to give rise to the cephalic shield, proximal to which the bud enlarges to form the trunk sac—the main parts of the zooid are thus delineated. Following the formation of the gut and the anus, the bud detaches from the parent stalk, being then able to crawl freely within the coenecium, and eventually assumes the mode of life of the parent.

Information on aspects of the reproduction and life cycle of *Rhabdopleura* is due mainly to the more recent work of Stebbing (1970a, b) and Dilly (1973) on *R. compacta*; the reviews by Dawydoff (1948) and Hyman (1959) are based on papers by Lankester (1884), Fowler (1904), Vaney and Conte (1906), and more specifically on Schepotieff's (1907a, b) observations on *R. normani*.

All evidence indicates that a colony of *Rhabdopleura* originates from a single individual, sexually produced. In the case of *R. compacta*, the development of the colony by budding ('astogeny') commences in the primary zooid, a small bud forming at the base of the contractile stalk. Later on the new zooid, as well as the primary zooid, produces buds which also become new zooids (Stebbing, 1970b).

The description of Schepotieff (1907b) of the origin of *Rhabdopleura normani* colonies indicates that they arise by one or more outgrowths from a ring of stolon, enclosed in a similarly shaped part of the parent coenecium. As already pointed out by Stebbing (1970b) and Hadfield (1975), it is clear that there are basic differences in the budding patterns of the two species: *R. compacta* buds only from the bases of pre-existing zooids while *R. normani* can bud at several points on the elongated growing stolons.

In *Rhabdopleura compacta* most of the buds are of two types: those that even at an early stage show the rudiments of arms and cephalic shields, develop into zooids, and thus contribute to the growth of the colony; and those, called 'dormant buds', which form ovoid or spherical masses, with no differentiation of tissues or organs (Stebbing, 1970b).

In *Rhabdopleura normani* such dormant buds were termed 'hybernacula' by Lankester (1884), who considered them to be normal buds whose development had become arrested. A different view was presented by Schepotieff (1907b), who called them 'sterile buds', since he considered them incapable of developing into zooids, like normal buds. As shown by Stebbing (1970b), in *R. compacta*, these dormant buds are capable of developing into zooids. The same author also calls our attention to a third type of bud, already described by Schepotieff (1907b) as a 'regenerative bud': it differs in appearance from a normal bud in that it develops in the tubes of adult zooids that have degenerated.

V. REPRODUCTIVE STRATEGIES IN THE HEMICHORDATA

It is generally accepted that the evolution of reproductive strategies in animals has proceeded, as a rule, along two different lines: (1) the development of fairly complex life cycles, or (2) the condensation of complex cycles towards direct development. Such complex cycles would include the free-living organisms that have developed one or more metamorphic stages with differing food requirements.

Complex cycles are likely to be quite vulnerable as a result of the potential for interruption at any point, due to factors like predation or the scarcity of a specific food. To compensate for such hazards there would be a very large production of ova, in order to ensure some minimal measure of success. As we shall see later, these two strategies are to be found, in different degrees, among the Hemichordata.

It has been reasoned by Cody (1966) that organisms have a limited amount of energy and that they allocate it to reproduction, competition, and avoidance of predation. According to him, in environments considered unstable (temperate, arctic), most energy would go into increasing the reproductive rate, thus increasing clutch size. In environments defined as stable (the tropics, islands, coastlines), we should find smaller clutches. However, as emphasized by Stearns (1976) in a excellent review on life-history tactics, "we have few, if any good data on how stable different environments appear to organisms and without an independent measure of stability, Cody's argument is circular".

Among the multitude of environmental factors those generally most relevant to life histories are food, temperature, breeding sites, refugia, competitors, and predators. However, although authors frequently invoke environmental instability to explain the trends they observe in life-history phenomena, no one has actually defined instability unambiguously and then measured it along with the relevant reproduction traits during the process of selection (Stearns, 1976).

The idea of r-selection and K-selection (or r-strategy and K-strategy), which originated with Dobzhansky (1950), was further developed by MacArthur (1962), Lewontin (1965), Skutch (1967), and others. A number of papers have been published on the subject in recent years; for a critical evaluation of the different ideas, see Stearns (1976, 1977) and various chapters in this volume.

The theory of r- and K-selection is qualitative, not quantitative, and admits comparisons only within limited groupings; it does, however, predict the association of the biological traits constituting life-history tactics into two groups: (1) r-selection: early age at first reproduction; large clutch size; semelparity; no parental care; a large reproductive effort; small, numerous, offspring; low assimilation efficiency; and a short generation time; (2) K-selection: delayed reproduction;

iteroparity; small clutches; parental care; smaller reproductive effort; a few large offspring; and high assimilation efficiency (Stearns, 1976).

In *Adaptation and Natural Selection*, published in 1966, Williams differentiated between the total effort the adult puts into its progeny, on the one hand, and the partitioning of that effort into a few large young or many small young. According to him, if progeny can grow faster as larvae outside the parent, when resources for the young are abundant and predation pressure is low, then many small progeny will be favoured; this would be the case for most Ptychoderidae and Spengelidae.

If resources for the young are scarce, or predator risk to small-size classes is high, then the parent will tend to produce a few large progeny: this could involve the cases of direct or abbreviated development as well as asexual propagation. The group producing a few large eggs would be called 'K-selected' and those producing many small eggs 'r-selected' (Stearns, 1976).

We are inclined to agree with Pianka (1974), who reasons that "no organism is completely r-selected or completely K-selected, but rather all must reach some compromise between the two extremes. We think of an r–K selection continuum, and an organism's position along it in a particular environment at a given instant in time." Such an idea seems to apply quite well to the situation in the Hemichordata, as can be appraised from the following general discussion.

The information available so far would seem to indicate that in the case of enteropneusts there are at least five major different life-history tactics: (1) transverse fission without 'regenerands', as in some populations of *Ptychodera flava*; (2) transverse fission plus 'regenerands', as in *Balanoglossus australiensis, B. capensis*, and *Glossobalanus crozieri*; (3) direct development, without a tornaria larva, as in *Saccoglossus*; (4) indirect development, with a tornaria larva, as in Ptychoderidae and Spengelidae; and (5) viviparity, as in *Xenopleura*. In relation to the Pterobranchia the possibilities known so far are sexual reproduction and asexual reproduction by budding.

For the case of transverse fission without regenerands it is relevant to remember that in his 1975 review Hadfield properly points out that when an enteropneust is broken into two halves, if each half regenerates the missing parts, reproduction has occurred. He emphasizes that his studies on Hawaiian population of *Ptychodera flava*, a species noted by Rao (1954) to have great restitutive powers, "have given no indication of any tendency toward regular reproduction via transverse fission at any time of the year". Quite a different picture emerges from the observations of Nishikawa (1977), who also worked on *P. flava*, but on animals collected at Kushimoto, on the coast of Japan. Such a study showed the presence not only of 'normal' animals but also of a fairly large number of 'branchiogenital' and 'hepatic' individuals, doubtless indicative of a process of asexual reproduction by transverse fission. In spite of repeated sampling and close observation of the sediment of the collection area, no 'regenerands' or 'larvae' were found at the Kushimoto site. The 'branchiogenital' individuals accounted for 10 to 20 per cent and the 'hepatic' individuals for 10 to 25 per cent of the total samples obtained on different occasions.

Another point to be stressed is the very high density of animals at this site, since on several occasions densities of up to 40 animals/100 cm^2 were quite common, and on one occasion up to 270 animals were collected in such an area; such high numbers would suggest a very high degree of success for this life-history tactic.

In relation to reproduction by transverse fission with the production of 'regenerands', the available information has already been presented and discussed. At this point a few more comments seem adequate for asexual reproduction as a life-history tactic. In his general discussion on the problem of vegetative reproduction in some animals (the idea being valid for the hemichordates), Williams (1975) points out that sexual reproduction can have several advantages over asexual reproduction, since it always generates variability through genetic recombination, and usually has evolved to permit dispersal and create a propagule that can be made resistant to harsh circumstances. It is to be noticed, however, that vegetative reproduction preserves successful genotypes, placing progeny in an environment that has been tested and found to be favourable, such a view assuming of course that conditions do not change. As already emphasized by Smith (1968), "if the genetic variance of the population is generated by mutation in a uniform environment sexual reproduction confers no advantage. But if the genetic variance has arisen because selection has favoured different genotypes in different environments, then sexual reproduction will accelerate adaptation to a new environment."

As pointed out by Stearns (1976), the question remains unresolved as to how much effort the individual should put into vegetative as against sexual reproduction at different ages and under different conditions; it would seem that the allocation of effort between vegetative and sexual reproduction could be strongly influenced by the probability of local extinction, competition, and the problems associated with dispersal. In the case of hemichordates the successful choice of substrata for settlement would be an added factor.

According to Williams (1975), "if the life cycle includes several asexual and one sexual generation the sexual reproduction will occur where ecologial differences will be greatest between two successive generations". He also notes that "where both asexual and sexual reproduction can occur simultaneously, the asexual offspring will develop immediately and near the parent, but dormant, widely dispersed propagules will be produced sexually." In the case of hemichordates this would imply the production of long-lived, free-swimming larvae. Another point emphasized by him is that "free-living lower plants and animals that regularly reproduce asexually and only rarely sexually always make the sexual mode a response to changed conditions or to stimuli predictive of changed conditions".

The process of direct development, without the formation of a tornaria larva, has been described in detail for *Saccoglossus* by Burdon-Jones (1952). Such a process includes, usually, a free-swimming larval stage, but these larvae swim about only for a day or so, after which they develop an equatorial constriction, marking the boundary between proboscis and collar; later on a second constriction appears, representing the collar-trunk boundary. During such changes the larva gradually elongates and takes on the typical worm-like appearance, assumes a benthonic life, and, by continued growth and elongation, changes into a young enteropneust. It is possible that a similar mode of development occurs also in *Balanoglossus australiensis*, aside from asexual propagation (Packard, 1968).

Indirect development, with the formation of a pelagic and planktotrophic tornaria larva, is found in the Ptychoderidae and Spengelidae. The knowledge on the successive developmental stages of tornaria has been summarized by Van der Horst (1932–1939), a number of stages being recognized (Muller, Heider, Metschnikoff, Krohn, Spengel, Agassiz, as well as metamorphosis); for further details, the reader is referred to Hadfield (1975).

The advantages of indirect development have already been discussed (see also Williams, 1975; Stearns, 1976). Regarding the problems of dispersal and successful choice of substrata for settlement, mention should be made of the phenomenon of delayed metamorphosis, as already emphasized by Bjornberg (1959) for a number of tornaria larvae.

The occurrence of viviparity in the Enteropneusta is based on information obtained from a single specimen, which, according to Gilchrist (1923), contained an advanced embryo in the trunk coelom; obviously, this would imply a case of not only very direct development but also internal fertilization, indicative of K-strategy.

VI. CONCLUDING REMARKS

Attempts to apply current concepts of reproductive strategies (including r- and K-selection) to hemichordates are hampered by a rather incomplete knowledge of the reproductive biology and ecology of representatives of this group; such a comment applies to both asexual and sexual reproduction.

An adequate understanding of the reproductive processes will require a much deeper knowledge on a number of basic aspects of hemichordate biology, such as maturation cycles and spawning (breeding periods), development in most genera, effect of nutrition on the formation of gametes at the expense of large yolk storage, factors which control the arrest of gamete development and final maturation just before spawning, possible effect of temperature and light on different species (summer and winter spawning), and finally viviparity, known only from a single case.

As far as asexual reproduction is concerned, the number of unanswered questions is legion; studies are badly needed on aspects such as histological and cellular processes involved in asexual propagation, the value of gonadal yolk reserves for the nourishment of the reconstituting fragments, dedifferentiation and/or redifferentiation of existing tissues, and the effect of environmental factors which may influence the triggering of production of new individuals by asexual means.

Adequate information on the ecology of hemichordates (including aspects like predation and competition) is almost non-existent. Knowledge on factors limiting the survival and growth of both adults and larvae, metamorphosis, settlement, energetics, and the like is fundamental for a proper evaluation of life- history tactics in hemichordates. It is hoped that in the near future a larger number of biologists may address themselves to the study of this poorly known but most interesting group.

REFERENCES

Assheton, R. (1908). 'A new species of *Dolichoglossus*', *Zool. Anz.*, **33**, 517–520.

Bjornberg, T.K.S. (1959). 'On Enteropneusta from Brazil', *Boll. Inst. Oceanogr. São Paulo, Brazil*, **10**, 1–104.

Burdon-Jones, C. (1952). 'Development and biology of the larva of *Saccoglossus horsti*', *Phil. Trans. roy. Soc. London*, **236B**, 553–590.

Cody, M. (1966). 'A general theory of clutch size', *Evolution*, **20**, 174–184.

Cori, C.I. (1902). 'Über das Vorkommen des *Polygordius* und *Balanoglossus* (Ptychodera) im Triester Golfe', *Zool. Anz.*, **25**, 361–365.

Damas, D., and Stiasny, G. (1961). 'Les larves planctoniques d'entéropneustes', *Mem. Acad. roy. Belgique Sci.*, **15**, 1–68.

Dawydoff, C. (1902). 'Über die Regeneration der Eichel bei den Enteropneusten', *Zool. Anz.*, **25**, 551–556.

Dawydoff, C. (1907a). 'Sur la morphologie des formations cardio-péricardiques des entéropneustes', *Zool. Anz.*, **31**, 352–362.

Dawydoff, C. (1907b). 'Sur le développement du néphridium de la trompe chez les entéropneustes', *Zool. Anz.*, **31**, 576–581.

Dawydoff, C. (1909). 'Beobachtung über den Regeneration-Prozess bei den Enteropneusten', *Z. wiss. Zool.*, **93**, 237–305.

Dawydoff, C. (1948). 'Classe des Entéropneustes', in *Traité de Zoologie* (Ed. P.P. Grassé), Vol. XI, Echinodermes, Stomochordés, Procordés, Masson et Cie, Paris, pp. 369–453.

Dilly, P.N. (1973). 'The larva of *Rhabdopleura compacta* (Hemichordata)', *Mar. Biol.*, **18**, 69–86.

Dobzhansky, T.H. (1950). 'Evolution in the tropics', *Amer. Sci.*, **38**, 209–221.

Fowler, G.H. (1904). 'Notes on *Rhabdopleura normani*', *Q. Jl microsc. Sci.*, **48**, 23.

Gilchrist, J.D.F. (1923). 'A form of dimorphism and asexual reproduction in *Ptychodera capensis*', *J. Linn. Soc.*, **35**, 393–398.

Hadfield, M.G. (1975). 'Hemichordata', in *Reproduction of Marine Invertebrates* (Eds. A.C. Giese and J.S. Pearse), Academic Press, New York, pp. 185–240.

Hadfield, M.G. (1978). 'Growth and metamorphosis of planktonic larvae of *Ptychodera flava* (Hemichordata, Enteropneusta)', in *Settlement and Metamorphosis of Marine Invertebrate Larvae* (Eds. Fu-Shiang Chia and M.E. Rice), Elsevier, New York, pp. 247–254.

Harmer, S.F. (1905). 'The Pterobranchia of the Siboga Expedition, with an account of other species', *Siboga Exped. Monogr.*, **26**, 1–132.

Hill, J.P. (1895). 'On a new species of Enteropneusta (*Ptychodera australiensis*) from the coast of New South Wales', *Proc. Linn. Soc. N.S.W.*, **10** (Ser. 2), 1–42.

Hyman, L.H. (1959). *The Invertebrates: Smaller Coelomate Groups*, Vol. V, McGraw-Hill, New York, 783 pp.

John, C.C. (1932). 'On the development of *Cephalodiscus*', *Discovery Rept.*, **6**, 191–204.

Kuwano, U. (1902). 'On a new enteropneust from Misaki, *Balanoglossus misakiensis*', *Ann. Zool. Japon.*, **4**, 77–84.

Lankester, E.R. (1884). 'A contribution to the knowledge of *Rhabdopleura*', *Q. Jl microsc. Sci.*, **24**, 622–647.

Lewontin, R.C. (1965). 'Selection for colonizing ability', in *The Genetics of Colonizing Species* (Eds. H.G. Baker and G.L. Stebbins), Academic Press, New York, pp. 79–94.

MacArthur, R.H. (1962). 'Some generalized theorems of natural selection', *Proc. natl Acad. Sci. U.S.A.*, **48**, 1893–1897.

Masterman, A.F. (1898). 'On the further anatomy and the budding processes of *Cephalodiscus dodecalophus*', *Trans. Roy. Soc. Edinb.*, **39**, 507–527.

Nishikawa, T. (1977). 'Preliminary report on the biology of the enteropneust, *Ptychodera flava* Eschscholtz, in the vicinity of Kushimoto, Japan', *Publ. Seto mar. Biol. Lab.*, **23**, 393–419.

Packard, A. (1968). 'Asexual reproduction in *Balanoglossus* (Stomochordata)', *Proc. roy. Soc. London*, **171B**, 261–272.

Petersen, J.A., and Ditadi, A.S.F. (1971). 'Asexual reproduction in *Glossobalanus crozieri* (Ptychoderidae, Enteropneusta, Hemichordata)', *Mar. Biol.*, **9**, 78–85.

Pianka, E. (1974). *Evolutionary Ecology*, Harper and Row, New York, 356 pp.

Rao, K.P. (1954). 'Bionomics of *Ptychodera flava* Eschscholtz', *J. Madras Univ.* (Section B), **24**, 1–5.

Ridewood, W.G. (1907). 'Pterobranchia: *Cephalodiscus*', *Nat. Antarct. Exped. 1901–1904 ("Discovery"), Nat. Hist. Rept.*, **2**, 1–67.

Schepotieff, A. (1907a). 'Die Anatomie von Rhabdopleura', *Zool. Jb. (Abt.) Anat. Ontog. Tiere*, **23**, 463–534.

Schepotieff, A. (1907b). 'Knospungsprozess und Gehäuse von *Rhabdopleura*', *Zool. Jb. (Abt.) Anat. Ontog. Tiere*, **24**, 193–238.

Schepotieff, A. (1908). 'Knospungsprocess von *Cephalodiscus*', *Zool. Jb. (Abt.) Anat. Ontog. Tiere*, **25**, 405–494.

Skutch, A.F. (1967). 'Adaptive limitation of the reproductive rate of birds', *Ibis*, **109**, 579–599.

Smith, J.M. (1968). 'Evolution in sexual and asexual populations', *Amer. Natur.*, **102**, 469–473.

Spengel, J.W. (1893). 'Die Enteropneusten des Golfes von Neapel', *Fauna Flora Golfes Neapel (Monogr.)*, **18**, 1–758.

Stearns, S.C. (1976). 'Life-history tactics: A review of the ideas', *Q. Rev. Biol.*, **51**, 3–47.

Stearns, S.C. (1977). 'The evolution of life history traits: A critique of the theory and a review of the data, *Ann. Rev. Ecol. Syst.*, **8**, 145–171.

Stebbing, A.R.D. (1970a). 'The status and ecology of *Rhabdopleura compacta* (Hemichordata) from Plymouth', *J. mar. biol. Ass. U.K.*, **50**, 209–221.

Stebbing, A.R.D. (1970b). 'Aspects of the reproduction and life cycle of *Rhabdopleura compacta (Hemichordata)*', *Mar. Biol.*, **5**, 205–212.

Tweedel, K.S. (1961). 'Regeneration of the enteropneust *Saccoglossus kowalevskii*', *Biol. Bull.*, **120**, 118–127.

Van der Horst, C.J. (1932–1939). 'Hemichordata', *Bronn's Kl. Ordn. Tierreichs*, **4**, Abt. 4, Buch 2, Teil 2, 1–737.

Vaney, A., and Conte, A. (1906). 'Recherches sur le *Rhabdopleura normani* Allman', *Rev. Suisse Zool.*, **14**, 143–183.

Willey, A. (1899). 'Enteropneusta from the South Pacific, with notes on West Indian species', in *Zoological Results Based on Material Collected in New Britain, New Guinea, Loyalty Islands and Elsewhere* (Ed. A. Willey), Part 3, Cambridge, pp. 223–334.

Williams, G.C. (1966). *Adaptation and Natural Selection*, Princeton University Press, Princeton.

Williams, G.C. (1975). *Sex and Evolution*, Princeton University Press, Princeton.

11. CEPHALOCHORDATA

JAYAPAUL AZARIAH
Department of Zoology, University of Madras, Guindy Campus, Madras 600 025, Tamil Nadu, India

I. INTRODUCTION

The cephalochordates are circumequatorial in distribution and are found in warm, shallow waters of the tropical, subtropical, and temperate zones. Being subjected to temperature differences over their areas of distribution, it is likely that different species of *Branchiostoma* may vary their reproductive strategies in relation to the prevailing temperature regime. That environmental factors may bring about changes in reproductive periodicities of *B. lanceolatum* is shown by the presence of an additional breeding season during from November through January in tropical waters (Sections III and IV). The reviews by Reverberi (1971) and Wickstead (1975), and more recently by Guraya (Volumes I and II of this series) summarize some aspects of the reproductive and developmental biology of cephalochordates. More recent studies on amphioxus oocytes are by Song and Wu (1986), Aizenstadt and Gabaeva (1987), Wang and Wang (1987), and Holland and Holland (1989, 1991); on spermatozoa and testis by Chen *et al.* (1988) and Holland and Holland (1989) respectively; and on development by Hirakow and Kajita (1990, 1991). In the present study, an attempt has been made to show the occurrence of variation in reproductive strategies among different species of *Branchiostoma* and to explore the possible influence of environment in shaping these strategies.

II. REGENERATION

Regeneration as a form of asexual reproduction is virtually unknown in amphioxus. Earlier studies on regenerative abilities of cephalochordates are very limited and all attempts to demonstrate conclusively that amphioxus is capable of regenerating lost parts have failed. The work of Biberhofer (1906) on amphioxus, collected from the Helgoland waters, showed that regeneration takes place only in lancelets freshly caught. If amputation is carried out after about seven days of capture, the lancelets failed even to effect wound healing, and frequently the myotomes were found to come out of their position, leaving bare the notochord. In such cases, ciliate attack, as observed by Chin (1941), was common. Azariah (unpublished) also observed a close relationship between the nerve cord and the notochord in the regeneration of the lost part. Interestingly, the amputated body bits, measuring a few millimetres, may be kept alive for a period of 30 days or more in sea water.

III. FECUNDITY

Amphioxus is one of the chief components of the level-bottom ecosystem, commonly designated as the 'amphioxus community' (Samuel, 1944). Pelagic lancelets and their diel vertical migration have been reviewed by Boshung and Shaw (1988). In a given area, the distribution of amphioxus is regional in sediments where about 65 per cent of the sand grains measure between 0.49 mm and 0.99 mm in diameter (Azariah, 1965). Lancelet larvae, being planktonic, drift away from the parent population and hence a knowledge of their abundance may provide an insight into the chances of maintaining a high concentration of the adult population. Information on the fecundity of amphioxus is available for only a few species. Azariah (1969), taking the number of eggs produced as a measure of fecundity, counted the number of eggs in 250 ovarian sacs of 12 lancelets (*Branchiostoma lanceolatum*) varying in length between 16 and 26 mm. The lancelets were in between stage III and stage V of maturity. In a given lancelet, the number of ova per ovarian pouch was found to vary widely. Normally, the ovary in the middle of the series contained the highest number of ova. In a specimen 25 mm long, a total of 22 ovaries were counted on the right side of the body: the tenth ovarian pouch contained the maximum number of 66 ova whereas the ovaries at the two ends contained between 0 and 10. The total number of ova in all ovarian pouches of the right side of the animal amounted to 576, suggesting that the fecundity of a single specimen of average length may well be over a thousand.

Besides egg counting, catch number of adults and larvae in plankton samples may serve as an index of fecundity. In the continental shelf of Western Louisiana, lancelet density reached a peak in January with a mean density of 4.6 specimens/100 m^3 (Boshung and Shaw, 1988). At the Lagos Harbour, as many as 400 larvae of *Branchiostoma nigeriense* per haul could be collected at 8 m depth from October through December (Webb, 1958). Gosselck and Kuehner (1973) reported 37,000 larvae per haul at a depth of 0 to 25 m from the bottom off the Northwest African coast. Gosselck (1975) calculated that the population of *B. senegalense* of Villa Cisneras, Spanish Sahara, produced about 45 $\times$ 10 eggs annually, which may amount to about 27 million tons of planktonic amphioxus larvae. Flood *et al.* (1976) have estimated the amount of planktonic larvae of *B. senegalense* which may pass beyond Cap Blanc (Northwest Africa) each year. In their computation of annual production of amphioxus larvae, they took into account the mean larval weight (0.25 mg), mean larval concentration (40,000 larvae/m^2), extension of pop-

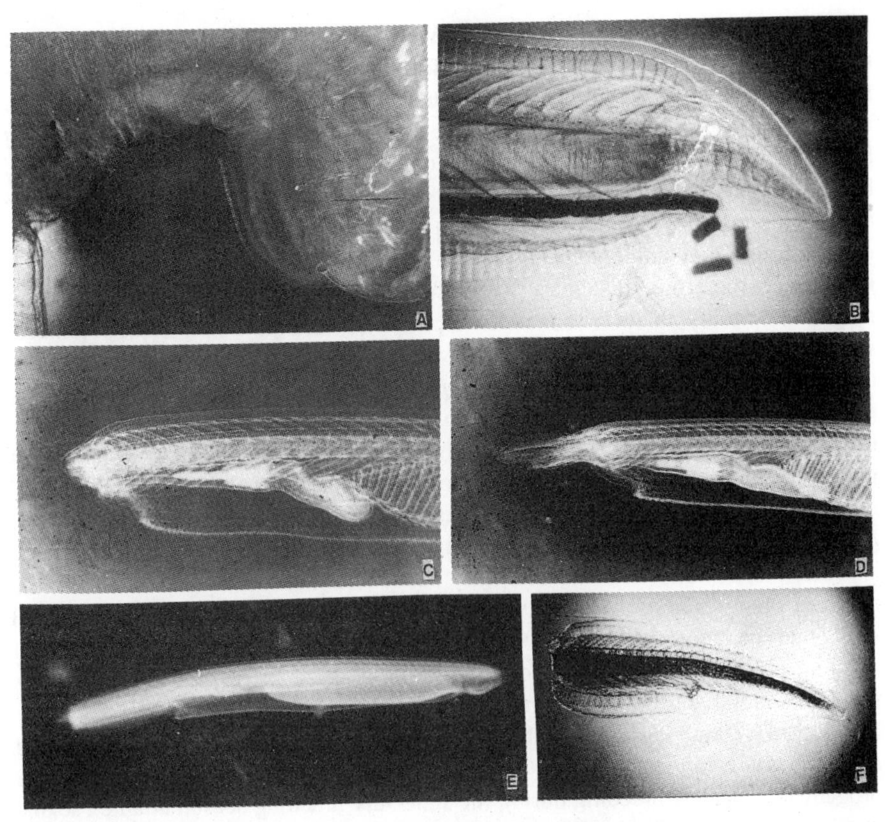

Fig. 1. Regeneration in *Branchiostoma lanceolatum*. **A**: Wound healing and resorption of the cephalic region. **B**: Caudal regeneration, post-ventral-storage chamber (ventral-fin) region. Note the formation of a new anus. **C** and **D**: Caudal regeneration, post-atrial region. **E**: An individual with a new regenerate. **F**: Amputated caudal portion.

ulated profile (110 km), the net translocation speed (0.1 m/s), and duration of larval season (four months). They calculated an estimated production of 1.1 million tons of amphioxus larvae per annum. It is thus clear that the reproductive capacity of amphioxus is indeed very high. The strategy of extravagance in production shown by lancelets may be related to their highly specific habitat selection in the face of depradation for settlement due to ocean currents.

Though we have succeeded in rearing individual amphioxus in laboratory aquaria for about one-and-a-half years, they failed to develop gonads in captivity without exception. What causes the sterility in laboratory cultures is not known. Likewise, since no regeneration is possible on laboratory acclimatization, it is presumed that some physiologically active substance may be lacking in the laboratory culture system. We have frequently observed that some of the gonadal pouches in *Branchiostoma lanceolatum* may not develop in some natural populations (Azariah,

1969). This confirms the observation of Webb (1955) in *B. nigeriense* that one or more of the gonadal pouches in the middle of the series may fail to develop or may only develop as a rudimentary structure. In this context, reports of the occurrence of hermaphroditism in amphioxus assume a new significance. Three types of hermaphroditism have been reported: (1) one or more of the gonadal pouches may be an ovary or a testis (Goodrich, 1912; Orton, 1914); (2) equal number of male and female gonads (Chen, 1931); and (3) both male and female gonadal elements may be found in a single gonadal pouch (Langerhans, 1875; Riddell, 1922). There is evidence, therefore, to suggest that in the development of each gonadal sac there may be flexibility to include the development of both male and female gonadal sac in a given series of gonads. The advantages of hermaphroditism as an effective reproductive strategy are too well known to be repeated here. It may also be rewarding to characterize the mechanism of sex determination in amphioxus from the viewpoint of the occurrence of hermaphroditism in nature and of sterility of laboratory-reared specimens. There is no consensus about the reported X and Y mechanism of sex determination in cephalochordates (*B. belcheri*: Nogusa, 1957; *B. floridae*: Howell and Boschung, 1971; *B. lanceolatum*: Colombera, 1974).

There appears to be some flexibility in the total number of gonads and in the range of their variation between sexes and in the number of gonad-bearing atrial myotomes in lancelets. Webb (1955) reported sexual differences in numbers of myotomes both anterior to and posterior to the gonads. In general the female has fewer gonadal pouches than the male in *Branchiostoma nigeriense*.

An interesting variation within the offspring of one parental population of *Branchiostoma lanceolatum* was noted by Azariah (1969). He found that the progeny of lancelets breeding during the June-July breeding season attained sexual maturity in about 12 months. During the next breeding season (November to January) the parental population, consisting of about 37 per cent of the total population, developed gonads and released a second set of offspring. The offspring of the earlier breeding season remained immature. The second set of offspring attained maturity in about six months and had only a life span of about 12 months. There were also differences in the rate of growth: the second set of offspring had a higher growth rate. Webb (1955) observed that the smallest specimen of *B. nigeriense* to bear gonad measured only about 7 mm. Azariah (1969) reported that, although the smallest gonad-bearing specimen of *B. lanceolatum* measured only 15 mm in length, maturity of the animal to bear gonads was not length dependent. Progeny of lancelets of both the breeding seasons, June to July and November to January, attained maturity for the first time when they were about 30 and 15 mm long respectively. Protracted spawning season and prolonged larval stage, shown by some lancelets (review in Boshung and Shaw, 1988), may ensure a steady and sustained recruitment over a long period and better prospects of larval survival to maturity. According to Webb (1955), some of the differences in the diagnostic features of West African lancelets may be purely due to environmental factors. In view of the above-mentioned variability in development of myotomal gonads between sexes and variation in the time of attainment of maturity, it may be of interest to carry out researches with a holistic approach, taking into consideration the influence, if any, of the environmental factors on the genetical basis of sex determination and on hormonal control of gonadal development. In lancelets evidence for endocrine influence on gonadal development and sexual maturity appears to be emerging: the ultrastructural study of the Hatschek's pit epithelial cells by Fang and Qi (1989) suggests that these cells may be the primitive gonadotropic cells of amphioxus.

IV. SEX RATIO

Information on sex ratio of populations of *Branchiostoma* is scanty. In the population of *B. lanceolatum* off the Madras coast, males were found to be more numerous than females, the (male: female) sex ratio being 3 : 2 during the June to July breeding season and 5 : 2 during the November to January breeding season.

V. SEXUAL PERIODICITY

In amphioxus the sexes are separate. The female can be distinguished externally by the honeycombed external outlines of the developing ova within. In males the testes are usually smooth without any external markings of the lobes (Azariah, 1969). In amphioxus gonads are only distinct during the breeding season, and in the absence of any other sex-dimorphic characters it is impossible to distinguish the sex during the non-breeding season. In view of the variability in total life span and the flexibility in gonadal development, it may be worthwhile to make a detailed study of sexual periodicity in *Branchiostoma* on a global scale.

VI. BREEDING CYCLE

Azariah (1969) deduced the pattern of breeding cycle of *Branchiostoma lanceolatum* by calculating the percentage of females in the population sample. On this basis, females were collected and counted during the months of May 1964 (40 per cent), June (33 per cent), July (51 per cent), August (6 per cent), December (7 per cent), and January 1965 (13 per cent). He noted that temporal variations occur in the onset and closure of the reproductive period. In temparate British waters, *B. lanceolatum* breeds only once in a year (Orton, 1914). However, the various populations of *B. lanceolatum* occurring in the English Channel and Mediterranean regions may have a short or an extended breeding season between spring and autumn (Willey, 1894; Orton, 1914; Conklin, 1932; Wickstead, 1967). As already mentioned (Section I), *B. lanceolatum* inhabiting tropical waters (e.g. off the Madras coast) has two breeding cycles a year. *Branchiostoma belcheri*, which has a life span of about three years, may attain sexual maturity when it is about one year old (Chin, 1941);

B. nigeriense, which has a life span of one year, spawns once in a year during August and September (Webb, 1958). In Helgoland, Courtney (1975a, b) found that *B. lanceolatum* matured after a period of two years and that 100 per cent of the population showed full gonadal development during the beginning of the third year. According to Courtney (1975a, b), spawning occurred annually through sexually dimorphic discrete areas of the gonadal envelope.

VII. CONCLUSION

From the foregoing brief account, it is clear that variability in the pattern of reproductive cycle is common among species of *Branchiostoma* and, in *B. lanceolatum*, between populations spread over wide geographical areas. In a study made in the author's laboratory (Azariah, unpublished) on morphometric variations in two latitudinally separated populations of *B. lanceolatum* inhabiting the Plymouth waters (U.K.) and the Indian waters, it was found that variations in biometric characters of myotomes may be attributed to temperature variations in accordance with the biogeographic size rule of Bergmann and Allen, a rule generally applied to homoeotherms. It is suggested that environmental factors, such as temperature and salinity, or synergistic effect of two or more parameters may control variations in reproductive strategies of lancelets. The question of genetic control, if any, of sexual maturation and reproductive periodicity in lancelets remains wide open.

REFERENCES

Aizenstadt, T.B., and Gabaeva, N.S. (1987). 'The perinuclear bodies (nuage) in the developing germ cells of the lancelet *Branchiostoma lanceolatum*', *Tsitologiya*, **29**, 137–141 (in Russian).

Azariah, J. (1965). 'Studies on the Cephalochordates of Madras coast. I. Taxonomic study', *J. mar. biol. Ass. India*, **7**, 348–363.

Azariah, J. (1969). 'Physiological and histochemical studies on some prochordates (Cephalochordates)', Ph.D. thesis, University of Madras, India.

Biberhofer, R. (1906). 'Über Regeneration bei *Amphioxus lanceolatus*', *Roux' Arch. Entwicklungsmech. Organismen*, **22**, 15–17.

Boshung, H.T., and Shaw, R.F. (1988). 'Occurrence of planktonic lancelets from Louisiana's (USA) continental shelf, with a review of pelagic *Branchiostoma* (order Amphioxi)', *Bull. mar. Sci.*, **43**, 229–240.

Chen, D., Zhao, X., Song, X., and Song, Y. (1988). 'The fine structure of the spermatozoa of the amphioxus (*Branchiostoma belcheri tsingtaoensis*', *Acta Zool. Sin.*, **34**, 106–109 (in Chinese).

Chen, T.Y. (1931). 'On a hermaphrodite specimen of the Chinese amphioxus', *Peking Nat. Hist. Bull.*, **5**, 11–16.

Chin, T.G. (1941). 'Studies on the biology of Amoy, amphioxus *Branchiostoma belcheri* Gray', *Philipp. J. Sci.*, **75**, 369–421.

Colombera, D. (1974). 'Male chromosomes in two populations of *Branchiostoma lanceolatum*', *Experientia*, **30**, 353–355.

Conklin, E.G. (1932). 'The embryology of amphioxus', *J. Morphol.*, **54**, 69–151.

Courtney, W.A.M. (1975a). 'Reproductive cycle and the occurrence of abnormal larvae in *Branchiostoma lanceolatum* at Helgoland', *Helgolander Wiss. Meersunters*, **27**, 19–27.

Courtney, W.A.M. (1975b). 'The temperature relationship and age structure of North Sea and Mediterranean populations of *Branchiostoma lanceolatum*', in *Protochordates* (Eds. E.J.W. Barrington and P.P.S. Jefferies), Symp. Zool. Soc. Lond., No. 36, pp. 213–233.

Flood, Per R., Braun, J.G., and De Leon, A.R. (1976). 'On the annual production of Amphioxus larvae (*Branchiostoma senegalense* Webb) off Cap Blanc, North West Africa', *Sarsia*, **61**, 63–70.

Fang, Yong-Qiang, and Qi, Xiang (1989). 'Ultrastructural study of the Hatschek's pit epithelial cells of amphioxus', *Sci. China*, Ser. B. *Chem. Life Sci., Earth Sci.*, **32**, 1465–1472.

Goodrich, E.S. (1912). 'A case of hermaphroditism in Amphioxus', *Zool. Anz.*, **30**, 443–448.

Gosselck, F. (1975). 'The distribution of *Branchiostoma senegalense* (Acrania Branchiostomidae) in the off-shore shelf region off North West Africa', *Int. Rev. ges. Hydrobiol. Hydrogr.*, **60**, 199–207.

Gosselck, F., and Kuehner, E. (1973). 'Investigations on the biology of *Branchiostoma senegalense* larvae off the Northwest African coast', *Mar. Biol.*, **22**, 67–73.

Hirakow, R., and Kajita, N. (1990). 'An electron microscopic study of the development of amphioxus, *Branchiostoma belcheri tsingtauense*: Cleavage', *J. Morphol.*, **203**, 331–344.

Hirakow, R., and Kajita, N. (1991). 'Electron microscopic study of the development of amphioxus, *Branchiostoma belcheri tsingtauense*: The gastrula', *J. Morphol.*, **207**, 37–52.

Holland, N.D., and Holland, L.Z. (1989). 'Fine structural study of the cortical reaction and formation of the egg coats in a lancelet (= amphioxus), *Branchiostoma floridae* (Phylum Chordata : Subphylum Caphalochordata = Acrania)', *Biol. Bull.*, **176**, 111–122.

Holland, N.D., and Holland, L.Z. (1989). 'The fine structure of the testis of a lancelet (= amphioxus), *Branchiostoma floridae* (Phylum Chordata : Subphylum Cephalochordata = Acrania),' *Acta Zool., Stockholm*, **70**, 221.

Holland, N.D., and Holland, L.Z. (1991). 'The fine structure of the growth stage of oocytes of a lancelet (= amphioxus), *Branchiostoma lanceolatum*', *Inv. Reprod. Dev.*, **19**, 107–122.

Howell, W.M., and Boshung, H.T., Jr. (1971). 'Chromosomes of the lancelet *Branchiostoma floridae* (Order Amphioxi)', *Experientia*, **27**, 1495–1496.

Langerhans, P. (1875). 'Zur Anatomie des *Amphioxus lanceolatus*', *Arch. mikrosk. Anat. Entwicklungsmech.*, **12**, 334–335.

Nogusa, S. (1957). 'The chromosomes of the Japanese lancelet *Branchiostoma belcheri* (Gray) with special reference to the sex chromosomes', *Ann. Zool. Japon.*, **30**, 42–46.

Orton, J.H. (1914). 'On a hermaphrodite specimen of amphioxus with notes on experiments in rearing amphioxus', *J. mar. biol. Ass. U.K.*, **10**, 506–512.

Probst, G. (1930). 'Regenerations Studien an Anneliden und *Branchiostoma lanceolatum* (Pallas)', *Rev. Suisse Zool.*, **37**, 343–351.

Riddell, W. (1922). 'On a hermaphrodite specimen of Amphioxus', *Ann. Mag. Nat. Hist.*, **1**, 613–617.

Reverberi, G. (1971). 'Amphioxus', in *Experimental Embryology of Marine and Freshwater Invertebrates* (Ed. G. Reverberi), North-Holland, Amsterdam, pp. 551–572.

Samuel, M. (1944). 'Preliminary observations on the animal communities of the level sea-bottom of the Madras coast', *J. Madras Univ.*, **15**, 45–71.

Song, Y.C., and Wu, S.C. (1986). 'An unknown ultrastructure of yolk granules in amphioxus egg', *Acta Zool. Sinica*, **32**, 32–34 (in Chinese).

Wang, D.H., and Wang, D.Y. (1987). 'The preliminary study of function of organelles in the vitellogenesis stage in the oocytes of amphioxus', *J. Xiamen Univ.* (*Nat. Sci.*), **26**, 615–619.

Webb, J.E. (1955). 'On the lancelets of West Africa', *Proc. Zool. Soc.* (*Lond.*), **125**, 421–443.

Webb, J.E. (1958). 'The ecology of Lagos Lagoon III. The life history of *Branchiostoma nigeriense* Webb', *Phil. Trans. roy. Soc. London*, **241B**, 335–353.

Wickstead, J.H. (1967). '*Branchiostoma lanceolatum* larvae: Some experiments on the effect of Thiouracil on metamorphosis', *J. mar. biol. Ass. U.K.*, **47**, 49–59.

Wickstead, J.H. (1975). 'Chordata: Acrania (Cephalochordata)', in *Reproduction of Marine Invertebrates* (Eds. A.C. Giese and J.S. Pearse), Vol. II, Academic Press, New York, pp. 283–319.

Willey, A. (1894). *Amphioxus and the Ancestry of the Vertebrates*, MacMillan, London, 316 pp.

SUBJECT INDEX

SPECIES INDEX